FUNDAMENTALS
OF
HEAT TRANSFER

Lindon C. Thomas

Professor of Mechanical Engineering
University of Petroleum and Minerals

FUNDAMENTALS
OF
HEAT TRANSFER

PRENTICE-HALL, INC., *Englewood Cliffs, New Jersey 07632*

Library of Congress Cataloging in Publication Data

Thomas, Lindon C.
   Fundamentals of heat transfer.

   Includes bibliographical references and index.
   1. Heat—Transmission. I. Title.
QC320.T49   536'.2   79-26073
ISBN 0-13-339903-6

Editorial/production supervision and interior design
  by Barbara A. Cassel
Manufacturing buyer: Gordon Osbourne

Printed in the United States of America

10 9 8 7 6 5 4 3 2 1

Prentice-Hall International, Inc., *London*
Prentice-Hall of Australia Pty. Limited, *Sydney*
Prentice-Hall of Canada, Ltd., *Toronto*
Prentice-Hall of India Private Limited, *New Delhi*
Prentice-Hall of Japan, Inc., *Tokyo*
Prentice-Hall of Southeast Asia Pte. Ltd., *Singapore*
Whitehall Books Limited, *Wellington, New Zealand*

To

Tisha and our children,
Sarah, Stephen, Mary, and *Mark*

And in memory of

R. N. Collins, who provided support and encouragement
during my first years as a teacher

CONTENTS

2 ONE-DIMENSIONAL HEAT TRANSFER 32

3 CONDUCTION HEAT TRANSFER: MULTIDIMENSIONAL 120

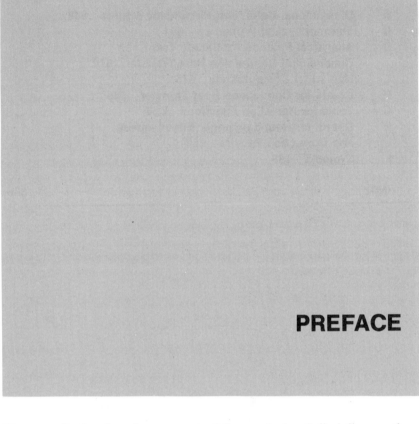

PREFACE

Heat transfer has long been a part of the curricula of disciplines such as aerospace, chemical, environmental, mechanical, and nuclear engineering, and physics. The significance of heat transfer, having heightened in recent years as a result of the world wide energy problem, is becoming more widely recognized in other fields such as architectural, civil, electrical, and petroleum engineering.

Fundamentals of Heat Transfer was primarily developed for use as a text in introductory undergraduate engineering heat transfer courses which are taught at the junior or senior level. The book should also serve as a useful reference for graduate students, practicing engineers, and other persons who are interested in energy related problems.

The most significant features that distinguish this book from other heat transfer texts are (1) the many practical examples; (2) the thorough presentation of fundamental concepts in the context of simple one-dimensional analyses; (3) the development of the numerical analysis approach, with emphasis on accuracy; (4) the practical treatment of radiation heat transfer; and (5) the coverage of convection heat transfer, with the practical lumped analysis approach set apart from the theory of convection.

The presentation throughout the book generally involves the use of brief developments and many examples. Elementary aspects of calculus and differential equations which provide background for an introductory study of heat transfer are reviewed in the Appendix. Fundamental concepts pertaining to the mechanisms of heat transfer are introduced in Chapter 1. One-dimensional analyses are developed in Chapter 2, with emphasis given to the importance of limiting criteria for common systems, such as fins, that are inherently multidimensional. Standard methods for solving multidimensional conduction problems are presented in Chapter 3, with attention focused primarily on the development of simple but accurate numerical finite difference solutions. Fundamentals of thermal radiation and the practical analysis of radiation heat transfer are presented in Chapter 4. The topic of convection heat transfer is introduced in Chapter 5, and a practical approach to the analysis of convection heat transfer which involves the use of modern convection correlations and simple lumped formulations is developed in Chapters 6 through 9 for forced and natural convection, boiling and condensation, and heat exchangers. Finally, the theory of convection is introduced in Chapter 10 for both laminar and turbulent flow.

The text contains sufficient material for coverage in two 3-hour semester courses. For use in a single course, the following material is suggested:

1-0—1-3	Introduction
2-1—2-8	One-dimensional analyses
3-1—3-6, 3-9	Multidimensional conduction
4-1—4-3-3, 4-4—4-5-3	Thermal radiation
5-0—5-2	Introduction to convection
6-1—6-4	Forced convection

This basic material can be coupled with selections from other sections in the book, depending on the specific objectives of the course, total number of contact hours, and background of the students.

Concerning the order of coverage of convection heat transfer, the format is such that the practical lumped approach of Chapters 6 through 9 can be studied before or after material in Chapter 10 which deals with the theory of convection.

Because we are in the transitional phase of metrication, both the international system of units (SI) and the English engineering system are used, with the SI system being used in the body of the text and in approximately 80 percent of the examples and problems.

It is hoped that this book will contribute to the effective study and teaching of the challenging subject of heat transfer. As indicated in the Acknowledgement, many colleagues and students have contributed to this work. Your suggestions, criticisms, and comments would also be very much appreciated.

Lindon C. Thomas

ACKNOWLEDGEMENT

This book was developed over a four- to five-year period, during which time I served on the faculties of the University of Petroleum and Minerals (UPM) and the University of Akron (UA). The broad base of support that these institutions have provided for my professional activities is gratefully acknowledged.

I am indebted to many colleagues who have contributed to this work. Professor E. M. Sparrow of the University of Minnesota, Professor F. M. White of the University of Rhode Island, and Professor R. D. Flack of the University of Virginia have reviewed the manuscript in its various stages of development over the past two to three years. Their suggestions, constructive criticisms, and general comments have been extremely helpful. In addition, feedback from Professors B. T. F. Chung and P. H. Gerhart of UA, Professor P. R. Clement of UPM, Professor E. G. Keshock of the University of Tennessee, Professor F. A. Kulacki of the Ohio State University, Professor B. G. Nimmo of UPM, and Professor R. Greif of the University of California, Berkeley, has enriched the book in many ways. I am also grateful to Mr. M. M. Al-Sharif of UPM for help with the computer programming and for checking much of the manuscript, and to Mr. D. J. Tandy of UPM for his many practical suggestions and for providing the photographs used in Figures E1-8, E2-5, and E2-14. All of these individuals have been most generous with their time and thoughts.

Because the book has been developed in an academic climate over a period of several years, a special acknowledgement is due to many other colleagues and students who have shared their perspectives with me. In particular, conversations with Professors N. Z. Azer and L. T. Fan of Kansas State University, Professor R. A. Granger of the U.S. Naval Academy, Professor E. N. Poulos of Cleveland State University, and Professor M. S. Selim of Texas Tech University, and active classroom participation by students at UPM and UA have made significant impacts on the book.

I also wish to acknowledge the responsiveness and cooperation of the many individuals and commercial firms that have provided illustrative material for the book. And I am pleased to acknowledge the help of Mrs. Nazli Ishaq who typed the manuscript which was used in classes at UPM over the past several years.

Finally, the love, patience, and forebearance of my wife, Tisha, and our children, Sarah, Stephen, Mary, and Mark, have been very important factors in making the book possible.

CHAPTER 1

INTRODUCTION

Heat transfer is defined as the transfer of energy across a system boundary caused solely by a temperature difference.

The study of heat transfer has long been a basic part of engineering curricula because of the significance of energy-related applications. For example, the transfer of heat in power plants from the energy source, be it fossil, nuclear, solar, or other, to the working fluid is one of the most basic processes in such systems. Similarly, the operation of refrigeration and air-conditioning units depend on the effective transfer of heat in condensers and evaporators. Other applications pertaining to environmental control which are of particular interest currently include the minimization of building-heat losses by means of improved insulating techniques and the use of supplemental energy sources, such as solar radiation, heat pumps, and fireplaces. Heat transfer is also very important in the operation of electrical machinery and transformers, and is often the controlling factor in the miniaturization of electronic systems.

Distillation, which is the basic separation process used in all refineries, is a particularly timely heat-transfer application. A crude oil distillation system consisting of a vertical column containing a number of trays and a furnace is shown in Fig. 1-1. The crude oil is piped through the heat exchanger (furnace) to the distillation column as shown in the figure. Within the column itself, the difference in the boiling temperatures of the various volatile substances is used to separate the crude oil into "cuts" by drawing off each component as it boils, and condensing the resulting cuts by cooling.

Although traditional fossil fuels such as oil will be utilized for many years to come, a new era has been brought upon us by rising costs of such exhaustible natural resources. As solutions are sought to our energy problems, there is no doubt that heat transfer will be a key factor. For example, the recovery of energy from waste streams, once dismissed as

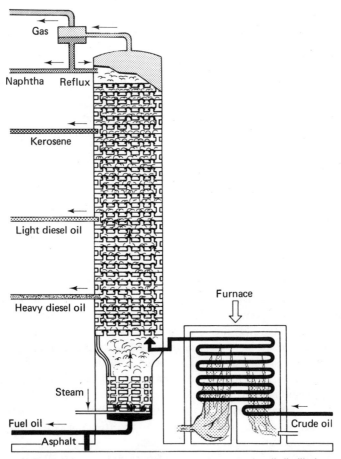

FIGURE 1-1   Example of heat transfer in a crude oil distillation column. (Courtesy of ARAMCO.)

being impractical, is now an economic necessity. Another example is the use of solar energy in heating, cooling, and energy generation. The schematic of a typical solar heating system is shown in Fig. 1-2(a). This arrangement is used in the Colorado State University Solar House shown in Fig. 1-2(b). In this and other solar energy applications, thermal radiation from the sun is captured in collectors, transferred to a working fluid such as water, and stored. The stored energy is then used for direct heating of the building during both day and night. Auxiliary electrical heating is provided for periods of inclement weather.

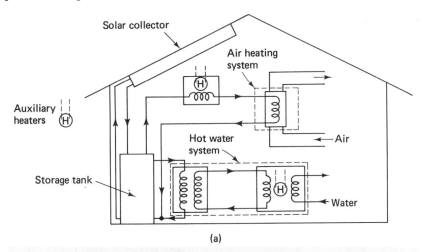

(a)

(b)

FIGURE 1-2 (a) Schematic of typical solar heating system. (b) Solar house No. 2 in CSU Solar Village. (Courtesy of Solar Laboratory, Colorado State University.)

The primary objective in the analysis of most heat-transfer problems is to (1) determine the temperature distribution within the system and the rate of heat transfer for specified operating conditions (the function of evaluation), or (2) prescribe the necessary operating conditions (size, shape, flow rates, etc.) in order to accomplish a given heat-transfer rate and/or temperatures (the function of design). Although emphasis in our study will be placed on the evaluation function, the concepts of thermal analysis that will be developed provide the foundation for the actual design of systems that involve heat transfer.

Because thermodynamics involves the study of heat and work for systems in equilibrium, a thermodynamic analysis can only provide us with predictions for the total quantity of heat transferred during a process in which a system goes from one equilibrium state (uniform temperature) to another. However, the length of time required for such processes to occur cannot be obtained by thermodynamics alone. On the other hand, the study of heat transfer involves a consideration of the mechanics of the transfer of thermal energy and is not restricted to equilibrium states. It is the science of heat transfer that enables us to perform the critical evaluation and design functions.

The analysis of heat-transfer processes requires the use of several fundamental laws, all of which are already familiar to the student. These fundamental principles are of a general nature and are independent of the mechanism by which heat is transferred. In addition, particular laws must be satisfied that pertain to each of the mechanisms by which heat transfer can be accomplished. These fundamental and particular laws are reviewed in the following two sections, after which brief consideration is given to the very useful analogy between heat transfer and the flow of electric current. A review of basic mathematical concepts, dimensions, and units that pertain to our study of heat transfer is presented in Appendices A and B.

1-1 FUNDAMENTAL LAWS

As in the case of thermodynamics, the *first law of thermodynamics* (conservation of energy) is the cornerstone of the science of heat transfer. This fundamental law takes the form (for nonrelativistic conditions)

$$\text{creation of energy} = 0$$
$$\Sigma E_o - \Sigma E_i + \Delta E_s = 0 \qquad (1\text{-}1)$$

where E_i and E_o represent the energy transfer into and out of the system, respectively, and ΔE_s is the change in energy stored within the system. The term ΔE_s includes internal energy U, kinetic energy KE, and potential energy PE, all of which are storable forms of energy; that is,

$$\Delta E_s = \Delta U + \Delta KE + \Delta PE \tag{1-2}$$

The first law of thermodynamics is written on a rate basis as

rate of creation of energy $= 0$

$$\Sigma \dot{E}_o - \Sigma \dot{E}_i + \frac{\Delta E_s}{\Delta t} = 0 \tag{1-3}$$

For a closed system, the terms ΣE_o and ΣE_i account for the transfer of heat and work across the boundaries. Thus, the first law for closed systems takes the familiar form

$$Q = W + \Delta U + \Delta KE + \Delta PE \tag{1-4}$$

where Q is the heat transferred into the system and W the work transferred out of the system. This equation is written on a rate basis as

$$q = \dot{W} + \frac{\Delta U}{\Delta t} + \frac{\Delta KE}{\Delta t} + \frac{\Delta PE}{\Delta t} \tag{1-5}$$

where $q(\equiv \dot{Q})$ is the rate of heat transferred into the system. The terms $\Delta U/\Delta t$, $\Delta KE/\Delta t$, and $\Delta PE/\Delta t$ are all zero for steady-state conditions.

For open systems, the terms ΣE_o and ΣE_i also account for the transfer of stored and flow energy (KE, PE, and H) which are associated with the fluid flow. [It should be recalled that the enthalpy H is equal to the sum of the internal energy U and flow energy PV (i.e., $H = U + PV$).] Therefore, the first law of thermodynamics takes the following more general form for open systems:

$$Q = W + H_o - H_i + KE_o - KE_i + PE_o - PE_i + \Delta U + \Delta KE + \Delta PE \tag{1-6}$$

or, on a rate basis,

$$q = \dot{W} + \dot{H}_o - \dot{H}_i + \dot{KE}_o - \dot{KE}_i + \dot{PE}_o - \dot{PE}_i + \frac{\Delta U}{\Delta t} + \frac{\Delta KE}{\Delta t} + \frac{\Delta PE}{\Delta t}$$

$$\tag{1-7}$$

For steady-state conditions, Eq. (1-7) reduces to

$$q = \dot{W} + \dot{H}_o - \dot{H}_i + \dot{KE}_o - \dot{KE}_i + \dot{PE}_o - \dot{PE}_i \tag{1-8}$$

Two other key fundamental laws are required in the analysis of heat transfer in fluids that are in motion: (1) the principle of *conservation of mass* (for nonrelativistic conditions),

$$\text{rate of creation of mass} = 0$$

$$\Sigma \dot{m}_o - \Sigma \dot{m}_i + \frac{\Delta m_s}{\Delta t} = 0 \tag{1-9}$$

and (2) *Newton's second law of motion* (with respect to the x direction for constant mass systems),

$$\text{rate of creation of momentum} = \text{sum of forces}$$

$$ma_x = \Sigma F_x \tag{1-10}$$

Three supplementary fundamental principles are required in the analysis of all heat-transfer processes: (1) the *second law of thermodynamics*, which provides us with the very critical conclusion that heat is transferred in the direction of decreasing temperature; (2) the *principle of dimensional continuity*, which requires that all equations be dimensionally consistent; and (3) *equations of state*, which provide information in equation, tabular, or graphical form pertaining to the thermodynamic properties at any state. Concerning the thermodynamic properties, information pertaining to the specific internal energy (U/m) and enthalpy (H/m) are tabulated for common substances (e.g., steam tables). In addition, U and H can be expressed in terms of the specific heats as follows:

$$c_v = \frac{\partial}{\partial T}\left(\frac{U}{m}\right)\bigg|_v \qquad c_P = \frac{\partial}{\partial T}\left(\frac{H}{m}\right)\bigg|_P \tag{1-11, 12}$$

For constant-volume and constant-pressure processes, U and H are given by

$$dU = mc_v\,dT \qquad dH = mc_P\,dT \tag{1-13, 14}$$

Further, for flow processes involving ideal gases or for small pressure drops, $\dot{H}$ can be written in terms of c_P and T as

$$d\dot{H} = \dot{m}c_P\,dT \tag{1-15}$$

1-2 BASIC MODES AND PARTICULAR LAWS

Conduction and thermal radiation represent the two fundamental mechanisms by which heat transfer is accomplished. These heat-transfer mechanisms occur in both solids and fluids. The transfer of heat by conduction (and sometimes by thermal radiation) from a solid surface to a moving fluid is known as *convection heat transfer*. These three modes of heat transfer and the particular laws that govern these phenomena are introduced in the following sections.

1-2-1 Conduction Heat Transfer

From the thermodynamic view, *temperature T* is a property which is an index of the kinetic energy possessed by molecules, atoms, and subatomic particles of a substance; the greater the agitation of these basic components of which physical matter is made, the higher the temperature. In this light, *conduction heat transfer* is simply the transfer of energy caused by physical interaction between adjacent molecules of a substance at different temperatures (levels of molecular kinetic energy).

Fourier law of conduction

On the basis of experimental observation, the rate of heat transferred by conduction in the x direction through a finite area A_x for the situation in which T is a function only of x can be expressed by

$$q_x = -kA_x \frac{dT}{dx} \tag{1-16}$$

where A_x is normal to the direction of transfer x, and k is the *thermal conductivity*. This equation was first used to analyze conduction heat transfer in 1822 by a French mathematical physicist named J. Fourier [1] and has come to be called the *Fourier law of conduction*. An example of a one-dimensional molecular conduction-heat-transfer problem for which this equation applies is illustrated in Fig. 1-3, which shows a plate with surface temperatures T_1 and T_2. Because no temperature differences occur in the y and z directions, q_y and q_z are both zero. For this case in which T_1 is greater than T_2, the temperature gradient is negative.

The consequence of the minus sign in Eq. (1-16) is that q_x is positive for situations such as this, in which the temperature gradient is negative. This result is consistent with the second law of thermodynamics, which stipulates that heat is transferred in the direction of decreasing temperature.

For situations in which the temperature is a function of time t and one space variable, such as x, the Fourier law of conduction is written as

$$q_x = -kA_x \frac{\partial T}{\partial x} \tag{1-17}$$

where q_x also is a function of t and x. Expressions of the form of Eqs.

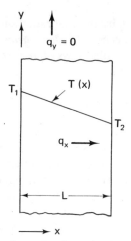

FIGURE 1-3 One-dimensional conduction heat transfer in a plate with $T_1 > T_2$; $q_y = 0$ and $q_z = 0$.

(1-16) and (1-17) can also be written for conduction heat transfer in the y, z, or r directions, as will be illustrated in Chap. 2. For situations in which the temperature is a function of more than one spatial dimension, the heat transfer in each direction must be accounted for. For example, for the case in which T is a function of x and y, we must write expressions for both q_x and q_y. To accomplish this task, we utilize a somewhat more general form of the Fourier law of conduction. This general Fourier law of conduction is presented in Chap. 3, which deals with multidimensional conduction heat transfer.

Thermal conductivity

The *thermal conductivity* k is a thermophysical property of the conducting medium which represents the rate of conduction heat transfer per unit area for a temperature gradient of 1°C/m (or 1°F/ft). The units for k are W/(m °C) [or Btu/(h ft °F)]. (Note that °C = 1 K and °F = 1° R.) The thermal conductivities for various common substances are listed in Table 1-1 for standard atmospheric conditions. More extensive tabulations of thermal conductivities and other properties are given in Tables A-C-1 through A-C-5 of the Appendix and in [2]–[6].

At room temperature, k ranges from values in the hundreds for good conductors of heat such as diamond and various metals to less than 0.01 W/(m °C) for some gases. Materials with values of k less than about 1 W/(m °C) are classified as insulators. As a rule of thumb, metals with good electrical conducting properties have higher thermal conductivities than do dielectric nonmetals or semiconductors. This is because the molecular interaction in good electrical conductors is enhanced by the move-

TABLE 1-1   Thermal conductivity of various substances at room temperature

	k	
Substance	W/(m °C)	Btu/(h ft °F)
Metals:		
Silver	420	240
Copper	390	230
Gold	320	180
Aluminum	200	120
Silicon	150	87
Nickel	91	53
Chromium	90	52
Iron (pure)	80	46
Germanium	60	35
Carbon steel (1% C)	54	31
Alloy steel (18% Cr, 8% Ni)	16	9.2
Nonmetal Solids:		
Diamond, type 2A	2300	1300
Diamond, type 1	900	520
Sapphire (Al_2O_3)	46	27
Limestone	1.5	0.87
Glass (Pyrex 7740)	1.0	0.58
Teflon (Duroid 5600)	0.40	0.23
Brick (B & W K-28)	0.25	0.14
Plaster	0.13	0.075
Cork	0.040	0.023
Liquids:		
Mercury	8.7	5.0
Water	0.6	0.35
Freon F-12	0.08	0.046
Gases:		
Hydrogen	0.18	0.10
Air	0.026	0.015
Nitrogen	0.026	0.015
Steam	0.018	0.01
Freon F-12	0.0097	0.0056

Source: From [3]–[5].
(1 W/(m °C) = 0.5778 Btu/(h ft °F), °C = 1 K, and °F = 1 °R.

ment of free electrons. Exceptions to this rule include dielectric crystals, such as diamond, sapphires, and quartz, and electrical semiconductors, such as silicon and germanium. A second rule is that solid phases of materials generally have higher thermal conductivities than do liquid phases. An exception to this rule is bismuth, which has a higher thermal conductivity for the liquid phase than for the solid phase.

The variation of k with temperature is shown in Fig. 1-4 for several representative substances and in Figs. A-C-1 through A-C-4 for various

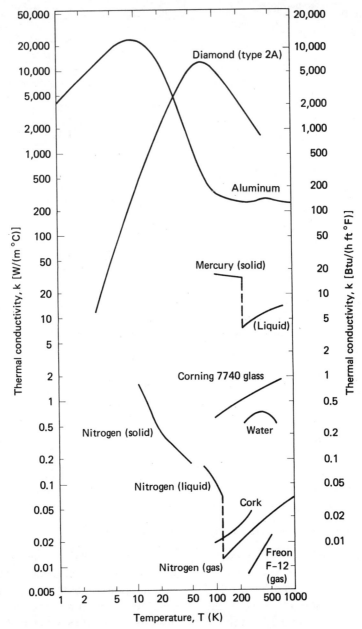

FIGURE 1-4 Variation of thermal conductivity k with temperature T for representative substances. (From [2]–[4].)

other common materials. The thermal conductivity of many of these substances varies by a factor of 10 or more for an order-of-magnitude change in temperature. On the other hand, the variation in k with temperature for some materials over certain temperature ranges is small enough to be neglected. We also note that exceptionally high thermal conductivities occur among the solid materials that were judged to be good conductors at room temperature. For example, the thermal conductivity of aluminum reaches a maximum value of about 20,000 W/(m °C) at 10 K. This is over 100 times as large as the value that occurs at room temperature. Substances under low-temperature conditions that have such exceedingly high thermal conductivities are known as *superconductors*.

In homogeneous materials, k can generally be assumed to be independent of direction (i.e., isotropic). However, some pure materials and laminates have thermal conductivities that are dependent upon the direction of heat flow. For example, the thermal conductivity of wood is different for heat conduction across the grain than for heat transfer parallel to the grain. Other materials with such nonisotropic characteristics include crystalline substances, laminated plastics, and laminated metals. For an introduction to the topic of conduction heat transfer in nonisotropic materials, the textbook by Eckert and Drake [7] is recommended.

Heat-transfer applications involving the more familiar metallic conductors such as copper and aluminum and insulators such as asbestos and cork are known to us all. To name only a few, tubes made of copper, stainless steel, aluminum, and other metals are used in boilers, evaporators, and condensers to transmit energy from one fluid to another; metal pans are used to cook food; double glass panes are used to minimize heat loss through windows; and asbestos, brick, and other insulative materials are used to reduce building heat losses in the winter and heat gains in the summer. Because of their high thermal conductivities and large electrical resistivities, silicon and diamond also find application, especially in the field of electronics. For example, silicon greases, pastes, and gaskets are often used in the construction of electronic systems in order to increase the rate of heat transfer while maintaining good electrical insulation between components. As another example, Fig. 1-5 shows a gold-plated diamond (type 2A) cube diode. Diodes such as these, ranging in size from below 0.1 mm to a few millimeters across, are used to generate high-frequency radio waves which relay telephone conversations and television broadcasts. These small diodes are characterized by very high power density operation, with operating temperatures in the range 150°C to 200°C. Because type 2A diamond has the highest thermal conductivity of all known materials in this temperature range, the diamond cube is used to remove the energy generated in the electronic semiconductor chip. Because of its effectiveness as a heat conductor, the diamond cube reduces the operating temperature of the diode, thereby increasing its lifetime and reliability.

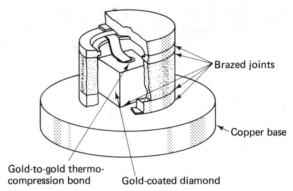

Gold-to-gold thermo-
compression bond   Gold-coated diamond

FIGURE 1-5   Microwave oscillator diode with diamond heat
sink. (Courtesy of Bell Telephone Laboratories.)

Analysis of conduction heat transfer

Consideration is now given to the analysis of conduction heat transfer in solids or stationary fluids. The important topic of conduction heat
transfer in moving fluids will be considered separately in Sec. 1-2-3, which
deals with convection.

The general theoretical analysis of conduction-heat-transfer problems
involves (1) the use of (a) the fundamental first law of thermodynamics
and (b) the Fourier law of conduction (particular law) in the development
of a mathematical formulation that represents the energy transfer in the
system; and (2) the solution of the resulting system of equations for the
temperature distribution. Once the temperature distribution is known, the
rate of heat transfer is obtained by use of the Fourier law of conduction.
The basic concepts involved in the theoretical analysis of conduction-
heat-transfer problems will be presented in Chap. 2 in the context of fairly
simple one-dimensional systems. These fundamentals will then be extended
to multidimensional systems in Chap. 3.

A simple practical approach to the analysis of basic steady-state
conduction-heat-transfer problems has been developed which involves the
use of an equation derived from the fundamental and particular laws. This
practical equation for conduction heat transfer takes the form (for systems
with uniform thermal conductivity)

$$q = kS(T_1 - T_2) \qquad (1\text{-}18)$$

where q is the rate of heat transfer conducted from a surface at temperature T_1 to a surface at T_2, and S is known as the *conduction shape factor*;
the unit for S is m (or ft). (The product kS is sometimes referred to as the
thermal conductance K.) The conduction shape factor S is dependent upon

geometry. For example, $S = A/L$ for one-dimensional conduction in a flat plate, $S = 2\pi L / \ln(r_2/r_1)$ for radial conduction in a hollow cylinder, and $S = 4\pi r_1 r_2 / (r_2 - r_1)$ for radial conduction in a hollow sphere. This practical approach to the analysis of conduction-heat-transfer problems is illustrated by several examples in this chapter. The theoretical basis for this simple method will be developed for one-dimensional systems in Chap. 2 and will be extended to more complex, multidimensional systems in Chap. 3.

EXAMPLE 1-1

Determine the rate of heat loss per unit area through a 10-cm-thick brick (B & W K-28) wall with surface temperatures of 15°C and 75°C.

Solution

Utilizing Eq. (1-18) with $S = A/L$, we write

$$q = kS(T_1 - T_2) = \frac{kA}{L}(T_1 - T_2)$$

Referring to Table 1-1, k is approximately equal to 0.25 W/(m °C). Thus, the rate of heat transfer is calculated to be

$$q = 0.25 \frac{W}{m\ °C} \left(\frac{A}{0.1\,m} \right) (75°C - 15°C)$$

It follows that the rate of heat transfer per unit area (i.e., heat flux) is

$$q'' = \frac{q}{A} = 150 \frac{W}{m^2}$$

1-2-2   Radiation Heat Transfer

The second fundamental mechanism of heat transfer involves the transfer of electromagnetic radiation emitted from a body as a result of vibrational and rotational movements of molecular, atomic, and subatomic particles. As we have already noted, temperature is an index of the level of agitation of these microscopic particles. Because molecules and their components are continuously in motion, thermal radiation is always emitted by physical matter. Furthermore, the rate at which this (internal) energy which is associated with molecular motion (i.e., temperature) is converted into thermal radiation increases with temperature.

The medium through which thermal radiation passes can be a vacuum or a transparent gas, liquid, or solid. Objects within the path *absorb*, *reflect*, and if they are transparent, *transmit* incident thermal radiation. The *absorptivity* α, *reflectivity* ρ, and *transmissivity* τ represent the fractions of incident thermal radiation that a body absorbs, reflects, and transmits, respectively. It follows that

$$\alpha + \rho + \tau = 1 \qquad (1\text{-}19)$$

These properties are primarily dependent upon the temperature of the emitting source and the nature of the surface that receives the thermal radiation. To illustrate, most incoming thermal radiation is absorbed by a surface coated with lampblack paint (i.e., $\alpha \simeq 0.97, \rho \simeq 0.03, \tau \simeq 0$), but is reflected from the surface of a polished aluminum plate (i.e., $\alpha \simeq 0.1, \rho \simeq 0.9, \tau \simeq 0$). As another example, a thin glass plate will transmit most of the thermal radiation from the sun, but will absorb much of the thermal radiation emitted from the low-temperature interior of a building, such as a greenhouse.

Thermal radiation generally passes through gases such as air with no significant absorption taking place. Such gases with $\tau \simeq 1$ are known as *nonparticipating gases*. Gases and transparent liquids that absorb significant quantities of the thermal radiation are known as *participating fluids*. Carbon dioxide and water vapor are examples of participating gases and water is a participating liquid. Of course, many liquids, such as mercury, are opaque to thermal radiation.

A body continually emits radiant energy in an amount which is related to its temperature and the nature of its surface. An object that absorbs all the radiant energy reaching its surface ($\alpha = 1$) is called a *blackbody*. Such ideal absorbers emit radiant energy at a rate that is proportional to the fourth power of the absolute temperature of the surface. The *Stefan–Boltzmann law* for blackbody thermal radiation takes the form

$$E_b = \sigma T_s^4 \qquad (1\text{-}20)$$

where the *total emissive power* E_b for a blackbody is the total rate of thermal radiation emitted by a perfect radiator per unit surface area, σ is the *Stefan–Boltzmann constant* [$\sigma = 5.670 \times 10^{-8}$ W/(m^2 K^4) $= 0.1714 \times 10^{-8}$ Btu/(h ft^2 °R^4)], and T_s is the *absolute* surface temperature. The experimental basis for this famous fourth-power law was first established by the Austrian scientist J. Stefan in 1879. This discovery was followed in 1884 by a theoretical development by another Austrian, L. Boltzmann. Equation (1-20) is shown in Fig. 1-6. Notice that the thermal radiation energy content rapidly falls from a very substantial level for temperatures of the order of 500 K to relatively small values for common environmental temperatures of the order of 50°C.

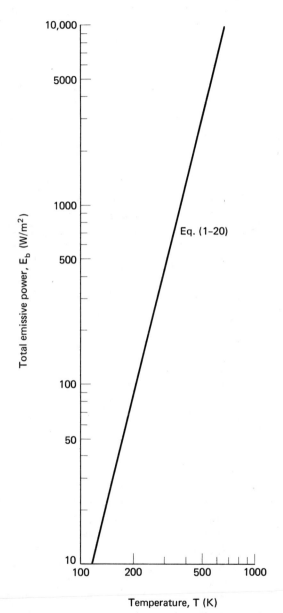

FIGURE 1-6 Stefan–Boltzmann law for blackbody thermal radiation.

For nonblackbody surfaces that absorb less than 100% of the incident radiant energy, the total emissive power E (i.e., the rate of thermal radiation emitted per unit surface area) is generally expressed by

$$E = \varepsilon E_b = \varepsilon \sigma T_s^4 \tag{1-21}$$

where the *emissivity* ε lies between zero and unity. For example, the emissivities of polished aluminum and lampblack paint at room temperature are of the order of 0.1 and 0.96, respectively. The emissivities of various substances are listed in Table A-C-7 of the Appendix. The relationship between the emissivity ε and the absorptivity α will be considered in Chap. 4.

By definition, the rate of radiation heat transfer q_R between two bodies is equal to the *net* rate of exchange of thermal radiation. For two infinite parallel blackbody plates that are separated by a vacuum or a nonparticipating gas, all the thermal radiation emitted from one surface reaches and is absorbed by the other body. The rate of thermal radiation leaving A_s that is absorbed by A_R is $A_s E_{bs}$, and the rate of radiant energy leaving A_R that is absorbed by A_s is $A_R E_{bR}$. With A_s and A_R being equal, it follows that the rate of radiation heat transfer q_R from A_s to A_R is simply

$$q_R = A_s(E_{bs} - E_{bR}) \tag{1-22}$$

or

$$q_R = \sigma A_s(T_s^4 - T_R^4) \tag{1-23a}$$

Taking the opposite orientation, the rate of radiation heat transfer from surface A_R to surface A_s is

$$q_R = \sigma A_s(T_R^4 - T_s^4) \tag{1-23b}$$

For situations such as the one illustrated in Fig. 1-7, only part of the thermal radiation leaving surface A_s reaches surface A_R, with the remainder of the radiation passing to the surroundings. This geometric effect is accounted for by the use of the *thermal radiation shape factor* F_{s-R}, which represents the fraction of thermal radiation leaving body s that reaches body R. For example, F_{s-R} for thermal radiation between two infinite parallel plates is unity, but F_{s-R} is only about 0.15 for the perpendicular plates shown in Fig. 1-7. In this connection, design charts for F_{s-R} for standard geometries are given in Chap. 4.

An expression for the rate of blackbody radiation heat transfer from

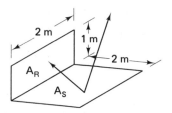

FIGURE 1-7 Thermal radiation from a plate surface A_s to a perpendicular plate A_R and to the surroundings.

A_s to A_R which accounts for this geometric consideration is given by

$$q_R = A_s F_{s-R}(E_{bs} - E_{bR}) = \sigma A_s F_{s-R}(T_s^4 - T_R^4) \qquad (1\text{-}24a)$$

Similarly, the net rate of exchange of thermal radiation from blackbody R to blackbody s is

$$q_R = A_R F_{R-s}(E_{bR} - E_{bs}) = \sigma A_R F_{R-s}(T_R^4 - T_s^4) \qquad (1\text{-}24b)$$

As will be seen in Chap. 4, the *principle of reciprocity* states that $A_s F_{s-R} = A_R F_{R-s}$. Hence, Eqs. (1-24a) and (1-24b) are identical, except for the sign.

To facilitate the practical analysis of systems involving thermal radiation, q_R is sometimes expressed in terms of the temperature difference $T_s - T_R$ by

$$q_R = \bar{h}_R A_s (T_s - T_R) \qquad (1\text{-}25)$$

where $\bar{h}_R$ is the *radiation-heat-transfer coefficient*. Comparing this defining equation for $\bar{h}_R$ with Eq. (1-24a) which was developed for thermal radiation between two blackbody surfaces, we obtain

$$\bar{h}_R = \sigma F_{s-R} \frac{T_s^4 - T_R^4}{T_s - T_R}$$

$$= \sigma F_{s-R}(T_s + T_R)(T_s^2 + T_R^2) \qquad (1\text{-}26)$$

Equations (1-24) and (1-25) will be utilized in the remainder of this chapter and in Chaps. 2 and 3 in the practical analysis of heat-transfer systems that involve ideal blackbody thermal radiation. This practical approach will then be extended to more realistic nonblackbody systems in Chap. 4.

EXAMPLE 1-2

The electrical cabinet shown in Fig. E1-2 is exposed to blackbody radiation with the surrounding walls, which are at 25°C. Determine the rate of radiation heat transfer if the cabinet temperature is 125°C.

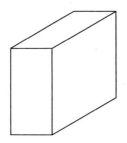

Surrounding walls
at 25°C

Cabinet dimensions:
0.418 m x 0.318 m x 0.160 m.
Total surface area for
radiation: $A_s = 0.268$ m².

FIGURE E1-2   Cabinet mounted on a vertical wall.

Solution

The energy emitted per unit area from the cabinet and the walls is calculated first. Utilizing the Stefan–Boltzmann law, Eq. (1-20), we have

$$E_{bs} = \sigma T_s^4 = 5.67 \times 10^{-8} \frac{W}{m^2\,K^4} (398\ K)^4$$

$$= 1420 \frac{W}{m^2}$$

$$E_{bR} = \sigma T_R^4 = \sigma (298\ K)^4 = 447 \frac{W}{m^2}$$

The rate of blackbody radiation heat transfer from A_s to the enclosure is given by Eq. (1-24a) with the radiation shape factor equal to unity.

$$q_R = A_s F_{s-R} (E_{bs} - E_{bR}) \tag{1-24a}$$

$$= 0.268\ m^2 (1) \left(1420 \frac{W}{m^2} - 447 \frac{W}{m^2} \right)$$

$$= 261\ W$$

If the space between the cabinet and the enclosure is evacuated, the total rate of heat transfer from the cabinet will be 261 W. By performing an energy balance on the cabinet for these conditions, we conclude that the power generated by the electrical system within the cabinet is 261 W. (If the cabinet is surrounded by air or another fluid, heat will also be removed by convection, which is the topic of the next section.)

1-2-3   Convection Heat Transfer

As we have already indicated, *convection* is the transfer of heat from a surface to a moving fluid. The conduction-heat-transfer mechanism always plays a primary role in convection. In addition, the thermal radiation-heat-transfer mechanism is also sometimes a factor. (But in our study of convection heat transfer, thermal radiation to the fluid will be neglected.) In addition to the transfer of energy via the basic heat-transfer mechanisms, energy is also transferred by macroscopic fluid motion in convection processes.

An introductory theoretical treatment of convection-heat-transfer processes is presented in Chap. 10. The theoretical analysis of convection requires that the fundamental laws of mass, momentum, and energy and the particular laws of viscous shear and conduction be utilized in the development of mathematical formulations for the fluid flow and energy transfer. The solution of these equations provides predictions for the velocity and temperature distributions within the fluid, after which predictions are developed for the rate of heat transfer into the fluid by the use of the Fourier law of conduction.

In regard to the theoretical analysis of convection-heat-transfer processes, the particular law for *Newtonian viscosity* can be written as

$$\tau = \frac{dF}{dA} = \mu\frac{\partial u}{\partial y} \tag{1-27}$$

where the viscosity μ is a property of the fluid with units kg/(m s) [or lb$_m$/(ft s)], y the distance from the wall, dA the differential area normal to y, dF the differential shear force acting on the area dA, τ the shear stress, and u the axial velocity. The law of Newtonian viscosity is supported by experimental observation, kinetic theory, and statistical mechanics. The viscosity is often coupled with the density and written as $\nu = \mu/\rho$, where ν is called the *kinematic viscosity*; the units for ν are m^2/s (or ft^2/s). The viscosity and kinematic viscosity of several Newtonian fluids are shown in Tables A-C-3 to A-C-5. The resistance to Newtonian fluid motion that results from shear stress is proportional to the viscosity, such that highly viscous fluids flow much less readily than do fluids with low viscosities such as water and air.

The engineer is generally primarily concerned about the rate of convection heat transfer rather than the temperature distribution within the fluid. Therefore, a practical approach to the analysis of convection heat transfer from surfaces such as the flat plate shown in Fig. 1-8 has been developed which employs an equation of the form

$$q_c = \bar{h}A_s(T_s - T_F) \tag{1-28}$$

where q_c is the rate of heat transferred from a surface at uniform temperature T_s to a fluid with reference temperature T_F, A_s is the surface area, and $\bar{h}$ is the *mean coefficient of heat transfer*; the units for $\bar{h}$ are W/(m^2 °C) [or

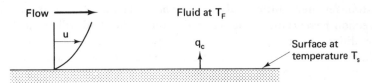

FIGURE 1-8   Convection heat transfer.

Btu/(h ft^2 °F)]. Equation (1-28) is often referred to as the *Newton law of cooling*. A more general form of Newton's law of cooling which applies to cases in which T_s and/or T_F are not uniform will be introduced in Chap. 2. The Newton law of cooling will be seen to be very useful in the analysis of heat-transfer processes involving convection combined with conduction and radiation in Sec. 1-2-4 and Chaps. 2 through 4, and in the evaluation and design of convection-heat-transfer systems in Chaps. 5 through 9.

Approximate ranges of $\bar{h}$ are shown in Table 1-2 for forced and natural convection in air and water. (For forced convection, the fluid motion is caused by mechanical means such as pumps and fans. On the other hand, natural convection is caused by temperature-induced density gradients within the fluid.) The actual value of $\bar{h}$ depends upon the hydrodynamic conditions as well as on the thermodynamic and thermophysical properties of the fluid. The details of evaluating $\bar{h}$ for standard fluid-flow systems will be considered in Chaps. 5 through 10.

TABLE 1-2   Convection-heat-transfer coefficients—range for representative applications

	$\bar{h}$	
System	$W/(m^2 \, °C)^a$	$Btu/(h \, ft^2 \, °F)$
Natural convection:		
Air	5–30	0.9–5
Water	200–600	30–100
Forced convection:		
Air	10–500	2–100
Water	$100–2\times10^4$	$20–4\times10^3$
Oil	$60–2\times10^3$	10–400
Boiling water at 1 atm	$2\times10^3–5\times10^4$	$300–9\times10^3$
Condensation of steam	$5\times10^3–10^5$	$900–2\times10^4$

^{a} 1 W/(m^2 °C)=0.1761 Btu/(h ft^2 °F).

EXAMPLE 1-3

Determine the rate of convection heat transfer from the cabinet in Example 1-2 if it is surrounded by air at 25°C with a coefficient of heat transfer of 6.9 W/(m^2 °C).

Solution

Utilizing the Newton law of cooling, we have

$$q_c = \bar{h} A_s (T_s - T_F) \tag{1-28}$$

$$= 6.9 \frac{W}{m^2 \, {}^\circ C} (0.268 \text{ m}^2)(125\,{}^\circ C - 25\,{}^\circ C)$$

$$= 185 \text{ W}$$

EXAMPLE 1-4

The temperature of a pan of water is maintained on a stove at 150°F with the coefficient of convection heat transfer to the water approximately equal to 50 Btu/(h ft² °F). The pan is made of thin copper plate. The area of the pan that is exposed to convective cooling by the surrounding air is 0.7 ft². Estimate the temperature of the pan if the air temperature is 70°F and the coefficient of heat transfer to the air is 2 Btu/(h ft² °F). See Fig. E1-4.

Solution

The temperature difference across the pan wall can be neglected because of the large thermal conductivity of copper and the small wall thickness. Assuming negligible thermal radiation effects, the first law of thermodynamics gives rise to

$$q_{c1} = q_{c2} + \frac{\Delta E_s}{\Delta t}$$

where $\Delta E_s / \Delta t$ is equal to zero for steady-state conditions. With q_{c1} and q_{c2} expressed in terms of the unknown wall temperature by Eq. (1-28),

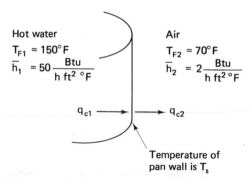

FIGURE E1-4 Convective heat loss from a pan of water.

we obtain

$$\bar{h}_1 A_s (T_{F1} - T_s) = \bar{h}_2 A_s (T_s - T_{F2})$$

$$T_s = \frac{\bar{h}_1 T_{F_1} + \bar{h}_2 T_{F_2}}{\bar{h}_2 + \bar{h}_1}$$

$$= \frac{\left[50\ \text{Btu}/(\text{h ft}^2\ {}^\circ\text{F})\right](150{}^\circ\text{F}) + \left[2\ \text{Btu}/(\text{h ft}^2\ {}^\circ\text{F})\right](70{}^\circ\text{F})}{(50 + 2)\ \text{Btu}/(\text{h ft}^2\ {}^\circ\text{F})}$$

$$= 147{}^\circ\text{F}$$

To calculate the rate of heat transfer, we write

$$q = q_{c1} = q_{c2} = \bar{h}_2 A_s (T_s - T_{F2})$$

$$= 2 \frac{\text{Btu}}{\text{h ft}^2\ {}^\circ\text{F}} (0.7\ \text{ft}^2)(147{}^\circ\text{F} - 70{}^\circ\text{F})$$

$$= 108 \frac{\text{Btu}}{\text{h}}$$

1-2-4   Combined Modes of Heat Transfer

Many heat-transfer processes encountered in practice involve combinations of conduction, thermal radiation, and convection. A situation in which combinations of these basic heat-transfer modes occur simultaneously is illustrated in Fig. 1-9. In this wall, which consists of two plates

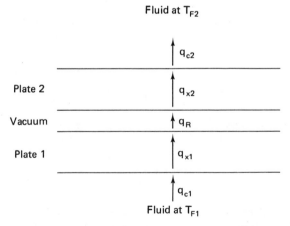

FIGURE 1-9 Heat-transfer process involving combined conduction, thermal radiation, and convection.

separated by a vacuum, heat is (1) convected from the fluid at T_{F1} to plate 1, (2) conducted through plate 1, (3) radiated from plate 1 to plate 2, (4) conducted through plate 2, (5) and convected from plate 2 to the fluid at T_{F2}. Similar composite walls have been developed for the storage of cryogenic liquids, which consist of multiple layers of highly reflective materials which are separated by evacuated spaces. These *superinsulations* provide very effective insulative walls with apparent thermal conductivities that are as low as 3×10^{-4} W/(m °C) [8].

EXAMPLE 1-5

An electronic cabinet made of anodized aluminum is cooled by natural convection and radiation. The surface area of the cabinet is 0.268 m², the ambient temperature is 25°C, and $\bar{h} = 6.9$ W/(m² °C). Estimate the rate of heat transfer from the cabinet if its surface temperature is to be maintained at 125°C.

Solution

The total rate of heat transfer q from the cabinet is

$$q = q_c + q_R$$

where q_c is given by Eq. (1-28). Because the cabinet is made of black anodized aluminum, we will utilize the blackbody approximation given by Eq. (1-24). Referring to Examples 1-2 and 1-3, we obtain

$$q_c = \bar{h} A_s (T_s - T_F) = 185 \text{ W}$$

$$q_R = A_s F_{s-R}(E_{bs} - E_{bR}) = \sigma A_s F_{s-R}(T_s^4 - T_R^4) = 261 \text{ W}$$

such that the total rate of heat transfer is

$$q = 446 \text{ W}$$

Note that the radiation heat transfer accounts for a very significant 58.5% of the total. Whereas radiation is important for systems such as this involving natural convection, radiation is often not a factor in forced convection systems. For example, if the cabinet is cooled by forced air with $\bar{h} = 500$ W/(m² °C), only about 1.9% of the heat transfer is accomplished by radiation. (Radiation is also often insignificant in systems involving bright metal surfaces with low emissivities.)

EXAMPLE 1-6

Liquid petroleum gas (LPG), which is widely used for domestic and industrial heating, consists of propane, butane, and mixtures of the two. LPG must be kept in liquid form because of its high volatility. The

liquification is often accomplished by refrigeration so that LPG can be
stored at atmospheric pressure. Butane is stored at about 0°C and
propane at -40°C.

With this brief background, we consider a cryogenic fluid that is
to be stored in the 2-m-diameter spherical chamber shown in Fig. E1-6.
A temperature of -40°C must be maintained by the refrigerant, which
flows within the 1-cm-thick shell surrounding the inner storage chamber. The outer part of the vessel consists of a 10-cm-thick insulating
material with $k = 0.60$ W/(m °C). The fluid surrounding the vessel is at
35°C with $\bar{h} = 150$ W/(m² °C). Determine the amount of refrigeration
that is required to maintain the fluid at -40°C.

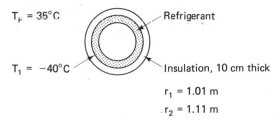

$T_F = 35$°C

Refrigerant

$T_1 = -40$°C

Insulation, 10 cm thick

$r_1 = 1.01$ m

$r_2 = 1.11$ m

FIGURE E1-6 Spherical vessel.

Solution

By applying the first law of thermodynamics to the outer surface of the
insulating wall, we obtain

$$\Sigma \dot{E}_i = \Sigma \dot{E}_o + \frac{\Delta E_s}{\Delta t}$$

$$q_c = q_r \tag{a}$$

The rate of heat transfer by convection q_c is given by Eq. (1-28),

$$q_c = \bar{h} A_s (T_F - T_2)$$

$$= \bar{h} 4\pi r_2^2 (35°C - T_2) = 150 \frac{W}{m^2 \ °C} (4\pi)(1.11 \text{ m})^2 (35°C - T_2)$$

$$= 2320 \frac{W}{°C} (35°C - T_2) \tag{b}$$

The rate of heat transfer by radial conduction q_r is given by Eq. (1-18) with $S = 4\pi r_1 r_2 / (r_2 - r_1)$; that is,

$$q_r = kS(T_2 - T_1)$$

$$= \frac{4\pi r_1 r_2 k}{r_2 - r_1}(T_2 + 40°C)$$

$$= 84.5 \frac{W}{°C}(T_2 + 40°C) \tag{c}$$

Similarly, the application of the first law of thermodynamics to the inside surface of the insulating wall gives

$$q_r = q_{ref} \tag{d}$$

such that the rate of heat transferred through the insulation by conduction q_r must be taken out by refrigeration.

To determine q_r or q_c, we first solve for the surface temperature T_2 by utilizing Eqs. (a) through (c).

$$84.5 \frac{W}{°C}(T_2 + 40°C) = 2320 \frac{W}{°C}(35°C - T_2)$$

$$T_2 = \frac{(2320 \text{ W}/°C)(35°C) - (84.5 \text{ W}/°C)(40°C)}{2320 \text{ W}/°C + 84.5 \text{ W}/°C}$$

$$= 32.4°C$$

Substituting this result back into Eq. (c), we have

$$q_{ref} = q_r = kS(T_2 - T_1)$$

$$= 84.5 \frac{W}{°C}(32.4°C + 40°C)$$

$$= 6120 \text{ W}$$

Hence, approximately 6.12 kW of refrigeration is required.

As suggested in Sec. 1-2-4, superinsulation-type arrangements involving at least one evacuated space are often used in such cryogenic applications. This type of arrangement is considered in Prob. 1-23.

1-3 ANALOGY BETWEEN HEAT TRANSFER AND THE FLOW OF ELECTRIC CURRENT

The basic laws of conduction, thermal radiation, and convection heat transfer often lead to relationships for the heat transfer rate q of the form

$$q = \frac{\Delta T}{R} \tag{1-29}$$

where the resistance R is a function of the thermal and geometric properties. The heat-transfer rate q and temperature T are analogous to the current I_e and voltage E_e in an electrical circuit. The thermal resistance R and the potential difference ΔT are dependent upon the heat-transfer mechanism considered. For example, Eq. (1-18) indicates that one-dimensional conduction heat transfer through a stationary medium with surface temperatures equal to T_1 and T_2 can be represented by a simple series circuit with resistance $R_k = 1/(kS)$ and potential difference $\Delta T = T_1 - T_2$. Similarly, $R_R = 1/(\bar{h}_R A_s)$ and $\Delta T = T_s - T_R$ for thermal radiation and $R_c = 1/(\bar{h}A_s)$ and $\Delta T = T_s - T_F$ for convection. As illustrated in Examples 1-7 and 1-8, heat-transfer problems can be represented by electrical networks. Such analogous electrical circuits often facilitate the solution of rather complex heat-transfer problems. As will be seen in Chaps. 2 and 3, electrical networks can also be set up for unsteady and general multidimensional systems.

EXAMPLE 1-7

100 W is generated by the flow of electric current in a 1-cm-diameter copper cable of 1 m length. The surrounding air temperature is 50°C and $\bar{h}$ is equal to 200 W/(m² °C). Determine the surface temperature of the cable for negligible thermal radiation losses.

Solution

Applying the first law of thermodynamics to the cable, we have

$$q = \dot{W} = 100 \text{ W}$$

The analogous thermal circuit is shown in Fig. E1-7 for the case in which the radiation loss from the cable is negligible.

To obtain T_s, we write

$$q = \dot{W} = \frac{T_s - T_F}{R_c}$$

or

$$T_s = \dot{W} R_c + T_F$$

$$= \frac{100 \text{ W}}{\left[200 \text{ W}/(\text{m}^2 \text{ °C})\right] \pi (0.01 \text{ m})(1 \text{ m})} + 50\text{°C}$$

$$= 65.9\text{°C}$$

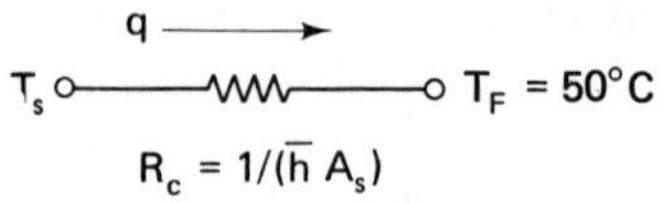

FIGURE E1-7 Thermal circuit.

EXAMPLE 1-8

Because excessive temperature shortens the life of transistors, the heat transfer in transistors and compact integrated circuits (ICs, often called chips) which contain transistors, resistors, and capacitors, is a critical design consideration. Consequently, manufacturers of transistors and ICs generally specify the maximum allowable internal junction temperature T_J and/or case temperature T_s, and the thermal resistance from the junction to case R_{JC} and from the case to ambient R_{CA}, assuming natural convection and radiative cooling. Whereas some silicon transistors and ICs can operate at temperatures as high as 200°C, the maximum temperature of components made of germanium is usually less than 100°C. In general, the life of a transistor or IC increases as its operating temperature decreases.

To illustrate, the thermal characteristics of the low-power silicon voltage regulator IC shown in Fig. El-8a are specified by the manufacturer for an ambient temperature of 25°C as follows:

> Maximum operating junction temperature: 150°C
> Thermal resistance: $R_{JC} = 60°C/W$
> Thermal resistance: $R_{CA} = 90°C/W$
> Maximum power rating: 600 mW

We want to calculate the junction and case temperatures of this device at its maximum power output of 600 mW.

FIGURE E1-8a Integrated circuit—voltage regulator ($\frac{1}{2} \times 1\frac{1}{2}$ cm).

Solution

The thermal circuit for this system is shown in Fig. El-8b. To calculate the junction temperature T_J for 600-mW output, we write

$$q = \frac{T_J - T_F}{R_{JC} + R_{CA}}$$

$$T_J = q(R_{JC} + R_{CA}) + T_F = 0.6 \ \text{W}\left(60 \ \frac{°C}{W} + 90 \frac{°C}{W}\right) + 25°C$$

$$= 115°C$$

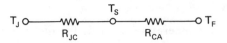

FIGURE E1-8b   Thermal circuit.

This temperature is well within the safe operating range for the unit.

The case temperature T_s at this maximum level of power dissipation is now calculated.

$$q = \frac{T_J - T_s}{R_{JC}}$$

$$T_s = T_J - qR_{JC}$$

$$= 115°C - 0.6 \text{ W}\left(60 \frac{°C}{W}\right)$$

$$= 79°C$$

PROBLEMS

1-1. List several heat-transfer processes that occur in the human body.

1-2. Describe two methods by which solar energy could be gathered during cloudless sunny days, stored, and utilized at some later time.

1-3. Describe a method by which electrical energy could be transformed into stored thermal energy during nonpeak night-time hours and utilized throughout the day for home-heating purposes.

1-4. Convert the following English units to SI units and check the results with Table A-B-1. (a) 1 Btu/h; (b) 1 Btu/(h ft²); (c) 1 Btu/(h ft² °F); (d) 1 Btu/(h ft °F); (e) 1 Btu/(h ft² °R⁴).

1-5. Determine the rate of heat transfer per unit area through a 5-in.-thick brick wall with surface temperatures equal to 35°F and −20°F; $k = 0.399$ Btu/(h ft °F).

1-6. Determine the rate of heat transfer in a 5-cm-long hollow copper rod with radii equal to 1 cm and 5 cm. The surfaces at r_1 and r_2 are maintained at $T_1 = 25°C$ and $T_2 = 75°C$, and the ends are insulated.

1-7. Determine the rate of heat transfer in the hollow cylinder of Prob. 1-6 if the surfaces at r_1 and r_2 are insulated and the ends are maintained at 25°C and 75°C, respectively.

1-8. Determine the rate of heat transfer in a copper spherical shell with radii equal to 3 cm and 6 cm if the surface temperatures are −120°C and 35°C.

1-9. An electrical device operating at 180°C with a 0.5-m² surface area is enclosed in a cabinet that has a surface temperature of 85°C. Determine the rate of radiation heat transfer between the two surfaces, assuming blackbody conditions.

1-10. A 1-mm-thick electrical heating plate generates 25 kW/m^3 per unit volume. One surface is insulated and the other is exposed to radiant exchange with a second surface at 70°C. Determine the surface temperature of the heater, assuming blackbody radiation.

1-11. A plate is exposed to a uniform heat flux of 100 Btu/(h ft^2) on one side. The other side exchanges thermal radiation with a surface at 0°F. Determine the temperature of the plate.

1-12. A very thin copper plate coated with carbon black lies between two blackbody plates at temperatures of 1000°C and 50°C. Determine the temperature of the intermediate plate and the rate of heat transfer per unit area.

1-13. Assume that a fluid at 10°C passes over a plate with surface temperature equal to 100°C and with $\bar{h}$ equal to 150 W/(m^2 °C). Determine the rate of heat transfer from 1 m^2 of the plate.

1-14. Water at atmospheric pressure and 30°F passes over a plate with 200 °F surface temperature and $\bar{h} = 500$ Btu/(h ft^2 °F). Sketch the analogous electrical circuit and determine the total quantity of heat transferred over a period of 30 s from a 1-ft^2 area.

1-15. Electricity passes through a 1-mm-diameter wire of 1 m length submerged in boiling water at a pressure of 20 psia. $\bar{h}$ is given as 4500 W/(m^2 °C) and the resistance of the wire is 0.5 Ω. Determine the current required to maintain the wire surface at 150°C. Also determine the rate of heat transferred from the wire.

1-16. A very thin copper plate coated with carbon black exchanges radiant energy with a blackbody plate at 1000°C. The other surface is exposed to a fluid with T_F equal to 50°C and $\bar{h}$ equal to 150 W/(m^2 °C). Determine the rate of heat transfer per unit area and the surface temperature.

1-17. Determine the rate of radiation heat transfer between two concentric spheres with 2-cm and 10-cm diameters. The surfaces are coated with carbon black and maintained at 0°C and 150°C, respectively.

1-18. Determine the rate of radiation heat transfer per unit length between two very long blackbody concentric cylinders with 2-in. and 10-in. diameters. The surfaces are maintained at 100°F and 32°F, respectively.

1-19. Show that the thermal resistance associated with blackbody thermal radiation is given by $R_R = [\sigma A_s F_{s-R}(T_s^2 + T_R^2)(T_s + T_R)]^{-1}$.

1-20. Solve Example 1-4 by the electrical analogy method.

1-21. Solve Example 1-6 by the electrical analogy method.

1-22. Forced-convection air cooling with $\bar{h} = 500$ W/(m^2 °C) and radiation are used to maintain the cabinet surface of Example 1-5 at 125°C. Estimate the rate of heat transfer and the relative importance of radiation in this problem.

1-23. A cryogenic fluid is maintained at −40°C in a 2-m diameter spherical chamber. The outer part of the vessel consists of a 1-cm thick shell containing refrigerant at $-40°C$, a 1-mm thick evacuated space, and a 10-cm thick layer of insulation [$k = 0.60$ W/(m °C)]. The surrounding fluid is at 35°C with $\bar{h} = 150$ W/(m² °C) and the surfaces can be considered to be blackbodies. Sketch the thermal circuit for this system and describe how the temperature of the inner radiative surface and the rate of heat transfer can be determined. (The actual solution to this type of problem is considered in Chap. 4.)

1-24. 0.05 kg/s of saturated liquid water at atmospheric pressure enters a tube (5 cm diameter, 0.5 m length) with surface temperature maintained at 150°C and $\bar{h} = 5000$ W/(m² °C). Determine the total rate of heat transfer to the fluid and the quality of the exiting steam.

1-25. 100 W is generated by the flow of electric current in a 1-m-long, 1-cm-diameter cable which has been heavily oxidized. The surrounding air temperature is 50°C, $\bar{h}$ is 200 W/(m² °C), and blackbody thermal radiation exchange occurs between the cable and the surroundings at 50°C. Determine the surface temperature of the cable. Are the effects of thermal radiation important in this problem?

1-26. A power transistor is to be operated at 2 W. The manufacturer rating for the thermal resistance from the case to the ambient is 30°C/W for natural convection cooling. Determine the temperature of the case if the air temperature is 25°C.

1-27. The junction-to-case thermal resistance of a germanium transistor is rated at 2°C/W and the case-to-ambient thermal resistance is 30°C/W. Determine the junction temperature if the power dissipation is 2 W and $T_F = 30°C$. (Note that if the junction temperature of germanium transistors exceeds 100°C, burnout will often occur.)

1-28. Certain high-reliability silicon power transistors can operate at junction temperatures as high as 200°C without burning out. Determine the maximum power at which such a transistor can operate if the junction-to-case thermal resistance is 2.5°C/W, the case-to-ambient thermal resistance is 75°C/W, and $T_F = 25°C$.

1-29. Estimate the level of importance of thermal radiation for the cable considered in Example 1-7.

REFERENCES

[1] FOURIER, J. B., *Théorie analytique de la Chaleur*. Paris, 1822; English translation by A. Freeman, Dover Publications, Inc., New York, 1955.

[2] TOULOUKIAN, Y. S., R. W. POWELL, C. Y. HO, and P. G. KLEMENS, *Thermophysical Properties of Matter*, Vol. 1, *Thermal Conductivity—Metallic Elements and Alloys*. New York: IFI/Plenum Data Corporation, 1970.

[3] TOULOUKIAN, Y. S., R. W. POWELL, C. Y. HO, and P. G. KLEMENS, *Thermophysical Properties of Matter*, Vol. 2, *Thermal Conductivity—Nonmetallic Solids*. New York: IFI/Plenum Data Corporation, 1970.

[4] TOULOUKIAN, Y. S., P. E. LILEY, and S. C. SAXENA, *Thermophysical Properties of Matter*, Vol. 3, *Thermal Conductivity—Nonmetallic Liquids and Gases*. New York: IFI/Plenum Data Corporation, 1970.

[5] POWELL, R. W., C. Y. HO, and P. E. LILEY, "Thermal Conductivity of Selected Materials," *NSRDS-NBS 8*. Washington, D.C.: U.S. Department of Commerce, National Bureau of Standards, 1966.

[6] HO, C. V., R. W. POWELL, and P. E. LILEY, *Thermal Conductivity of Elements*, Vol. 1, *First Supplement to Journal of Physical and Chemical Reference Data*. Washington, D.C.: American Chemical Society, 1972.

[7] ECKERT, E. R. G., and R. M. DRAKE, *Analysis of Heat and Mass Transfer*. New York: McGraw-Hill Book Company, 1972.

[8] BARRON, R, *Cryogenic Systems*. New York: McGraw-Hill Book Company, 1967.

CHAPTER 2

ONE-DIMENSIONAL
HEAT TRANSFER

2-1 INTRODUCTION

The primary objective of this chapter is to establish the fundamental
concepts involved in analyzing basic one-dimensional heat-transfer problems. These principles first will be demonstrated in the context of steady-
state conduction heat transfer in a plane wall. One-dimensional analyses
will then be developed for several other representative problems. This
study will provide the foundation for our analysis of more complex
multi-dimensional conduction, thermal radiation, and convection heat
transfer in Chaps. 3 through 10.

2-2 CONDUCTION HEAT TRANSFER IN A FLAT PLATE

2-2-1 Introduction

We consider one-dimensional steady conduction heat transfer in the plane
wall shown in Fig. 2-1(a). The classic differential approach to solving this
problem is presented for several standard boundary conditions, after which
an efficient but less general short method will be introduced.

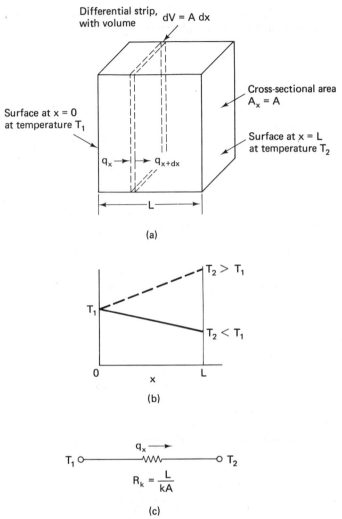

FIGURE 2-1   One-dimensional conduction heat transfer in a flat plate. (a) System and differential control volume. (b) Temperature distribution. (c) Thermal circuit.

2-2-2   Differential Formulation

To develop mathematical equations that represent the energy transfer within the wall, attention is focused upon the differential strip $A\,dx$ shown in Fig. 2-1(a). First, we recognize that $\Delta E_s/\Delta t$ is zero because of steady-state conditions, and conduction heat transfer in the y and z directions is zero because no temperature gradients occur in these directions. Thus, the temperature is a function of x alone. With these points in mind, the application of the first law of thermodynamics given by Eq. (1-3) to the element $A\,dx$ gives

Step 1

$$q_x = q_{x+dx} \tag{2-1}$$

This equation indicates that q_x does not change with x.

The next step in the development of the differential formulation requires the use of a basic relationship between q_x and q_{x+dx}. This relationship is established on the basis of the definition of the derivative as follows:

$$\frac{dq_x}{dx} = \lim_{\Delta x \to 0} \frac{q_{x+\Delta x} - q_x}{\Delta x} \tag{2-2}$$

By replacing Δx by the infinitesimal quantity dx, this equation gives rise to

$$q_{x+dx} = q_x + \frac{dq_x}{dx}\,dx \tag{2-3}$$

Utilizing this expression, we eliminate q_{x+dx} in Eq. (2-1) to obtain

Step 2

$$\frac{dq_x}{dx} = 0 \tag{2-4}$$

The quantity q_x is now expressed in terms of the temperature distribution T by introducing the Fourier law of conduction, Eq. (1-16); that is,

Step 3

$$\frac{d}{dx}\left(-kA_x \frac{dT}{dx}\right) = 0 \tag{2-5}$$

Because the cross-sectional area A_x is independent of x for this application ($A_x = A$), this equation reduces to

$$\frac{d}{dx}\left(k \frac{dT}{dx}\right) = 0 \tag{2-6}$$

For situations in which the thermal conductivity is essentially uniform, this expression takes the simple form

$$\frac{d^2 T}{dx^2} = 0 \tag{2-7}$$

The differential formulation is completed by writing the boundary conditions. As the term implies, a boundary condition is a mathematical statement pertaining to the behavior of the dependent variable (T in our case) at the system boundary. It should be recalled that the number of boundary conditions is equal to the highest-order derivative in an ordinary differential equation. Hence, two boundary conditions must be specified in order to solve steady one-dimensional conduction heat-transfer problems. For the moment, the surface temperatures are simply identified by T_1 and T_2,

$$T(0) = T_1, \qquad T(L) = T_2 \qquad (2\text{-}8,9)$$

The treatment of several common types of boundary conditions will be considered later in this section.

2-2-3   Solution

The solution to Eq. (2-7) is written on the basis of a simple double integration as

$$T = C_1 x + C_2 \qquad (2\text{-}10)$$

By applying the boundary conditions, the constants of integration C_1 and C_2 are written in terms of the surface temperatures as follows:

$$T_1 = C_2 \qquad (2\text{-}11)$$
$$T_2 = C_1 L + C_2 \qquad (2\text{-}12)$$

or

$$C_1 = \frac{T_2 - T_1}{L} \qquad (2\text{-}13)$$

Therefore, the temperature distribution takes the form

$$T = (T_2 - T_1)\frac{x}{L} + T_1$$

or

$$\frac{T - T_1}{T_2 - T_1} = \frac{x}{L} \qquad (2\text{-}14)$$

This linear relationship is shown in Fig. 2-1(b) for situations in which $T_1 > T_2$ and $T_1 < T_2$. Because Eq. (2-14) does not involve k, we conclude that the temperature distribution is totally independent of the material (steel, wood, asbestos, etc.).

An expression is obtained for the rate of heat transfer in the wall by utilizing the Fourier law of conduction together with Eq. (2-14); that is,

$$q_x = -kA \frac{dT}{dx}\bigg|_x = \frac{kA}{L}(T_1 - T_2) \qquad (2\text{-}15)$$

Consistent with Eq. (2-1), this equation indicates that the rate of heat transfer in the plate is independent of x.

Equation (2-15) provides the basis for the practical expressions given by Eqs. (1-18) and (1-29). For example, the thermal resistance R_k is simply

$$R_k = \frac{L}{kA} \qquad (2\text{-}16)$$

for a flat plate with uniform thermal conductivity. The thermal circuit associated with one-dimensional conduction heat transfer in a plate is shown in Fig. 2-1(c).

EXAMPLE 2-1

Determine the temperature distribution and rate of heat transfer across a copper plate with cross-sectional area 1 m² and thickness 5 cm and with surface temperatures of 130°C and 15°C.

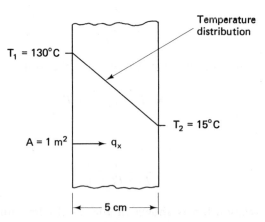

FIGURE E2-1   Temperature distribution in a plane wall.

Solution

The temperature distribution is given by Eq. (2-14),

$$T = (15°C - 130°C)\frac{x}{0.05\ m} + 130°C$$

This linear distribution is shown in Fig. E2-1. The rate of heat transfer obtained from Eq. (2-15) is

$$q_x = 380\frac{W}{m\ °C}\left(\frac{1\ m^2}{0.05\ m}\right)(130°C - 15°C)$$

$$= 874\ kW$$

2-2-4 Boundary Conditions

For situations in which the surface temperatures T_1 and T_2 are known, the boundary conditions given by Eqs. (2-8) and (2-9), together with the differential equation, provide a complete mathematical model of the problem. However, for the many applications encountered in practice in which one or both surface temperatures are not known, the actual heat transfer at the boundaries must be accounted for. For such cases, the solutions for T and q developed above still apply, but T_1 and T_2 must be evaluated. Examples of other common types of boundary conditions include convection, thermal radiation, specified heat flux, and combinations of these. These types of boundary conditions will be considered later in this section, as will another important type of boundary condition which involves interfacial conduction in composites.

Problems involving these other types of boundary conditions can be solved by any of several approaches. One way is to replace Eq. (2-8) and/or Eq. (2-9) by the formal mathematical statement of the boundary condition(s) and to then solve for the constants of integration. Thus, for a flat plate with uniform thermal conductivity, the boundary conditions are utilized to evaluate C_1 and C_2 in Eq. (2-10). In another somewhat simpler approach, Eqs. (2-8) and (2-9) are retained with the unknown surface temperatures T_1 and T_2 being determined by the use of the electrical analogy concept. For example, R_k is given by Eq. (2-16) for a flat plate with uniform thermal conductivity. Both the formal and the network approaches will be demonstrated in the sections that follow.

Convection boundary condition

For the case illustrated in Fig. 2-2(a), in which the surface at $x = 0$ is exposed to fluid with temperature T_F, the heat convected from the fluid is conducted into the wall. Consequently, the boundary condition at $x = 0$

is written as

$$q_c = q_x$$

$$\bar{h}[\,T_F - T(0)\,] = -k \frac{dT}{dx}\Big|_0 \tag{2-17}$$

where the surface temperature $T(0) = T_1$, is unknown. The other surface at $x = L$ is maintained at a known temperature T_2, such that the second boundary condition is given by Eq. (2-9).

At this point in our study, $\bar{h}$ is assumed to be known. Details concerning the evaluation of $\bar{h}$ are presented in Chaps. 5 through 10, which deal specifically with the study of convection heat transfer.

Substituting the solution for T given by Eq. (2-10) into Eqs. (2-9) and (2-17), we obtain Eq. (2-12),

$$T_2 = C_1 L + C_2 \tag{2-12}$$

and

$$-kC_1 = \bar{h}(T_F - C_2) \tag{2-18}$$

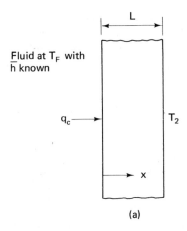

(a)

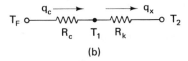

(b)

FIGURE 2-2 One-dimensional conduction heat transfer in a plane wall with convection at one surface. (a) System. (b) Thermal circuit.

The solution of these two equations for the two constants of integration gives rise to an expression for T of the form

$$\frac{T-T_2}{T_F-T_2} = \frac{L-x}{L+k/\bar{h}} \tag{2-19}$$

Hence, the unknown surface temperature T_1 is given by

$$\frac{T_1-T_2}{T_F-T_2} = \frac{L}{L+k/\bar{h}} = \frac{1}{1+k/(\bar{h}L)} \tag{2-20}$$

With T_1 now known, the temperature distribution within the wall can also be conveniently expressed by Eq. (2-14),

$$\frac{T-T_1}{T_2-T_1} = \frac{x}{L} \tag{2-14}$$

A more efficient approach to the solution of this convection-heat-transfer problem involves the use of the electrical analog. The series circuit is shown in Fig. 2-2(b). An analysis of this elementary circuit indicates that the rate of heat transfer can be expressed as

$$q = \frac{T_F-T_2}{R_c+R_k} \tag{2-21}$$

where $R_c = 1/(\bar{h}A)$, $R_k = L/(kA)$, and $q = q_x = q_c$. This circuit can also be solved for the surface temperature T_1 by writing

$$q = q_c = q_x$$
$$\frac{T_F-T_2}{R_c+R_k} = \frac{T_F-T_1}{R_c} = \frac{T_1-T_2}{R_k} \tag{2-22}$$

By equating q and q_c, we have

$$\frac{T_1-T_F}{T_2-T_F} = \frac{R_c}{R_c+R_k} = \frac{1}{1+L\bar{h}/k} \tag{2-23}$$

Alternatively, by equating q and q_x, we obtain Eq. (2-20), and by equating q_c and q_x, we get

$$T_1 = \frac{T_F/R_c+T_2/R_k}{1/R_c+1/R_k} = \frac{T_F\bar{h}+T_2 k/L}{\bar{h}+k/L} \tag{2-24}$$

All three of these equations for T_1 are equivalent. Note that Eq. (2-24) can also be obtained directly from Eq. (2-17) by replacing q_x by $(T_1 - T_2)/R_k$.

It should also be noted that the convective resistance is negligible for large values of $\bar{h}$. For this situation, Eq. (2-23) indicates that the surface temperature T_1 is simply

$$T_1 = T_F \qquad (2\text{-}25)$$

Other representative problems involving one-dimensional steady conduction heat transfer in flat plates with convection, thermal radiation, and uniform heat flux boundary conditions are presented in Examples 2-2 through 2-4.

EXAMPLE 2-2

Two fluids are separated by a 2-in.-thick stainless steel plate [$k = 45$ Btu/(h ft °F)] with an area of 10 ft². The fluid temperatures and mean coefficient of heat transfer are $T_{F1} = 50°F$, $T_{F2} = 0°F$, $\bar{h}_1 = 200$ Btu/(h ft² °F), and $\bar{h}_2 = 150$ Btu/(h ft² °F). Determine the surface temperatures and the rate of heat transfer through the plate for negligible thermal radiation at the surfaces. Utilize the electrical analogy approach.

Solution

The analogous thermal circuit is shown in Fig. E2-2. The thermal resistances are

$$R_{c1} = \frac{1}{\bar{h}_1 A} \qquad R_k = \frac{L}{kA} \qquad R_{c2} = \frac{1}{\bar{h}_2 A}$$

Solving for q, we obtain

$$q = \frac{T_{F1} - T_{F2}}{R_{c1} + R_k + R_{c2}} = \frac{(50°F - 0°F)A}{\left[\dfrac{1}{200} + \dfrac{2}{45(12)} + \dfrac{1}{150} \right] \dfrac{\text{h ft}^2\ °F}{\text{Btu}}}$$

$$q = 32{,}500 \frac{\text{Btu}}{\text{h}} \qquad q'' = \frac{q}{A} = 3250 \frac{\text{Btu}}{\text{h ft}^2}$$

The temperature distribution within the plate is given by Eq. (2-14),

$$\frac{T - T_1}{T_2 - T_1} = \frac{x}{L}$$

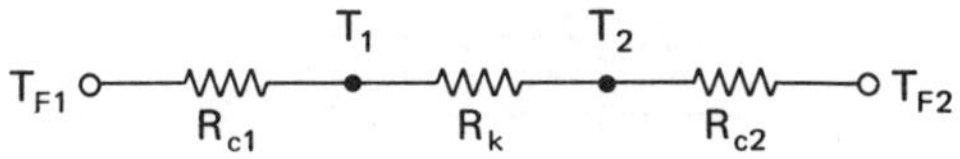

FIGURE E2-2 Thermal circuit.

T_1 and T_2 are obtained as follows:

$$q = \frac{T_{F1} - T_1}{R_{c1}} \qquad T_1 = T_{F1} - R_{c1}q$$

$$T_1 = 50°F - \frac{32{,}500 \text{ Btu/h}}{\left[200 \text{ Btu}/(\text{h ft}^2 \text{ °F})\right]10 \text{ ft}^2} = 33.8°F$$

$$q = \frac{T_2 - T_{F2}}{R_{c2}} \qquad T_2 = T_{F2} + R_{c2}q$$

$$T_2 = 0°F + \frac{32{,}500 \text{ Btu/h}}{\left[150 \text{ Btu}/(\text{h ft}^2 \text{ °F})\right]10 \text{ ft}^2} = 21.7 \text{ °F}$$

To check our assumption that the thermal radiation is small, we compute the total emissive power for surface A_1.

$$E_1 = \varepsilon \sigma T_1^4 = \varepsilon \left(0.171 \times 10^{-8} \frac{\text{Btu}}{\text{h ft}^2 \text{ °R}^4}\right)(494 \text{ °R})^4$$

$$= \varepsilon \left(101 \frac{\text{Btu}}{\text{h ft}^2}\right)$$

Referring to Table A-C-7, we find that the emissivity for stainless steel is of the order of 0.25. It follows that E_1 is less than 1% of q''. Thus, the assumption that thermal radiation effects are small is reasonable.

EXAMPLE 2-3

A 1-in.-thick carbon steel plate with a 1-ft^2 cross-sectional area is exposed to a uniform heat flux of 5000 Btu/(h ft^2). The other surface of the plate is maintained at a temperature of 212°F. Determine the unknown surface temperature.

Solution

The situation in which a surface is exposed to a specified uniform rate of heat transfer per unit area $q_0''(\equiv q_0/A)$ is illustrated in Fig. E2-3a on page 42. (This type of boundary condition can be achieved by heating the surface with an electrical heating plate or by thermal radiation with

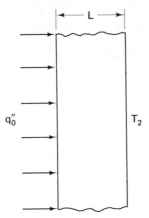

FIGURE E2-3a   Plane wall with uniform heat flux.

$T_R \gg T_s$.) The uniform heat-flux boundary condition is written as

$$-k \frac{dT}{dx}\bigg|_0 = q_0''$$

The thermal circuit for this problem is shown in Fig. E2-3b, where $q_x = q_0$ and where $T(0) = T_1$ is unknown. Based on this simple circuit, we write

$$q_0'' = \frac{T_1 - T_2}{L/k}$$

such that T_1 is given by

$$T_1 = \frac{L}{k} q_0'' + T_2$$

Substituting into this equation, we obtain

$$T_1 = \frac{1\ \text{ft}}{12} \frac{1}{31\ \text{Btu}/(\text{h ft}^2\ {}^\circ\text{F})} \frac{5000\ \text{Btu}}{\text{h ft}^2} + 212\,^\circ\text{F}$$

$$= 225\,^\circ\text{F}$$

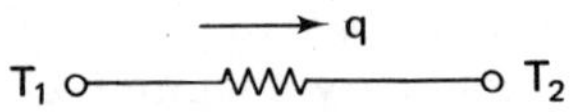

$T_1 \circ\!\!-\!\!\!\text{WW}\!\!\!-\!\!\circ T_2$

FIGURE E2-3b   Thermal circuit.

EXAMPLE 2-4

One surface of the plate shown in Fig. E2-4a is exposed to blackbody thermal radiation with $T_R = 1000°C$. The other surface is maintained at a temperature $T_2 = 15°C$. Obtain a solution for the unknown surface temperature at the radiating surface and the heat transfer rate. Assume that $F_{s-R} \simeq 1$.

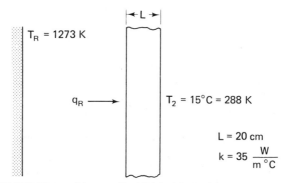

FIGURE E2-4a Plane wall with blackbody thermal radiation.

Solution

The boundary condition at the surface which is exposed to thermal radiation is given by

$$q_R = q_x$$

$$\sigma A_s F_{s-R}(T_R^4 - T_1^4) = -kA \frac{dT}{dx}\bigg|_0 \qquad (a)$$

where $T(0)$ is denoted by T_1 and $A_s = A_x = A$. The thermal circuit for this problem is shown in Fig. E2-4b, where $R_R = [\sigma A_s F_{s-R}(T_R^2 + T_1^2)(T_R + T_1)]^{-1}$.

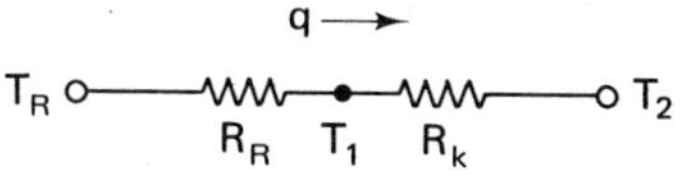

$$q \longrightarrow$$

$$T_R \circ\!\!-\!\!\wedge\!\!\wedge\!\!\wedge\!\!-\!\!\bullet\!\!-\!\!\wedge\!\!\wedge\!\!\wedge\!\!-\!\!\circ T_2$$
$$R_R \quad T_1 \quad R_k$$

FIGURE E2-4b Thermal circuit.

We have shown that the rate of heat transfer through the wall by conduction q_x can be written as

$$q_x = \frac{kA}{L}(T_1 - T_2) \qquad (b)$$

Thus, Eq. (a) reduces to

$$\sigma F_{s-R}\left(T_R^4 - T_1^4\right) = \frac{k}{L}\left(T_1 - T_2\right) \tag{c}$$

An iterative approach is suggested for solving this nonlinear equation for the unknown $T(0) = T_1$. This simple approach is outlined as follows: (1) rewrite Eq. (c) in the form (to four significant figures)

$$T_1 = T_2 + F_{s-R}\frac{\sigma L}{k}\left(T_R^4 - T_1^4\right)$$

$$= \frac{T_2 + F_{s-R}(\sigma L/k)T_R^4}{1 + F_{s-R}(\sigma L/k)T_1^3}$$

$$= \frac{1139 \text{ K}}{1 + 3.24 \times 10^{-10}T_1^3/\text{K}^3} \tag{d}$$

(2) compute an approximate value of T_1 by substituting an assumed value of T_1 into the right-hand side of Eq. (d), (3) compute a new approximate value of T_1 by substituting the value computed in step 2 into the right-hand side of Eq. (d), (4) continue this procedure until the calculations for T_1 converge satisfactorily. With T_1 known, the rate of heat transfer can be calculated from Eq. (b).

For our particular problem, we start the iteration sequence with T_1 set equal to 1000 K. The convergence is slow, with the first several values of T_1 being

$$T_1^0 = 1000 \text{ K} \qquad T_1^1 = 860.3 \text{ K} \qquad T_1^2 = 944.2 \text{ K} \qquad T_1^3 = 894.9 \text{ K}$$

Continuing this iteration procedure, our calculations for T_1 converge to 913.4 K (to within four significant figures) after 18 iterations.

The rate of heat transfer per unit area is

$$q_x'' = \frac{q_x}{A} = \frac{k}{L}\left(T_1 - T_2\right) = \frac{35\text{W}/(\text{m } °\text{C})}{0.2 \text{ m}}\left(913.4 \text{ K} - 288 \text{ K}\right)$$

$$= 1.094 \times 10^5 \frac{\text{W}}{\text{m}^2} \tag{e}$$

To double-check, we calculate q_R''.

$$q_R'' = \frac{q_R}{A} = \sigma F_{s-R}\left(T_R^4 - T_1^4\right)$$

$$= 5.67 \times 10^{-8}\frac{\text{W}}{\text{m}^2 \text{ K}^4}\left[(1273 \text{ K})^4 - (913.4 \text{ K})^4\right] = 1.094 \times 10^5 \frac{\text{W}}{\text{m}^2} \tag{f}$$

The agreement between q_x'' and q_R'' assures us of a proper solution.

Returning briefly to the iterative solution for T_1, Eq. (d) can be written in the alternative forms

$$T_1 = 1139 \text{ K} - 3.24 \times 10^{-10} T_1^4 / \text{K}^3 \tag{g}$$

and

$$T_1 = \left(\frac{1139 \text{ K} - T_1}{3.24 \times 10^{-10} / \text{K}^3} \right)^{1/4} \tag{h}$$

The use of Eq. (g) gives rise to a convergent iterative solution for T_1 of 913.4 K. But the convergence is much slower than for Eq. (d). On the other hand, the use of Eq. (h) produces iterative calculations for T_1 that diverge! A formal criterion for the iterative convergence or divergence of equations such as Eqs. (d), (g), and (h) is given in [1]. However, it is generally a simple matter to perform a few calculations to determine whether convergence is being achieved.

It should also be mentioned that more efficient higher-order iterative schemes are available. For example, the Newton–Raphson method which is introduced in Prob. 2-15 only requires two iteration steps.

Composite systems

We now consider the interfacial condition associated with conduction heat transfer within a wall that is composed of several layers of different materials. This situation is illustrated in Figs. 2-3(a) and 2-4(a) for two types of joints. We designate the temperature distributions in these two materials by T_I and T_{II}. Because this composite actually involves two systems, materials I and II, we must write two boundary conditions for each material. Consequently, because the interface is a part of both materials, two interfacial boundary conditions are prescribed. Hence, for composite solids with perfect thermal contact, one interfacial boundary condition is

$$T_I(0) = T_{II}(0) \tag{2-26}$$

This situation in which the temperature distribution in the entire composite is continuous is illustrated in Fig. 2-3(b). The second boundary condition is written on the basis of the first law of thermodynamics, which states that the rate of energy conducted into the interface must be equal to the rate of

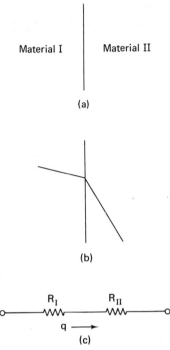

FIGURE 2-3 Composite with perfect thermal contact. (a) System interface. (b) Representative temperature distribution. (c) Thermal circuit.

energy conducted out.

$$q_{\mathrm{I}x} = q_{\mathrm{II}x}$$

$$-k_1 \frac{dT_{\mathrm{I}}}{dx}\bigg|_0 = -k_{\mathrm{II}} \frac{dT_{\mathrm{II}}}{dx}\bigg|_0 \tag{2-27}$$

This statement is synonymous with the Kirchhoff law for electrical current flow at a junction. The thermal circuit for this composite is shown in Fig. 2-3(c).

For situations in which an imperfect mechanical joint is made because of surface roughness, a discontinuity occurs in the temperature distribution at the interface, as shown in Fig. 2-4(b). For such cases, the interfacial temperatures in materials I and II are related empirically through an equation of the form

$$q_{tc} = h_{tc} A \left[T_{\mathrm{I}}(0) - T_{\mathrm{II}}(0) \right] \tag{2-28}$$

where h_{tc} is called the *thermal contact coefficient*. The first law of thermodynamics still applies, such that we obtain the following two interfacial

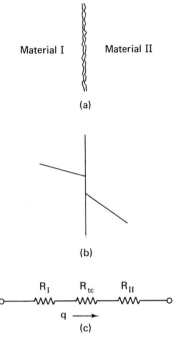

(a)

(b)

(c)

FIGURE 2-4 Composite with imperfect thermal contact. (a) System interface. (b) Representative temperature distribution. (c) Thermal circuit.

boundary conditions:

$$-k_I A \left.\frac{dT_I}{dx}\right|_0 = h_{tc} A \left[T_I(0) - T_{II}(0) \right] \tag{2-29}$$

$$h_{tc} A \left[T_I(0) - T_{II}(0) \right] = -k_{II} A \left.\frac{dT_{II}}{dx}\right|_0 \tag{2-30}$$

The thermal circuit for this type of problem is shown by Fig. 2-4(c); the thermal contact resistance R_{tc} is equal to $1/(h_{tc}A)$. As R_{tc} becomes small, this circuit reduces to the circuit shown in Fig. 2-3(c).

The thermal contact coefficient h_{tc} is dependent upon the material, surface roughness, contact pressure, and temperature. Experimental data for h_{tc} are available in references [1]–[3] for standard materials such as aluminum, copper, and stainless steel. For example, h_{tc} ranges from 5 to 50 kW/(m² °C) for various aluminum surfaces at 200°C with contact pressure between 1 and 30 atm. Thermal contact coefficients are generally much smaller for stainless steel [order of 3 kW/(m² °C)] and much larger for copper [order of 150 kW/(m² °C)].

A practical means of reducing the thermal contact resistance is to insert a material of good thermal conductivity between the two surfaces. Thermal greases containing silicon have been developed for this purpose. Thin soft metal foil can also be used for certain applications.

EXAMPLE 2-5

A germanium power transistor is capable of operating at up to 5 W. The manufacturer's rating for the case to ambient thermal resistance of the transistor is 30°C/W and the case temperature is not to exceed 80°C. In order to lower its operating temperature for a given power input, the transistor is to be mounted to a black anodized aluminum frame which serves as a heat sink, as shown in Fig. E2-5a. The frame provides an additional 1600 mm^2 surface area for cooling. To minimize the thermal contact resistance between the transistor and the heat sink and at the same time maintain proper electrical insulation, the surfaces are first cleaned and coated with a silicon grease (such as Dow-Corning Silicon Heat Sink Compound 340). A special electrical insulating gasket made of mica, beryllium oxide, or anodized aluminum is then inserted and the transistor is bolted tightly to the frame. As a rule of thumb, for proper

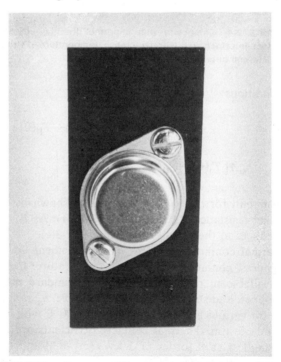

FIGURE E2-5a   Power amplifier mounted on aluminum frame.

thermal contact the thermal contact resistance is of the order of 0.5°C/W.

Determine the maximum power at which the transistor can safely be operated if the surrounding walls and air are at 25°C and $\bar{h} = 10$ W/(m² °C).

Solution

With the transistor case temperature T_s equal to 80°C, the rate of heat transferred directly to the surroundings is

$$q_T = \frac{T_s - T_F}{R_T} = \frac{80°C - 25°C}{30°C/W}$$

$$= 1.83 \text{ W}$$

In addition, heat is dissipated through the heat sink. The thermal circuit for the heat transfer through the heat sink (assuming negligible resistance to conduction) is shown in Fig. E2-5b. Assuming that blackbody conditions are approximated, R_R is given by

$$R_R = \frac{1}{\sigma A_s F_{s-R}(T_0 + T_R)(T_0^2 + T_R^2)}$$

Setting T_0 equal to T_s as a first approximation, we have

$$R_R = \frac{1}{\sigma(1600 \times 10^{-6} \text{ m}^2)(1)(353 \text{ K} + 298 \text{ K})\left[(353 \text{ K})^2 + (298 \text{ K})^2\right]}$$

$$= 79.3 \frac{°C}{W}$$

R_c is given by

$$R_c = \frac{1}{\bar{h} A_s} = \frac{1}{\left[10 \text{ W/(m}^2 \text{ °C)}\right](1600 \times 10^{-6} \text{ m}^2)}$$

$$= 62.5 \frac{°C}{W}$$

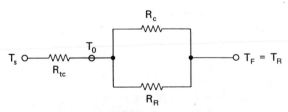

FIGURE E2-5b Thermal circuit.

To calculate the equivalent resistance R_{HS} for the heat transfer from the heat sink, we write

$$\frac{1}{R_{HS}} = \frac{1}{R_R} + \frac{1}{R_c}$$

$$= \frac{1}{79.3\ °C/W} + \frac{1}{62.5\ °C/W}$$

$$R_{HS} = 35\ \frac{°C}{W}$$

It follows that the rate of heat transferred from the power amplifier through the heat sink is

$$q_{HS} = \frac{T_s - T_F}{R_{tc} + R_{HS}} = \frac{80°C - 25°C}{0.5°C/W + 35°C/W}$$

$$= 1.55\ W$$

Based on this result, the temperature of the heat sink is calculated.

$$q_{HS} = \frac{T_s - T_0}{R_{tc}}$$

$$T_0 = T_s - q_{HS} R_{tc}$$

$$= 80°C - 1.55\ W\left(0.5\frac{°C}{W}\right)$$

$$= 79.2°C$$

Because the difference between T_0 and T_s is so small, our approximation for R_R is quite adequate.

Thus, the total power that can be safely dissipated from the transistor/heat-sink unit is estimated to be

$$q = q_T + q_{HS}$$

$$= 1.83\ W + 1.55\ W = 3.38\ W$$

Without the heat sink, the transistor could only be operated at 1.83 W. For operation at the 5-W level, a larger more efficient heat sink would be required.

To compare the rate of heat transfer by radiation and convection from the heat sink, we write

$$q_R = \frac{T_0 - T_F}{R_R} = \frac{79.2°C - 25°C}{79.3°C/W} = 0.683\ W$$

$$q_c = \frac{T_0 - T_F}{R_c} = \frac{79.2°C - 25°C}{62.5°C/W} = 0.867\ W$$

Thus, radiation accounts for about 44% of the heat transfer from the heat sink.

The remainder of our study of composites will be restricted to cases in which the thermal contact resistance is assumed to be negligible. Three basic situations to be considered include series, parallel, and combined series–parallel arrangements.

Series Arrangements A composite wall that is composed of two materials in series is shown in Fig. 2-5(a). For situations in which the surface temperatures are uniform, the heat transfer in composite walls that consist of two or more materials in series is always one-dimensional. The analogous electrical circuit for this two-wall series arrangement is shown in Fig. 2-5(b). The heat-transfer rate is simply

$$q_x = \frac{T_1 - T_3}{R_{\text{I}} + R_{\text{II}}} \tag{2-31}$$

where $R_{\text{I}} = L_1/(k_{\text{I}}A)$ and $R_{\text{II}} = L_2/(k_{\text{II}}A)$. The unknown interfacial temperature T_2 can be expressed as

$$T_1 - T_2 = q_x R_{\text{I}} = (T_1 - T_3)\frac{R_{\text{I}}}{R_{\text{I}} + R_{\text{II}}} \tag{2-32}$$

Equations can also be written for the temperature profile within each

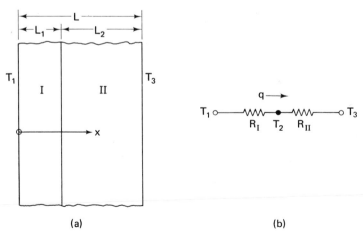

(a)   (b)

FIGURE 2-5 One-dimensional conduction heat transfer in a composite plane wall: series arrangement. (a) System. (b) Thermal circuit.

material as

$$\frac{T_{Ix} - T_1}{T_2 - T_1} = \frac{x}{L} \tag{2-33}$$

$$\frac{T_{IIx} - T_2}{T_3 - T_2} = \frac{x - L_1}{L_2} \tag{2-34}$$

These equations both come directly from Eq. (2-14).

Parallel Arrangements   A simple parallel composite of two materials is shown in Fig. 2-6(a). For this situation, the heat transfer is one-dimensional and can be represented by the electrical circuit shown in Fig. 2-6(b). The solution for the total heat-transfer rate in this parallel network can be written as

$$q_x = \frac{T_1 - T_2}{R_k} \tag{2-35}$$

where the equivalent parallel resistance is

$$\frac{1}{R_k} = \frac{1}{R_I} + \frac{1}{R_{II}} = \frac{k_I A_I}{L} + \frac{k_{II} A_{II}}{L} \tag{2-36}$$

This same result is achieved by recognizing that q_x is equal to the sum of the heat-transfer rates in the individual materials; that is,

$$q_x = q_{Ix} + q_{IIx}(T_1 - T_2)\left(\frac{1}{R_I} + \frac{1}{R_{II}}\right) \tag{2-37}$$

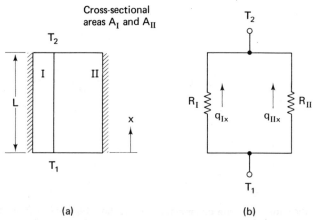

FIGURE 2-6   Heat transfer in a composite plane wall: parallel arrangement. (a) System. (b) Thermal circuit.

The temperature profile in each material is given by [from Eq. (2-14)]

$$T_{Ix} - T_1 = (T_2 - T_1)\frac{x}{L} \qquad T_{IIx} - T_1 = (T_2 - T_1)\frac{x}{L} \qquad (2\text{-}38,39)$$

The fact that these equations indicate that $T_{Ix} = T_{IIx}$ is consistent with the observation that this is a one-dimensional heat-transfer problem.

Combined Series–Parallel Arrangements A composite wall that provides combined series and parallel paths for the heat transfer is illustrated in Fig. 2-7(a). Such heat-transfer problems are almost always two- or three-dimensional. However, approximate solutions to these types of problems can be obtained under certain conditions by assuming one-dimensional heat transfer. For approximate one-dimensional heat transfer, the thermal circuit is given by Fig. 2-7(b). For this situation, the total rate of heat transfer can be expressed as

$$q_x = q_{Ix} + q_{IIx} \qquad (2\text{-}40)$$

where q_{Ix} and q_{IIx} are given by

$$q_{Ix} = q_{IIIx} = \frac{T_1 - T_2}{R_I + R_{III}} \qquad (2\text{-}41)$$

$$q_{IIx} = q_{IVx} = \frac{T_1 - T_2}{R_{II} + R_{IV}} \qquad (2\text{-}42)$$

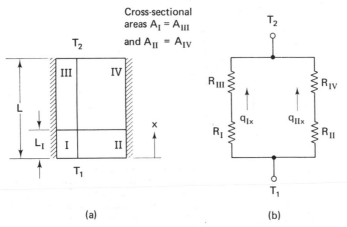

(a)   (b)

FIGURE 2-7 Heat transfer in a composite plane wall: series–parallel arrangement. (a) System. (b) Approximate one-dimensional thermal circuit.

An approximate criterion for which a one-dimensional analysis is appropriate is given in Fig. 2-8 for a representative system involving four thermal resistances. This figure indicates that the heat transfer is essentially one-dimensional for situations in which $R_I/R_{III} \simeq R_{II}/R_{IV}$. In addition, this criterion is satisfied for small values of both R_I/R_{III} and R_{II}/R_{IV}, and for large values of both R_I/R_{III} and R_{II}/R_{IV}. For example, with $R_{II}/R_{IV} = 10$, the error is of the order of 1% for $R_I/R_{III} > 3$ and about 10% for $R_I/R_{III} > 0.4$. For the case in which R_I/R_{III} and R_{II}/R_{IV} are both very

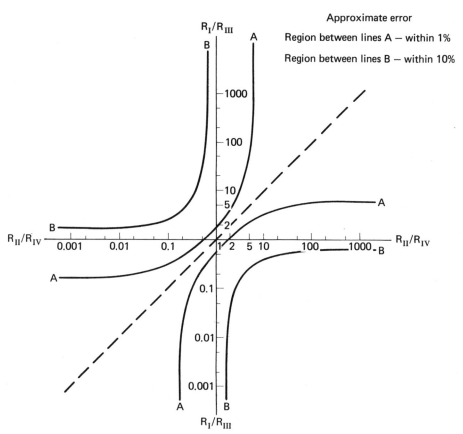

Note: This figure is based on a numerical finite difference solution to Prob. 3-55 which involves a parallel composite with convection at one surface. The numerical technique introduced in Chap. 3 can be used to develop criterion for other arrangements.

FIGURE 2-8   Criterion for applicability of one-dimensional analysis to composite wall with four series–parallel thermal resistances.

small, the analogous electrical circuit essentially reduces to a simple parallel circuit involving R_{III} and R_{IV} alone. Similarly, for the case in which R_I/R_{III} and R_{II}/R_{IV} are very large, the circuit reduces to a parallel network which only involves R_I and R_{II}. As indicated earlier, parallel heat-transfer systems involving only two thermal resistances are one-dimensional.

EXAMPLE 2-6

One surface of the 1-m-thick composite plate shown in Fig. E2-6a is kept at 0°C. The other surface is exposed to a fluid with $T_F = 100°C$ and $\bar{h} = 1000$ W/(m² °C). Estimate the rate of heat transfer through this wall.

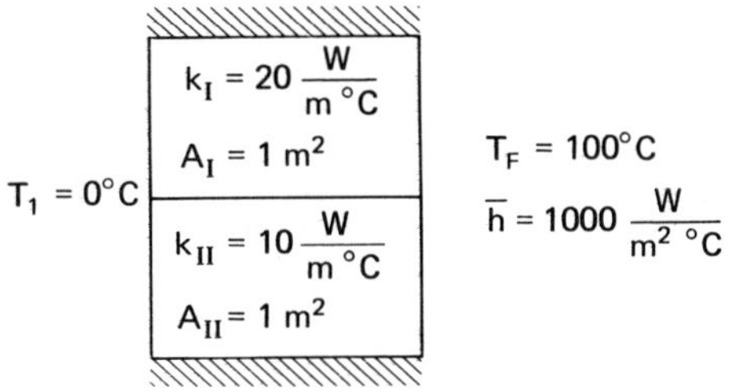

FIGURE E2-6a Composite plane wall.

Solution

Because this composite wall/fluid system provides both series and parallel paths, this heat-transfer problem is actually two-dimensional. Assuming for the moment that the two-dimensional effects are secondary, we sketch the approximate thermal circuit in Fig. E2-6b. Calculating R_I, R_{II}, R_{III}, and R_{IV}, we have

$$R_I = \frac{L}{k_I A_I} = \frac{1 \text{ m}}{[20 \text{ W}/(\text{m °C})](1 \text{ m}^2)} = 0.05 \frac{°C}{W}$$

$$R_{II} = \frac{L}{k_{II} A_{II}} = \frac{1 \text{ m}}{[10 \text{ W}/(\text{m °C})](1 \text{ m}^2)} = 0.1 \frac{°C}{W}$$

$$R_{III} = \frac{1}{\bar{h} A_{III}} = \frac{1}{[1000 \text{ W}/(\text{m}^2 \text{ °C})](1 \text{ m}^2)} = 0.001 \frac{°C}{W}$$

$$R_{IV} = R_{III}$$

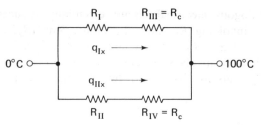

FIGURE E2-6b   Thermal circuit.

and

$$\frac{R_I}{R_{III}} = \frac{0.05°C/W}{0.001°C/W} = 50$$

$$\frac{R_{II}}{R_{IV}} = \frac{0.1°C/W}{0.001°C/W} = 100$$

Referring to Fig. 2-8, we find that the error for a one-dimensional analysis for this system is less than 1%.

The rate of heat transfer is therefore approximated by

$$q = \frac{T_1 - T_F}{R_I + R_{III}} + \frac{T_1 - T_F}{R_{II} + R_{IV}}$$

$$= (0°C - 100°C)\left(\frac{1}{0.05 + 0.001} + \frac{1}{0.1 + 0.001}\right)\frac{W}{°C}$$

$$= -100°C\left(19.6\frac{W}{°C} + 9.9\frac{W}{°C}\right) = -2950 \ W$$

Other types of boundary conditions

Several other types of boundary conditions are sometimes required in the analysis of heat-transfer problems. Examples include energy dissipation caused by relative interfacial motion, and a moving interface associated with phase change. Reference [4] is suggested as an introduction to the analysis of systems with moving boundaries.

2-2-5   Differential Formulation/Solution: Summary

To recapitulate, the development of the differential formulation/solution for conduction heat transfer involves (1) the application of the first law of thermodynamics to a differential element within the system, (2) the utilization of the definition of the derivative in relating the energy entering the differential element to the energy exiting, (3) the use of the appropriate particular law(s) (in this case the Fourier law of conduction), (4) a consideration of the conditions that exist at the boundaries, and (5) the solution of the resulting system of equations for the temperature distribution and the rate of heat transfer.

2-2-6 Short Method

The full differential formulation/solution approach outlined above can be utilized to analyze any steady one-dimensional conduction-heat-transfer problem. However, a much shorter method can be developed for one-dimensional problems in which q_x is constant. This approach involves the direct integration of the one-dimensional Fourier law of conduction, Eq. (1-16). To develop this short method in the context of the flat-plate geometry, we merely separate and integrate Eq. (1-16) as follows:

$$\int \frac{q_x}{A_x} dx = - \int k \, dT \tag{2-43}$$

Because q_x and A_x are constant for this problem, this equation reduces to

$$\frac{q_x}{A} \int dx = - \int k \, dT \tag{2-44}$$

which can be integrated. For the case in which the thermal conductivity is uniform, Eq. (2-44) can be written as

$$\frac{q_x}{A} \int dx = - k \int dT \tag{2-45}$$

The integration of this equation from $x=0$ to L and $T=T_1$ to T_2 gives rise to Eq. (2-15),

$$q_x = \frac{kA}{L}(T_1 - T_2) \tag{2-15}$$

On the other hand, the integration from $x=0$ to x and $T=T_1$ to T provides a relationship for the temperature distribution of the form

$$q_x = \frac{kA}{x}(T_1 - T) \tag{2-46}$$

Replacing q_x from Eq. (2-15), we obtain Eq. (2-14),

$$\frac{T - T_1}{T_2 - T_1} = \frac{x}{L} \tag{2-14}$$

This short method can be generalized for any steady one-dimensional conduction heat-transfer system by writing

$$q_\xi \int \frac{d\xi}{A_\xi} = - \int k\,dT \qquad (2\text{-}47)$$

where ξ is equal to x, y, or z in Cartesian coordinates and r in cylindrical or spherical coordinates; q_ξ is constant but A_ξ can be a function of ξ. The integration from one boundary to the other provides us with an expression for q_ξ. The integration from one boundary to an intermediate location ξ gives rise to a prediction for the temperature distribution. Because of its directness, this method will be used in the following sections whenever possible. However, the more general differential formulation/solution approach or numerical techniques will be required later when problems involving more complex situations are encountered.

2-3   CONDUCTION HEAT TRANSFER IN RADIAL SYSTEMS

2-3-1   Introduction

Whereas the cross-sectional area A_x for conduction heat transfer in a plane wall is constant, many situations are encountered in which the area through which the heat is transferred is dependent upon the spatial coordinate. For example, for radial conduction heat transfer in the hollow cylinder shown in Fig. 2-9(a), A_r is equal to $2\pi rL$. This classic problem is analyzed in this section, after which an interesting concept known as the critical radius will be introduced.

2-3-2   Analysis

For one-dimensional radial conduction heat transfer in a hollow cylinder, the Fourier law of conduction takes the form

$$q_r = - kA_r \frac{dT}{dr} \qquad (2\text{-}48)$$

Because $q_r = q_{r+dr}$ for this problem, the heat transfer can be obtained by the short method as follows:

$$q_r \int_{r_1}^{r_2} \frac{dr}{2\pi rL} = - k \int_{T_1}^{T_2} dT \qquad (2\text{-}49)$$

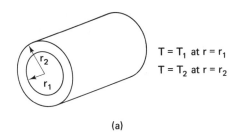

$T = T_1$ at $r = r_1$

$T = T_2$ at $r = r_2$

(a)

$r_2/r_1 = 1$ $r_2/r_1 = 2$ $r_2/r_1 = 4$

1.0

$\dfrac{T - T_1}{T_2 - T_1}$

0

1.0   2.0   3.0   4.0

r/r_1

(b)

FIGURE 2-9   One-dimensional radial heat transfer in a hollow cylinder. (a) System. (b) Temperature distribution, Eq. (2-52).

for uniform thermal conductivity, or

$$q_r = \frac{2\pi Lk}{\ln(r_2/r_1)}(T_1 - T_2) \qquad (2\text{-}50)$$

Thus, the thermal resistance to radial conduction in a hollow cylinder is

$$R_k = \frac{\ln(r_2/r_1)}{2\pi Lk} \qquad (2\text{-}51)$$

Similar to the analysis of heat transfer in a plane wall, the temperature distribution in this hollow cylinder can be obtained by integrating from r_1 to r and T_1 to T. The resulting expression for T is given by

$$\frac{T - T_1}{T_2 - T_1} = \frac{\ln(r/r_1)}{\ln(r_2/r_1)} \qquad (2\text{-}52)$$

This profile is shown in Fig. 2-9(b) for three values of r_2/r_1. It is observed that the temperature distribution is nearly linear for values of r_2/r_1 of the order of unity, but decidedly nonlinear for the larger values of r_2/r_1. This nonlinearity is caused by the variation in A_r with respect to r.

For cases in which the surface temperatures are not specified, T_1 and T_2 are determined by utilizing the appropriate thermal boundary conditions. This point is illustrated in Example 2-7.

For one-dimensional radial heat transfer in a hollow sphere, A_r is equal to $4\pi r^2$. Utilizing the short method, the rate of heat transfer in a hollow sphere with surface temperatures T_1 and T_2 is easily shown to be

$$q_r = \frac{4\pi r_1 r_2 k}{r_2 - r_1}(T_1 - T_2) \tag{2-53}$$

and the temperature profile is given by

$$\frac{T - T_1}{T_2 - T_1} = \frac{r - r_1}{r_2 - r_1}\frac{r_2}{r} \tag{2-54}$$

Note that the thermal resistance for this system is

$$R_k = \frac{r_2 - r_1}{4\pi r_1 r_2 k} \tag{2-55}$$

EXAMPLE 2-7

Refrigerant flows in a 1.9-in.-O.D. copper tube with 0.281-in. wall thickness. The inside surface temperature is 5°F and the room temperature is 70°F. Determine the thickness of insulative pipe covering [$k_i = 0.428$ Btu/(h ft °F)] required to reduce the heat gain to the pipe by 25% for the case in which forced convection heat transfer occurs with $\bar{h} = 10$ Btu/(h ft² °F). Assume that the thermal radiation effects are negligible.

Solution

The thermal circuit is shown in Fig. E2-7.

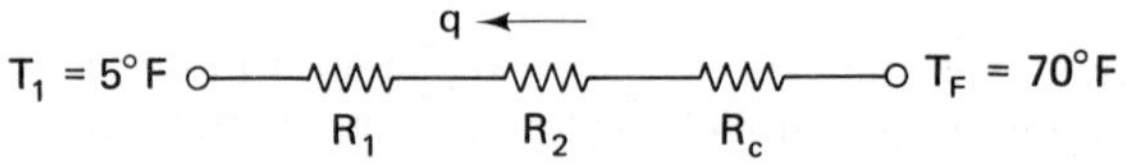

$$T_1 = 5°F \; \circ\!\!-\!\!\!\bigwedge\!\!\!\bigwedge\!\!\!-\!\!\!\bigwedge\!\!\!\bigwedge\!\!\!-\!\!\!\bigwedge\!\!\!\bigwedge\!\!\!-\!\!\circ \; T_F = 70°F$$
$$\overset{q\;\longleftarrow}{} \quad R_1 \qquad R_2 \qquad R_c$$

FIGURE E2-7 Thermal circuit.

$$R_1 = \frac{\ln(r_2/r_1)}{2\pi L k} = \frac{1}{L}\frac{\ln(0.95/0.669)}{2\pi}\frac{\text{h ft °F}}{230 \text{ Btu}} = \frac{2.43 \times 10^{-4}}{L}\frac{\text{h ft}}{\text{Btu}}$$

$$R_2 = \frac{\ln(r_0/r_2)}{2\pi L k_i} \qquad R_c = \frac{1}{\bar{h} A_s} = \frac{0.1}{A_s}\frac{\text{h ft}^2\text{ °F}}{\text{Btu}}$$

With no insulation, q_r is given by

$$q_r = \frac{5°F - 70°F}{\dfrac{2.43 \times 10^{-4}}{L}\ \dfrac{\text{h ft °F}}{\text{Btu}} + \dfrac{0.1}{2\pi L(0.95/12)\text{ft}}\ \dfrac{\text{h ft}^2\ °F}{\text{Btu}}} = q_c$$

$$\frac{q_r}{L} = \frac{-65°F}{(2.43 \times 10^{-4} + 0.201)\ \text{h ft °F/Btu}} = -323\frac{\text{Btu}}{\text{h ft}} \tag{a}$$

Note that the thermal resistance R_1 of the pipe wall is negligible.

A 25% reduction in the rate of heat transfer results in $q_r/L = -242$ Btu/(h ft). It follows that

$$q_r \simeq \frac{5°F - 70°F}{R_2 + R_c} = \frac{-65°F}{\dfrac{\ln(r_0/r_2)}{2\pi k_i L} + \dfrac{1}{\bar{h}2\pi r_0 L}}$$

$$\ln\left(\frac{r_0}{r_2}\right) + \frac{k_i}{\bar{h} r_0} = \frac{-65°F(2\pi k_i)}{q_r/L}$$

$$= \frac{-65°F(2\pi)[0.428\ \text{Btu}/(\text{h ft °F})]}{-242\ \text{Btu}/(\text{h ft})} = 0.722$$

This nonlinear equation is solved for r_0 by iteration as follows:

$$\ln\left(\frac{r_0}{r_2}\right) = 0.722 - \frac{0.428}{10}\frac{\text{ft}}{r_0}$$

$$\frac{r_0}{r_2} = \exp\left(0.722 - \frac{0.0428}{r_0}\text{ft}\right) \tag{b}$$

Starting with an assumed value for r_0/r_2 of 2, we have

$$\frac{r_0}{r_2} = \exp\left[0.722 - \frac{0.0428}{2(0.95/12)}\right] = 1.57$$

Substituting this value back into Eq. (b), we obtain

$$\frac{r_0}{r_2} = 1.46$$

Continuing this iteration sequence, we arrive at

$$\frac{r_0}{r_2} = (1.4)$$

after only three more steps. Thus, the thickness of insulation required to reduce the rate of heat loss by 25% is

$$\delta = 1.4r_2 - r_2 = 0.4(0.95 \text{ in.})$$
$$= 0.38 \text{ in.}$$

EXAMPLE 2-8

Utilize the differential formulation approach to obtain expressions for the temperature distribution, rate of heat transfer, and thermal resistance for the cylindrical section shown in Fig. E2-8.

FIGURE E2-8   Cylindrical section.

Solution

Because no heat is transferred across the surfaces at $\theta = 0$ and θ_1 and because T_1 and T_2 are uniform, the heat transfer in this system is one-dimensional. This one-dimensional (r direction) conduction-heat-transfer problem can be analyzed by either the differential formulation approach or the short method.

The differential formulation is developed as follows:

Step 1

$$q_r = q_{r+dr}$$

Step 2

$$q_r = q_r + \frac{dq_r}{dr} dr \qquad \text{or} \qquad \frac{dq_r}{dr} = 0$$

Step 3

$$\frac{d}{dr}\left(-kA_r \frac{dT}{dr}\right) = \frac{d}{dr}\left(-k\theta_1 rL \frac{dT}{dr}\right) = 0$$

or, for uniform thermal conductivity,

$$\frac{d}{dr}\left(r \frac{dT}{dr}\right) = 0 \qquad\qquad\qquad (a)$$

Finally, the boundary temperatures can be designated by $T(r_1) = T_1$ and $T(r_2) = T_2$.

Next, Eq. (a) is solved for the temperature distribution. Integrating, we have

$$r\frac{dT}{dr} = C_1$$

Continuing,

$$\int_{T_1}^{T} dT = \int_{r_1}^{r} \frac{C_1\,dr}{r}$$

$$T - T_1 = C_1 \ln\left(\frac{r}{r_1}\right)$$

Utilizing the boundary condition at r_2, C_1 is given by

$$C_1 = \frac{T_2 - T_1}{\ln(r_2/r_1)}$$

and the temperature profile becomes

$$\frac{T - T_1}{T_2 - T_1} = \frac{\ln(r/r_1)}{\ln(r_2/r_1)}$$

To obtain an expression for the rate of heat transfer, we utilize the Fourier law of conduction as follows:

$$q_r = -k\theta_1 rL\frac{dT}{dr} = -k\theta_1 LC_1$$

$$= k\theta_1 L\frac{T_1 - T_2}{\ln(r_2/r_1)}$$

or

$$q_r = \frac{T_1 - T_2}{R_k}$$

where $R_k = \ln(r_2/r_1)/(\theta_1 kL)$. These results are seen to be identical to those which were obtained on the basis of the short method for the case of a hollow circular cylinder with $\theta_1 = 2\pi$.

2-3-3 Critical Radius

As shown in Example 2-7, convection and composite wall boundary conditions associated with geometries for which the area is not constant are handled in the same way that these complications were treated for the plane-wall geometry. For the case of a plane wall exposed to a fluid, an increase in the thickness of the wall results in an increase in the internal resistance $R_k[\equiv L/(kA)]$ but does not change the surface resistance R_c. Hence, such an increase in the thickness of a plane wall always reduces the rate of heat transfer through the wall. Of course, a reduction in heat transfer is most easily accomplished by the use of an insulating material of low thermal conductivity. On the other hand, an increase in the wall thickness or the addition of an insulating material does not always bring about a decrease in the heat-transfer rate for geometries with nonconstant cross-sectional area.

To see this point, consider the hollow cylinder of radii r_1 and r_2 shown in Fig. 2-10(a). The inside surface is maintained at temperature T_1, and the temperature T_F of the surrounding fluid is specified. The thermal circuit for this problem is shown in Fig. 2-10(b). Now we ask, what will be the effect of increasing the outside radius r_2, with T_1, T_F, and $\bar{h}$ held constant? The rate of heat transfer from the outside surface to the fluid is

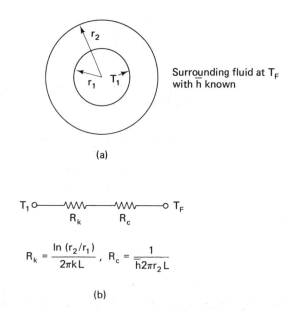

(a)

$$T_1 \circ\!\!-\!\!\bigwedge\!\!\bigwedge\!\!-\!\!\bigwedge\!\!\bigwedge\!\!-\!\!\circ T_F$$
$$\quad\quad R_k \quad\quad R_c$$

$$R_k = \frac{\ln\,(r_2/r_1)}{2\pi k L}, \quad R_c = \frac{1}{\bar{h}2\pi r_2 L}$$

(b)

FIGURE 2-10 Convection heat transfer from a hollow circular cylinder. (a) System. (b) Thermal circuit.

given by

$$q_r = \frac{T_1 - T_F}{R_k + R_c} = \frac{T_1 - T_F}{\dfrac{\ln(r_2/r_1)}{2\pi Lk} + \dfrac{1}{\bar{h}2\pi Lr_2}} \tag{2-56}$$

An increase in r_2 is seen to increase R_k but to decrease R_c. Therefore, the addition of material can either decrease or increase the rate of heat transfer, depending upon the change in the total resistance $R_c + R_k$ with r_2.

To see the effect of r_2 on q_r, this equation is plotted in Fig. 2-11 for various values of $k/(\bar{h}r_1)$, with r_2/r_1 taken as the independent variable. For $k/(\bar{h}r_1)$ less than unity, the rate of heat transfer q_r continuously

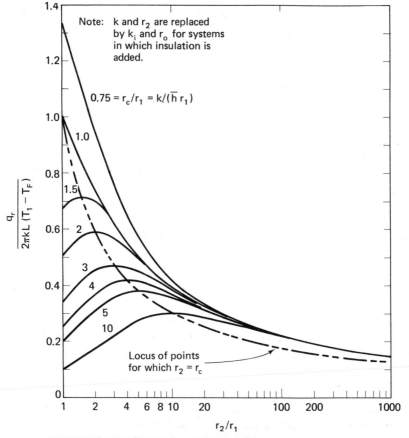

FIGURE 2-11 Effect of increase in outside radius r_2 on heat transfer from a pipe.

decreases as r_2 increases from a value of r_1. But, for $k/(\bar{h}r_1)$ greater than unity, q_r increases to a maximum and then decreases. To determine the radius r_c at which q_r is maximized for $k/(\bar{h}r_1)$ greater than unity, we set dq_r/dr_2 equal to zero,

$$\left.\frac{dq_r}{dr_2}\right|_{r_c} = -(T_1 - T_F)2\pi Lk\left.\left(\frac{1}{r_2} - \frac{k}{hr_2^2}\right)\right|_{r_c} = 0 \qquad (2\text{-}57)$$

with the result

$$r_c = \frac{k}{\bar{h}} \qquad (2\text{-}58)$$

r_c is referred to as the *critical radius*. If $r_2 > r_c$, the addition of material to the outside surface will decrease the rate of heat transfer, as in the case of Example 2-7, where $r_c = 0.514$ in. and $r_2 = 0.95$ in. This is often the case for conditions of forced convection for which $\bar{h}$ is large and r_c is small. But if $r_2 < r_c$, the addition of material will increase the heat-transfer rate until $r_2 = r_c$, after which additional increases in r_2 will decrease q_r. This situation occurs more often for natural convection than for forced convection because of the low values of $\bar{h}$, especially in gases.

In contrast, if insulation is added to the inside surface, both R_c and R_k increase. Hence, the addition of insulation to the inside surface always reduces the heat-transfer rate and the critical radius concept has no significance.

Heat transfer in a sphere is affected by the addition of insulation to the outside or inside surface in much the same way as in a cylinder. However, for a sphere, the critical radius is given by

$$r_c = \frac{2k}{\bar{h}} \qquad (2\text{-}59)$$

Likewise, a critical radius would be expected to be found for thermal radiation heat transfer or combined convection and radiation from the outside surface of a cylinder or sphere.

EXAMPLE 2-9

Determine the thickness of insulative pipe covering [$k_i = 0.428$ Btu/(h ft °F)] required to reduce the convective heat loss from the 1.9-in.-O.D. pipe of Example 2-7 by 25% for the case in which natural convection cooling occurs with $\bar{h} = 3.6$ Btu/(h ft^2 °F).

Solution

The critical radius for this situation is

$$r_c = \frac{k_i}{\bar{h}} = \frac{0.428 \text{ Btu/(h ft °F)}}{3.6 \text{ Btu/(h ft}^2 \text{ °F)}} = 0.119 \text{ ft} = 1.43 \text{ in.}$$

Thus, we have

$$\frac{r_c}{r_2} = \frac{1.43 \text{ in.}}{0.95 \text{ in.}} = 1.5$$

Because r_c/r_2 is greater than unity, the effect of insulation for r_0 less than r_c will be to increase the rate of heat transfer q_r. In contrast, for the forced convection conditions of Example 2-7, r_c/r_2 is equal to 0.541, such that the effect of insulation is to reduce q_r for all values of r_0.

Referring to Fig. 2-11, the dimensionless heat flux $q_r/[2\pi k_i L(T_1 - T_F)]$ increases from a value of 0.66 at $r_0/r_2 = 1$ to a maximum value of approximately 0.71 at r_0/r_2 equal to r_c/r_2 ($\equiv 1.5$), and then begins to fall toward zero. Note that the rate of heat transfer for $r_0/r_2 = 1$ is given by the simple Newton law of cooling.

$$q_r = 2\pi r_2 L \bar{h}(T_1 - T_F)$$

or

$$\frac{q_r}{2\pi k_i L(T_1 - T_F)} = \frac{\bar{h} r_2}{k_i} = \frac{r_2}{r_c} = \frac{1}{1.5} = 0.667$$

Our problem calls for a 25% reduction in the rate of heat transfer; that is,

$$\left. \frac{q_r}{2\pi k_i L(T_1 - T_F)} \right|_{r_0} = 0.5$$

Utilizing Fig. 2-11, we find that the dimensionless heat flux reaches this value for r_0/r_2 equal to approximately 5. Thus, the thickness of insulation is

$$\delta = r_0 - r_2 = 5(0.95 \text{ in.}) - 0.95 \text{ in.}$$
$$= 3.8 \text{ in.}$$

2-4 VARIABLE THERMAL CONDUCTIVITY

2-4-1 Introduction

As indicated in Chap. 1, the thermal conductivity of most materials is at least somewhat dependent upon temperature. The variation of thermal conductivity with temperature is shown in Fig. 2-12 for several common metals for the temperature range $-100°C$ to $300°C$. Whereas the assumption of uniform thermal conductivity is generally acceptable for problems involving small temperature differences, many situations are encountered in which the variation in k with T cannot be neglected. Referring to Fig. 2-12, we see that a linear approximation for k can be utilized over limited temperature ranges for these materials; that is,

$$k(T) = k_0(1 + \beta_T T) \qquad (2\text{-}60)$$

where β_T is known as the *temperature coefficient of thermal conductivity*. The units for β_T are $1/°C$ (or $1/°F$).

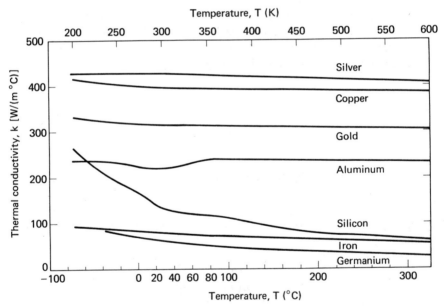

FIGURE 2-12 Variation of thermal conductivity k with temperature for several metals.

2-4-2 Analysis

Utilizing the short method, k must be retained within the integral in Eq. (2-47) for cases in which the variation of the thermal conductivity with temperature is significant. By integrating from one boundary to the other,

Eq. (2-47) gives

$$q_\xi \int_{\xi_1}^{\xi_2} \frac{d\xi}{A_\xi} = -\int_{T_1}^{T_2} k\, dT \tag{2-61}$$

where $\xi_1 = 0$ and $\xi_2 = L$ for a flat plate and $\xi_1 = r_1$ and $\xi_2 = r_2$ for a hollow cylinder or sphere.

Introducing the mean thermal conductivity $\bar{k}$,

$$\bar{k} = \frac{1}{T_2 - T_1} \int_{T_1}^{T_2} k\, dT \tag{2-62}$$

the rate of heat transfer q_ξ is given by

$$q_\xi = \frac{\bar{k}(T_1 - T_2)}{\int_{\xi_1}^{\xi_2} d\xi / A_\xi} \tag{2-63}$$

Based on this expression, equations can be written for the thermal resistances for flat plates, and hollow cylinders and spheres with variable thermal conductivity of the forms

$$R_k = \frac{L}{A\bar{k}} \qquad \text{flat plate} \tag{2-64}$$

$$R_k = \frac{\ln(r_2/r_1)}{2\pi L \bar{k}} \qquad \text{hollow cylinder} \tag{2-65}$$

$$R_k = \frac{r_2 - r_1}{4\pi r_1 r_2 \bar{k}} \qquad \text{hollow sphere} \tag{2-66}$$

EXAMPLE 2-10

The surfaces of a 10-cm-thick plate are maintained at 0°C and 100°C. The thermal conductivity varies with temperature according to $k = k_0(1 + \beta_T T)$ with $k = 50$ W/(m °C) at 0°C and $k = 100$ W/(m °C) at 100°C. Determine the heat-transfer flux and the temperature distribution in the plate.

Solution

We first want to evaluate k_0 and β_T. Setting k equal to 50 W/m °C) at 0°C and 100 W/(m °C) at 100°C, we find that $k_0 = 50$ W/(m °C) and $\beta_T = 0.01/°C$. (The absolute temperature scale can also be used, for which case $k_0 = -87$ W/(m °C) and $\beta_T = -5.78 \times 10^{-3}/°C$.)

To calculate the rate of heat transfer, we utilize Eq. (2-63), with the result

$$q = \frac{\bar{k}A}{L}(T_1 - T_2) \tag{a}$$

where

$$\bar{k} = \frac{1}{T_2 - T_1} \int_{T_1}^{T_2} k\,dT$$

$$= \frac{k_0}{100°C} \int_{0°C}^{100°C} (1 + \beta_T T)\,dT$$

$$= \frac{k_0}{100°C} \left(T + \frac{\beta_T}{2} T^2 \right) \Big|_{0°C}^{100°C}$$

$$= \frac{k_0}{100°C} \left[100°C + \frac{10^{-2}}{2°C}(100°C)^2 \right]$$

$$= 1.5k_0$$

Following through with the calculation, we have

$$q'' = 1.5\frac{k_0}{L}(T_1 - T_2) = \frac{1.5}{0.1\text{ m}}\left(50\frac{\text{W}}{\text{m °C}} \right)(0°C - 100°C)$$

$$= -75\frac{\text{kW}}{\text{m}^2}$$

To obtain the temperature distribution, the Fourier law of conduction is integrated from 0 to x and from 0°C to T as follows:

$$q = -kA\frac{dT}{dx} \tag{b}$$

$$\frac{q}{A}\int_0^x dx = -\int_{0°C}^T k\,dT$$

$$\frac{q}{A}x = -k_0\left(T + \frac{\beta_T}{2} T^2 \right)$$

where $q/A = q'' = 1.5k_0/L(T_1 - T_2) = -75$ kW/m². It follows that

$$T + \frac{\beta_T}{2} T^2 + 1.5(T_1 - T_2)\frac{x}{L} = 0 \tag{c}$$

Solving this quadratic equation for T, we have

$$T = \frac{-1 \pm \sqrt{1 - 4(\beta_T/2)1.5(T_1 - T_2)(x/L)}}{\beta_T} \tag{d}$$

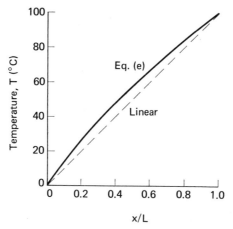

FIGURE E2-10 Temperature distributions in plane wall with variable thermal conductivity.

where the positive sign must be utilized to satisfy the boundary conditions. Substituting in for β_T, T_1, and T_2, T becomes

$$T = \frac{-1 + \sqrt{1 + 3(x/L)}}{0.01} \, °C \qquad (e)$$

This temperature distribution is compared with the linear profile associated with uniform thermal conductivity in Fig. E2-10. Notice the distinct nonlinear behavior of T for this case in which k is a function of temperature.

2-5 INTERNAL ENERGY SOURCES

2-5-1 Introduction

Heat-transfer systems involving internal energy sources include chemical and nuclear reaction in which energy is generated by chemical reaction or by the interaction of nuclear particles. Other important heat-transfer applications involving internal energy sources occur in electrical and electronic circuits. Examples include electrical resistance heaters made of nickel/chromium alloys, incandescent electric lamps with tungsten filaments, and semiconductor chips and transistors made of silicon or germanium. The dissipation of heat is also important in the operation of electric motors, generators, transformers, and relays.

The strength of an internal energy source is generally represented by the rate of energy generated per unit volume $\dot{q}$. For electrical and electronic systems, $\dot{q}$ is given by

$$\dot{q} = \frac{I_e^2 R_e}{V} = \frac{I_e^2 \rho_e}{A^2} \qquad (2\text{-}67)$$

where the current I_e is assumed to be uniform and $\rho_e(\equiv R_e A/L)$ is the *electrical resistivity*. The resistivity is generally a linear function of temperature; that is,

$$\rho_e = \rho_0(1 + \alpha_0 T) \qquad (2\text{-}68)$$

where ρ_0 is the resistivity at a reference temperature such as 0°C, and α_0 is the temperature *coefficient of resistance* at the reference temperature. Accordingly, the power dissipation per unit volume associated with the flow of electric current can be approximated by an equation of the form

$$\dot{q} = \dot{q}_0(1 + \alpha_0 T) \qquad (2\text{-}69)$$

where $\dot{q}_0 = I_e^2 \rho_0/A^2$. Data for ρ_0 and α_0 are tabulated in reference [6] for several common metals and alloys. For example, $\rho_0 = 1.8 \times 10^{-8}$ Ω m and $\alpha_0 = 0.004/°C$ for copper wire, and $\rho_0 = 20 \times 10^{-8}$ Ω m and $\alpha_0 = 0.005/°C$ for carbon steel. In comparison, the electrical resistivities of electrical insulators such as mica and electrical semiconductors such as germanium are of the order of 10^{15} Ω m and 0.4 Ω m, respectively. In general, α_0 is quite small for metals and alloys, such that the internal energy generation can be assumed to be uniformly distributed (i.e., independent of the spatial coordinates) for moderate temperatures. However, the variation of $\dot{q}$ becomes an important factor for high-temperature operation.

2-5-2   Analysis

Application of the first law of thermodynamics to the problem of conduction heat transfer in a flat plate with internal energy generation (see Fig. 2-13) leads to

Step 1

$$\dot{q}\,dV + q_x = q_{x+dx} \qquad (2\text{-}70)$$

(The energy generation can be considered as energy brought into the differential element.) Unlike the case of steady-state conduction heat transfer with no energy generation [represented by Eq. (2-1)], Eq. (2-70) indicates that q_x is not equal to q_{x+dx} for energy-generation problems. Therefore, the short method, which involves the mere integration of the Fourier law, cannot be utilized, unless the dependence of q_x on x is known. Hence, we proceed with the development of the differential formulation.

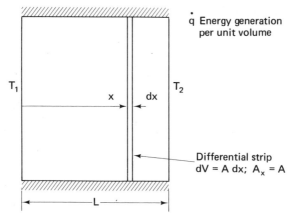

FIGURE 2-13 Conduction heat transfer in a plane wall with internal energy generation.

The substitution of Eq. (2-3) into Eq. (2-70) gives

Step 2

$$\dot{q}A\,dx = \frac{dq_x}{dx}\,dx \tag{2-71}$$

where A_x is constant $(A_x = A)$ and $dV = A\,dx$. With q_x given by the Fourier law of conduction, Eq. (2-71) becomes

Step 3

$$\frac{d}{dx}\left(k\frac{dT}{dx}\right) + \dot{q} = 0 \tag{2-72}$$

For the case of uniform thermal conductivity and energy generation, this equation reduces to the form

$$\frac{d^2T}{dx^2} + \frac{\dot{q}_0}{k} = 0 \tag{2-73}$$

Notice that Eq. (2-73) is identical to Eq. (2-7) for $\dot{q}_0 = 0$.

Because this is a second-order differential equation, two boundary conditions are required; they are

$$T(0) = T_1 \qquad T(L) = T_2 \tag{2-74, 75}$$

This simple differential equation can be solved by two integrations as follows:

$$\frac{dT}{dx} = -\frac{\dot{q}_0}{k}x + C_1 \qquad T = \frac{\dot{q}_0}{k}\frac{x^2}{2} + C_1 x + C_2 \tag{2-76, 77}$$

The boundary conditions require that $C_2 = T_1$ and $C_1 = (T_2 - T_1)/L + \dot{q}_0 L/(2k)$, such that Eq. (2-77) becomes

$$T - T_1 = (T_2 - T_1)\frac{x}{L} + \frac{\dot{q}_0}{k}\frac{L^2}{2}\left[\frac{x}{L} - \left(\frac{x}{L}\right)^2\right] \qquad (2\text{-}78)$$

Focusing attention on the case in which symmetrical cooling takes place with $T_2 = T_1$, the temperature distribution is given by

$$T - T_1 = \frac{2\dot{q}_0\ell^2}{k}\left[\frac{x}{L} - \left(\frac{x}{L}\right)^2\right] \qquad (2\text{-}79)$$

or

$$T - T_1 = \frac{1}{2}\frac{\dot{q}_0\ell^2}{k}\left[1 - \left(\frac{\xi}{L/2}\right)^2\right] \qquad (2\text{-}80)$$

where $\ell = V/A_s = L/2$ and $\xi = x - L/2$. This temperature distribution is shown in Fig. 2-14 in terms of both x and ξ. The maximum temperature within the plate for symmetrical cooling occurs at the center where $dT/dx = 0$.

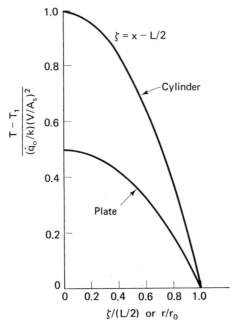

FIGURE 2-14   Temperature distributions for plate and cylinder with uniform energy generation.

Setting $x = L/2$ in Eq. (2-79) or $\xi = 0$ in Eq. (2-80), we have

$$T_{max} = T_1 + \frac{\dot{q}_0}{k} \frac{\ell^2}{2} = T_1 + \frac{\dot{q}_0}{k} \frac{L^2}{8} \qquad (2\text{-}81)$$

Because materials break down above certain temperatures, this maximum temperature represents a critical design consideration.

To obtain the rate of heat transfer for the case in which $T_2 = T_1$, we apply the Fourier law of conduction as follows:

$$q_x = -kA \frac{dT}{dx}\bigg|_x = \dot{q}_0 \frac{V}{2}\left(\frac{2x}{L} - 1\right) \qquad (2\text{-}82)$$

Hence, the rates of heat transfer at the surfaces $x = 0$ and $x = L$ are

$$q_0 = -\dot{q}_0 \frac{V}{2} \quad \text{and} \quad q_L = \dot{q}_0 \frac{V}{2} \qquad (2\text{-}83,84)$$

such that the total rate of energy generated within the plate $\dot{q}_0 V$ is transferred into the fluid from the two surfaces.

For the more general case in which $T_2 \neq T_1$, Eq. (2-78) is coupled with the Fourier law of conduction to obtain

$$q_x = \frac{kA}{L}(T_1 - T_2) + \frac{\dot{q}_0 V}{2}\left(\frac{2x}{L} - 1\right) \qquad (2\text{-}85)$$

It follows that q_0 and q_L are

$$q_0 = \frac{kA}{L}(T_1 - T_2) - \frac{\dot{q}_0 V}{2} \qquad (2\text{-}86)$$

and

$$q_L = \frac{kA}{L}(T_1 - T_2) + \frac{\dot{q}_0 V}{2} \qquad (2\text{-}87)$$

Thus, we see that the total rate of heat transfer is found by simply superimposing the heat transfer through a nongenerating plate with surface temperature T_1 and T_2 upon the heat transfer in a generating plate with both surfaces at T_1.

Solutions are also available for the temperature distribution in heat-generating cylinders and spheres. For example, the temperature distribution in a solid circular cylinder with uniform heat generation takes the form

$$T - T_1 = \frac{\dot{q}_0 \ell^2}{k} \left[1 - \left(\frac{r}{r_1} \right)^2 \right]$$
(2-88)

where $\ell = V/A_s = r_1/2$. This expression is shown in Fig. 2-14. The maximum temperature in the cylinder occurs at the center.

$$T_{max} = T_1 + \frac{\dot{q}_0 \ell^2}{k} = T_1 + \frac{\dot{q}_0}{k} \frac{r_1^2}{4}$$
(2-89)

Notice that for the same volume over surface area ratio V/A_s and $\dot{q}_0/k$ ratio, the circular cylinder operates at twice the relative temperature $T - T_1$ as the flat plate. To discover why this is true, Prob. 2-41 should be considered.

EXAMPLE 2-11

Four amperes of current flow in a 1-mm-diameter copper wire [$k = 360$ W/(m °C)] with resistivity equal to about 30 $\mu\Omega$ cm. The insulation is 2 mm thick and has a thermal conductivity of 0.05 W/(m °C). Determine the surface temperature of the insulation and the maximum operating temperature of the wire if the system is cooled by blackbody thermal radiation with $T_R = -200°C$ and $F_{s-R} = 1.0$.

Solution

The power generated within the wire is

$$\dot{W} = I_e^2 R_e = I_e^2 \rho_e \frac{L}{A}$$

$$\frac{\dot{W}}{L} = (4 \text{ A})^2 (30 \times 10^{-6} \ \Omega \ 0.01 \text{ m}) \frac{4}{\pi (10^{-3} \text{ m})^2} = 6.11 \frac{\text{W}}{\text{m}}$$

Thus, the energy generated per unit volume is

$$\dot{q}_0 = \frac{\dot{W}}{LA} = \frac{6.11 \text{ W/m}}{\pi (10^{-3} \text{ m})^2 / 4} = 7.78 \times 10^6 \frac{\text{W}}{\text{m}^3}$$

Because this energy must be radiated away,

$$\dot{W} = q_R = \sigma A_s F_{s-R} (T_s^4 - T_R^4)$$

or

$$T_s^4 = T_R^4 + 6.11 L \frac{W}{m} \left(\frac{1}{\sigma A_s F_{s-R}} \right)$$

such that the surface temperature T_s is found to be 288 K ($= 15.1°C$). To find the interfacial temperature T_1, we write

$$q_r = \frac{T_1 - T_s}{\dfrac{\ln(r_0/r_1)}{2\pi k_i L}} = q_R$$

$$T_1 = T_s + \frac{q_R}{L} \frac{\ln(r_0/r_1)}{2\pi k_i}$$

$$= 15.1°C + 6.11 \frac{W}{°C} \frac{\ln(0.0025/0.0005)}{2\pi [0.05 W/(m \ °C)]}$$

$$= 46.4°C$$

Finally, utilizing Eq. (2-89), T_{max} is calculated.

$$T_{max} = T_1 + \frac{\dot{q}_0}{k} \left(\frac{r_1}{2} \right)^2$$

$$= 46.4°C + \frac{7.78 \times 10^6 \ W/m^3}{360 \ W/(m \ °C)} \left(\frac{0.5 \times 10^{-3} \ m}{2} \right)^2$$

$$= 46.4°C + 1.35 \times 10^{-3} \ °C \approx 46.4°C$$

Thus, we see that the temperature throughout the wire is essentially uniform.

EXAMPLE 2-12

Determine the effect of electrical insulation on the maximum operating temperature T_{max} of a wire with radius r_1 in which uniform energy generation occurs. The wire is surrounded by a fluid with T_F and $\bar{h}$ specified and radiation effects are to be neglected.

Solution

The maximum wire temperature is given by Eq. (2-89) in terms of the surface temperature of the wire T_1,

$$T_{max} = T_1 + \frac{\dot{q}_0 (r_1/2)^2}{k} \tag{a}$$

With $\dot{q}_0$, r_1, and k for the wire specified, T_{max} depends solely on T_1.

To see the effect of insulation on T_1, we write an expression for the heat transfer from the surface of the wire to the fluid as follows:

$$q = \frac{T_1 - T_F}{R_{ki} + R_c} = \dot{q}_0 V \tag{b}$$

Solving for T_1, we have

$$T_1 = T_F + \dot{q}_0 V \left[\frac{\ln(r_0/r_1)}{2\pi k_i L} + \frac{1}{\bar{h} 2\pi r_0 L} \right] \tag{c}$$

or

$$T_1 = T_F + \frac{\dot{q}_0 V}{2\pi k_i L} \left(\ln \frac{r_0}{r_1} + \frac{k_i}{\bar{h} r_1} \frac{r_1}{r_0} \right) \tag{d}$$

$T_1 - T_F$ can be plotted as a function of r_0/r_1 and $\bar{h} r_1 / k$. For values of $k_i/(\bar{h} r_1)$ greater than unity, $T_1 - T_F$ falls toward a minimum, after which it increases. The location of the minimum in $T_1 - T_F$ occurs at the critical radius

$$r_c = \frac{k_i}{\bar{h}}$$

Based on Eq. (a), T_{max} is also minimized at this location. This is seen by setting dT_1/dr_0 equal to zero. Thus, we reach the somewhat surprising conclusion that the maximum operating temperature of the wire can actually be reduced by adding electrical insulation if the wire radius is less than the critical radius.

2-6 EXTENDED SURFACES

2-6-1 Introduction

Situations often arise in which means are sought for increasing the heat convected from a surface. A consideration of the Newton law of cooling, Eq. (1-28),

$$q_c = \bar{h} A_s (T_s - T_F) \tag{2-90}$$

suggests that q_c can be increased by increasing $\bar{h}$, $T_s - T_F$, or A_s. As already indicated, $\bar{h}$ is a function of the geometry, fluid properties, and flow rate. The modulation of $\bar{h}$ through the control of these factors provides a means by which q_c can be increased or decreased, as will be discussed in Chaps. 5

through 10. With regard to the effect of $T_s - T_F$ on the rate of heat transfer, difficulties are often encountered in automobile cooling systems in very hot weather because T_F is too high. Concerning the third factor, which is the object of this section, the area of a surface that is exposed to the fluid is often "extended" by the use of fins or spines, as illustrated in Fig. 2-15. Familiar applications of such extended surface heat-transfer devices include automobile radiators, power transistors, and high-voltage electrical transformers. In addition, extended surfaces are commonly used to increase the rate of thermal radiation from surfaces.

Referring to the surface extension of the plane wall illustrated in Fig. 2-16, heat is transferred from the wall into the fin itself by conduction and from the fin surface by convection. Hence, the decrease in the surface convection resistance R_c brought about by the increase in surface area A_s is accompanied by an increase in the conduction resistance R_k. In order for the rate of heat transfer from the wall to be increased by the use of a surface extension, the decrease in R_c must be greater than the increase in

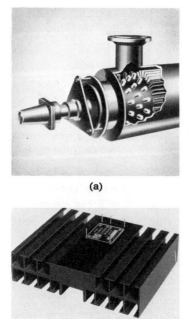

(a)

(b)

FIGURE 2-15 Typical applications of extended surfaces. (a) Longitudinal fins and spines in a fired heater. Fluid flowing in the annulus is heated by a gas- or oil-fired flame inside the fintube. (Courtesy of Brown Fintube Company.) (b) Fins on a high-power voltage regulator. (Courtesy of RS Components Ltd.)

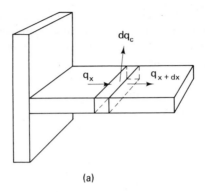

(a)

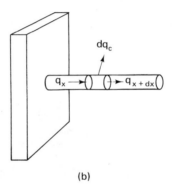

(b)

FIGURE 2-16 Extended surfaces with uniform cross section.
(a) Fin with rectangular profile. (b) Fin with circular profile
(spine).

R_k. As a matter of fact, the surface resistance must be the controlling
factor ($R_k < R_c$, or preferably $R_k \ll R_c$) in practical fin applications.

In order to provide a guide for the design of extended surface
systems, we introduce the *Biot number* Bi,

$$\mathrm{Bi} = \frac{\bar{h}\ell}{k} \tag{2-91}$$

The characteristic length ℓ is equal to V/A_s, where V is the fin volume.
The Biot number is generally taken as a rule-of-thumb approximation for
the ratio between the conduction resistance and the surface resistance; that
is,

$$\mathrm{Bi} \simeq \frac{R_k}{R_c} \tag{2-92}$$

Because R_k must be considerably smaller than R_c for a fin to be effective, we conclude that the Biot number should be small for fin applications. In practice, the following criterion is generally maintained in the design of convective surface extensions:

$$\text{Bi} = \frac{\bar{h}\ell}{k} \lesssim 0.1 \tag{2-93}$$

In keeping with this criterion, we conclude that fins should generally be considered for applications involving small values of $\bar{h}$ and ℓ and large values of k. Referring back to Table 1-2, we see that likely fin applications occur for natural convection, forced convection of gases and, to a lesser extent, forced convection of liquids. However, because of the very high values of $\bar{h}$ associated with two-phase fluids, fins are generally not useful for applications involving boiling or condensation.

2-6-2 Analysis

Returning to the fin with uniform cross section which is shown in Fig. 2-16, the heat transfer in this system is clearly two-dimensional. However, for situations in which the Biot number is much less than unity, the temperature is essentially a function of x alone, such that an approximate one-dimensional analysis of the heat-transfer process can be developed. For values of the Biot number less than 0.1, the use of a one-dimensional analysis generally leads to an error less than about 5%, as will be demonstrated in Chap. 3. However, for values of the Biot number much greater than 0.1, the multidimensionality must be accounted for in the analysis. For such situations in which the Biot number is not small, the concept of using fins to increase the heat transfer must be questioned.

Assuming that the temperature distribution in the extended surface shown in Fig. 2-16 is essentially one-dimensional, the first law of thermodynamics is applied to the differential volume $A\,dx$, with the following result:

Step 1

$$q_x = q_{x+dx} + dq_c \tag{2-94}$$

Step 2

$$\frac{dq_x}{dx}\,dx + dq_c = 0 \tag{2-95}$$

Whereas q_x is given by the one-dimensional Fourier law of conduction, Eq. (1-16), dq_c requires the use of the *general Newton law of cooling*, which is given by

$$dq_c = h\,dA_s(T_s - T_F) \tag{2-96}$$

h is known as the *local* coefficient of heat transfer. Incidentally, the integration of this equation for situations in which T_s and T_F are uniform gives rise to the simpler form of the Newton law of cooling, Eq. (2-90), where

$$\bar{h} = \frac{1}{A_s} \int_{A_s} h\,dA_s \tag{2-97}$$

Utilizing the one-dimensional Fourier law of conduction and the general Newton law of cooling, Eq. (2-95) takes the form

Step 3

$$\frac{d}{dx}\left(kA\frac{dT}{dx}\right)dx - hp\,dx(T - T_F) = 0 \tag{2-98}$$

where T_s is set equal to T in this approximate one-dimensional analysis. For our particular application, the cross-sectional area A, perimeter p, and k are uniform such that Eq. (2-98) reduces to

$$\frac{d^2T}{dx^2} - \frac{hp}{kA}(T - T_F) = 0 \tag{2-99}$$

For the case in which the base temperature T_0 is specified, one boundary condition is written as

Step 4

$$T = T_0 \qquad \text{at } x = 0 \tag{2-100}$$

The second boundary condition is written by recognizing that the fin loses heat from the tip by convection; that is,

$$-k\frac{dT}{dx} = h(T - T_F) \qquad \text{at } x = L \tag{2-101}$$

where the coefficient h at the tip is not necessarily equal to the coefficient along the perimeter. For the case in which the fin is very long, the temperature of the fin approaches T_F as x increases, such that Eq. (2-101) reduces to

$$\frac{dT}{dx} = T - T_F = 0 \qquad \text{as } x \to \infty \tag{2-102}$$

Of course, other boundary conditions can be written at $x=0$ or at $x=L$, depending upon the dictates of the actual problem under consideration.

Equation (2-102) will now be used to obtain predictions for the temperature distribution and heat transfer in a very long fin. With h approximated by $\bar{h}$ and with T_F assumed to be uniform, Eq. (2-99) takes the form

$$\frac{d^2\psi}{dx^2} - m^2\psi = 0 \tag{2-103}$$

where $\psi = T - T_F$ and $m^2 = \bar{h}p/(kA)$. Recognizing that an exponential function satisfies Eq. (2-103), the substitution of

$$\psi = Ce^{cx} \tag{2-104}$$

into this differential equation gives

$$c^2 - m^2 = 0 \tag{2-105}$$

Hence, $c = \pm m$, and the solution is

$$\psi = C_1 e^{mx} + C_2 e^{-mx} \tag{2-106}$$

The constants C_1 and C_2 are evaluated on the basis of the boundary conditions. Based on Eq. (2-100), $\psi = T_0 - T_F$ at $x = 0$ such that

$$T_0 - T_F = C_1 + C_2 \tag{2-107}$$

As suggested above, the second boundary condition can be written in the form of Eq. (2-102) for cases in which L is very long. For this situation, $\psi = 0$ as $x \to \infty$ and Eq. (2-106) gives

$$0 = \lim_{x \to \infty} (C_1 e^{mx} + C_2 e^{-mx}) \tag{2-108}$$

This equation requires that C_1 be equal to zero, such that C_2 is equal to $T_0 - T_F$ [from Eq. (2-107)]. Therefore, the solution for this case is

$$\psi = T - T_F = (T_0 - T_F)e^{-mx} \tag{2-109}$$

To obtain the total rate of heat transfer from the fin into the fluid q_F, we perform a lumped energy balance on the entire fin. It follows that $q_F = q_0$, where the rate of heat transfer at the base q_0 is obtained from the Fourier law of conduction,

$$q_0 = -kA \left.\frac{dT}{dx}\right|_0 \tag{2-110}$$

With the temperature distribution given by Eq. (2-109), our prediction for q_F becomes

$$q_F = q_0 = \sqrt{\bar{h}pkA} \; (T_0 - T_F) \qquad (2\text{-}111)$$

Parenthetically, this same result can also be obtained by equating q_F to the total rate of heat convected from the surface,

$$q_F = \int dq_c$$
$$= \int_0^\infty hp(T - T_F)\,dx \qquad (2\text{-}112)$$

where h is again approximated by $\bar{h}$. Substituting for $T - T_F$ and integrating, we arrive at Eq. (2-111).

To determine the minimum fin length for which this solution applies, we merely require that T be approximately equal to T_F at x equal to L; that is,

$$\frac{T_L - T_F}{T_0 - T_F} = e^{-mL} < \varepsilon \qquad (2\text{-}113)$$

where ε is a small number. With ε equal to 0.01, mL must be greater than a value of about 4.6 in order for our analysis to be reasonable.

For the case in which mL is significantly less than 4.6, the heat transfer through the tip must be accounted for by the use of Eq. (2-101). The use of this more general boundary condition for "short" convecting fins gives rise to an expression for the temperature distribution of the form (see Prob. 2-51)

$$\frac{T - T_F}{T_0 - T_F} = \frac{\cosh[m(L-x)] + \bar{h}/(km)\sinh[m(L-x)]}{\cosh(mL) + \bar{h}/(km)\sinh(mL)} \qquad (2\text{-}114)$$

To obtain q_F for this type of fin, we apply the Fourier law of conduction as follows:

$$q_F = -kA\frac{dT}{dx}\bigg|_0 = -kA\left[\frac{-m\sinh(mL) - m\bar{h}/(km)\cosh(mL)}{\cosh(mL) + \bar{h}/(km)\sinh(mL)}\right](T_0 - T_F)$$

$$= \sqrt{\bar{h}pkA}\left[\frac{\sinh(mL) + \bar{h}/(km)\cosh(mL)}{\cosh(mL) + \bar{h}/(km)\sinh(mL)}\right](T_0 - T_F) \qquad (2\text{-}115)$$

Other types of boundary conditions are sometimes encountered in fin applications. For example, for a fin with insulated tip, T and q_F are given by

$$\frac{T - T_F}{T_0 - T_F} = \frac{\cosh\left[m(L-x)\right]}{\cosh(mL)} \qquad (2\text{-}116)$$

and

$$q_F = \sqrt{hpkA} \; \tanh(mL)(T_0 - T_F) \qquad (2\text{-}117)$$

Expressions can also be obtained for T and q_F for fins with specified tip temperature; that is (see Prob. 2-52),

$$T - T_F = (T_0 - T_F)\frac{\sinh\left[m(L-x)\right]}{\sinh(mL)} + (T_1 - T_F)\frac{\sinh(mx)}{\sinh(mL)} \qquad (2\text{-}118)$$

and

$$q_F = \sqrt{hpkA} \left[\coth(mL) - \frac{1}{\sinh(mL)}\right](T_0 - T_F + T_1 - T_F) \qquad (2\text{-}119)$$

But in this case it should be noted that $q_F \neq q_0$.

2-6-3 Fin Resistance

To express the rate of heat transfer from a fin in terms of a thermal resistance, we write

$$q_F = \frac{T_0 - T_F}{R_F} \qquad (2\text{-}120)$$

For example, by introducing Eq. (2-111), the thermal resistance for a very long fin with small Biot number and negligible radiation effects is

$$R_F = \frac{1}{\sqrt{hpkA}} \qquad (2\text{-}121)$$

The thermal resistance of several standard fin geometries are summarized in Table 2-1. It should be noted that the manufacturers of fin units for electronic devices and other systems generally provide a rating for the thermal resistance of the unit, which accounts for conduction, natural convection, and radiation effects. Table 2-2 gives the thermal resistance rating for several standard heat-sink fin units for electronic applications.

TABLE 2-1 Thermal resistance: fins with small Biot number

System	*Thermal resistance R_F* *
Fin with uniform cross section: Very long with $Bi < 0.1$ 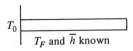 T_0 T_F and $\bar{h}$ known	$\dfrac{1}{\sqrt{\bar{h}pkA}}$
Fin with uniform cross section: Insulated tip with $Bi < 0.1$ T_0 $\longrightarrow q_c = 0$ T_F and $\bar{h}$ known	$\dfrac{1}{\sqrt{\bar{h}pkA}\,\tanh(mL)}$
Fin with uniform cross section: Tip at T_1 with $Bi < 0.1$ T_0 $\longleftarrow T_1$ T_F and $\bar{h}$ known	$\dfrac{1}{\sqrt{\bar{h}pkA}\left[\coth(mL) - \dfrac{1}{\sinh(mL)}\right]\left(1 + \dfrac{T_1 - T_F}{T_0 - T_F}\right)}$
Fin with uniform cross section: Convection from tip with $Bi < 0.1$ T_0 $\longrightarrow q_x = q_c$ T_F and $\bar{h}$ known	$\dfrac{1}{\sqrt{\bar{h}pkA}\,\dfrac{\sinh(mL) + \bar{h}/(mk)\cosh(mL)}{\cosh(mL) + \bar{h}/(mk)\sinh(mL)}}$
Blackbody fin with convection and radiation with uniform cross section: Very long with $Bi < 0.1$ T_0 $T_F = T_R = T_\infty$ $\bar{h}$ and F_{s-R} known	$\left(\bar{h}pkA + \dfrac{2}{5}kA\sigma p F_{s-R}\,\dfrac{T_0^5 - 5T_0 T_\infty^4 + 4T_\infty^5}{(T_0 - T_\infty)^2}\right)^{-\frac{1}{2}}$

*$m = \sqrt{\bar{h}p/(kA)}$

TABLE 2-2 Thermal resistance for natural convection and radiation from fin heat-sink units*

Thermal resistance R_F	Description
14°C/W	RS 401–778 Predrilled to accept T0–3 semiconductor case. 44.5 mm x 31.7 mm x 13.7 mm
19°C/W	RS 401–863 Predrilled to accept variety of plastic packaged semiconductor cases. 30 mm x 25 mm x 12.5 mm
10.5°C/W	RS 401–964 Predrilled to accept variety of plastic packaged semiconductor cases. 38 mm x 27 mm x 22.5 mm
4°C/W (with fins vertical)	RS401–497 100 mm length, overall cross section 64.5 mm x 15 mm
2.1°C/W (with fins vertical)	RS401–403 100 mm length, overall cross section 123.8 mm x 26.7 mm
1.1°C/W (with fins vertical)	RS 401–807 152 mm length, overall cross section 130 mm x 32 mm
0.5°C/W (with fins vertical)	RS 401–958 115 mm length, overall cross section 120 mm x 120 mm

*Courtesy of RS Components Ltd.

2-6-4 Fin Efficiency

Traditionally, the rate of heat transfer from surface extensions is generally presented in the literature in terms of the *fin efficiency* η_F. Fin efficiency is defined by

$$\eta_F = \frac{q_F}{q_{max}} \tag{2-122}$$

where q_{max} is the rate of heat transfer for the idealistic situation in which the Biot number is equal to zero (i.e., $R_k \simeq 0$) and the entire surface area of the fin A_s is at the base temperature T_0; that is,

$$q_{max} = \bar{h} A_s (T_0 - T_F) \tag{2-123}$$

The fin efficiency is easily expressed in terms of the fin resistance by writing

$$\eta_F = \frac{T_0 - T_F}{R_F} \frac{1}{\bar{h} A_s (T_0 - T_F)}$$

$$R_F = \frac{1}{\eta_F \bar{h} A_s} \tag{2-124}$$

Utilizing Eq. (2-111), we see that the fin efficiency for a very long fin is given by

$$\eta_F = \frac{\sqrt{\bar{h} p k A}}{\bar{h} A_s} = \frac{A}{p L \sqrt{Bi}} \tag{2-125}$$

Because of its usefulness as a design criterion, η_F is given in Fig. 2-17 for several standard fin geometries. Notice that parabolic and triangular profiles are more efficient than rectangular profiles. What is more, these nonuniform profiles contain less material than fins with rectangular profiles, which cuts down on weight and cost.

It should also be observed that η_F decreases with fin length. This happens because the temperature of a fin approaches T_F as x increases. Of course, the local rate of heat transfer dq_c convected from a fin decreases as $T - T_F$ falls. Therefore, an analysis should be made in the design of any fin unit which accounts for this decrease in fin efficiency with length and the cost of material.

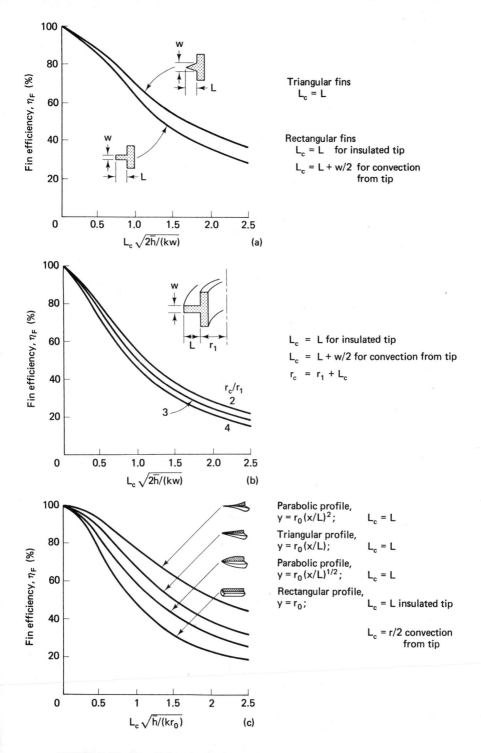

FIGURE 2-17 Fin efficiencies for approximate one-dimensional heat transfer in various extended surfaces for small Biot number. (a) Rectangular and triangular fins. (b) Circumferential fins. (c) Spines. (From Gardner [7].)

EXAMPLE 2-13

A 1-cm-diameter 3-cm-long steel fin $[k = 43 \text{ W}/(\text{m }°\text{C})]$ transfers heat from a wall at 200°C to a fluid at 25°C with $\bar{h} = 120 \text{ W}/(\text{m}^2 \text{ }°\text{C})$. Determine the rate of heat transfer from the fin for the case in which the tip is insulated and thermal radiation effects are negligible.

Solution

The system is shown in Fig. E2-13. The Biot number is

$$\text{Bi} = \frac{\bar{h}\ell}{k} = \frac{\bar{h}}{k}\frac{D}{4} = 0.00698$$

Because the Biot number is much less than 0.1, we are justified in using an approximate one-dimensional analysis.

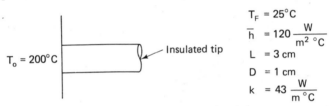

$T_F = 25°\text{C}$

$\bar{h} = 120 \dfrac{\text{W}}{\text{m}^2 \text{ }°\text{C}}$

$L = 3 \text{ cm}$

$D = 1 \text{ cm}$

$k = 43 \dfrac{\text{W}}{\text{m }°\text{C}}$

FIGURE E2-13   Fin with insulated tip.

For this situation in which the tip is insulated, q_F is given by Eq. (2-117),

$$q_F = \sqrt{\bar{h}pkA} \ \tanh(mL)(T_0 - T_F)$$

where

$$m = \sqrt{\frac{\bar{h}p}{kA}} = \frac{33.4}{\text{m}}$$

and

$$\sqrt{\bar{h}pkA} = 0.113 \frac{\text{W}}{°\text{C}}$$

Following through with the calculation, we obtain

$$q_F = 15.1 \text{ W}$$

EXAMPLE 2-14

The power transistor of Example 2-5 is to be mounted on an RS 401-778 anodized aluminum heat-sink fin unit, as shown in Fig. E2-14a. Determine the maximum power that can be dissipated safely by this transistor for a case temperature of 80°C, ambient temperature of 25°C, and $\bar{h} = 10$ W/(m² °C).

FIGURE E2-14a   Power transistor mounted on RS 401-778 heat sink.

Solution

Neglecting radiation effects for the moment and focusing attention on one of the two fins, we have

$$m = \sqrt{\frac{\bar{h}p}{kA}} = \sqrt{\frac{[10 \text{ W}/(\text{m}^2 \text{ °C})](64 \times 10^{-3} \text{ m})}{[200 \text{ W}/(\text{m °C})](31 \times 10^{-6} \text{ m}^2)}} = \frac{10.2}{\text{m}}$$

$$mL = \frac{10.2}{\text{m}}(12 \times 10^{-3} \text{ m}) = 0.122$$

Because mL is much less than 4.6, we must account for the heat loss through the tip. Therefore, Eq. (2-115) is utilized.

$$q_F = \sqrt{\bar{h}pkA}\left[\frac{\sinh(mL) + \bar{h}/(km)\cosh(mL)}{\cosh(mL) + \bar{h}/(km)\sinh(mL)}\right](T_0 - T_F)$$

where

$$\sqrt{\bar{h}pkA} = \left[10\frac{W}{m^2\,°C}(64\times 10^{-3}\,m)\left(200\frac{W}{m\,°C}\right)(31\times 10^{-6}\,m^2)\right]^{1/2}$$

$$= 6.30\times 10^{-2}$$

$$\frac{\bar{h}}{km} = \frac{10\,W/(m^2\,°C)}{[200\,W/(m\,°C)](10.2/m)} = 4.90\times 10^{-3}$$

It follows that

$$q_F = 7.95\times 10^{-3}(T_0 - T_F)\frac{W}{°C}$$

After mounting the power transducer, the area of the primary surface of the fin unit which is exposed to convective cooling is approximately 800 m². It follows that the rate of heat transfer from the primary surface is

$$q_p = \bar{h}A_p(T_0 - T_F)$$

$$= 10\frac{W}{m^2\,°C}(800\times 10^{-6}\,m^2)(T_0 - T_F)$$

$$= 8\times 10^{-3}\frac{W}{°C}(T_0 - T_F)$$

The total rate of heat convected from the heat-sink fin unit is given by

$$q_c = 2q_F + q_p$$

$$= 2(0.00795)(T_0 - T_F)\frac{W}{°C} + 0.008(T_0 - T_F)\frac{W}{°C}$$

$$= 0.0239(T_0 - T_F)\frac{W}{°C}$$

Thus, the thermal resistance to convection for the heat sink is

$$R_c = \frac{1}{0.0239}\frac{°C}{W} = 41.8\frac{°C}{W}$$

(Note that R_c for this heat-sink fin unit is about 49% lower than for the simple flat-plate frame system of Example 2-5.)

Assuming that the power transistor is properly attached to the heat sink (see Example 2-5), the thermal circuit for heat transfer through the heat sink is shown in Fig. E2-14b. It follows that the power dissipated

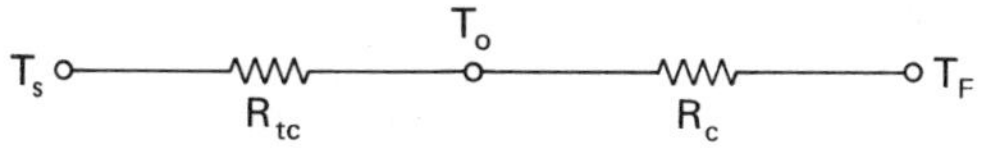

FIGURE E2-14b Thermal circuit.

through the heat sink is

$$q_{HS} = \frac{T_s - T_F}{R_{tc} + R_{HS}}$$

$$= \frac{80°C - 25°C}{0.5°C/W + 41.8°C/W} = 1.3 \text{ W}$$

This coupled with the 1.83 W transferred directly from the transistor to the surroundings gives a total power dissipation rate of

$$q = q_T + q_{HS}$$
$$= 1.83 \text{ W} + 1.3 \text{ W} = 3.13 \text{ W}$$

Because we have neglected the radiation losses, this will be a conservative estimate. The combined effects of convection and radiation heat transfer from a short fin such as this can be fairly easily analyzed by means of the numerical techniques introduced in Chap. 3. However, for design purposes, the manufacturers of heat-sink fin units generally specify the approximate thermal resistance for combined natural convection and radiation. Referring to Table 2-2, we find that the resistance of the fin unit in this example is rated by the manufacturer at 14°C/W. But it should be noted that this rating is based on laboratory measurements in free air (at about 25°C) at an unspecified transistor frame temperature T_s. In actuality, R_{HS} is strongly dependent upon T_s. Therefore, this value of R_{HS} should be used as a rough estimate in design work. Replacing R_c in Fig. E2-14b by this value, the rate of heat transfer from the heat sink becomes

$$q_{HS} = \frac{80°C - 25°C}{14.5°C/W} = 3.79 \text{ W}$$

such that the thermal limit on the total power dissipation from the transistor/heat-sink unit is

$$q = q_T + q_{HS}$$
$$= 1.83 \text{ W} + 3.79 \text{ W}$$
$$= 5.62 \text{ W}$$

Therefore, we conclude that the power transistor could be operated safely near its maximum power rating of 5 W.

EXAMPLE 2-15

Determine the rate of heat transfer from the long anodized aluminum fin shown in Fig. E2-15. The base temperature T_0 is 80°C, the air and surrounding walls are 25°C, and the coefficient of heat transfer due to natural convection cooling is 10 W/(m² °C).

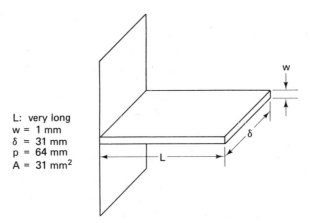

L: very long
w = 1 mm
δ = 31 mm
p = 64 mm
A = 31 mm²

FIGURE E2-15   Convecting and radiating fin.

Solution

The heat transfer in this fin can be approximated by a one-dimensional analysis if the Biot number R_k/R_s is very small. Assuming approximate blackbody conditions, we estimate the Biot number for combined convection and radiation as follows:

$$\text{Bi} = \frac{R_k}{R_s} = \frac{\ell/k}{\dfrac{1}{\bar{h}} + \dfrac{1}{\sigma F_{s-R}(T_s + T_R)(T_s^2 + T_R^2)}} \tag{a}$$

where ℓ is set equal to V/A_s of the fin. As a conservative measure, T_s and T_R are both set equal to 353 K($= 80°C$). The Biot number is then found to be 1.21×10^{-5}, such that a one-dimensional analysis can be safely used.

Referring to the lumped-differential element shown in Fig. E2-15, the application of the first law of thermodynamics gives

$$q_x = q_{x+dx} + dq_c + dq_R \tag{b}$$

It follows that

$$0 = \frac{dq_x}{dx} dx + dq_c + dq_R \tag{c}$$

Because the temperature along the fin is a function of x, we utilize the general Newton law of cooling as well as a generalized form of Eq. (1-24); that is,

$$dq_R = \sigma F_{s-R} dA_s (T_s^4 - T_R^4) \tag{d}$$

Substituting these particular laws into Eq. (c), we obtain

$$\frac{d}{dx}\left(kA\frac{dT}{dx}\right)dx = hp\,dx(T - T_F) + \sigma p\,dx F_{s-R}(T^4 - T_R^4)$$

or

$$\frac{d^2T}{dx^2} = \frac{hp}{kA}(T - T_F) + \frac{\sigma p F_{s-R}}{kA}(T^4 - T_R^4) = m^2(T - T_F) + m_R^2(T^4 - T_R^4) \tag{e}$$

where h is approximated by $\bar{h}$, $m = \overline{hp}/(kA)$, and $m_R^2 = \sigma p F_{s-R}/(kA)$. The boundary condition at the base is

$$T = T_0 \qquad \text{at } x=0 \tag{f}$$

For the case in which T_F and T_R are both equal to T_∞ and the fin is very long we also have

$$T = T_\infty \quad \text{or} \quad \frac{dT}{dx} = 0 \qquad \text{as } x \to \infty \tag{g}$$

The simplest way to solve this nonlinear system of equations for T is to use the numerical finite-difference approach introduced in Chap. 3. However, we can obtain an analytical solution for the temperature gradient by making the substitution

$$\psi = \frac{dT}{dx} \tag{h}$$

This puts our nonlinear differential equation into the form

$$\frac{d\psi}{dx} = m^2(T - T_\infty) + m_R^2(T^4 - T_\infty^4) \tag{i}$$

Based on Eq. (h), $dx = dT/\psi$. Therefore, Eq. (i) can be written as

$$\psi\frac{d\psi}{dT} = m^2(T - T_\infty) + m_R^2(T^4 - T_\infty^4) \tag{j}$$

We now separate the variables and integrate to obtain

$$\frac{\psi^2}{2} = m^2\left(\frac{T^2}{2} - T_\infty T\right) + m_R^2\left(\frac{T^5}{5} - TT_\infty^4\right) + C_1 \tag{k}$$

$$\psi = \frac{dT}{dx} = \pm\sqrt{2}\left[m^2\left(\frac{T^2}{2} - T_\infty T\right) + m_R^2\left(\frac{T^5}{5} - TT_\infty^4\right) + C_1\right]^{1/2} \tag{l}$$

where the negative sign is retained because the gradient is known to be negative for the case in which $T_0 > T$.

The constant C_1 can be evaluated by introducing the boundary condition given by Eq. (g); that is,

$$\frac{dT}{dx}\bigg|_\infty = 0 = \sqrt{2}\left[m^2\left(\frac{T_\infty^2}{2} - T_\infty^2\right) + m_R^2\left(\frac{T_\infty^5}{5} - T_\infty^5\right) + C_1\right]^{1/2} \tag{m}$$

where $T(\infty) = T_\infty$. Hence, C_1 is given by

$$C_1 = m^2\frac{T_\infty^2}{2} + \frac{4}{5}m_R^2 T_\infty^5$$

and Eq. (l) becomes

$$\frac{dT}{dx} = -\sqrt{2}\left[m^2\left(\frac{T^2}{2} - T_\infty T + \frac{T_\infty^2}{2}\right) + m_R^2\left(\frac{T^5}{5} - T_\infty^4 T + \frac{4}{5}T_\infty^5\right)\right]^{1/2} \tag{n}$$

An expression can now be written for the rate of heat transfer from the fin.

$$q_F = -kA\frac{dT}{dx}\bigg|_0$$

$$= kA\sqrt{2}\left[\frac{m^2}{2}(T_0^2 - 2T_\infty T_0 + T_\infty^2) + \frac{m_R^2}{5}(T_0^5 - 5T_\infty^4 T_0 + 4T_\infty^5)\right]^{1/2}$$

$$= \left[\bar{h}pkA(T_0 - T_\infty)^2 + \frac{2}{5}kA\sigma pF_{s-R}(T_0^5 - 5T_\infty^4 T_0 + 4T_\infty^5)\right]^{1/2} \tag{o}$$

or

$$q_F = \sqrt{q_c + q_R} \tag{p}$$

where

$$q_c = \sqrt{\bar{h}pkA}\ (T_0 - T_\infty) \tag{q}$$

and

$$q_R = \left[\frac{2}{5} kA\sigma p F_{s-R} \left(T_0^5 - 5T_\infty^4 T_0 + 4T_\infty^5 \right) \right]^{1/2} \qquad \text{(r)}$$

Calculating q_c and q_R, we have

$$q_c = \left[\left(10\frac{W}{m^2\,°C} \right)(64 \times 10^{-3} m)\left(200\frac{W}{m\,°C} \right)(31 \times 10^{-6}\ m^2) \right]^{1/2}$$

$$\times (353°C - 298°C)$$

$$= 3.46\ W$$

$$q_R = \left\{ \frac{2}{5}\left(200\frac{W}{m\,°C} \right)(31 \times 10^{-6}\ m^2)\left(5.67 \times 10^{-8}\frac{W}{m^2\ K^4} \right)(64 \times 10^{-3}\ m)(1) \right.$$

$$\left. \times \left[(353\ K)^5 - 5(298\ K)^4(353\ K) + 4(298\ K)^5 \right] \right\}^{1/2}$$

$$= 2.94\ W$$

Thus, the total rate of heat transfer from the fin is

$$q_F = \sqrt{(3.46\ W)^2 + (2.94\ W)^2}$$
$$= 4.54\ W$$

The fact that q_F is about 31% greater than q_c reinforces our earlier conclusion that radiation can play a very significant role in heat transfer from fins with near blackbody surfaces.

2-7   UNSTEADY HEAT-TRANSFER SYSTEMS

2-7-1   Introduction

We now turn our attention to unsteady heat-transfer processes such as the cooling of a billet, which is illustrated in Fig. 2-18. The heat transfer in unsteady systems like this is actually multidimensional because the temperature within the body is a function of time t and at least one space dimension. However, for problems such as this that involve convection and/or radiation, approximate one-dimensional analyses can be utilized if the Biot number (R_k/R_s) is small (i.e., $\text{Bi} \gtrsim 1$). Under these circumstances, the variation in temperature with the spatial coordinates will be very slight, such that the temperature can be taken as a function of time alone. Approximate one-dimensional lumped analyses are developed for

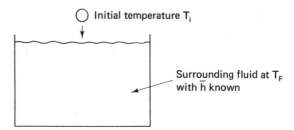

FIGURE 2-18   Convective cooling of a billet initially at temperature T_i which is dropped into a bath at temperature T_F.

representative unsteady convection heat-transfer processes in this section.
Unsteady heat-transfer systems involving multidimensions and radiation
will be considered in Chaps. 3 and 4.

2-7-2 Analysis

We consider the situation illustrated in Fig. 2-18 in which a body with
uniform initial temperature T_i is suddenly exposed to an environmental
temperature T_F. Applying the first law of thermodynamics to the lumped
volume V, we obtain

$$q_c + \frac{\Delta E_s}{\Delta t} = 0 \tag{2-126}$$

$$\bar{h} A_s (T - T_F) + \frac{dU}{dt} = 0 \tag{2-127}$$

The internal energy U can be expressed in terms of the specific heat at
constant volume; that is,

$$c_v = \frac{\partial}{\partial T} \left(\frac{U}{m} \right) \Big|_0 \tag{2-128}$$

Since the volume is essentially constant for heat transfer in solids, dU can
be written as

$$dU = m c_v \, dT = \rho V c_v \, dT \tag{2-129}$$

Therefore, Eq. (2-127) takes the form

$$\frac{dT}{dt} + \frac{\bar{h} A_s}{\rho V c_v} (T - T_F) = 0 \tag{2-130}$$

The initial condition associated with this equation is

$$T = T_i \qquad \text{at } t = 0 \tag{2-131}$$

This completes our formulation for situations in which the mass of the surrounding fluid is large and T_F is essentialy independent of time. However, for systems in which the mass of the surrounding fluid is not large, the variation in T_F with time must be accounted for. This is done by applying the first law of thermodynamics to the surrounding fluid itself, as illustrated in Example 2-18.

For the case in which T_F is constant, Eq. (2-130) can be written in the form

$$\frac{d\psi}{dt} + \frac{\bar{h}A_s\psi}{\rho V c_v} = 0 \tag{2-132}$$

where $\psi = T - T_F$. With $\bar{h}$ approximated by a constant, this homogeneous first-order equation can be separated and integrated to obtain

$$\int_{\psi_i}^{\psi} \frac{d\psi}{\psi} = \frac{-\bar{h}A_s}{\rho V c_v} \int_0^t dt \tag{2-133}$$

where $\psi_i = T_i - T_F$. Continuing, we have

$$\ln\left(\frac{\psi}{\psi_i}\right) = -\frac{\bar{h}A_s t}{\rho V c_v} \tag{2-134}$$

Hence, the temperature history is represented by

$$\frac{T - T_F}{T_i - T_F} = \exp\left(\frac{-\bar{h}A_s t}{\rho V c_v}\right) \tag{2-135}$$

The rate of heat convected from the surface at any instant t can be written as

$$q_c = \bar{h}A_s(T - T_F)$$

$$= \bar{h}A_s(T_i - T_F)\exp\left(\frac{-\bar{h}A_s t}{\rho V c_v}\right) \tag{2-136}$$

To obtain the total heat transferred from the surface over a length of time τ, we simply integrate as follows:

$$Q_c = \int_0^\tau q_c \, dt = \bar{h}A_s(T_i - T_F)\int_0^\tau \exp\left(\frac{-\bar{h}A_s t}{\rho V c_v}\right) dt$$

$$= \rho V c_v(T_i - T_F)\left[1 - \exp\left(\frac{-\bar{h}A_s \tau}{\rho V c_v}\right)\right] \tag{2-137}$$

The maximum total amount of energy that can be convected to the fluid is obtained by merely allowing τ to become very large; that is,

$$Q_{max} = \rho V c_v (T_i - T_F) \tag{2-138}$$

Notice that Q_{max} is equal to the relative internal energy possessed by the body at time $t = 0$. The dimensionless ratio Q_c / Q_{max} is given by

$$\frac{Q_c}{Q_{max}} = 1 - \exp\left(\frac{-\bar{h} A_s \tau}{\rho V c_v}\right) \tag{2-139}$$

EXAMPLE 2-16

A metallic ball $[k = 200 \text{ W}/(\text{m } ^\circ\text{C}), \rho = 2500 \text{ kg}/\text{m}^3, c_v = 0.8 \text{ kJ}/(\text{kg } ^\circ\text{C})]$ 0.5 cm in diameter initially at 250°C is dropped into a large tank of fluid which is maintained at 25°C and 1 atm pressure. The mean coefficient of heat transfer is approximately equal to 3000 W/(m² °C) for boiling and 250 W/(m² °C) for nonboiling. Determine the temperature history of the ball. Also estimate the instantaneous rate of heat transfer.

Solution

Calculating the Biot number and assuming negligible radiation effects, we find that $\text{Bi} \ll 0.1$ for the entire process. Hence, the lumped differential formulation is given by Eqs. (2-130) and (2-131).

The solution to Eq. (2-130) with the initial condition $T(0) = 250°C$ is given by Eq. (2-135) with $\bar{h} = 3000 \text{ W}/(\text{m}^2 \text{ °C})$; that is,

$$\frac{T - T_F}{T_i - T_F} = \frac{T - 25°C}{250°C - 25°C} = \exp\left(-\frac{1.8t}{\text{s}}\right) \tag{a}$$

This equation applies to the period of time for which $T \geqslant 100°C$. To find the time t_1 at which boiling ceases, T is set equal to 100°C in this equation, with the result

$$t_1 = \frac{\text{s}}{1.8} \ln\left(\frac{225}{75}\right) = 0.61 \text{ s} \tag{b}$$

The solution to Eq. (2-130) for the condition $T(t_1) = 100°C$ and $\bar{h} = 250$ W/(m² °C) is given by

$$\frac{T - 25°C}{100°C - 25°C} = \exp\left(-0.15\frac{t - t_1}{\text{s}}\right) \tag{c}$$

Equations (a) and (b) are shown in Fig. E2-16.

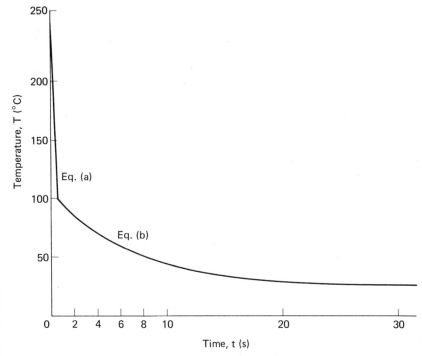

FIGURE E2-16 Temperature history.

To calculate the rate of convection from the surface, we write

$$q_c = \bar{h} A_s (T - T_F)$$

$$q_c'' = 3000 \frac{W}{m^2 \, °C} (225°C) \exp\left(-\frac{1.8t}{s}\right) \qquad (d)$$

for $t \leqslant 0.61$ s, and

$$q_c'' = 250 \frac{W}{m^2 \, °C} (75°C) \exp\left(-0.15 \frac{t - 0.61}{s}\right) \qquad (e)$$

for $t \geqslant 0.61$ s.

To estimate the levels of radiation heat transfer in this problem, one can easily calculate the instantaneous rate of thermal radiation emitted by a blackbody ball with temperature given by Eqs. (a) and (b). This calculation indicates that the maximum rate of radiation from the ball is less than 1% of the rate of convection heat transfer (see Prob. 2-67).

EXAMPLE 2-17

Electric current of 5 A is suddenly passed through a 1-mm-diameter copper wire initially at temperature 25°F. Develop an expression for the instantaneous temperature of the wire for uniform internal energy generation, constant surrounding fluid temperature T_F of 25°C, coefficient of heat transfer $\bar{h}$ of 25 W/(m² °C), and negligible radiation effects.

Solution

Because the Biot number for this system is much less than 0.1, an approximate one-dimensional formulation is developed. Utilizing the first law of thermodynamics, we obtain

$$\Sigma \dot{E}_o - \Sigma \dot{E}_i + \frac{\Delta E_s}{\Delta t} = 0$$

$$\bar{h} A_s (T - T_F) - \dot{q}_0 V + \rho V c_v \frac{dT}{dt} = 0$$

or

$$\frac{d\psi}{dt} + \frac{\bar{h} A_s \psi}{\rho V c_v} - \frac{\dot{q}_0}{\rho c_v} = 0 \qquad (a)$$

where $\psi = T - T_F$. The initial condition is

$$\psi(0) = 0 \qquad (b)$$

To obtain the homogeneous solution ψ_H, we write

$$\frac{d\psi_H}{\psi_H} = - \frac{\bar{h} A_s}{\rho V c_v} dt$$

and integrate.

$$\ln\left(\frac{\psi_H}{C_1}\right) = - \frac{\bar{h} A_s t}{\rho V c_v}$$

or

$$\psi_H = C_1 \exp\left(\frac{-\bar{h} A_s t}{\rho V c_v}\right) \qquad (c)$$

To obtain the particular solution ψ_p, we assume that

$$\psi_p = C_3 + C_4 t$$

Substituting this expression into Eq. (a),

$$C_4 + \frac{\bar{h}A_s}{\rho V c_v}(C_3 + C_4 t) = \frac{\dot{q}_0}{\rho c_v}$$

Hence, we see that $C_4 = 0$ and $C_3 = \dot{q}_0 V/(\bar{h}A_s)$; that is,

$$\psi_p = \frac{\dot{q}_0 V}{\bar{h}A_s} \tag{d}$$

Thus, ψ takes the form

$$\psi = C_1 \exp\left(\frac{-\bar{h}A_s t}{\rho V c_v}\right) + \frac{\dot{q}_0 V}{\bar{h}A_s} \tag{e}$$

Utilizing the initial condition,

$$C_1 = -\frac{\dot{q}_0 V}{\bar{h}A_s}$$

and

$$\psi = T - T_F = \frac{\dot{q}_0 V}{\bar{h}A_s}\left[1 - \exp\left(\frac{-\bar{h}A_s t}{\rho V c_v}\right)\right] \tag{f}$$

The electrical resistivity ρ_e is approximately 1.8×10^{-8} Ω m. Consequently, the rate of internal energy generation per unit volume $\dot{q}_0$ is

$$\dot{q}_0 = \frac{I_e^2 \rho_e}{A^2} = \frac{(5\text{A})^2(1.8 \times 10^{-8}\Omega \text{ m})}{\left[\pi(0.001 \text{ m})^2/4\right]^2}$$

$$= 730 \frac{\text{kW}}{\text{m}^3}$$

The other pertinent properties for copper are $\rho = 8950$ kg/m³, $c_v = 0.383$ kJ/(kg °C), and $k = 386$ W/(m °C). It follows that

$$\frac{\dot{q}_o V}{\bar{h}A_s} = \frac{730 \text{ kW/m}^3}{25 \text{ W/(m}^2 \text{ °C)}} \frac{0.001 \text{ m}}{4} = 7.3°\text{C}$$

$$\frac{\bar{h}A_s}{\rho V c_v} = \frac{25 \text{ W/(m}^2 \text{ °C)}}{(8950 \text{ kg/m}^3)[0.383 \text{ kJ/(kg °C)}]} \frac{4}{0.001 \text{ m}}$$

$$= \frac{2.92 \times 10^{-2}}{\text{s}}$$

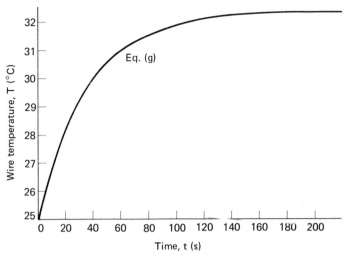

FIGURE E2-17   Temperature history.

Hence, the instantaneous temperature of the wire is given by

$$T = 25°C + 7.3°C\left[1 - \exp\left(-\frac{2.92 \times 10^{-2}t}{s}\right)\right] \tag{g}$$

This equation is shown in Fig. E2-17. Notice that the wire temperature approaches a steady-state value of 32.3°C after about 160 s.

The instantaneous convection heat-transfer flux is obtained by coupling the Newton law of cooling and Eq. (g).

$$q_c'' = \bar{h}(T_s - T_F)$$

$$= 25\frac{W}{m^2\,°C}(7.3°C)\left[1 - \exp\left(-\frac{2.92 \times 10^{-2}t}{s}\right)\right]$$

$$= 183\frac{W}{m^2}\left[1 - \exp\left(-\frac{2.92 \times 10^{-2}t}{s}\right)\right] \tag{h}$$

The heat flux increases from zero at time zero to a maximum steady-state value of 183 W/m².

EXAMPLE 2-18

The metallic ball of Example 2-16 is dropped into a 0.0012-kg oil bath which is initially at 25°C; the boiling point of the oil is 400°C and the properties are $\rho = 82.5$ kg/m³, $c_v = 2.2$ kJ/(kg °C). The oil is well stirred such that $\bar{h} = 350$ W/(m² °C) and the temperature throughout the bath is uniform, but time-dependent. The container is insulated. Determine the final steady-state temperature of the ball and oil and the time required to reach steady state for the case in which radiation effects are small.

Solution

We develop lumped energy balances on both the ball and the oil as follows:

$$\text{Ball:} \quad 0 = \bar{h} A_s (T_I - T_{II}) + (\rho V c_v)_I \frac{dT_I}{dt} \tag{a}$$

$$\text{Oil:} \quad \bar{h} A_s (T_I - T_{II}) = (\rho V c_v)_{II} \frac{dT_{II}}{dt} \tag{b}$$

$$T_I(0) = T_{Ii} = 250°C \tag{c}$$

$$T_{II}(0) = T_{IIi} = 25°C \tag{d}$$

These differential equations are put into the operator format

$$(D + K_I) T_I = K_I T_{II} \tag{e}$$

$$(D + K_{II}) T_{II} = K_{II} T_I \tag{f}$$

where $K_I = \bar{h} A_s / (\rho V c_v)_I = 0.21/s$ and $K_{II} = \bar{h} A_s / (\rho V c_v)_{II} = 1 \times 10^{-2}/s$. We now eliminate T_{II} by combining Eqs. (e) and (f).

$$(D + K_I) T_I = K_I K_{II} T_I \frac{1}{D + K_{II}}$$

or

$$\frac{d}{dt} \left(\frac{dT_I}{dt} \right) + (K_I + K_{II}) \frac{dT_I}{dt} = 0 \tag{g}$$

Separating the variables, we obtain

$$\frac{d(dT_I / dt)}{dT_I / dt} = -(K_I + K_{II}) dt \tag{h}$$

A first integration gives

$$\ln \left(\frac{dT_I / dt}{C_I} \right) = -(K_I + K_{II}) t \tag{i}$$

or

$$\frac{dT_I}{dt} = C_1 \exp \left[-(K_I + K_{II}) t \right] \tag{j}$$

A second integration gives

$$T_I = \frac{C_1}{K_I + K_{II}} \left\{ 1 - \exp\left[-(K_I + K_{II})t \right] \right\} + C_2 \tag{k}$$

Referring to Eq. (a), we see that

$$\frac{dT_I}{dt} = -K_I(T_{Ii} - T_{IIi}) \qquad \text{at } t = 0 \tag{1}$$

Thus, $C_1 = -K_I(T_{Ii} - T_{IIi})$. To evaluate C_2, we employ the initial condition given by Eq. (c); that is, $C_2 = T_{Ii}$. Hence, our solution for T_I is

$$\frac{T_I - T_{Ii}}{T_{Ii} - T_{IIi}} = \frac{K_I}{K_I + K_{II}} \left\{ \exp\left[-(K_I + K_{II})t \right] - 1 \right\}$$

The final steady-state temperature T_{ss} of the system is

$$T_{ss} = 250°C - 225°C \frac{K_I}{K_I + K_{II}} = 35.2°C$$

The time for the temperature to reach 99% of its steady-state value is obtained as follows:

$$0.01 = \exp\left[-(K_I + K_{II})t_{ss} \right]$$

$$t_{ss} = \frac{\ln 0.01}{-(K_I + K_{II})} = \frac{\ln 0.01}{0.22/s} = 20.9 \text{ s}$$

2-7-3   Electrical Analogy

As suggested in Chap. 1, an analogy exists between unsteady one-dimensional heat transfer and the unsteady flow of electric current. For unsteady heat-transfer systems, we represent the thermal capacity $C(\equiv mc_v)$ of a mass by an electrical capacitor C_e. The electric current passing through a capacitor is proportional to the time rate of change of voltage,

$$I_e = -C_e \frac{dE_e}{dt} \tag{2-140}$$

With this in mind, the thermal network for the problem under consideration is shown in Fig. 2-19. The thermal capacitor is initially charged at the potential T_i with the switch in position 1. The process is then initiated by throwing the switch to position 2, with the energy stored in the capacitor being dissipated through the resistance. With the product $R_e C_e$ set equal to $R_e C [\equiv \rho V c_v / (\bar{h} A_s)]$ and with $E_{ei} - E_{eo}$ set equal to $T_i - T_F$, the flow of electric current in this circuit is perfectly analogous to the flow of heat in

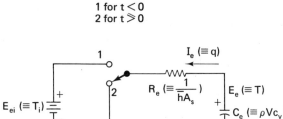

Switch position

1 for t < 0
2 for t > 0

FIGURE 2-19 Analogous electrical circuit for unsteady one-dimensional convection heat transfer.

the thermal system. Hence, instantaneous measurements in I_e and $E_e - E_{eo}$ correspond to the instantaneous rate of heat transfer q and temperature difference $T - T_F$, respectively. This extended electrical analogy concept will be utilized in Chap. 3 in the analysis of more complex multidimensional heat-transfer processes.

2-8 SUMMARY: PRACTICAL SOLUTION RESULTS

In this chapter we have dealt with classic one-dimensional heat-transfer systems. These include conduction in flat plates, hollow cylinders and spheres, and composites, conduction with convection and other types of boundary conditions, conduction with variable thermal conductivity, and conduction with internal energy generation. In addition, approximate one-dimensional analyses have been developed for heat transfer in extended surfaces and unsteady lumped heat-transfer processes with small values of the Biot number.

For systems without internal energy generation, we found that the rate of heat transfer can be expressed in terms of practical equations in any of the three formats

$$q = \frac{\Delta T}{R} = K\Delta T = kS\Delta T \qquad (2\text{-}141)$$

where R is the thermal resistance, K the thermal conductance, S the conduction shape factor, and ΔT the temperature difference that characterizes the particular problem. The thermal resistances and temperature profiles for the key steady one-dimensional nongenerating systems treated in this chapter are summarized in Table 2-3. Effects of convection, thermal

TABLE 2-3 Thermal resistance: steady one-dimensional systems

System	Thermal resistance, R
Plane wall: surfaces at $x = 0$ and $x = L$ at T_1 and T_2.	$\dfrac{L}{kA}$

T_1 T_2

$\leftarrow L \rightarrow$

| Hollow Circular Cylinder: surfaces at $r = r_1$ and $r = r_2$ at T_1 and T_2. | $\dfrac{\ln (r_2/r_1)}{2\pi L k}$ |

T_1 T_2

| Hollow Circular Cylinder Section: surfaces at $r = r_1$ and $r = r_2$ at T_1 and T_2; surfaces at $\theta = 0$ and $\theta = \theta_1$ insulated. | $\dfrac{\ln (r_2/r_1)}{\theta_1 L k}$ |

T_1
T_2
θ_1

| Hollow Sphere: surfaces at $r = r_1$ and $r = r_2$ at T_1 and T_2. | $\dfrac{r_2 - r_1}{4\pi r_1 r_2 k}$ |

T_1 T_2

radiation, and composite materials on steady-state heat transfer in these systems can be accounted for by the network approach. The thermal resistances for several fin units are given in Tables 2-1 and 2-2.

In regard to the problem of heat transfer with internal energy generation, the maximum temperature T_{max} is a very important design parameter. T_{max} is given in Table 2-4 for two standard geometries.

For unsteady one-dimensional convection heat transfer, we found that the instantaneous rate q and total Q are given by

$$q_c = \bar{h} A_s (T_i - T_F) \exp\left(\frac{-\bar{h} A_s t}{\rho V c_v} \right) \qquad (2\text{-}136)$$

$$Q_c = \rho V c_v (T_i - T_F) \left[1 - \exp\left(\frac{-\bar{h} A_s \tau}{\rho V c_v} \right) \right] \qquad (2\text{-}137)$$

For other one-dimensional heat-transfer problems involving variable properties, nonuniform internal energy generation, nonstandard geometries, and other types of boundary conditions, the reader can refer back to

TABLE 2-4 T_{max} for systems with uniform internal energy generation

System	T_{max}
Flat Plate	$T_1 + \dfrac{\dot{q}_0}{k} \dfrac{L^2}{8}$
Solid Circular Cylinder with radius r_1.	$T_1 + \dfrac{\dot{q}_0}{k} \dfrac{r_1^2}{4}$

previous sections in Chap. 2. The principles set forth in this chapter can be readily adapted to the analysis of any one-dimensional heat-transfer problem for which solutions are not already available. However, the more complex problems such as those involving thermal radiation and nonuniform cross-sectional areas may require the use of numerical techniques introduced in Chap. 3.

PROBLEMS

2-1. Measurements for the axial temperature distribution in a 1-in.-diameter copper rod of 9-in. length are given in Fig. P2-1. The ends are at $T_1 = 345°F$ and 158°F, and the perimeter is insulated. Compare the theoretical temperature distribution with these data and determine the rate of heat transfer.

2-2. One surface of a stainless steel plate [$k = 10$ W/(m K)] of 10-cm thickness is at a temperature of 80°C. The other surface is exposed to a uniform heat

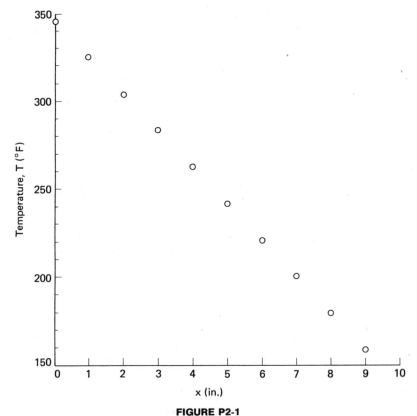

FIGURE P2-1

flux of 500 W/m^2. Determine the temperature distribution in the plate and the unknown surface temperature.

2-3. The surface temperatures of a glass window with dimensions $\frac{1}{32}$ in. by 1 ft by 1 ft are 100°F and 68°F. Determine the rate of heat conducted through the window and the total heat loss by conduction over a 12-h period.

2-4. An electrical heating plate of 1-mm thickness is sandwiched between two identical materials of 1-cm thickness and 0.5-m^2 cross-sectional area. The ends are insulated and $\dot{q}$ is equal to 500 kW/m^3. The surface temperatures of one plate are found to be equal to 85°C and 150°C. Determine the thermal conductivity of the material.

2-5. Solve Prob. 2-2 using the electrical analog approach.

2-6. Solve Prob. 2-3 using the electrical analog approach.

2-7. A 1-mm-thick glass window is exposed to fluids at temperatures of 100°C and 68°C. The coefficients of heat transfer are 250 W/(m^2 °C) and 60 W/(m^2 °C), respectively. Determine the rate of heat transfer per unit area and the surface temperatures. Neglect effects of radiation.

2-8. The inside of a furnace wall 1 ft thick constructed of fire clay brick is 1000°F. The room temperature is 50°F. Determine the rate of heat transfer through a 20-ft^2 area and the surface temperature if $\bar{h} = 150$ Btu/(h ft^2 °F).

2-9. A 10-cm-thick copper plate is exposed to a fluid at 5°C with $\bar{h} = 150$ W/(m^2 °C). The rate of heat transfer per unit area into the other surface of the plate is 1000 W/m^2. Determine the surface temperatures of the plate.

2-10. A 10-cm-thick plate [$k = 16$ W/(m °C)] coated with carbon black is exposed to radiation from a blackbody source at 200°C. The rate of heat transfer per unit area into the other surface of the plate is 1000 W/m^2. Determine the surface temperatures of the plate.

2-11. If the nonradiating surface of the plate described in Prob. 2-10 is maintained at 15°C, determine the unknown surface temperature and the rate of heat transfer per unit area.

2-12. A 10-cm-thick aluminum plate coated with carbon black is exposed to radiation on both sides from blackbody sources at 200°C and 2000°C. Determine the surface temperatures of the plate and the rate of heat transfer per unit area through the plate. [*Hint:* Assume a value of $T(L)$ and iterate for $T(0)$. Then compute a new value of $T(L)$. Continue this double iteration until $T(0)$ and $T(L)$ converge.]

2-13. A furnace wall is constructed of 5-in.-thick brick. The radiating surfaces and the hot gases within the furnace are at a temperature of 2000°F with $\bar{h}$ equal to 500 Btu/(h ft^2 °F). The air temperature outside the furnace wall is at 85°F with $\bar{h}$ equal to 10 Btu/(h ft^2 °F). Determine the surface temperatures of the wall and the rate of heat transfer per unit area.

2-14. The inside wall temperature of a 5-cm-thick brick furnace wall is 1000°C. The outside air temperature is 30°C and $\bar{h} = 10$ W/(m^2 °C). (a) Determine the outside surface temperature. (b) Determine the thickness of asbestos needed to reduce the surface temperature by 50%.

2-15. The general formula for the Newton–Raphson iteration method for solving an equation of the form $f(x) = 0$ is given in [1] as $x_{i+1} = x_i - f(x_i)/f'(x_i)$, where i is the iteration index. Demonstrate that Eq. (d) in Example 2-4 can be solved in only two steps by this approach, starting with $T_1 = 1000$ K.

2-16. A composite wall consists of a series arrangement of aluminum (5 cm thick), copper (10 cm thick), and stainless steel (2 cm thick). The surface temperatures are 15°C and 100°C. Determine the rate of heat transfer per unit area, the temperature distribution, and the interfacial temperatures.

2-17. Two aluminum plates 10 cm thick are bolted together with a contact pressure of 15 atm. The surfaces of the composite plate are at temperatures 100°C and 0°C. Determine the temperature distribution and rate of heat transfer for the cases in which the interfacial surfaces are (a) polished and silicon grease is used; (b) ground with roughness of about 2.5 μm, for which case $h_{tc} \simeq 11.4$ kW/(m^2 °C).

2-18. A 10-cm-thick composite wall consists of a parallel arrangement of aluminum with 0.1-m^2 cross-sectional area and stainless steel with 0.5-m^2 cross-sectional area. The surface temperatures are 15°C and 100°C. Determine the rate of heat transfer and the temperature distribution in this one-dimensional system.

2-19. One surface of the composite wall of Prob. 2-18 is exposed to a fluid at 100°C with $\bar{h} = 5$ W/(m^2 °C) and the other surface is held at 15°C. Estimate the rate of heat transfer in this two-dimensional system.

2-20. The thermal conductivity of a nonhomogeneous wall of thickness L is given as a function of x. The surfaces at $x = 0$ and $x = L$ are maintained at T_1 and T_2. This system is a generalization of a series composite and is one-dimensional. Develop expressions for the rate of heat transfer q_x for (a) $k = k_1$ $(1 + \gamma x)$, and (b) $k = k_1$ for $0 \leqslant x < L/2$ and $k = k_2$ for $L/2 < x \leqslant L$; $k_1 = 15$ W/(m °C), $k_2 = 100$ W/(m °C), and $\gamma = 0.1/$m. Show that your solution to part (b) is equivalent to Eq. (2-31).

2-21. The thermal conductivity of a nonhomogeneous wall of thickness L is given as a function of y. The surfaces at $x = 0$ and $x = L$ are held at T_1 and T_2. This system is a generalization of a parallel composite and is one-dimensional. Develop expressions for the rate of heat transfer for (a) $k = k_1(1 + \gamma y)$, and (b) for $k = k_1$ for $0 \leqslant y < w/2$ and $k = k_2$ for $w/2 < y \leqslant w$; $k_1 = 15$ W/(m °C), $k_2 = 100$ W/(m °C), and $\gamma = 0.1/$m. Show that your solution to part (b) is equivalent to Eq. (2-35).

2-22. Complex structures such as the particulate composite shown in Fig. P2-22 are sometimes encountered. A system such as this in which k is a function of both x and y represents a generalization of combined series/parallel composites and is multidimensional. A practical approach which can sometimes be used in handling these types of problems is to simply establish upper and lower bounds on the heat transfer. As pointed out by Zinsmeister [8], an upper bound for this system can be established by distributing the volumes of the individual phases in a simple parallel arrangement, and a lower bound can be obtained from a simple series arrangement. Assuming

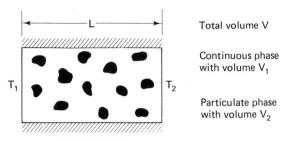

FIGURE P2-22

that we have a particulate composite wall of thickness L and total volume V, with V_1 and V_2 being the volume of the continuous and particulate phases, show that upper and lower bounds on the rate of heat transfer are given by

$$q = \frac{V_1 k_1 + V_2 k_2}{L^2}(T_1 - T_2) \qquad \text{upper bound}$$

$$q = \left(\frac{V}{L}\right)^2 \frac{k_1 k_2}{V_1 k_2 + V_2 k_1}(T_1 - T_2) \qquad \text{lower bound}$$

2-23. The inside and outside surface temperatures of a hollow copper cylinder with 1-cm and 5-cm radii are 90°C and 15°C, respectively. Determine the rate of heat transfer. Also plot the temperature distribution and compare your results with the temperature distribution in a flat plate of 4 cm thickness.

2-24. Saturated water at a pressure of 1 atm flows inside a hollow stainless steel cylinder of 5-cm and 6-cm radii. The outside air temperature is 50°C with $\bar{h}_1$ and $\bar{h}_2$ equal to 500 W/(m² °C) and 30 W/(m² °C), respectively. Determine the rate of heat transfer over a length of 1 m and determine the surface temperatures.

2-25. The inside surface temperature of a hollow aluminum sphere with r_1 and r_2 equal to 10 cm and 15 cm, respectively, is 50°C. The outside surface, which is coated with carbon black, exchanges radiant energy with the walls of an enclosure which are maintained at 500°C. Determine the outside surface temperature of the sphere and the rate of heat transfer.

2-26. The inside surface temperature of a hollow aluminum sphere of 1-in. and 6-in. radii is 0°F. The outside surface is exposed to air at a temperature of 50°F with $\bar{h}$ equal to 15 Btu/(h ft² °F). Determine the unknown surface temperature and the rate of heat transfer.

2-27. To reduce the rate of heat transfer in the sphere described in Prob. 2-26, a layer of asbestos insulation is used. Determine the thickness of insulation required to reduce the heat loss by 50%. Is the critical radius of importance for this problem?

2-28. Steam at 150°C passes through a 2-cm-O.D. pipe which is insulated with asbestos. The outside air temperature is 35°C and $\bar{h}$ is equal to 6 W/(m²

°C). Determine the thickness of asbestos for which the heat loss is a maximum. What is the rate of heat transfer for this condition?

2-29. (a) Show that the differential equation for one-dimensional (r direction) conduction heat transfer in a hollow sphere takes the form

$$\frac{d}{dr}\left(r^2 \frac{dT}{dr}\right) = 0$$

(b) Based on the differential-formulation approach, develop expressions for T and q_r for the case in which the surface temperatures at r_1 and r_2 are maintained at T_1 and T_2.

2-30. Develop the differential formulation for one-dimensional conduction heat transfer in a hollow cylinder with the temperature at r_1 maintained at T_1 and with the surrounding fluid at temperature T_F with $\bar{h}$ known.

2-31. Show that the critical radius r_c for convection from a hollow sphere is equal to $2k/\bar{h}$.

2-32. Develop a system of curves similar to Fig. 2-11 for the rate of convection heat transfer from a hollow sphere. The temperature at r_1 is T_1 and the surrounding fluid temperature is T_F.

2-33. Determine the critical radius for blackbody thermal radiation between a hollow cylinder with $T_1 = 0°C$ at $r = r_1 = 1$ cm, $T_R = 1000°C$, $F_{s-R} = 1.0$, and for three different values of $k/(\sigma F_{s-R} r_1)(10^{-8}$ K^3, 10^{-10} K^3, 10^{-12} K^3). To see the effect of r_2/r_1 on the radiation heat transfer for the system, plot $q_R/[2\pi k L(T_1 - T_R)]$ vs. r_2/r_1 for these three values of $k/(\sigma F_{s-R} r_1)$.

2-34. Determine the thickness of insulative pipe covering [$k_i = 0.428$ Btu/(h ft °F)] required to reduce the convective heat loss by 25% from a 1.9-in.-O.D. pipe with surface temperature equal to 200°F. The surrounding fluid is at 70°F and $\bar{h} = 3.6$ Btu/(h ft^2 °F).

2-35. Determine the critical thickness for blackbody radiation from a hollow stainless steel cylinder [$k = 45$ W/(m °C)] with the inside surface at $r_1 = 1$ cm at a temperature $T_1 = 0°C$. The cylinder is surrounded by a vacuum with $F_{s-R} = 1$ and $T_R = 1000°C$.

2-36. Determine the thickness of insulative covering [$k_i = 0.428$ Btu/(h ft °F)] required to reduce the total heat loss (both convective and radiative) from a 1.9-in.-O.D. pipe with a surface temperature of 200°F by 25%. Assume blackbody conditions with the surrounding wall and air at 70°F and with $\bar{h} = 3.6$ Btu/(h ft^2 °F).

2-37. Determine the temperature distribution and rate of heat transfer through a 5-cm-thick wall in which the thermal conductivity varies linearly with temperature; that is, $k = k_0(1 + \beta_T T)$, where $k = 50$ W/(m °C) at 0°C and 75 W/(m °C) at 100°C. The surface temperatures are 15°C and 55°C. Utilize the short method and the Celsius temperature scale. Plot the temperature distribution and compare it with the distribution in a wall with uniform thermal conductivity. Also show that the same results are obtained by use of absolute temperatures.

2-38. Develop the differential formulation for Prob. 2-37 and solve these equations for the temperature distribution and the rate of heat transfer through the wall.

2-39. The wall described in Prob. 2-37 is exposed to a uniform heat flux equal to 500 W/m^2 on one surface and to a fluid at 50°C with $\bar{h}$ equal to 40 W/(m^2 °C) on the other side. Determine the surface temperatures of the wall.

2-40. Determine the heat transfer through a 10-cm-thick iron wall with surface temperatures equal to 2 K and 10 K. Refer to Fig. A-C-2 for the variation in thermal conductivity of iron at cryogenic temperatures.

2-41. For uniform internal energy generation in a circular cylinder and a flat plate, show that $q_r(r_0)$ is twice as large as $q_x(L)$ for the same V/A_s. Does this explain why the dimensionless operating temperature of the cylinder is twice as large for the plate? (See Fig. 2-14.)

2-42. For uniform energy generation in a flat plate with both surfaces maintained at T_1, q_x can be easily deduced. With this point in mind, use the short method to solve for the temperature distribution in this system.

2-43. A 5-mm-diameter copper wire of 1 m length conducts 100 A of electric current. Determine the rate of heat transfer from the wire and determine the maximum temperature within the wire if the surrounding fluid temperature is 50°C and $\bar{h}$ is equal to 200 W/(m^2 °C).

2-44. Show that the energy equation for one-dimensional conduction heat transfer in a hollow sphere with internal energy generation takes the form

$$\frac{1}{r^2}\frac{d}{dr}\left(r^2\frac{dT}{dr}\right) + \frac{\dot{q}}{k} = 0$$

2-45. Show that the volume of a differential spherical shell is $dV = 4\pi r^2 dr$.

2-46. If the wire described in Prob. 2-43 is insulated with 1 cm of material with a thermal conductivity of 0.1 W/(m °C), determine the maximum temperature in the wire and the outside surface temperature of the insulation.

2-47. Heat transfer in electric motors is important for two reasons. First, energy lost through heat transfer obviously reduces the efficiency of the machine. Second, because of deterioration of the insulation with increasing temperature, the life expectancy of electric motors is shortened by overheating. (As a rule of thumb, the time to failure of organic insulation is halved for each 8 to 10°C rise.) Because of its importance in the operation of electric motors, the operating temperature should be carefully monitored. Describe how a measurement of the electrical resistance of the motor coil can be used in conjunction with Eq. (2-69) to estimate the temperature.

2-48. A 1-cm-diameter stainless steel rod connects two walls at 0°C and 25°C which are 5 cm apart at 0°C and 25°C. The surrounding fluid temperature is 15°C and $\bar{h}$ is equal to 120 W/(m^2 °C). Determine the temperature distribution in the rod and the heat-transfer rate into the rod at each wall. What is the net rate of heat transfer between the fluid and the rod?

2-49. A 5-mm-diameter copper fin of 1 cm length is attached to a wall at 200°C. The air temperature is 5°C and $\bar{h}$ is equal to 25 W/(m² °C). Calculate the Biot number and determine the rate of heat transfer from this short fin, the temperature at the end, and the fin efficiency.

2-50. Air and water are separated by a thin plane wall made of carbon steel. It is proposed to increase the heat transfer rate between these two fluids by adding straight rectangular carbon steel fins $\frac{1}{8}$ in. thick, 1 in. long, and spaced 1 in. between centers, to the wall. What percent increase in heat transfer can be realized by adding fins to (a) the air side and (b) the water side of the plane wall? The air and water side coefficients may be taken as 40 and 250 Btu/(h ft² °F), respectively.

2-51. Develop Eqs. (2-114) and (2-115).

2-52. Develop Eqs. (2-118) and (2-119).

2-53. One surface of a very thin plate [$k = 250$ W/(m °C)] has a surface temperature T_s equal to 100°C. The other surface is exposed to a fluid with $T_F = 20°C$ and $\bar{h} = 75$ W/(m² °C). A circular fin ($D = 2$ mm, $L = 1$ cm) is attached to the surface. Determine the rate of heat transfer from the fin, and the number of fins required per square meter of primary surface in order to increase the rate of heat transfer by 100%.

2-54. Assuming that the emissivity of the fin in Example 2-13 is equal to 0.2, show that the effects of thermal radiation from the fin are small.

2-55. Estimate the rate of heat transfer in cylindrical fins with diameter of 2 cm and length 5 cm which are exposed to a convection environment with $\bar{h}$ equal to 20 W/(m² °C), for two fin materials (copper and 1.5% carbon steel); $T_0 = 100°C$ and $T_F = 0°C$. Utilize the one-dimensional assumption in these calculations. Also calculate the Biot number and comment on the appropriateness of the one-dimensional analysis for these fins.

2-56. A 2-cm-diameter 10-cm-long copper heating rod is connected to two surfaces which are at 50°C. The energy generation uniformly within the rod is equal to 100 kW. The surrounding fluid is also at 50°C with $\bar{h} = 150$ W/(m² °C). Determine the temperature distribution in the rod and the rate of heat transfer to the fluid and to the walls.

2-57. Estimate the increase in heat-transfer rate which can be obtained from a cylindrical wall by using one aluminum pin-shaped fin per square inch, each having a diameter of $\frac{3}{16}$ in. and a length of 1 in. Assume that the heat-transfer coefficient is 25 Btu/(h ft² °F), $T_0 = 600°F$, and $T_F = 70°F$.

2-58. Estimate the operating case temperature of the power amplifier (without heat sink) of Example 2-5 for a power output of 5 W. Assume that $T_R = T_F = 25°C$, and $\bar{h} = 10$ W/(m² °C).

2-59. A silicon high-power transistor with TO-3 case is to operate at 50 W. The thermal resistances are rated at $R_{JC} = 2°C/W$ and $R_{CA} = 38°C/W$. The junction temperature must not exceed 150°C. Estimate the maximum power that this transistor can be operated at if the ambient temperature is 25°C and it is to be mounted (a) without a heat sink; (b) on a RS 401-778 heat sink; (c) on a RS 401-403 heat sink.

2-60. The high-power voltage regulator unit shown in Fig. 2-15(b) is to operate at 30 W. Its thermal resistance when standing upright in free air is rated by the manufacturer at 1.3°C/W. Determine the maximum air temperature if the case temperature is to be held to 80°C.

2-61. In order to operate a silicon power transistor with a TO-3 case at its maximum power rating of 115 W, the large RS 401-958 heat-sink fin unit is to be used. Determine the operating temperature of the case and the junction temperature for this application, assuming $R_{JC} = 1°C/W$, $R_{CA} = 38°C/W$, and the ambient temperature is 25°C.

2-62. The case to ambient thermal resistance for a silicon transistor housed in a TO-5 case is rated at 150°C/W. Determine the power at which the case temperature will reach an operating temperature of 175°C for an ambient temperature of 25°C.

2-63. Determine the power at which the case temperature of the transistor of Prob. 2-62 will reach the maximum value if the transistor is mounted on the RS 401-964 heat-sink fin unit.

2-64. Determine the maximum power at which a germanium transistor can be operated if the junction to case thermal resistance is rated at 2°C/W. The transistor is housed in a TO-3 case with case to ambient thermal resistance of 30°C/W. The air temperature is 25°C and the junction temperature must not exceed 100°C.

2-65. Determine the maximum power dissipation for Prob. 2-64 if the transistor is mounted on a RS 401-497 heat-sink fin unit.

2-66. A copper ball of 1 in. diameter initially at 75°F is dropped into an ice bath which is maintained at 32°F. The bath is stirred such that the coefficient of heat transfer is 250 Btu/(h ft^2 °F). Determine the temperature history of the ball and the length of time required to reach steady state. Also determine the rate of heat transfer to the fluid as a function of time.

2-67. Show that the effects of thermal radiation are small for the unsteady process in Example 2-16. Assume that the emissivity of the surface is 0.02.

2-68. The sole plate of a household iron has a surface area of 0.05 m^2 and is made of stainless steel with a total mass of 1.5 kg and specific heat of 0.46 kJ/(kg °C). The iron is rated at 500 W and is initially at the ambient temperature which is 30°C. The coefficient of heat transfer is 10 W/(m^2 °C). After the iron is turned on, how long will it take to reach 110°C?

2-69. A 1-mm-diameter copper wire initially at a temperature of 30°F is suddenly exposed to a fluid with sinusoidal temperature given by $T_F = (100 + 50 \sin 2\pi t)$ °F with $\bar{h}$ equal to 5 Btu/(h ft^2 °F). Determine the temperature of the wire.

2-70. Measurements for the axial temperature distribution in the copper rod shown in Fig. P2-70a are given in Fig. P2-70b. To estimate the significance of two-dimensional effects, develop an approximate one-dimensional analysis for the axial temperature distribution and compare your results with the data. What is the maximum percent difference between one-dimensional theory and the data?

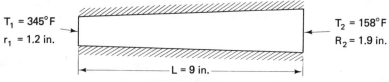

FIGURE P2-70a

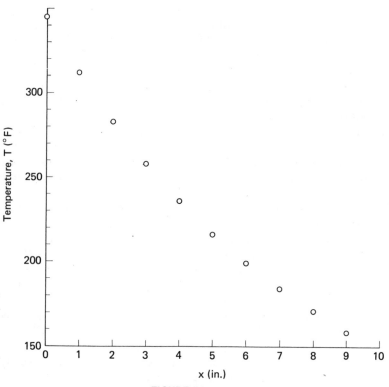

FIGURE P2-70b

REFERENCES

[1] KREYSZIG, E., *Advanced Engineering Mathematics*, 3rd ed. New York: John Wiley & Sons, Inc., 1972.

[2] BARZELAY, M. E., K. N. TONG, AND G. F. HOLLOWAY, "Effect of Pressure on Thermal Conductance of Contact Joints," *NACA Tech. Note 3295*, May 1955.

[3] CLAUSING, A. M., "Heat Transfer at the Interface of Dissimilar Metals—The Influence of Thermal Strain," *Int. J. Heat Mass Transfer*, **9**, 1966, 791–801.

[4] MOORE, C. J., JR., H. A. BLUM, AND H. ATKINS, "Studies Classification Bibliography for Thermal Contact Resistance Studies," *ASME Paper 68-WA/HT-18*, December 1968.

[5] ARPACI, V. S., *Conduction Heat Transfer.* Reading, Mass.: Addison-Wesley Publishing Company, Inc., 1966.

[6] SCHNEIDER, P. J., *Conduction Heat Transfer.* Reading, Mass.: Addison-Wesley Publishing Company, Inc., 1955.

[7] GARDNER, K. A., "Efficiency of Extended Surfaces," *Trans. ASME,* **67**, 1945, 621–631.

[8] ZINSMEISTER, G. E., "Teaching Heat Conduction in Composites—I," *ME News*, **16**, 1979, 22–26.

CHAPTER 3

CONDUCTION HEAT TRANSFER: MULTIDIMENSIONAL

3-1 INTRODUCTION

As we have seen, one-dimensional heat transfer sometimes occurs in flat-plate, cylindrical, and spherical geometries. We have also seen that the heat transfer in multidimensional thermal systems such as series–parallel composites, extended surfaces, and unsteady convective or radiative heating and cooling of bodies, can sometimes be approximated by one-dimensional analyses. But many heat-transfer systems encountered in practice are distinctly multidimensional and cannot be approximated by one-dimensional mathematical models. Therefore, we now turn our attention to the analysis of multidimensional conduction heat transfer. Our first order of business will be to present the general Fourier law of conduction for multidimensional conduction heat transfer. Mathematical formulations and solutions will then be developed for representative systems, with emphasis placed on the modern numerical finite-difference approach. In addition, practical solution results will be given to aid in the design of standard multidimensional conduction-heat-transfer systems.

3-2 GENERAL FOURIER LAW OF CONDUCTION

As indicated in Chap. 1, a more general form of the Fourier law must be utilized in the analysis of heat transfer when more than one spatial dimension is involved. For such multidimensional problems, the Fourier law of conduction for isotropic materials takes the form (in Cartesian coordinates; see Fig. 3-1)

$$dq_x = -k \, dA_x \frac{\partial T}{\partial x} \qquad (3\text{-}1a)$$

$$dq_y = -k \, dA_y \frac{\partial T}{\partial y} \qquad (3\text{-}1b)$$

$$dq_z = -k \, dA_z \frac{\partial T}{\partial z} \qquad (3\text{-}1c)$$

$dA_x = dy \, dz$
$dA_y = dx \, dz$
$dA_z = dx \, dy$

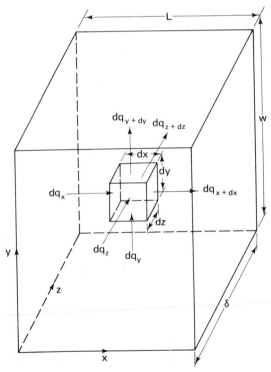

FIGURE 3-1 Differential volume for rectangular solid.

where dq_x, dq_y, and dq_z are the differential rates of heat transfer through the differential areas dA_x $(=dy\,dz)$, dA_y $(=dx\,dz)$, and dA_z $(=dx\,dy)$, respectively. (The general Fourier law of conduction is written in vector form in Prob. 3-1.) It follows that the total rate of heat transfer through a finite area, say A_x $(=w\delta$ in Fig. 3-1), is given by

$$q_x = \int_{A_x} -k\frac{\partial T}{\partial x}\, dA_x = \int_0^\delta \int_0^w -k\frac{\partial T}{\partial x}\, dy\, dz \qquad (3\text{-}2)$$

Notice that the term $\partial T/\partial x$ must be retained within the integral(s) in Eq. (3-2) for situations in which T is a function of two or three spatial dimensions. However, for the case in which T is a function of t and/or x only, Eq. (3-2) reduces to the simple form of the Fourier law given by Eq. (1-17).

$$q_x = -kA_x\frac{\partial T}{\partial x} \qquad (3\text{-}3)$$

3-3   MATHEMATICAL FORMULATIONS

3-3-1   Introduction

Of the various methods that have been developed for the mathematical modeling of energy transfer in conduction-heat-transfer systems, the differential and finite-difference formulations are the most frequently used. Whereas the differential formulation approach, which was introduced in Chap. 2, provides the basis for establishing exact analytical solutions to many fundamental problems, the finite-difference formulation method lends itself to the analysis of the more complex problems often encountered in practice for which exact solutions are difficult, if not impossible, to obtain. Both of these methods of formulation will be developed in this section.

   Two other methods that are useful in the analysis of conduction heat transfer are the finite-element and the integral formulations. The finite-element formulation approach, which is the newest of these four methods, has gained in popularity over the last few years. However, this numerical approach is based on the calculus of variations, which is developed in higher-level texts [1]. Consequently, this approach will not be developed in our study. The simpler integral formulation approach will be introduced in the context of the theoretical analysis of convection heat transfer in Chap. 10.

3-3-2 Differential Formulation

Differential formulations are developed in this section for multidimensional conduction heat transfer in both Cartesian and cylindrical coordinate systems, with consideration given to spherical systems in Example 3-5 and Prob. 3-11. The differential formulation for a multidimensional system is developed by following the same steps outlined in Sec. 2-2-5 for one-dimensional systems.

Cartesian coordinate system

Consider heat transfer in the rectangular solid shown in Fig. 3-1. The initial temperature is T_i and the internal energy generation per unit volume is given by $\dot{q}$. The conditions at the six surfaces are, of course, prescribed by the boundary conditions. Several different boundary conditions will be considered in this chapter. Applying the first law of thermodynamics to the differential element $dx\,dy\,dz$, we obtain
Step 1

$$dq_x + dq_y + dq_z + \dot{q}\,dV = dq_{x+dx} + dq_{y+dy} + dq_{z+dz} + \frac{\partial U}{\partial t} \qquad (3\text{-}4)$$

where $\partial U/\partial t = \partial(\rho\,dV\,c_v T)/\partial t$ with $dV = dx\,dy\,dz$. (Note that differential rates of heat are transferred through the differential areas dA_x, dA_y, and dA_z.) Utilizing the definition of the partial derivative, we have [see Eq. (A-6) in the Appendix]

$$dq_{x+dx} = dq_x + \frac{\partial(dq_x)}{\partial x}\,dx \qquad (3\text{-}5)$$

plus similar expressions relating dq_y and dq_z to dq_{y+dy} and dq_{z+dz}. Thus, Eq. (3-4) becomes
Step 2

$$\dot{q}\,dV = \frac{\partial(dq_x)}{\partial x}\,dx + \frac{\partial(dq_y)}{\partial y}\,dy + \frac{\partial(dq_z)}{\partial z}\,dz + dV\frac{\partial}{\partial t}(\rho c_v T) \qquad (3\text{-}6)$$

Introducing the Fourier law of conduction, we obtain
Step 3

$$\frac{\partial}{\partial x}\left(k\,dA_x\frac{\partial T}{\partial x}\right)dx + \frac{\partial}{\partial y}\left(k\,dA_y\frac{\partial T}{\partial y}\right)dy + \frac{\partial}{\partial z}\left(k\,dA_z\frac{\partial T}{\partial z}\right)dz + \dot{q}\,dV$$
$$= dV\frac{\partial}{\partial t}(\rho c_v T) \qquad (3\text{-}7)$$

Dividing through by the differential volume, this equation reduces to

$$\frac{\partial}{\partial x}\left(k\frac{\partial T}{\partial x}\right) + \frac{\partial}{\partial y}\left(k\frac{\partial T}{\partial y}\right) + \frac{\partial}{\partial z}\left(k\frac{\partial T}{\partial z}\right) + \dot{q} = \frac{\partial}{\partial t}(\rho c_v T) \qquad (3\text{-}8)$$

For constant properties k, ρ, and c_v, we have

$$\frac{\partial^2 T}{\partial x^2} + \frac{\partial^2 T}{\partial y^2} + \frac{\partial^2 T}{\partial z^2} + \frac{\dot{q}}{k} = \frac{1}{\alpha}\frac{\partial T}{\partial t} \qquad (3\text{-}9)$$

where the *thermal diffusivity* α is equal to $k/(\rho c_v)$.

The formulation is completed by writing

$$T = T_i(x,y,z) \qquad \text{at } t = 0 \qquad (3\text{-}10)$$

plus six conditions in x, y, and z. These boundary conditions are expressed in the same way as those that were developed for one-dimensional systems in Chap. 2, except for the fact that partial derivatives are used instead of total derivatives. This point is illustrated in Examples 3-1 through 3-3.

With $\dot{q} = 0$, Eq. (3-9) reduces to the *Fourier equation*,

$$\frac{\partial^2 T}{\partial x^2} + \frac{\partial^2 T}{\partial y^2} + \frac{\partial^2 T}{\partial z^2} = \frac{1}{\alpha}\frac{\partial T}{\partial t} \qquad (3\text{-}11)$$

For conditions of steady state, this equation reduces further to the *Laplace equation*,

$$\frac{\partial^2 T}{\partial x^2} + \frac{\partial^2 T}{\partial y^2} + \frac{\partial^2 T}{\partial z^2} = 0 \qquad (3\text{-}12)$$

If we have steady-state conditions with internal energy generation, Eq. (3-9) reduces to the *Poisson equation*,

$$\frac{\partial^2 T}{\partial x^2} + \frac{\partial^2 T}{\partial y^2} + \frac{\partial^2 T}{\partial z^2} + \frac{\dot{q}}{k} = 0 \qquad (3\text{-}13)$$

The steady one-dimensional form of Eq. (3-9) or (3-13) takes the form

$$\frac{d^2 T}{dx^2} + \frac{\dot{q}}{k} = 0 \qquad (3\text{-}14)$$

This equation was solved for various conditions in Chap. 2.

EXAMPLE 3-1

Write the differential formulation for the energy transfer in the rectangular plate shown in Fig. E3-1.

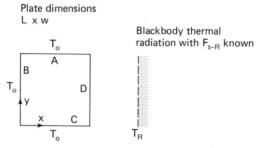

Plate dimensions
L x w

Blackbody thermal
radiation with F_{s-R} known

FIGURE E3-1 Conduction in rectangular plate with blackbody thermal radiation.

Solution

Because heat is conducted from radiating surface D to surfaces A and C as well as to surface B, the temperature distribution within this plate is a function of both x and y. Thus, the two-dimensional Laplace equation applies; that is,

$$\frac{\partial^2 T}{\partial x^2} + \frac{\partial^2 T}{\partial y^2} = 0$$

The boundary conditions are

$$T = T_0 \qquad \text{at} \quad x = 0$$

$$-k\frac{\partial T}{\partial x} = \sigma F_{s-R}(T^4 - T_R^4) \qquad \text{at} \quad x = L$$

$$T = T_0 \qquad \text{at} \quad y = 0$$

$$T = T_0 \qquad \text{at} \quad y = w$$

Because of symmetry, either one of the y conditions can be replaced by

$$\frac{\partial T}{\partial y} = 0 \qquad \text{at} \quad y = \frac{w}{2}$$

EXAMPLE 3-2

The flat plate shown in Fig. E3-2 is initially at a temperature T_i. It is suddenly exposed to a fluid at temperature T_F with h known. Write the

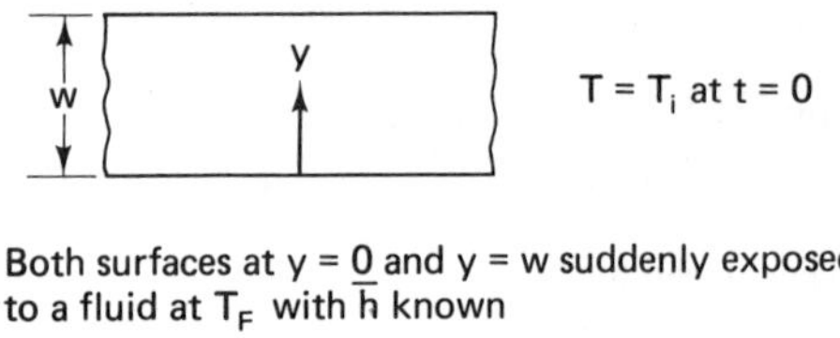

Both surfaces at y = 0 and y = w suddenly exposed
to a fluid at T_F with $\bar{h}$ known

FIGURE E3-2   Unsteady conduction in plane wall.

differential formulation for the energy transfer, assuming uniform properties.

Solution

The temperature distribution for this problem is a function of both distance y and time t. The energy equation for this unsteady two-dimensional conduction heat-transfer system takes the form

$$\alpha \frac{\partial^2 T}{\partial y^2} = \frac{\partial T}{\partial t}$$

This is the two-dimensional form of the Fourier equation. The initial and boundary conditions are

$$T = T_i \qquad\qquad \text{at} \quad t = 0$$

$$h(T_F - T) = -k\frac{\partial T}{\partial y} \qquad \text{at} \quad y = 0$$

$$-k\frac{\partial T}{\partial y} = h(T - T_F) \qquad \text{at} \quad y = w$$

Note that because of symmetry, either one of the y conditions can be replaced by

$$\frac{\partial T}{\partial y} = 0 \qquad \text{at} \quad y = \frac{w}{2}$$

Strictly speaking, the coefficient of heat transfer h is a function of both time and location for problems such as this. However, approximate solutions are generally obtained by taking the average steady-state value of h over the entire surface (i.e., $h = \bar{h}$). For the case in which h is very large, the boundary conditions reduce to

$$T = T_F \qquad \text{at} \quad y = 0 \quad \text{and} \quad y = w$$

For steady-state conditions, the energy equation reduces to

$$\frac{\partial^2 T}{\partial y^2} = 0$$

such that

$$T = T_F$$

The rate of heat transfer through the plate is zero for this case.

Another limiting condition occurs for small values of the Biot number. As indicated in Chap. 2, the temperature is nearly independent of location for values of the Biot number less than 0.1, such that the energy equation can be approximated by

$$hA_s(T - T_F) + \rho V c_v \frac{dT}{dt} = 0$$

with

$$T = T_i \qquad \text{at} \quad t = 0$$

EXAMPLE 3-3

The composition of material in the flat plate shown in Fig. E3-3 varies with x and y. Write the differential formulation for energy transfer in this system.

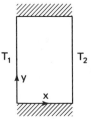

FIGURE E3-3 System with composition and thermal conductivity dependent on both x and y.

Solution

Even though the surface temperatures are uniform in this system, the temperature distribution is two-dimensional because the thermal conductivity is a function of both x and y. The energy equation is therefore given by Eq. (3-8) with $\dot{q}=0$, $\partial T/\partial z=0$ and $\partial U/\partial t=0$; that is,

$$\frac{\partial}{\partial x}\left(k\frac{\partial T}{\partial x}\right)+\frac{\partial}{\partial y}\left(k\frac{\partial T}{\partial y}\right)=0 \tag{a}$$

The boundary conditions take the form

$$
\begin{array}{llll}
T=T_1 & \text{at} & x=0 & \text{(b)} \\
T=T_2 & \text{at} & x=L & \text{(c)} \\
\dfrac{\partial T}{\partial y}=0 & \text{at} & y=0 \text{ and } y = w & \text{(d,e)}
\end{array}
$$

These equations represent the general mathematical formulation for combined series-parallel composites with perfect thermal contact at the boundaries. (An approximate approach to the solution of this type of problem is given in Prob. 2-22. In addition, solutions for the rate of heat transfer in systems in which k is a function of only one space dimension are developed in Probs. 2-20 and 2-21.)

Cylindrical coordinate system

Attention is given next to systems that lend themselves to the cylindrical coordinates shown in Fig. 3-2. The coordinates r and θ are related to x and y by

$$x=r\cos\theta \tag{3-15}$$

$$y=r\sin\theta \tag{3-16}$$

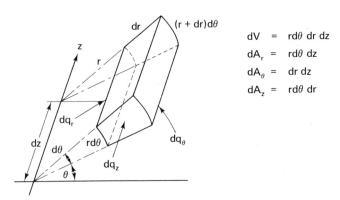

$$
\begin{aligned}
dV &= r d\theta\, dr\, dz \\
dA_r &= r d\theta\, dz \\
dA_\theta &= dr\, dz \\
dA_z &= r d\theta\, dr
\end{aligned}
$$

FIGURE 3-2 Differential volume for cylindrical system.

The general Fourier law of conduction for heat transfer in the r, θ, and z directions takes the form

$$dq_r = -k\,dA_r\frac{\partial T}{\partial r} \tag{3-17a}$$

$$dq_\theta = -k\,dA_\theta\frac{1}{r}\frac{\partial T}{\partial \theta} \tag{3-17b}$$

$$dq_z = -k\,dA_z\frac{\partial T}{\partial z} \tag{3-17c}$$

where $dA_r = r\,d\theta\,dz$, $dA_\theta = dr\,dz$, and $dA_z = r\,d\theta\,dr$.

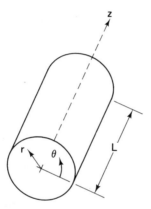

FIGURE 3-3 Cylindrical body.

Consider, for example, the cylindrical body shown in Fig. 3-3. The initial temperature is given by T_i and the conditions at the boundaries $r = r_0$, $z = 0$, and $z = L$ are assumed to be known. Applying the first law of thermodynamics to a differential volume dV ($= r\,d\theta\,dr\,dz$) such as that shown in Fig. 3-2, we obtain

Step 1

$$dq_r + dq_\theta + dq_z + \dot{q}\,dV = dq_{r+dr} + dq_{\theta+d\theta} + dq_{z+dz} + \frac{\partial U}{\partial t} \tag{3-18}$$

where $U = \rho\,dV\,c_v T$. Utilizing the definition for the partial derivative, Eq. (3-18) takes the form

Step 2

$$\dot{q}\,dV = \frac{\partial}{\partial r}(dq_r)\,dr + \frac{\partial}{\partial \theta}(dq_\theta)\,d\theta + \frac{\partial}{\partial z}(dq_z)\,dz + dV\frac{\partial}{\partial t}(\rho c_v T) \tag{3-19}$$

Then,

Step 3

$$\frac{\partial}{\partial r}\left(kr\,d\theta\,dz\,\frac{\partial T}{\partial r}\right)dr + \frac{\partial}{\partial \theta}\left(k\,dr\,dz\,\frac{1}{r}\frac{\partial T}{\partial \theta}\right)d\theta$$
$$+ \frac{\partial}{\partial z}\left(kr\,d\theta\,dr\,\frac{\partial T}{\partial z}\right)dz + \dot{q}\,dV = dV\frac{\partial}{\partial t}(\rho c_v T) \quad (3\text{-}20)$$

In simplifying Eq. (3-20), the first term must be handled carefully when differentiating because $dA_r(=r\,d\theta\,dz)$ is a function of r. Taking note of this point, Eq. (3-20) reduces to

$$\frac{1}{r}\frac{\partial}{\partial r}\left(kr\frac{\partial T}{\partial r}\right) + \frac{1}{r^2}\frac{\partial}{\partial \theta}\left(k\frac{\partial T}{\partial \theta}\right) + \frac{\partial}{\partial z}\left(k\frac{\partial T}{\partial z}\right) + \dot{q} = \frac{\partial}{\partial t}(\rho c_v T) \quad (3\text{-}21)$$

or, for uniform properties,

$$\frac{1}{r}\frac{\partial}{\partial r}\left(r\frac{\partial T}{\partial r}\right) + \frac{1}{r^2}\frac{\partial^2 T}{\partial \theta^2} + \frac{\partial^2 T}{\partial z^2} + \frac{\dot{q}}{k} = \frac{1}{\alpha}\frac{\partial T}{\partial t} \quad (3\text{-}22)$$

The formulation is completed by writing

$$T = T_i(r,\theta,z) \qquad \text{at } t=0 \quad (3\text{-}23)$$

together with six conditions in r, θ, and z.

The steady one-dimensional form of Eq. (3-22),

$$\frac{1}{r}\frac{d}{dr}\left(r\frac{dT}{dr}\right) + \frac{\dot{q}}{k} = 0 \quad (3\text{-}24)$$

was solved in Chap. 2 for several conditions.

EXAMPLE 3-4

Write the differential formulation for steady heat transfer in the cylindrical fin with circular cross section shown in Fig. E3-4.

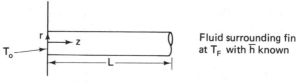

FIGURE E3-4   Circular fin.

Solution

The differential formulation for this two-dimensional problem takes the form

$$\frac{\partial^2 T}{\partial z^2} + \frac{1}{r}\frac{\partial}{\partial r}\left(r\frac{\partial T}{\partial r}\right) = 0$$

$$T = T_0 \qquad \text{at} \quad z = 0$$

$$-k\frac{\partial T}{\partial z} = \bar{h}(T - T_F) \qquad \text{at} \quad z = L$$

$$\frac{\partial T}{\partial r} = 0 \qquad \text{at} \quad r = 0$$

$$-k\frac{\partial T}{\partial r} = \bar{h}(T - T_F) \qquad \text{at} \quad r = r_0$$

As indicated in Chap. 2, for small values of the Biot number the variation of temperature with r is very slight, such that an approximate one-dimensional formulation can be utilized. The approximate one-dimensional energy equation takes the form

$$\frac{d^2 T}{dz^2} - \frac{\bar{h}p}{kA}(T - T_F) = 0$$

(The axial coordinate is sometimes designated by x instead of z.)

EXAMPLE 3-5

A spherical ball initially at temperature T_i is suddenly exposed to a convecting fluid with temperature T_F and with known h. Develop the differential formulation.

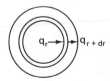

FIGURE E3-5 Differential volume for unsteady heat transfer in sphere.

Solution

Referring to Fig. E3-5, the energy balance gives
Step 1

$$q_r = q_{r+dr} + \frac{\Delta E_s}{\Delta t}$$

[where $dV = 4\pi r^2 dr$ (Prob. 2-45)].
Step 2

$$q_r = q_r + \frac{\partial q_r}{\partial r} dr + \frac{\partial}{\partial t}(\rho dV c_v T)$$

Step 3

$$0 = \frac{\partial}{\partial r}\left(-kr^2 \frac{\partial T}{\partial r}\right) + r^2 \frac{\partial}{\partial t}(\rho c_v T)$$

For uniform properties, this equation reduces to the form

$$\frac{1}{r^2} \frac{\partial}{\partial r}\left(r^2 \frac{\partial T}{\partial r}\right) = \frac{1}{\alpha} \frac{\partial T}{\partial t}$$

or

$$\frac{\partial^2 T}{\partial r^2} + \frac{2}{r} \frac{\partial T}{\partial r} = \frac{1}{\alpha} \frac{\partial T}{\partial t}$$

The initial boundary conditions are

$$T = T_i \qquad\qquad \text{at} \quad t = 0$$

$$\frac{\partial T}{\partial r} = 0 \qquad\qquad \text{at} \quad r = 0$$

$$-k \frac{\partial T}{\partial r} = h(T - T_F) \qquad\qquad \text{at} \quad r = r_0$$

As indicated in Chap. 2, for small values of the Biot number, the temperature is essentially independent of r such that the energy equation can be approximated by

$$\frac{dT}{dt} + \frac{hA}{\rho V c_v}(T - T_F) = 0$$

3-3-3 Numerical Finite-Difference Formulation

The numerical finite-difference formulation for multidimensional conduction heat transfer will be introduced in the framework of steady two-dimensional (x,y) and unsteady three-dimensional (t,x,y) systems with internal energy generation. The basic concepts developed in our study will be seen to apply to other multidimensional conduction-heat-transfer systems as well.

Steady-state systems

To develop a numerical finite-difference formulation for steady conduction heat transfer in a rectangular solid such as is shown in Fig. 3-4(a), we subdivide the entire system into a number Z_s of subvolumes with dimensions $\Delta x \, \Delta y \, \delta$. The resulting network of subvolumes is traditionally diagrammed by the use of nodes, with the mass of each subvolume assumed to be concentrated at its node. Each subvolume is treated as a lumped subsystem, with the temperature of a node assumed to represent

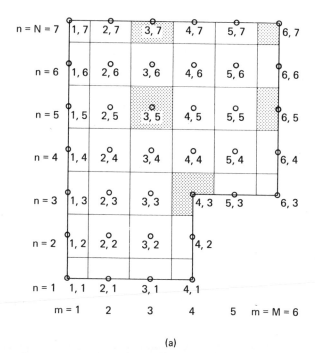

(a)

FIGURE 3-4 (a) Representation of a rectangular solid by network of subvolumes and nodes. Shading indicates representative interior and exterior nodes. Plate thickness is δ.

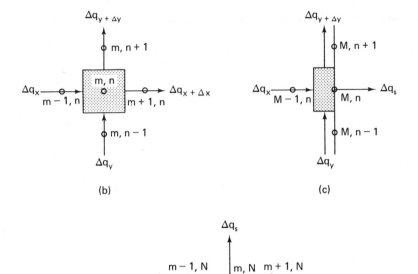

(b)   (c)

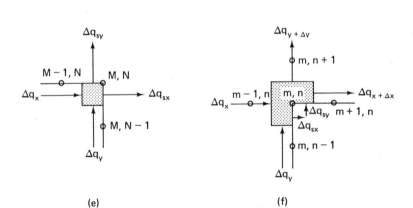

(d)

(e)   (f)

FIGURE 3-4   (b) Interior node (m,n). (c) Regular exterior node
(M,n). (d) Regular exterior node (m,N). (e) Outer corner node
(M,N). (f) Inner corner node $(4,3)$.

the mean temperature of its subvolume. The distances between adjacent
nodes is Δx or Δy. The x and y location of each node is given by
$(m-1)\Delta x$ and $(n-1)\Delta y$, respectively; the values of m and n take on
integer values with m ranging between 1 and M, and n taking values

between 1 and N. The temperature at the location x,y is designated by $T_{m,n}$.

The mechanics of developing a finite-difference network is quite straightforward. We first sketch in horizontal and vertical construction lines that are Δy and Δx apart, with Δy generally set equal to Δx. These construction lines also include the boundaries of the system. The nodes are then located at all intersections of the construction lines. The subvolumes are formed by sketching in horizontal and vertical lines that lie halfway between adjacent construction lines. This procedure will produce the desired system of nodes and subvolumes.

Once the finite-difference grid is established, the numerical finite-difference formulation is developed by merely performing energy balances on each interior and exterior node. This approach will be seen to produce a system of Z_s algebraic equations which represent the energy transfer within the system.

Energy Balance—Interior Nodes Applying the first law of thermodynamics to an interior subvolume such as the one shown in Fig. 3-4(b), we obtain

$$\Delta q_x + \Delta q_y + \dot{q}\,\Delta V = \Delta q_{x+\Delta x} + \Delta q_{y+\Delta y} + \frac{\Delta E_s}{\Delta t} \tag{3-25}$$

where $\Delta E_s/\Delta t = 0$ for steady-state conditions and $\Delta V = \delta\,\Delta x\,\Delta y$. Using simple difference approximations for the Fourier law of conduction (see Example 3-6), Eq. (3-25) takes the form

$$-k\delta\,\Delta y\,\frac{T_{m,n} - T_{m-1,n}}{\Delta x} - k\delta\,\Delta x\,\frac{T_{m,n} - T_{m,n-1}}{\Delta y} + \dot{q}\,\Delta V$$

$$= -k\delta\,\Delta y\,\frac{T_{m+1,n} - T_{m,n}}{\Delta x} - k\delta\,\Delta x\,\frac{T_{m,n+1} - T_{m,n}}{\Delta y}$$

$$\tag{3-26}$$

Rearranging this equation, we obtain (for uniform k)

$$\frac{T_{m-1,n} + T_{m+1,n} - 2T_{m,n}}{\Delta x^2} + \frac{T_{m,n-1} + T_{m,n+1} - 2T_{m,n}}{\Delta y^2} + \frac{\dot{q}}{k} = 0 \tag{3-27}$$

or, with Δy set equal to Δx,

$$T_{m+1,n} + T_{m-1,n} + T_{m,n+1} + T_{m,n-1} - 4T_{m,n} + \frac{\dot{q}}{k}\Delta x^2 = 0 \tag{3-28}$$

This equation applies to each internal node, such that we now have a system of simple algebraic equations which replaces the original partial differential equation.

Parenthetically, Eq. (3-28) can also be developed by transforming the differential energy equation given by Eq. (3-13) (with $\partial T/\partial z = 0$),

$$\frac{\partial^2 T}{\partial x^2} + \frac{\partial^2 T}{\partial y^2} + \frac{\dot{q}}{k} = 0 \qquad (3\text{-}29)$$

into the finite-difference format. This approach is presented in [2] and [3].

EXAMPLE 3-6

Write finite-difference approximations for the two-dimensional (x,y) Fourier law of conduction.

Solution

The Fourier law of conduction for two dimensions (x,y) takes the form

$$dq_x = -k\, dA_x \frac{\partial T}{\partial x} \qquad dq_y = -k\, dA_y \frac{\partial T}{\partial y}$$

Referring to Fig. E3-6, we see that the differential areas are approxi-

$$dq_{x+dx} \simeq \Delta q_{x+\Delta x} = -k\delta\, (T_{m+1,n} - T_{m,n})$$

$$dq_{y+dy} \simeq \Delta q_{y+\Delta y} = -k\delta\, (T_{m,n+1} - T_{m,n})$$

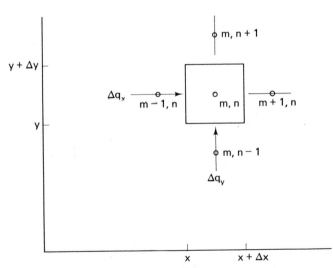

FIGURE E3-6   Finite-difference subvolume.

mated by $dA_x = \delta \Delta y$ and $dA_y = \delta \Delta x$, where $\Delta y = \Delta x$. The gradient $\partial T/\partial x$ at x can be approximated by the simple central difference (with respect to the face at x)

$$\frac{\partial T}{\partial x} = \frac{T_{m,n} - T_{m-1,n}}{\Delta x}$$

Similarly, $\partial T/\partial y$ at y can be approximated by

$$\frac{\partial T}{\partial y} = \frac{T_{m,n} - T_{m,n-1}}{\Delta y}$$

Thus, the steady two-dimensional Fourier law can be approximated by finite-difference equations of the form

$$dq_x \simeq \Delta q_x = -k\delta \left(T_{m,n} - T_{m-1,n} \right)$$
$$dq_y \simeq \Delta q_y = -k\delta \left(T_{m,n} - T_{m,n-1} \right)$$

Finite-difference approximations can also be written for dq_{x+dx} and dq_{y+dy} of the form

$$dq_{x+dx} \simeq \Delta q_{x+\Delta x} = -k\delta \left(T_{m+1,n} - T_{m,n} \right)$$
$$dq_{y+dy} \simeq \Delta q_{y+\Delta y} = -k\delta \left(T_{m,n+1} - T_{m,n} \right)$$

Energy Balance—Exterior Nodes Following our approach to the development of the finite-difference energy equation for an interior node, we perform energy balances on subvolumes associated with exterior nodes of the system shown in Fig. 3-4(a). Because the node of each exterior subvolume lies on the system boundary, these exterior subvolumes are smaller in size than are the interior subvolumes. For example, the corner node $(6,7)$ has a volume ΔV equal to one-fourth the size of an interior subvolume.

Referring to the regular exterior node shown in Fig. 3-4(c), we apply the first law of thermodynamics, to obtain

$$\Delta q_x + \Delta q_y + \dot{q}\,\Delta V = \Delta q_s + \Delta q_{y+\Delta y} + \frac{\Delta E_s}{\Delta t} \tag{3-30}$$

where $\Delta E_s/\Delta t = 0$, $\Delta V = \delta \Delta y \Delta x/2$, and Δq_s is the rate of heat transfer out of the subvolume at the surface. Employing the Fourier law of conduction, we obtain

$$
-k\delta \Delta y \frac{T_{M,n}-T_{M-1,n}}{\Delta x} - k\delta \frac{\Delta x}{2} \frac{T_{M,n}-T_{M,n-1}}{\Delta y} + \dot{q}\,\Delta V
$$

$$
= \Delta q_s - k\delta \frac{\Delta x}{2} \frac{T_{M,n+1}-T_{M,n}}{\Delta y} \qquad (3\text{-}31)
$$

Rearranging this equation with $\Delta q_s = \delta\,\Delta y\,q_s''$ and $\Delta y = \Delta x$, we write

$$
2T_{M-1,n} + T_{M,n+1} + T_{M,n-1} - 4T_{M,n} + \left(\frac{\dot{q}}{k} - \frac{2}{\Delta x}\frac{q_s''}{k}\right)\Delta x^2 = 0 \quad (3\text{-}32)
$$

Similar expressions can be developed for regular exterior nodes at $y = N\Delta y$ [see Fig. 3-4(d)] and for corner nodes [see Figs. 3-4(e) and (f)]. The energy balance for these types of nodes takes the general form

$$
\Delta q_x + \Delta q_y + \dot{q}\,\Delta V = \Delta q_{x+\Delta x} + \Delta q_{y+\Delta y} + \Sigma\,\Delta q_s + \frac{\Delta E_s}{\Delta t} \qquad (3\text{-}33)
$$

where $\Delta E_s/\Delta t = 0$ for steady-state conditions, each term depends upon the particular geometry, and $\Sigma\,\Delta q_s$ represents the total rate of energy transfer from the surfaces.

For convection, thermal radiation or specified heat flux boundary conditions, the unknown exterior nodal temperatures satisfy equations such as Eqs. (3-32) and (3-33) for steady-state conditions with q_s'' ($\equiv \Delta q_s/\Delta A$) specified by

$$q_s'' = h(T_{m,n} - T_F) \qquad \text{convection} \qquad\qquad (3\text{-}34)$$
$$q_s'' = \sigma F_{s-R}(T_{m,n}^4 - T_R^4) \qquad \text{blackbody thermal radiation} \qquad (3\text{-}35)$$
$$q_s'' = f(x,y) \qquad \text{specified heat flux} \qquad\qquad (3\text{-}36)$$
$$q_s'' = 0 \qquad \text{insulated surface} \qquad\qquad (3\text{-}37)$$

Once the surface temperature is predicted, Eqs. (3-34) and (3-35) provide the means by which the rate of heat transfer from the surface Δq_s can be determined for convection and thermal radiation boundary conditions. On the other hand, for specified wall-temperature boundary conditions, the exterior nodal temperatures are known a priori, with nodal equations such as Eqs. (3-32) and (3-33) providing the basis for predicting the rate of heat transfer from the surface. For example, for an isothermal boundary condition at the surface $x = (M-1)\Delta x$, $T_{M,n} = T_{M,n+1} = T_{M,n-1}$

$= T_s$ and Eq. (3-32) reduces to

$$q_s'' = \frac{k}{\Delta x}(T_{M-1,n} - T_s) + \frac{\Delta x\,\dot{q}}{2} \tag{3-38}$$

Thus, after $T_{M-1,n}$ has been determined, the rate of heat transfer across the surface can be obtained.

To summarize, the full numerical finite-difference formulation for steady conditions consists of the Z_s algebraic equations produced at the interior and exterior nodes. These expressions take the form of Eqs. (3-28), (3-32), and (3-33).

EXAMPLE 3-7

Develop a numerical finite-difference formulation for two-dimensional heat transfer in a square with 5-cm-long sides if three faces are maintained at 30 °C and the other face exchanges blackbody thermal radiation with a surface at 500 °C with $F_{s-R} = 0.25$. Utilize a grid spacing with $L/\Delta x = 3$. The thermal conductivity k and thickness δ of the square are assumed to be known. (The differential formulation for this problem is given in Example 3-1.)

Solution

The finite-difference grid with $L/\Delta x = 3$ is shown in Fig. E3-7. This network involves the six unknown nodal temperatures T_1, T_2, T_3, T_4, T_5, and T_6. However, because of symmetry, we know that $T_4 = T_1$, $T_5 = T_2$,

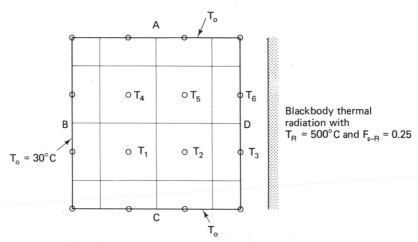

FIGURE E3-7 Finite-difference grid for rectangular solid.

and $T_6 = T_3$. Therefore, we need only develop the nodal equations for T_1, T_2, and T_3. These nodal equations are developed as follows:

Interior nodes at T_1 and T_2

$$\Delta q_x + \Delta q_y = \Delta q_{x+\Delta x} + \Delta q_{y+\Delta y}$$

$$-k\,\Delta y\,\delta\frac{T_{m,n}-T_{m-1,n}}{\Delta x} - k\,\Delta x\,\delta\frac{T_{m,n}-T_{m,n-1}}{\Delta y} = -k\,\Delta y\,\delta\frac{T_{m+1,n}-T_{m,n}}{\Delta x}$$

or

$$T_{m-1,n} + T_{m,n-1} + T_{m+1,n} = 3T_{m,n}$$

We therefore have

$$T_0 + T_0 + T_2 = 3T_1 \tag{a}$$

for the node at T_1, and

$$T_1 + T_0 + T_3 = 3T_2 \tag{b}$$

for the node at T_2.

Exterior Node at T_3

$$\Delta q_x + \Delta q_y = \Delta q_R$$

$$-k\,\Delta y\,\delta\frac{T_3-T_2}{\Delta x} - k\frac{\Delta x}{2}\delta\frac{T_3-T_0}{\Delta y} = \sigma\,\Delta y\,\delta F_{s-R}\big(T_3^4 - T_R^4\big)$$

or

$$T_2 + \frac{1}{2}T_0 + \frac{\sigma\,\Delta y\,F_{s-R}T_R^4}{k} = \frac{3}{2}T_3 + \frac{\sigma\,\Delta y F_{s-R}T_3^4}{k} \tag{c}$$

Because Eqs. (a) through (c) can be solved for the temperatures T_1, T_2, and T_3, this completes our numerical finite-difference formulation for the temperature distribution with $L/\Delta x = 3$. Once these temperatures are known, predictions can be developed for the rate of heat transfer at any surface. To illustrate, the rate of heat transfer from radiating surface D is given by

$$q_R = 2\left[\sigma\,\Delta y\,\delta F_{s-R}\big(T_3^4 - T_R^4\big) + \sigma\frac{\Delta y}{2}\delta F_{s-R}\big(T_0^4 - T_R^4\big)\right]$$

To obtain the rate of heat transfer at the other surfaces, energy balances must be developed at each node along the surface of interest. For example, the rate of heat transfer at surface C is approximated by

$$q_C = -k\,\Delta x\,\delta\big[(T_1 - T_0) + (T_2 - T_0)\big] - k\frac{\Delta x}{2}\,\delta(T_3 - T_0)$$

As we shall see momentarily, the rather course grid spacing used in this illustration gives rise to an error of the order of 10%. To obtain higher levels of accuracy, a finer-grid mesh must be used.

Unsteady systems

To develop a numerical finite-difference formulation for unsteady conduction heat transfer in a rectangular solid with internal energy generation, we subdivide the system into Z_s subvolumes, with the temperature at time t and location x,y designated by $T^\tau_{m,n}$; the time index τ takes on integer values $0,1,2,\ldots$, and time t is equal to $\tau\,\Delta t$. Following the pattern established in our finite-difference formulation for steady-state systems, the first law of thermodynamics is then applied to each node at an instant of time t. For example, the development of an energy balance on an interior subvolume such as the one shown in Fig. 3-4(b) gives

$$\Delta q_x + \Delta q_y + \dot{q}\,\Delta V = \Delta q_{x+\Delta x} + \Delta q_{y+\Delta y} + \frac{\Delta E_s}{\Delta t} \tag{3-39}$$

where $\Delta E_s/\Delta t$ is no longer equal to zero and the heat-transfer rates are specified by the Fourier law of conduction at the instant t.

Based on our previous study, we know that $\Delta E_s/\Delta t$ can be expressed in terms of the specific heat and instantaneous temperature difference by (for uniform properties)

$$\frac{\Delta E_s}{\Delta t} = \frac{\Delta U}{\Delta t} = \rho\,\Delta V c_v \frac{\Delta T}{\Delta t} \tag{3-40}$$

This relationship for $\Delta E_s/\Delta t$ together with our finite-difference approximation for the Fourier law of conduction are substituted into Eq. (3-39), to obtain

$$-k\delta\,\Delta y\,\frac{T^\tau_{m,n} - T^\tau_{m-1,n}}{\Delta x} - k\delta\,\Delta x\,\frac{T^\tau_{m,n} - T^\tau_{m,n-1}}{\Delta y} + \dot{q}\,\Delta V$$

$$= -k\delta\,\Delta y\,\frac{T^\tau_{m+1,n} - T^\tau_{m,n}}{\Delta x} - k\delta\,\Delta x\,\frac{T^\tau_{m,n+1} - T^\tau_{m,n}}{\Delta y} + \rho\,\Delta V c_v \frac{\Delta T}{\Delta t} \tag{3-41}$$

With Δy set equal to Δx, this equation can be written as

$$\frac{\Delta T}{\Delta t} = \frac{\alpha}{\Delta x^2}(T_{m+1,n}^\tau + T_{m-1,n}^\tau + T_{m,n+1}^\tau + T_{m,n-1}^\tau - 4T_{m,n}^\tau) + \frac{\dot{q}}{k}\alpha \quad (3\text{-}42)$$

Similar equations can be written for the exterior subvolumes.

To complete our formulation, $\Delta T/\Delta t$ is usually expressed in terms of the *forward time difference,*

$$\frac{\Delta T}{\Delta t} = \frac{T_{m,n}^{\tau+1} - T_{m,n}^\tau}{\Delta t} \quad\quad (3\text{-}43)$$

or the *backward time difference,*

$$\frac{\Delta T}{\Delta t} = \frac{T_{m,n}^\tau - T_{m,n}^{\tau-1}}{\Delta t} \quad\quad (3\text{-}44)$$

Both of these approximations are utilized in the numerical solution of heat-transfer problems.

Utilizing the forward time difference, Eq. (3-42) takes the form

$$T_{m,n}^{\tau+1} - T_{m,n}^\tau = \frac{\alpha\,\Delta t}{\Delta x^2}(T_{m+1,n}^\tau + T_{m-1,n}^\tau$$

$$+ T_{m,n+1}^\tau + T_{m,n-1}^\tau - 4T_{m,n}^\tau) + \frac{\dot{q}}{k}\alpha\,\Delta t \quad (3\text{-}45)$$

This forward-time-difference equation expresses the nodal temperature at one instant $T_{m,n}^{\tau+1}$ in terms of the nodal temperature distribution $T_{m,n}^\tau$, $T_{m+1,n}^\tau$, $T_{m-1,n}^\tau$, $T_{m,n+1}^\tau$, $T_{m,n-1}^\tau$ at the earlier time $\tau\,\Delta t$. Because the nodal temperature distribution is known at some initial instant of time, this type of equation is *explicit* in that the unknown temperature at the next instant of time can be calculated directly.

On the other hand, the use of the backward time difference gives rise to

$$T_{m,n}^\tau - T_{m,n}^{\tau-1} = \frac{\alpha\,\Delta t}{\Delta x^2}(T_{m+1,n}^\tau + T_{m-1,n}^\tau + T_{m,n+1}^\tau$$

$$+ T_{m,n-1}^\tau - 4T_{m,n}^\tau) + \frac{\dot{q}}{k}\alpha\,\Delta t \quad (3\text{-}46)$$

or

$$T_{m,n}^{\tau+1} - T_{m,n}^\tau = \frac{\alpha\,\Delta t}{\Delta x^2}(T_{m+1,n}^{\tau+1} + T_{m-1,n}^{\tau+1} + T_{m,n+1}^{\tau+1}$$

$$+ T_{m,n-1}^{\tau+1} - 4T_{m,n}^{\tau+1}) + \frac{\dot{q}}{k}\alpha\,\Delta t \quad (3\text{-}47)$$

This backward-time-difference type of equation is *implicit* because the nodal temperature at any instant $T_{m,n}^{\tau+1}$ is given in terms of the unknown nodal temperature distribution at that same instant of time.

EXAMPLE 3-8

A plate of 3 mm thickness is initially at 50 °C. One side of the plate is suddenly exposed to blackbody thermal radiation with $F_{s-R} = 0.5$ and $T_R = 500$ °C. The other side of the plate is maintained at 50 °C. Develop an explicit finite-difference formulation for this unsteady problem with $\Delta x = L/3$.

Solution

The finite-difference grid network for this problem is shown in Fig. E3-8. Note that we have the three unknowns: T_2^τ, T_3^τ, and T_4^τ. The energy balance on each node with unknown temperature is developed as follows, assuming uniform properties:

Interior Nodes m = 2 and 3

$$q_x = q_{x+\Delta x} + \frac{\Delta E_s}{\Delta t}$$

$$-kA \frac{T_m^\tau - T_{m-1}^\tau}{\Delta x} = -kA \frac{T_{m+1}^\tau - T_m^\tau}{\Delta x} + \rho A \, \Delta x c_v \frac{T_m^{\tau+1} - T_m^\tau}{\Delta t}$$

or

$$T_m^{\tau+1} - T_m^\tau = \frac{\alpha \Delta t}{\Delta x^2} (T_{m+1}^\tau + T_{m-1}^\tau - 2T_m^\tau)$$

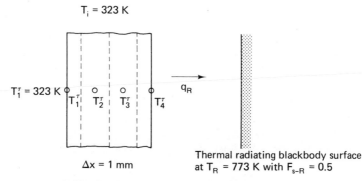

FIGURE E3-8 Finite-difference grid for plane wall.

Hence, we obtain

$$T_2^{\tau+1} - T_2^{\tau} = \frac{\alpha\,\Delta t}{\Delta x^2}(T_3^{\tau} + T_1^{\tau} - 2T_2^{\tau}) \tag{a}$$

and

$$T_3^{\tau+1} - T_3^{\tau} = \frac{\alpha\,\Delta t}{\Delta x^2}(T_4^{\tau} + T_2^{\tau} - 2T_3^{\tau}) \tag{b}$$

Exterior Node m = M = 4

$$q_x = q_R + \frac{\Delta E_s}{\Delta t}$$

$$-kA\frac{T_4^{\tau} - T_3^{\tau}}{\Delta x} = Aq_R'' + \rho A\frac{\Delta x}{2}c_v\frac{T_4^{\tau+1} - T_4^{\tau}}{\Delta t}$$

or

$$T_4^{\tau+1} - T_4^{\tau} = \frac{2\alpha\,\Delta t}{\Delta x^2}\left(-\frac{\Delta x\, q_R''}{k} + T_3^{\tau} - T_4^{\tau}\right) \tag{c}$$

where $q_R'' = \sigma F_{s-R}[(T_4^{\tau})^4 - T_R^4]$. To complete the formulation, we write the initial condition.

$$T_2^0 = T_3^0 = T_4^0 = T_i \tag{d}$$

Utilizing this initial condition, Eqs. (a) through (c) can be solved for T_2^1, T_3^1, and T_4^1. These values of T_m^{τ} at $\tau = 1$ can, in turn, be utilized to obtain T_m^2. Following through in this fashion, predictions can be developed for T_m^{τ} for successively larger and larger values of τ.

Once T_m^{τ} is known, predictions can be developed for the rate of heat transfer. For example, the heat transfer flux at the radiating surface is simply

$$q_R'' = \sigma F_{s-R}\left[(T_4^{\tau})^4 - T_R^4\right] \tag{c}$$

To obtain the instantaneous rate of heat transfer at the other face, we merely perform an energy balance on exterior node (1) as follows:

$$q_0 = q_{\Delta x/2} + \frac{\cancel{\Delta E_s}}{\cancel{\Delta t}}$$

$$q_0 = -kA\frac{T_2^{\tau} - T_1^{\tau}}{\Delta x} \tag{f}$$

where $\Delta E_s/\Delta t$ is zero because T_1^{τ} is independent of time.

Accuracy

The differential formulation is generally the standard against which the numerical finite-difference formulation of a problem is compared. The error introduced by using finite-difference approximations of the derivatives decreases toward zero as the increments approach infinitesimal proportions. Consequently, the error of a finite-difference formulation can be reduced by decreasing the volume ΔV ($=\Delta x \Delta y \Delta z$) and the time increment Δt. For instance, Example 3-9 indicates that the use of forward or backward difference approximations for $\partial T/\partial t$ involves an error of the order of Δt. Thus, a 50% decrease in the increment Δt cuts the forward or backward difference approximation error for $\partial T/\partial t$ roughly in half. On the other hand, the error associated with the central difference approximations for $\partial T/\partial x$ is of the order of Δx^2. Hence, a 50% reduction in Δx would be expected to reduce the error by a factor of about four.

However, the use of smaller volumetric increments also brings about an increase in the number of subvolumes Z_s and algebraic equations required to represent a system. Of course, the length of time required to obtain a solution increases as the number of equations increases. Likewise, the solution time for unsteady problems increases with decreasing increment in time Δt. Consequently, a compromise concerning element size and time increment must be made on the basis of the accuracy required and the cost of computation.

It should also be mentioned that improved accuracy can be achieved by the use of higher-order finite-difference approximations for derivatives in space and time. For example, the central difference approximation for $\partial T/\partial t$ presented in Example 3-9 has an error of the order of Δt^2.

The question remains: What grid size Δx and time increment Δt must be utilized to achieve reasonable accuracy? The best way to obtain an accurate solution is to actually solve the problem for successively smaller and smaller increments in Δx and Δt. For example, with $L/\Delta x = M$ and $\alpha \Delta t/\Delta x^2$ held constant, solutions can be obtained for $M = 2, 3, 4, \ldots$. If the solution for the temperature distribution and heat-transfer rate converges for increasing M, one can be assured of a reasonably accurate solution.

Alternatively, the accuracy of a finite-difference formulation/solution can be estimated by obtaining the numerical solution for simplified conditions or similar problems for which exact solutions are available. For instance, if the steady-state solution to an unsteady problem is known, the general unsteady finite-difference formulation can be solved for steady-state conditions and a comparison can be made. However, it must be emphasized that this approach does not guarantee the same solution accuracy for the general problem as for the simplified problem.

EXAMPLE 3-9

Utilize the Taylor theorem [4] to evaluate the error of standard finite-difference approximations for $\partial T / \partial t$, where T is a function of t, x, and y.

Solution

Based on the Taylor theorem, T at $t + \Delta t$ can be expanded in terms of T at t by

$$T(t+\Delta t,x,y) = T(t,x,y) + \Delta t \frac{\partial T}{\partial t} + \frac{1}{2}\Delta t^2 \frac{\partial^2 T}{\partial t^2} + \frac{1}{6}\Delta t^3 \frac{\partial^3 T}{\partial t^3}$$
$$+ \frac{1}{24}\Delta t^4 \frac{\partial^4 T}{\partial t^4} + \dots \qquad (a)$$

Rearranging this equation, we obtain

$$\frac{\partial T}{\partial t} = \frac{T(t+\Delta t,x,y) - T(t,x,y)}{\Delta t} + O(\Delta t) \qquad (b)$$

where $O(\Delta t)$ designates terms containing first- and higher-order powers of Δt. This equation provides the basis for the forward difference approximation,

$$\frac{\partial T}{\partial t} \simeq \frac{T_{m,n}^{\tau+1} - T_{m,n}^{\tau}}{\Delta t} \qquad (c)$$

which has an accuracy of the order of Δt. Hence, the error is reduced by approximately one-half by halving the increment Δt. Similarly, $T(t - \Delta t, x, y)$ can be written as

$$T(t-\Delta t,x,y) = T(t,x,y) - \Delta t \frac{\partial T}{\partial t} + \frac{1}{2}\Delta t^2 \frac{\partial^2 T}{\partial t^2} - \frac{1}{6}\Delta t^3 \frac{\partial^3 T}{\partial t^3}$$
$$+ \frac{1}{24}\Delta t^4 \frac{\partial^4 T}{\partial t^4} - \dots \qquad (d)$$

This equation takes the form

$$\frac{\partial T}{\partial t} = \frac{T(t,x,y) - T(t-\Delta t,x,y)}{\Delta t} + O(\Delta t) \qquad (e)$$

Thus, we conclude that the backward difference approximation,

$$\frac{\partial T}{\partial t} \simeq \frac{T_{m,n}^{\tau} - T_{m,n}^{\tau-1}}{\Delta t} \tag{f}$$

also has an accuracy of the order of Δt.

A higher-order approximation can be developed for $\partial T / \partial t$ by subtracting Eq. (d) from Eq. (a); that is,

$$\frac{\partial T}{\partial t} = \frac{T(t+\Delta t, x, y) - T(t-\Delta t, x, y)}{2\Delta t} + O(\Delta t^2) \tag{g}$$

This equation provides the basis for the central difference approximation,

$$\frac{\partial T}{\partial t} \simeq \frac{T_{m,n}^{\tau+1} - T_{m,n}^{\tau-1}}{2\,\Delta t} \tag{h}$$

which has an improved accuracy of the order of Δt^2.

Similar Taylor series expansions can be developed for $\partial T / \partial x$ at the interface between two nodes. Based on this approach, we can show that the central difference approximation for $\partial T / \partial x$ at the interface between two subvolumes,

$$\frac{\partial T}{\partial x} \simeq \frac{T_{m,n}^{\tau} - T_{m-1,n}^{\tau}}{\Delta x} \tag{i}$$

has an error of the order of Δx^2 (see Prob. 3-17).

3-4 SOLUTION TECHNIQUES—INTRODUCTION

Several techniques are available for the solution of the equations produced by our differential and finite-difference formulations. These solution techniques include *analytical*, *numerical*, *analogical*, and *graphical* methods. Brief consideration will first be given to the analytical solution approach. Attention will then be turned to the three other methods, with emphasis given to the numerical approach.

Solutions obtained by these various methods for standard two-dimensional conduction-heat-transfer problems are available in formula and chart form as a convenience for design calculations. These practical results will be presented in Sec. 3-9.

3-5 ANALYTICAL SOLUTIONS

3-5-1 Introduction

Among the various methods that can be utilized to develop analytical solutions for multidimensional heat-transfer problems, the *product solution* technique and the *approximate integral* approach are the simplest. The classical product solution approach is introduced in this section. Although the integral approach can be used to solve many conduction-heat-transfer problems, this popular method will be introduced in Chap. 10 in the context of convection heat transfer. Other analytical solution techniques for multidimensional conduction-heat-transfer problems are available in the literature.

3-5-2 Product Solution Method

The product solution technique is particularly well suited to the solution of steady and unsteady multidimensional conduction-heat-transfer problems associated with simple linear boundary conditions. To introduce this solution concept, we consider the steady-state two-dimensional conduction-heat-transfer problem illustrated in Fig. 3-5.

The differential formulation for this problem for the case in which k is uniform, $\dot{q}=0$, $\partial T/\partial z=0$, and $\partial T/\partial t=0$ takes the form

$$\frac{\partial^2 T}{\partial x^2} + \frac{\partial^2 T}{\partial y^2} = 0 \tag{3-48}$$

$$T=T_1 \quad \text{at} \quad x=0 \qquad T=T_1 \quad \text{at} \quad x=L \tag{3-49, 50}$$

$$T=T_1 \quad \text{at} \quad y=0 \qquad T=f(x) \quad \text{at} \quad y=w \tag{3-51, 52}$$

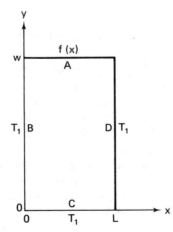

FIGURE 3-5 Steady two-dimensional conduction heat transfer in a rectangular solid.

The use of the product solution technique requires that only one nonhomogeneous term appear in the linear differential formulation. Therefore, we eliminate three of the nonhomogeneous boundary conditions by utilizing the substitution $\psi = T - T_1$. This simple substitution transforms Eqs. (3-48) through (3-52) into the form

$$\frac{\partial^2 \psi}{\partial x^2} + \frac{\partial^2 \psi}{\partial y^2} = 0 \qquad (3\text{-}53)$$

$$\psi = 0 \quad \text{at } x = 0 \qquad \psi = 0 \qquad \text{at } x = L \qquad (3\text{-}54, 55)$$

$$\psi = 0 \quad \text{at } y = 0 \qquad \psi = F(x) \qquad \text{at } y = w \qquad (3\text{-}56, 57)$$

where $F(x) = f(x) - T_1$. Here x is referred to as the homogeneous direction and y is the nonhomogeneous direction.

The solution to Eqs. (3-53) through (3-57) is developed in Appendix D by assuming a product solution of the form

$$\psi(x,y) = X(x) Y(y) \qquad (3\text{-}58)$$

where $X(x)$ is a function of x and $Y(y)$ is a function of y. The general solution takes the final form

$$T - T_1 = \sum_{n=1}^{\infty} c_n \frac{\sinh(n\pi y / L)}{\sinh(n\pi w / L)} \sin \frac{n\pi x}{L} \qquad (3\text{-}59)$$

where

$$c_n = \frac{2}{L} \int_0^L F(x) \sin \frac{n\pi x}{L} \, dx \qquad (3\text{-}60)$$

For the case in which $f(x)$ is equal to a uniform temperature T_2, we have

$$c_n = \frac{2}{L} \int_0^L (T_2 - T_1) \sin \frac{n\pi x}{L} \, dx = -\frac{2}{L}(T_2 - T_1)\frac{L}{n\pi}\left[\cos(n\pi) - 1\right]$$

$$= (T_2 - T_1)\frac{4}{n\pi} \qquad n = 1, 3, 5, \ldots \qquad (3\text{-}61)$$

$$= 0 \qquad n = 2, 4, 6, \ldots$$

Hence, Eq. (3-59) reduces to

$$\frac{T - T_1}{T_2 - T_1} = \frac{2}{\pi} \sum_{n=1}^{\infty} \frac{1 - (-1)^n}{n} \frac{\sinh(n\pi y / L)}{\sinh(n\pi w / L)} \sin \frac{n\pi x}{L} \qquad (3\text{-}62)$$

This series is uniformly convergent [4] in the region $0 \leqslant y < w$ for all values

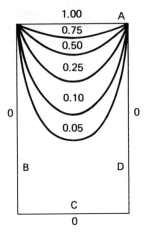

FIGURE 3-6 Dimensionless isotherms $(T-T_1)/(T_2-T_1)$ for steady two-dimensional conduction heat transfer in a rectangular solid; $L/w=0.6$.

of x. Isotherms obtained on the basis of Eq. (3-62) are shown in Fig. 3-6.

For the case in which $f(x)=T_1+T_2 \sin(\pi x/L)$ [i.e., $F(x)=T_2 \sin(\pi x/L)$], we find that (see Prob. 3-22)

$$c_1 = T_2 \tag{3-63}$$
$$c_n = 0 \qquad n=2,3,4,\ldots$$

such that the temperature distribution is given by

$$T-T_1 = T_2 \frac{\sinh(\pi y/L)}{\sinh(\pi w/L)} \sin \frac{\pi x}{L} \tag{3-64}$$

Once the temperature distribution for the problem of interest is known, the rate of heat transfer can be obtained by utilizing the appropriate particular law. To illustrate, we select the situation for which $f(x)=T_1+T_2 \sin(\pi x/L)$. To obtain the rate of heat transfer in the y direction, the general Fourier law of conduction is utilized; that is,

$$dq_y = -k\, dA_y \frac{\partial T}{\partial y} = -k\delta\, dx\, T_2 \frac{\pi}{L} \frac{\cosh(\pi y/L)}{\sinh(\pi w/L)} \sin \frac{\pi x}{L} \tag{3-65}$$

The total rate of heat transfer q_y is now obtained by integrating from x

equal to 0 to L as follows:

$$q_y = \int_0^L -k\frac{\partial T}{\partial y}\,\delta\,dx$$

$$= -k\delta T_2\frac{\pi}{L}\frac{\cosh(\pi y/L)}{\sinh(\pi w/L)}\left(-\frac{L}{\pi}\cos\frac{\pi x}{L}\right)\Big|_0^L$$

$$= -2k\delta T_2\frac{\cosh(\pi y/L)}{\sinh(\pi w/L)} \tag{3-66}$$

Thus, the rate of heat transfer across surface A at $y = w$ is

$$q_A = -2k\delta T_2\coth\frac{\pi w}{L} \tag{3-67}$$

and the rate of heat transfer across surface C at $y = 0$ is

$$q_C = -2k\delta T_2\frac{1}{\sinh(\pi w/L)} \tag{3-68}$$

Similar expressions can be developed for the rate of heat transfer in the x direction.

As illustrated in Examples 3-10 and 3-11, the same approach is used to develop expressions for the rate of heat transfer for the more complex steady and unsteady systems in which the temperature field is expressed in terms of infinite series.

The product solution approach has been utilized in the solution of a large number of standard steady and unseady conduction heat-transfer problems. The results of some of these analyses will be presented in Sec. 3-9.

EXAMPLE 3-10

Consider the two-dimensional problem shown in Fig. 3-5 for the case in which $f(x) = T_2 = 100°C$, $T_1 = 0°C$, $k = 100$ W/(m °C) and $\delta = 1$ cm. The solution to this problem is given by Eq. (3-62). First, demonstrate that this equation satisfies the energy equation. Then utilize this result to obtain predictions for the rates of heat transfer across surfaces A and C of the plate for the case in which $L = w$.

Solution

As already mentioned, this infinite series converges for all values of x and y within the domain of the body. Predictions for the temperature at various locations within the body are shown in Fig. 3-6. Because Eq.

(3-62) is uniformly convergent in the region $0 \leqslant y < w$ for all values of x, this equation can be differentiated term by term in this domain.

To demonstrate that Eq. (3-62) satisfies the energy equation, Eq. (3-48), we obtain $\partial^2 T/\partial x^2$ and $\partial^2 T/\partial y^2$ as follows:

$$\frac{\partial T}{\partial x} = (T_2 - T_1) \frac{2}{\pi} \sum_{n=1}^{\infty} \frac{1-(-1)^n}{n} \frac{\sinh(n\pi y/L)}{\sinh(n\pi w/L)} \frac{n\pi}{L} \cos \frac{n\pi x}{L} \qquad \text{(a)}$$

$$\frac{\partial^2 T}{\partial x^2} = (T_2 - T_1) \frac{2}{\pi} \sum_{n=1}^{\infty} \frac{1-(-1)^n}{n} \frac{\sinh(n\pi y/L)}{\sinh(n\pi w/L)} \left(\frac{-n^2\pi^2}{L^2} \right) \sin \frac{n\pi x}{L} \quad \text{(b)}$$

and

$$\frac{\partial T}{\partial y} = (T_2 - T_1) \frac{2}{\pi} \sum_{n=1}^{\infty} \frac{1-(-1)^n}{n} \frac{n\pi}{L} \frac{\cosh(n\pi y/L)}{\sinh(n\pi w/L)} \sin \frac{n\pi x}{L} \qquad \text{(c)}$$

$$\frac{\partial^2 T}{\partial y^2} = (T_2 - T_1) \frac{2}{\pi} \sum_{n=1}^{\infty} \frac{1-(-1)^n}{n} \left(\frac{n\pi}{L} \right)^2 \frac{\sinh(n\pi y/L)}{\sinh(n\pi w/L)} \sin \frac{n\pi x}{L} \quad \text{(d)}$$

Substituting these results for $\partial^2 T/\partial x^2$ and $\partial^2 T/\partial y^2$ into Eq. (3-48),

$$\frac{\partial^2 T}{\partial x^2} + \frac{\partial^2 T}{\partial y^2} = 0$$

we find that this equation is indeed satisfied.

To determine the rate of heat transfer in the y direction, we apply the general Fourier law of conduction.

$$dq_y = -k\, dA_y \frac{\partial T}{\partial y}$$

$$= -k(T_2 - T_1) \left\{ \frac{2}{L} \sum_{n=1}^{\infty} \left[1-(-1)^n \right] \frac{\cosh(n\pi y/L)}{\sinh(n\pi w/L)} \sin \frac{n\pi x}{L} \right\} \delta\, dx$$

$$\text{(e)}$$

To obtain an expression for the total rate of heat transfer over the region $0 \leqslant x \leqslant L$, we integrate with respect to x; i.e.,

$$q_y = k\delta(T_2 - T_1) \frac{2}{L} \sum_{n=1}^{\infty} \left[1-(-1)^n \right] \frac{\cosh(n\pi y/L)}{\sinh(n\pi w/L)} \frac{L}{n\pi} \left[\cos(n\pi) - 1 \right]$$

$$= -k\delta(T_2 - T_1) \frac{4}{\pi} \sum_{n=1}^{\infty} \frac{1-(-1)^n}{n} \frac{\cosh(n\pi y/L)}{\sinh(n\pi w/L)} \qquad \text{(f)}$$

Setting y equal to zero, the total rate of heat transfer across face C is given by

$$q_C = -k\delta(T_2 - T_1)\frac{4}{\pi}\sum_{n=1}^{\infty}\frac{1-(-1)^n}{n}\frac{1}{\sinh(n\pi)} \qquad (g)$$

where $w = L$ for the square body. The infinite series

$$\frac{4}{\pi}\sum_{n=1}^{\infty}\frac{1-(-1)^n}{n}\frac{1}{\sinh(n\pi)} \qquad (h)$$

converges fairly rapidly to approximately 0.221. Hence, q_C is given by

$$q_C = -0.221k\delta(T_2 - T_1) = -22.1 \text{ W} \qquad (i)$$

To obtain the rate of heat transfer across face A, we allow y to approach w in Eq. (f).

$$q_A = -k\delta(T_2 - T_1)\frac{4}{\pi}\sum_{n=1}^{\infty}\frac{1-(-1)^n}{n}\coth(n\pi) \qquad (j)$$

The infinite series

$$\sum_{n=1}^{\infty}\frac{1-(-1)^n}{n} = 2\left(1 + \frac{1}{3} + \frac{1}{5} + \frac{1}{7} + \dots\right) \qquad (k)$$

is divergent [4]. This, together with the fact that $\coth(n\pi)$ is always greater than unity, indicates that q_A is unbounded. This rather startling result stems from the fact that we have the drastic situation in which the two surfaces A and B at different temperatures T_2 and T_1 are in direct contact. If we were to attempt to set up an electrical analog for this problem, we would have burn out because of shorting at the corner sections.

EXAMPLE 3-11

The surfaces of a flat plate initially at uniform temperature T_i are suddenly brought to a temperature T_1. The differential formulation takes the form

$$\alpha\frac{\partial^2\psi}{\partial x^2} = \frac{\partial\psi}{\partial t}$$

$$\psi = T_i - T_1 \quad \text{at} \quad t = 0$$
$$\psi = 0 \qquad \quad \text{at} \quad x = 0$$
$$\psi = 0 \qquad \quad \text{at} \quad x = I$$

where $\psi = T - T_1$. These equations can be solved by means of the product solution approach [5]–[7][1], with the result

$$\frac{T - T_1}{T_i - T_1} = \frac{2}{\pi} \sum_{n=1}^{\infty} \frac{1 - (-1)^n}{n} \exp(-n^2\pi^2\alpha t / L^2) \sin\frac{n\pi x}{L}$$

which is uniformly convergent for $t > 0$ over the entire plate $0 \leqslant x \leqslant L$. Develop expressions for the instantaneous rate of heat transfer from the plate and the total heat transferred over a period of time τ.

Solution

Taking q_x in the x direction as positive,

$$q = -q_0 + q_L$$

and because of symmetry

$$q = -2q_0$$

Introducing the Fourier law of conduction,

$$q_0 = -kA_x \frac{\partial T}{\partial x}\bigg|_0$$

$$= -kA(T_i - T_1)\frac{2}{\pi} \sum_{n=1}^{\infty} \frac{1 - (-1)^n}{n} \exp(-n^2\pi^2\alpha t / L^2)\frac{n\pi}{L}\cos\frac{n\pi x}{L}\bigg|_0$$

$$= -kA(T_i - T_1)\frac{2}{L} \sum_{n=1}^{\infty} \left[1 - (-1)^n\right]\exp(-n^2\pi^2\alpha t / L^2) \qquad \text{(a)}$$

To obtain an expression for the accumulated heat transfer Q_0 over the span of time τ, we write

$$Q_0 = \int_0^{\tau} q_0\, dt$$

$$= -kA(T_i - T_1)\frac{2}{L} \sum_{n=1}^{\infty} \left[1 - (-1)^n\right]\left(\frac{-L^2}{n^2\pi^2\alpha}\right)\exp(-n^2\pi^2\alpha t / L^2)\bigg|_0^{\tau}$$

$$= -kA(T_i - T_1)\frac{2L}{\alpha\pi^2} \sum_{n=1}^{\infty} \frac{1 - (-1)^n}{n^2}\left[1 - \exp(-n^2\pi^2\alpha\tau / L^2)\right] \qquad \text{(b)}$$

---

[1]This system of equations can also be conveniently solved by means of the approximate integral technique and the Laplace transform method.

The accumulated heat transfer for the entire plate Q is

$$Q = -2Q_0$$

The infinite series in Eqs. (a) and (b) are both convergent. These infinite series are generally evaluated by digital computer, except for large values of $\alpha t / L^2$ or $\alpha \tau / L^2$ for which cases hand calculations can be quickly made. Calculations for T, q, and Q for this problem are given in chart form in Sec. 3-9.

EXAMPLE 3-12

Solutions to heat-transfer problems involving more than one nonhomogeneous term can be developed by applying the *principle of superposition* if the equations are linear (see Appendix A). In this approach, solutions to more involved problems are obtained by summing up or superimposing solutions of sets of simpler problems. To demonstrate this important superposition concept, consider steady two-dimensional conduction heat transfer in the rectangular solid shown in Fig. 3-5 for the case in which surfaces A and D are both at T_2.

Solution

The differential formulation for this problem takes the form

$$\frac{\partial^2 \psi}{\partial x^2} + \frac{\partial^2 \psi}{\partial y^2} = 0 \tag{a}$$

$$\psi = 0 \qquad \text{at} \quad x = 0 \tag{b}$$
$$\psi = T_2 - T_1 \quad \text{at} \quad x = L \tag{c}$$
$$\psi = 0 \qquad \text{at} \quad y = 0 \tag{d}$$
$$\psi = T_2 - T_1 \quad \text{at} \quad y = w \tag{e}$$

where the use of $\psi (\equiv T - T_1)$ has created homogeneous conditions at $x = 0$ and $y = 0$. This mathematical formulation is represented by Fig. E3-12a. However, we are left with two nonhomogeneous boundary conditions. Because the use of the product solution technique requires that only one nonhomogeneous condition occur, we turn to the principle of superposition to achieve a solution.

Utilizing the superposition concept, ψ is assumed to be equal to the sum of two simpler solutions; that is,

$$\psi = \psi_I + \psi_{II} \tag{f}$$

Our differential formulation can be written in terms of ψ_I and ψ_{II} by

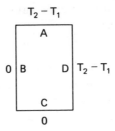

FIGURE E3-12a   Rectangular plate with two nonhomogeneous boundary conditions.

merely substituting Eq. (f) into Eqs. (a) through (e).

$$\frac{\partial^2 \psi_I}{\partial x^2} + \frac{\partial^2 \psi_{II}}{\partial x^2} + \frac{\partial^2 \psi_I}{\partial y^2} + \frac{\partial^2 \psi_{II}}{\partial y^2} = 0$$

$$
\begin{aligned}
\psi_I + \psi_{II} &= 0 & \text{at} \quad x &= 0 \\
\psi_I + \psi_{II} &= T_2 - T_1 & \text{at} \quad x &= L \\
\psi_I + \psi_{II} &= 0 & \text{at} \quad y &= 0 \\
\psi_I + \psi_{II} &= T_2 - T_1 & \text{at} \quad y &= w
\end{aligned}
\qquad \text{(g)}
$$

Based on this system of equations, the following two simpler problems are formulated such that the differential formulation for $\psi(= \psi_I + \psi_{II})$ remains satisfied:

$$\frac{\partial^2 \psi_I}{\partial x^2} + \frac{\partial^2 \psi_I}{\partial y^2} = 0 \qquad \frac{\partial^2 \psi_{II}}{\partial x^2} + \frac{\partial^2 \psi_{II}}{\partial y^2} = 0$$

$$
\begin{aligned}
\psi_I &= 0 & \text{at} \quad x = 0 & \qquad \psi_{II} = 0 \\
\psi_I &= 0 & \text{at} \quad x = L & \qquad \psi_{II} = T_2 - T_1 \\
\psi_I &= 0 & \text{at} \quad y = 0 & \qquad \psi_{II} = 0 \\
\psi_I &= T_2 - T_1 & \text{at} \quad y = w & \qquad \psi_{II} = 0
\end{aligned}
\qquad \text{(h)}
$$

The formulations for ψ_I and ψ_{II} are represented by Figs. E3-12b and E3-12c. Note that Fig. E3-12a is equivalent to superimposing Fig. E3-12b on Fig. E3-12c. The solution for ψ_I is given by Eq. (3-62),

$$\psi_I = T_I - T_1 = (T_2 - T_1) \frac{2}{\pi} \sum_{n=1}^{\infty} \frac{1 - (-1)^n}{n} \frac{\sinh(n\pi y / L)}{\sinh(n\pi w / L)} \sin \frac{n\pi x}{L} \qquad \text{(i)}$$

By inspection, we see that the solution for ψ_{II} can be obtained from Eq. (3-62) by merely interchanging y and x and w and L; that is,

$$\psi_{II} = T_{II} - T_1 = (T_2 - T_1) \frac{2}{\pi} \sum_{n=1}^{\infty} \frac{1 - (-1)^n}{n} \frac{\sinh(n\pi x / w)}{\sinh(n\pi L / w)} \sin \frac{n\pi y}{w} \qquad \text{(j)}$$

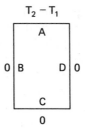

FIGURE E3-12b Rectangular plate with one nonhomogeneous boundary condition at $y = w$.

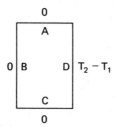

FIGURE E3-12c Rectangular plate with one nonhomogeneous boundary condition at $x = L$.

Returning to Eq. (f), our full solution is

$$T - T_1 = \psi = \psi_{\text{I}} + \psi_{\text{II}}$$

$$= (T_2 - T_1) \frac{2}{\pi} \sum_{n=1}^{\infty} \frac{1 - (-1)^n}{n} \left[\frac{\sinh(n\pi y/L)}{\sinh(n\pi w/L)} \sin\frac{n\pi x}{L} \right.$$

$$\left. + \frac{\sinh(n\pi x/w)}{\sinh(n\pi L/w)} \sin\frac{n\pi y}{w} \right] \tag{k}$$

The principle of superposition can be utilized to solve a wide range of steady and unsteady heat-transfer problems, as long as the differential equation and boundary–initial conditions are linear.

3-6 NUMERICAL APPROACH

3-6-1 Introduction

The numerical finite-difference formulation of heat-transfer problems, no matter how complex, consists of a system of algebraic equations. Because of the widespread use of digital computers which can quickly solve the large numbers of algebraic equations that are often produced by finite-difference formulations, the numerical solution approach is today the primary method of solution of multidimensional heat-transfer problems. This powerful solution approach is introduced in this section for both steady and unsteady multidimensional conduction-heat-transfer problems. In both of these cases, the mechanics involved in producing predictions for the temperature distribution and rate of heat transfer will be introduced in the context of rather coarse finite-difference grids which result in small numbers of equations that are solved by hand. Then, more accurate solutions will be developed for formulations involving finer grids by the use of FORTRAN computer programs. Throughout this section, the question of solution accuracy should be kept foremost in mind.

3-6-2   Steady-State Systems

The finite-difference equations for steady two-dimensional (x,y) conduction heat transfer are summarized in Table 3-1, with the equations for the exterior nodes expanded for several common boundary conditions. Similar equations can be developed for steady one-dimensional and three-dimensional systems by applying the first law of thermodynamics and the appropriate particular law to each subvolume with unknown nodal temperature. (Nodal equations are also available for curved boundaries in [9] and [10].)

Finite-difference equations are written for each interior node and for those exterior nodes for which the temperature is unspecified. For the moment, we represent these equations for a steady two-dimensional system as follows:

$$a_{11}T_1 + a_{12}T_2 + \ldots + a_{1j}T_j + \ldots + a_{1Z}T_Z = C_1 \qquad (3\text{-}69\text{-}1)$$
$$a_{21}T_1 + a_{22}T_2 + \ldots + a_{2j}T_j + \ldots + a_{2Z}T_Z = C_2 \qquad (3\text{-}69\text{-}2)$$
$$\vdots \qquad\qquad\qquad\qquad\qquad \vdots$$
$$a_{j1}T_1 + a_{j2}T_2 + \ldots + a_{jj}T_j + \ldots + a_{jZ}T_Z = C_j \qquad (3\text{-}69\text{-}j)$$
$$\vdots \qquad\qquad\qquad\qquad\qquad \vdots$$
$$a_{Z1}T_1 + a_{Z2}T_2 + \ldots + a_{Zj}T_j + \ldots + a_{ZZ}T_Z = C_Z \qquad (3\text{-}69\text{-}Z)$$

where Z is equal to Z_s minus the number of external nodes for which the temperature is specified. This system of equations must be solved for the Z unknown individual nodal temperatures. Once the solution for the temperature distribution is determined, the rate of heat transfer within the body or at its surface can be obtained.

For finite-difference formulations involving a small number of subvolumes, the nodal equations can generally be conveniently solved by

TABLE 3-1 Summary of finite-difference nodal equations for steady-state two-dimensional (x,y) conditions with heat generation

Type node	Nodal equation
1. Interior node	$T_{m+1,n} + T_{m-1,n} + T_{m,n+1} + T_{m,n-1}$ (1)

$$-4T_{m,n} + \Delta x^2 \, \frac{\dot{q}}{k} = 0$$

2. Exterior nodes

  a. General equations

    i. Regular exterior node (M, n)

$$2T_{M-1,n} + T_{M,n+1} + T_{M,n-1} - 4T_{M,n} \qquad (2i)$$
$$+ \Delta x^2 \, \frac{\dot{q}}{k} = 2\,\Delta x \, \frac{q_s''}{k}$$

    ii. Outer corner node (M, N)

$$T_{M-1,N} + T_{M,N-1} - 2T_{M,N} + \frac{\Delta x^2}{4}\frac{\dot{q}}{k} = 2\,\Delta x \, \frac{q_s''}{k} \quad (2ii)$$

    iii. Inner corner node (m, n)

$$2T_{m-1,n} + 2T_{m,n+1} + T_{m+1,n} + T_{m,n-1} - 6T_{m,n} \qquad (2iii)$$
$$+ \frac{3}{4}\,\Delta x^2 \, \frac{\dot{q}}{k} = 2\,\Delta x \, \frac{q_s''}{k}$$

  b. Boundary conditions[a]

    i. Specified surface temperature

$T_{m,n}$ known for $m = M$ and for $n = N$

    ii. Specified heat flux

q_s'' known

    iii. Insulated surface

$q_s'' = 0$

    iv. Convection

$q_s'' = h(T_{m,n} - T_F)$ at $m = M$ and/or $n = N$

    v. Blackbody thermal radiation

$q_s'' = F_{s-R}\,(T_{m,n}^4 - T_R^4)$ at $m = M$ and/or $n = N$

[a]These boundary conditions apply to each type of exterior node.

159

standard algebraic manipulation. This approach is demonstrated in Examples 3-13 and 3-14. However, the accuracy of such course finite-difference formulations is often inadequate. The number of equations required to accurately model multidimensional heat-transfer systems usually range from 10 or so to many thousands. For problems that require a large number of small subvolumes to obtain the desired accuracy, the use of standard algebraic techniques is prohibitive. Therefore, more efficient solution techniques must be used in solving such problems. Both iterative and direct methods are available for the systematic solution of simultaneous algebraic equations. Several of the more simple and commonly used methods are discussed in the following sections.

EXAMPLE 3-13

A 1-cm-diameter 3-cm-long steel fin ($k = 43$ W/(m °C)) transfers heat from the wall of a heat exchanger at 200°C to a fluid at 25°C with $\bar{h} = 120$ W/(m² °C). Develop a numerical finite-difference solution for the rate of heat transfer for the case in which the tip is insulated and thermal radiation effects are negligible. Utilize a grid spacing of $\Delta x = L/4$.

Solution

As shown in Example 2-13, the Biot number is equal to 0.00698, such that the heat transfer in this two-dimensional system can be approximated by a one-dimensional analysis.

A one-dimensional finite-difference grid is shown in Fig. E3-13 for $\Delta x = L/4$. Because T_1 is specified, this network only involves four unknown nodal temperatures. Therefore, four algebraic equations must be written to solve for T_2, T_3, T_4, and T_5. Focusing attention on node

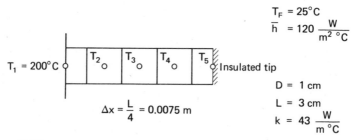

FIGURE E3-13 Finite-difference grid for fin with small Biot number.

(2), an energy balance is developed as follows:

Node (2)
$$q_x = q_{x+\Delta x} + \Delta q_c$$
$$-kA\frac{T_2 - T_1}{\Delta x} = -kA\frac{T_3 - T_2}{\Delta x} + \bar{h}p\,\Delta x(T_2 - T_F)$$

where $\bar{h}p\Delta x^2/(kA) = 0.0628$ for this problem. Rearranging, we obtain

$$T_1 - \left(2 + \frac{\bar{h}p\,\Delta x^2}{kA}\right)T_2 + T_3 + \frac{\bar{h}p\,\Delta x^2}{kA}T_F = 0$$

Similar equations can be written for nodes (3) and (4) as follows:

Node (3)
$$T_2 - \left(2 + \frac{\bar{h}p\,\Delta x^2}{kA}\right)T_3 + T_4 + \frac{\bar{h}p\,\Delta x^2}{kA}T_F = 0$$

Node (4)
$$T_3 - \left(2 + \frac{\bar{h}p\,\Delta x^2}{kA}\right)T_4 + T_5 + \frac{\bar{h}p\,\Delta x^2}{kA}T_F = 0$$

The energy balance for node (5) takes the form

Node (5)
$$q_x = \Delta q_c$$
$$-kA\frac{T_5 - T_4}{\Delta x} = \bar{h}p\frac{\Delta x}{2}(T_5 - T_F)$$

or

$$T_4 - \left(1 + \frac{\bar{h}p}{kA}\frac{\Delta x^2}{2}\right)T_5 + \frac{\bar{h}p}{kA}\frac{\Delta x^2}{2}T_F = 0$$

Substituting for the various parameters, our four nodal equations are summarized as follows:

$$200°C - 2.06T_2 \quad + T_3 \qquad\qquad +1.57\ °C = 0 \qquad \text{(a)}$$
$$T_2 - 2.06T_3 \quad + T_4 \qquad\qquad +1.57\ °C = 0 \qquad \text{(b)}$$
$$T_3 - 2.06T_4 \quad + T_5 + 1.57\ °C = 0 \qquad \text{(c)}$$
$$T_4 - 1.03T_5 + 0.785°C = 0 \qquad \text{(d)}$$

These equations can be easily solved by standard algebraic manipulation. For example, Eqs. (a) and (b) can be combined to eliminate T_2 and Eqs. (c) and (d) can be coupled to remove T_5. The solution of the resulting two equations gives

$$T_3 = 155°C \qquad T_4 = 145°C$$

T_2 and T_5 are then found from Eqs. (a) and (d); that is,

$$T_2 = 173°C \qquad T_5 = 141°C$$

By comparing these predictions for the nodal temperatures with the exact analytical solution given by Eq. (2-116), we find a maximum error of only about 2%.

To obtain an approximation for the rate of heat transfer from the fin, we develop an energy balance at node (1) as follows:

Node (1) $q_F = q_{\Delta x/2} + \Delta q_c$

$$q_F = -kA \frac{T_2 - T_1}{\Delta x} + \bar{h}p \frac{\Delta x}{2}(T_1 - T_F) = 14.6 \text{ W}$$

Referring back to Example 2-13, we find that the analytical solution for the rate of heat transfer q_F is

$$q_F = 15.1 \text{ W}$$

Thus, the error in our finite-difference prediction for q_F is about 3.3%.

To improve the accuracy of our solution, a finer grid can be utilized. [To get a better feel for the effect of grid size on the accuracy of finite-difference solutions, it is suggested that this rather simple problem be solved for $\Delta x = L$, $\Delta x = L/2$, and $\Delta x = L/3$ (see Prob. 3-36).]

EXAMPLE 3-14

Develop an approximate numerical solution to the two-dimensional heat-transfer problem shown in Fig. E3-14a by utilizing a grid with $\Delta x = L/3$. Also evaluate the accuracy of the solution.

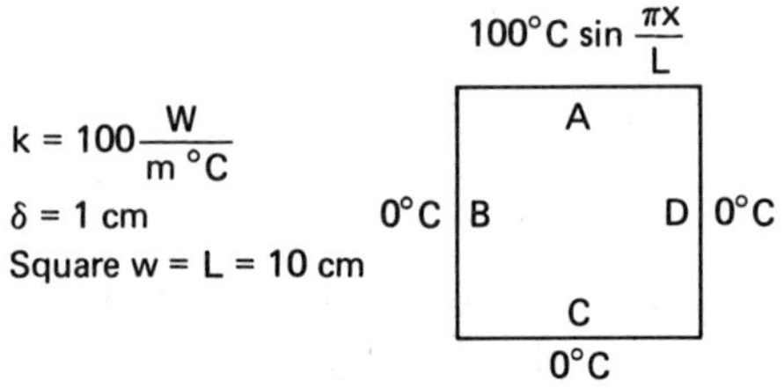

FIGURE E3-14a   Two-dimensional heat transfer in rectangular plate.

Solution

The finite-difference grid for this problem is shown in Fig. E3-14b for $\Delta x = L/3$. This grid produces 16 nodes, but only four of the nodal temperatures are unknown; these are T_1, T_2, T_3, and T_4. Referring to Fig. E3-14b and utilizing Eq. (1) in Table 3-1, or by developing energy balances on each interior node, the nodal equations for this system are written as

Node (2,2)	$-4T_1 + T_2 + T_3 = 0$	(a)
Node (2,3)	$T_1 - 4T_2 + T_4 + 86.6°C = 0$	(b)
Node (3,2)	$T_1 - 4T_3 + T_4 = 0$	(c)
Node (3,3)	$+ T_2 + T_3 - 4T_4 + 86.6°C = 0$	(d)

These four equations can be solved for the unknown nodal temperatures. However, our computational work can be eased by recognizing that system symmetry requires $T_3 = T_1$ and $T_4 = T_2$. Consequently, we really only have two unknown nodal temperatures to concern ourselves with. Utilizing this fact, Eqs. (a) and (b) become

Node (2,2)	$-3T_1 + T_2 = 0$	(e)
Node (2,3)	$T_1 - 3T_2 \qquad 86.6°C = 0$	(f)

Equations (e) and (f) are easily solved to obtain

$$T_1 = \frac{1}{8} 86.6°C = 10.8°C \tag{g}$$

$$T_2 = \frac{3}{8} 86.6°C = 32.4°C \tag{h}$$

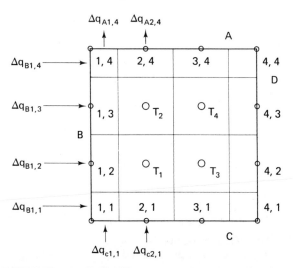

FIGURE E3-14b Finite-difference grid for rectangular plate.

A comparison of these predictions with the exact analytical solution developed in Sec. 3-5 reveals an error of 15.3% in T_1 and 8.06% in T_2.

Referring to Fig. E3-14b, the rate of heat transfer across the heating surface is

$$q_A = 2(\Delta q_{A1,4} + \Delta q_{A2,4})$$

To approximate the rate of heat transfer at nodes $(1,4)$ and $(2,4)$, energy balances are written for these nodes. Considering node $(2,4)$ first, we write

$$\Delta q_{A2,4} = \Delta q_x + \Delta q_y$$
$$= -k\delta \frac{\Delta y}{2} \frac{86.6°C - 0°C}{\Delta x} - k\delta \Delta x \frac{86.6°C - T_2}{\Delta y}$$

Setting $T_2 = 32.4°C$ and utilizing the values of k and δ for this problem (i.e., $k\delta = 1$ W/°C), this equation reduces to

$$\Delta q_{A2,4} = -k\delta \left(\frac{86.6°C - 0°C}{2} + 86.6°C - 32.4°C \right) = -97.5 \text{ W}$$

The negative sign indicates that heat is transferred into the surface at node $(2,4)$.

The energy balance for the corner node $(1,4)$ takes the form

$$\Delta q_{B1,4} + \Delta q_y = \Delta q_{A1,4} + \Delta q_x$$

where

$$\Delta q_x = -k\delta \frac{\Delta y}{2} \frac{86.6°C - 0°C}{\Delta x} = -43.3 \text{ W}$$
$$\Delta q_y = -k\delta \frac{\Delta x}{2} \frac{0°C - 0°C}{\Delta y} = 0 \text{ W}$$

Thus, we have the result

$$\Delta q_{A1,4} = \Delta q_{B1,4} + 43.3 \text{ W} \tag{i}$$

where $\Delta q_{B1,4}$ is unknown. Because we have one equation but two unknowns, both $\Delta q_{A1,4}$ and $\Delta q_{B1,4}$ are indeterminant. About the best that we can do in this situation is to simply neglect the corner node, that is, assume that $\Delta q_{A1,4}$ is negligible.

Summing up, q_A is approximated by

$$q_A = 2(0 \text{ W} - 97.5 \text{ W}) = -195 \text{ W}$$

The exact solution for q_A is found from Eq. (3-67), which is developed in Sec. 3-5, to be -200 W. Thus, the error in q_A is only 2.5%. However, upon closer inspection we find that the exact solutions for $\Delta q_{A1,4}$ and $\Delta q_{A2,4}$ are -13.4 W and -86.6 W, respectively. The error in $\Delta q_{A2,4}$ is a substantial 12.6%, with the contribution of $\Delta q_{A1,4}$ and $\Delta q_{A4,4}$ being about 13% of q_A. These errors are of the order of the errors in the predictions for the temperature distribution. It just so happens that these errors compensate each other. We are not always so fortunate.

As seen in this example, the inherent indeterminant nature of the heat transfer through corner nodes for surfaces with specified temperatures is one of the main sources of error in this type of numerical finite-difference formulation. (This type of error does not occur for convection, thermal radiation, or specified heat flux boundary conditions.) This error can be reduced by decreasing the grid size. (This feat is sometimes accomplished by using variable grid spacing, with the smallest subvolumes utilized in the region where the accuracy is most critical.)

Following the pattern established in approximating q_A, the following predictions are obtained for q_B and q_C:

$$q_B = -86.5 \text{ W} \qquad q_C = -21.6 \text{ W}$$

Because of symmetry, we write

$$q_D = -q_B = 86.5 \text{ W}$$

To check our calculations, we perform an energy balance on the entire system.

$$q_A = q_B + q_C + q_D = (-86.5 - 21.6 - 86.5) \text{ W}$$
$$= -194.6 \text{ W}$$

Because this value is within 0.3% of the direct calculation for $q_A (= -195$ W), we conclude that no calculation errors have been made. This small difference is a consequence of the use of three significant figures in our calculations and should not be confused with the actual error that results from the use of finite-difference approximations. Such an energy balance can not be used to determine the accuracy of a numerical finite-difference analysis!

As indicated in Sec. 3-3 the error associated with our finite-difference approximations should be of the order of Δx^2. Thus, with $\Delta x / L$ equal to one-third in this formulation, we expect an error of the order of 11%. In checking back over our results, we find that this is fairly representative of the order of error in our calculations for the temperatures and rates of heat transfer. Thus, we would expect that a grid spacing of $\Delta x / L = 0.1$ would be required to reduce the errors to the order of 1%. This is indeed the case, as is shown in Example 3-17.

Iterative method

Iterative techniques have been devised for the simultaneous solution of systems of equations of the form of Eqs. (3-69-1) through (3-69-Z) which are similar to the simple iterative scheme introduced in Chap. 2. This approach involves the development of successive approximations for the unknowns that converge toward the exact solution.

The *Gauss–Seidel method* is one of the more popular iterative schemes for solving systems of algebraic equations. In this approach, Eqs. (3-69-1) through (3-69-Z) are rewritten as follows:

$$T_1 = \frac{1}{a_{11}} \left[C_1 - (a_{12}T_2 + \dots + a_{1j}T_j + \dots + a_{1Z}T_Z) \right] \qquad (3\text{-}70\text{-}1)$$

$$T_2 = \frac{1}{a_{22}} \left[C_2 - (a_{21}T_1 + a_{23}T_3 + \dots + a_{2j}T_j + \dots + a_{2Z}T_Z) \right] \qquad (3\text{-}70\text{-}2)$$

$$\vdots$$

$$T_j = \frac{1}{a_{jj}} \left[C_j - (a_{j1}T_1 + \dots + a_{jj-1}T_{j-1} + a_{jj+1}T_{j+1} + \dots + a_{jZ}T_Z) \right]$$

$$(3\text{-}70\text{-}j)$$

$$\vdots$$

$$T_Z = \frac{1}{a_{ZZ}} \left[C_Z - (a_{Z1}T_1 + \dots + a_{Zj}T_j + \dots + a_{ZZ-1}T_{Z-1}) \right] \qquad (3\text{-}70\text{-}Z)$$

By introducing approximations for the unknown temperatures ($T_1 = T_1^0$, $T_2 = T_2^0$, $T_j = T_j^0$, etc.) into Eqs. (3-70-1) through (3-70-Z), new first-order calculations are developed for the unknowns, T_1^1, T_2^1, and so on. The first-order calculations are then substituted into Eqs. (3-70-1) through (3-70-Z) to produce second-order calculations. Assuming convergence, this procedure is continued until the ith iteration produces sufficient accuracy for all values of T_j; that is,

$$\left| \frac{T_j^{i+1} - T_j^i}{T_j^i} \right| < \varepsilon \qquad (3\text{-}71)$$

where $i = 1, 2, \dots$, and ε is a small number which determines the accuracy of the solution of Eqs. (3-69). Formal convergence criteria for the Gauss–Seidel method is presented in [4].

The mechanics involved in the use of this iterative method is illustrated in Example 3-15. Notice that new temperature values are utilized in the iteration pattern as soon as they are calculated. This approach is adapted to digital computation in Example 3-15 and in several other examples.

It should be mentioned that a hand-calculation iterative scheme known as the relaxation method is also available. However, because it does not lend itself to efficient digital computation, the relaxation method is no longer of real practical significance.

EXAMPLE 3-15

Develop a more accurate numerical solution for the fin problem of Example 3-13 by employing the Gauss–Seidel iteration method.

Solution

A finite-difference grid network is shown in Fig. E3-15a for M nodes. By applying the first law of thermodynamics to each node, we obtain

$m = 2, 3, \ldots, M-1$

$$q_x = q_{x+\Delta x} + \Delta q_c$$

$$T_{m-1} - \left(2 + \frac{\bar{h}p}{kA}\Delta x^2\right)T_m + T_{m+1} + \frac{\bar{h}p}{kA}\Delta x^2\, T_F = 0 \qquad \text{(a)}$$

and $m = M$

$$q_x = \Delta q_c$$

$$T_{M-1} - \left(1 + \frac{\bar{h}p}{kA}\frac{\Delta x^2}{2}\right)T_M + \left(\frac{\bar{h}p}{kA}\frac{\Delta x^2}{2}\right)T_F = 0 \qquad \text{(b)}$$

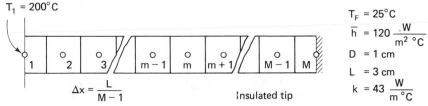

$T_1 = 200°C$

$T_F = 25°C$

$\bar{h} = 120\,\dfrac{W}{m^2\,°C}$

$\Delta x = \dfrac{L}{M-1}$ Insulated tip

$D = 1$ cm

$L = 3$ cm

$k = 43\,\dfrac{W}{m\,°C}$

FIGURE E3-15a Finite-difference grid for fin with small Biot number.

These $M-1$ equations are written in the Gauss–Seidel format as follows:

$m = 2, 3, \ldots, M-1$

$$T_m = \frac{1}{2 + \bar{h}p\,\Delta x^2/(kA)}\left(T_{m-1} + T_{m+1} + \frac{\bar{h}p}{kA}\Delta x^2\,T_F\right) \qquad \text{(c)}$$

$m = M$

$$T_M = \frac{1}{1 + \bar{h}p\,\Delta x^2/(2kA)}\left(T_{M-1} + \frac{\bar{h}p}{kA}\frac{\Delta x^2}{2}\,T_F\right) \qquad \text{(d)}$$

For example, for the $M=5$ grid of Example 3-13, we have (to four significant figures)

$$
\begin{aligned}
T_2 &= \frac{1}{2.063}(200°C + T_3 + 1.57°C) \\[4pt]
T_3 &= \frac{1}{2.063}(T_2 + T_4 + 1.57°C) \\[4pt]
T_4 &= \frac{1}{2.063}(T_3 + T_5 + 1.57°C) \\[4pt]
T_5 &= \frac{1}{1.031}(T_4 + 0.785°C)
\end{aligned}
\qquad \text{(e)}
$$

Starting our iterative calculations with $T_m = 200°C$, first-round calculations are obtained as follows:

$$
\begin{aligned}
T_2 &= \frac{1}{2.063}(200°C + 200°C + 1.57°C) = 194.7°C \\[4pt]
T_3 &= \frac{1}{2.063}(194.7°C + 200°C + 1.57°C) = 192.1°C \\[4pt]
T_4 &= \frac{1}{2.063}(192.1°C + 200°C + 1.57°C) = 190.8°C \\[4pt]
T_5 &= \frac{1}{1.031}(190.8°C + 0.785°C) = 185.8°C
\end{aligned}
\qquad \text{(f)}
$$

A second iteration gives

$$
\begin{aligned}
T_2 &= \frac{1}{2.063}(200°C + 192.1°C + 1.57°C) = 190.8°C \\[4pt]
T_3 &= \frac{1}{2.063}(190.8°C + 190.8°C + 1.57°C) = 185.7°C \\[4pt]
T_4 &= \frac{1}{2.063}(185.7°C + 185.8°C + 1.57°C) = 180.8°C \\[4pt]
T_5 &= \frac{1}{1.031}(180.8°C + 0.785°C) = 176.1°C
\end{aligned}
\qquad \text{(g)}
$$

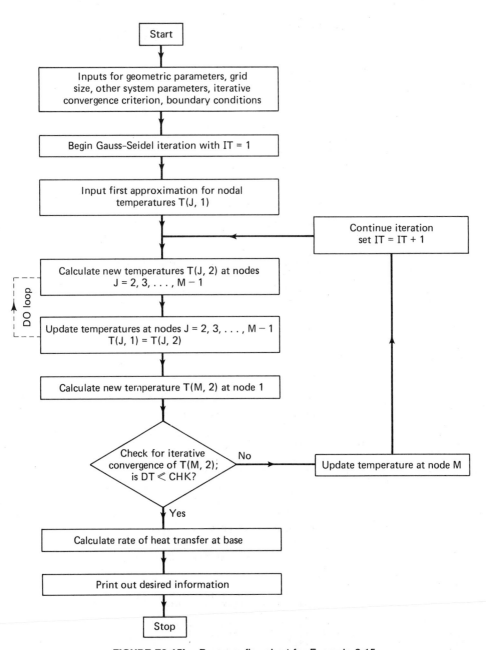

FIGURE E3-15b Program flowchart for Example 3-15.

```
C           EXAMPLE 3-15
0001        REAL K
0002        REAL L
0003        DIMENSION T(100,2)
C
C           INPUTS FOR GEOMETRIC PARAMETERS.
0004        L=0.03
0005        D=0.01
0006        A=3.1416*D*D/4.
0007        P=3.1416*D
C
C           INPUTS FOR GRID SIZE.
0008        DO 50 M=3,20
0009        M1=M-1
0010        DX=L/M1
C
C           INPUTS FOR PROPERTIES.
0011        K=43.
C
C           INPUTS FOR OTHER SYSTEM PARAMETERS.
0012        H=120.
0013        CC1=H*P*DX*DX/(K*A)
0014        T1=200.
0015        TF=25.
C
C           INPUT FOR ITERATIVE CONVERGENCE CRITERION.
0016        CHK=1.E-5
C
C           INPUT FOR BOUNDARY CONDITIONS.
0017        T(1,1)=T1
0018        T(1,2)=T1
C
C           BEGIN GAUSS SEIDEL ITERATION WITH IT=1.
0019        IT=1
C
C           INPUT FIRST APPROXIMATION FOR NODAL TEMPERATURES.
0020        DO 1 J=2,M
0021      1 T(J,1)=T1
C
C           CALCULATE NEW TEMPERATURES T(J,2) AT NODES J=2,3,......,M-1.
0022      2 DO 3 J=2,M1
0023        T(J,2)=(T(J-1,1)+T(J+1,1)+CC1*TF)/(2.+CC1)
C
C           UPDATE TEMPERATURES AT NODES J=2,3,...,M-1.  T(J,1)=T(J,2).
0024      3 T(J,1)=T(J,2)
C
C           CALCULATE NEW TEMPERATURE T(M,2) AT NODE M.
0025        T(M,2)=(T(M-1,1)+CC1/2.*TF)/(1.+CC1/2.)
C
C           CHECK FOR ITERATIVE CONVERGENCE OF T(M,2).
0026        DT=ABS((T(M,2)-T(M,1))/T(M,1))
0027        IF(DT .LE. CHK) GO TO 4
C
C           UPDATE TEMPERATURE AT NODE M.
0028        T(M,1)=T(M,2)
C
C           CONTINUE ITERATION.  IT=IT+1.
0029        IT=IT+1
0030        GO TO 2
C
C           CALCULATE RATE OF HEAT TRANSFER AT BASE.
0031      4 QF=-K*A/DX*(T(2,2)-T(1,2))+H*P*DX/2.*(T(1,2)-TF)
C
C           PRINT OUT DESIRED INFORMATION.
0032     90 FORMAT(1X,2(I5,3X),3(F8.3,3X))
0033        PRINT 90,M,IT,T(M,2),QF
0034     50 CONTINUE
0035        STOP
0036        END
```

FIGURE E3-15c FORTRAN program for Example 3-15.

The iteration calculations for T_2, T_3, T_4, and T_5 converge within about 2% of the values obtained in Example 3-13 after 22 iterations. But because of the slow convergence, hand calculation is impractical for this problem. Therefore, in order to develop more accurate finite-difference solutions with larger values of M, numerical digital computation is now employed.

Utilizing FORTRAN computer language, Eqs. (c) and (d) are written as

$$T(J,2) = (T(J-1,1) + T(J+1,1) + (CC1*TF)/(2. + CC1)) \qquad (h)$$

for $J = m = 2, 3, \dots, M - 1$, and

$$T(M,2) = (T(M-1,1) + CC1/2.*TF)/(1. + CC1/2) \qquad (i)$$

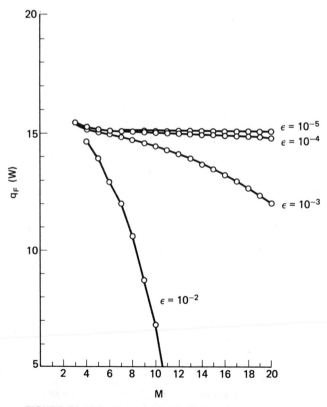

FIGURE E3-15d Numerical calculations for q_F.

where CC1 = H*P*DX*DX/(K*A), H = $\bar{h}$, P = p, DX = Δx, K = k, and TF = T_F. The second index represents the new (index 2) and old (index 1) iteration values for the nodal temperatures. The rate of heat transfer q_F from the fin is obtained by performing an energy balance on node (1); q_F is expressed in FORTRAN by

$$QF = -K*A/DX*(T(2,2) - T(1,2)) + H*P*DX/2.*(T(1,2) - TF) \qquad (j)$$

A simple flowchart and FORTRAN program are given in Figs. E3-15b and E3-15c which solves the $M - 1$ nodal equations by the Gauss–Seidel method. The program also calculates the rate of heat transfer from the fin. This FORTRAN program requires that the temperature at node M satisfy the specified iterative convergence criterion ε (designated by CHK in the program). (Logic that permits us to check and satisfy convergence for all nodal temperatures is developed in Example 3-17.)

Calculations for the rate of heat transfer from the fin are shown in Fig. E3-15d for $M = 3, 4, \ldots, 20$, and for $\varepsilon = 10^{-2}, 10^{-3}, 10^{-4}$, and 10^{-5}. Note that the calculations are dependent upon both M and ε! The calculations for q_F essentially converge to approximately 15.1 W for $M \geqslant 6$ (i.e., $\Delta x \gtrsim 0.005$ m) and $\varepsilon = 10^{-5}$. This value is in agreement with the analytical solution developed in Example 2-13.

EXAMPLE 3-16

Determine the rate of heat transfer and temperature distribution for the anodized aluminum fin shown in Fig. E3-16a. The base is at 80°C, the air and surrounding walls are at 25°C, and the coefficient of heat transfer due to natural convection cooling is 10 W/(m² °C).

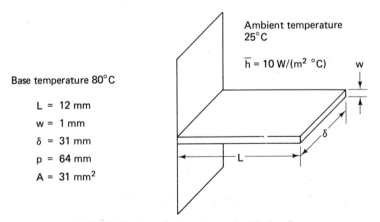

Base temperature 80°C

L = 12 mm
w = 1 mm
δ = 31 mm
p = 64 mm
A = 31 mm²

Ambient temperature 25°C

$\bar{h}$ = 10 W/(m² °C)   w

FIGURE E3-16a   Convecting and radiating fin.

Solution

As shown in Example 2-15, a one-dimensional analysis is warranted for this problem because the Biot number is much less than 0.1. Because of the complexity of the problem, we will utilize the numerical approach.

Utilizing the finite-difference grid of Fig. E3-15a in Example 3-15, the nodal equations are given by

$m = 2, 3, \ldots, M - 1$

$$q_x = q_{x+\Delta x} + \Delta q_c + \Delta q_R$$

$$-kA \frac{T_m - T_{m-1}}{\Delta x} =$$

$$-kA \frac{T_{m+1} - T_m}{\Delta x} + \bar{h}p \, \Delta x (T_m - T_F) + \sigma p \, \Delta x \, F_{s-R}(T_m^4 - T_R^4)$$

(a)

$m = M$

$$q_x = \Delta q_c + \Delta q_R + q_c + q_R$$

$$-kA \frac{T_M - T_{M-1}}{\Delta x} = \bar{h}p \frac{\Delta x}{2}(T_M - T_F) + \sigma p F_{s-R} \frac{\Delta x}{2}(T_M^4 - T_R^4)$$

$$+ \bar{h}A(T_M - T_F) + \sigma A F_{s-R}(T_M^4 - T_R^4)$$

(b)

where q_c and q_R account for the heat transfer from the tip. To put Eqs. (a) and (b) into the Gauss–Seidel iteration form, we write these equations as follows:

$m = 2, 3 \ldots, M - 1$

$$T_m = \frac{T_{m-1} + T_{m+1} + CC_1 T_F + CR_1 T_R^4}{2 + CC_1 + CR_1 T_m^3}$$

(c)

$m = M$

$$T_M = \frac{T_{M-1} + (CC_1/2 + CC_2)T_F + (CR_1/2 + CR_2)T_R^4}{1 + CC_1/2 + CC_2 + (CR_1/2 + CR_2)T_M^3}$$

(d)

where $CC_1 = \bar{h}p \, \Delta x^2 / (kA)$, $CC_2 = \bar{h} \, \Delta x / k$, $CR_1 = \sigma p F_{s-R} \, \Delta x^2 / (kA)$, and $CR_2 = \sigma F_{s-R} \, \Delta x / k$. Notice that T_m appears on both sides of these nonlinear equations. (A similar iterative formulation was developed in Example 2-4 for one-dimensional radiation heat transfer from a flat plate.)

Equations (c) and (d) are written in FORTRAN language as follows:

$$T(J,2) = (T(J-1,1) + T(J+1,1) + CC1*TF + CR1*TR**4.)$$
$$/(2. + CC1 + CR1*T(J,1)**3.) \qquad (e)$$

for $J = 2, 3, \ldots, M-1$, and

$$T(M,2) = (T(M-1,1) + (CC1/2. + CC2)*TF + (CR1/2. + CR2)*TR**4.)$$
$$/(1. + CC1/2. + CC2) + (CR1/2. + CR2)*T(M,1)**3.) \qquad (f)$$

for $J = M$, where $CC1 = H*P*DX*DX/(K*A)$, $CC2 = H*DX/K$, CR1 $= SIGMA*P*DX*DX*FSR/(K*A)$, and $CR2 = SIGMA*DX*FSR/K$. By applying the first law of thermodynamics to node (1), an expression is obtained for the rate of heat transfer q_F from the fin of the form

$$QF = -K*A/DX*(T(2,2) - T(1,2)) + H*P*DX/2.*(T(1,2) - TF)$$
$$+ SIGMA*P*DX/2.*FSR*(T(1,2)**4. - TR**4.) \qquad (g)$$

The FORTRAN program presented in Example 3-15 is employed to solve these equations by utilizing the following inputs for geometric parameters, properties, and other system parameters:

Inputs for geometric parameters

$$L = 12.E - 3$$
$$P = 64.E - 3$$
$$A = 31.E - 6$$

Inputs for properties

$$K = 200.$$

Inputs for other system parameters

$$H = 10.$$
$$SIGMA = 5.67E - 8$$
$$FSR = 1.$$
$$CC1 = H*P*DX*DX/(K*A)$$
$$CC2 = H*DX/K$$
$$CR1 = SIGMA*P*DX*DX*FSR/(K*A)$$
$$CR2 = SIGMA*DX*FSR/K$$
$$T1 = 353.$$
$$TF = 298.$$
$$TR = 298.$$

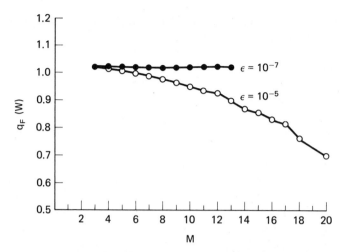

FIGURE E3-16b Numerical calculations for q_F.

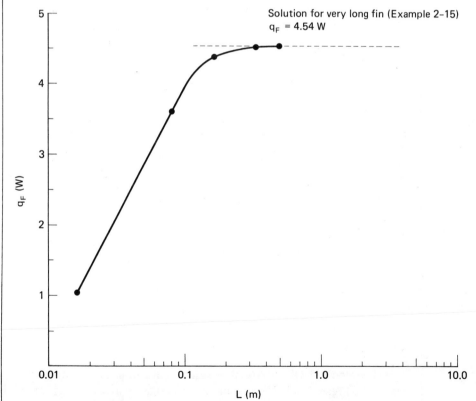

FIGURE E3-16c Numerical calculations for q_F for various lengths.

Of course, the equations for $T(J,2)$, $T(M,2)$, and QF must be specified in accordance with Eqs. (e), (f), and (g).

Calculations are shown in Fig. E3-16b for the rate of heat transfer from the fin for $M = 3,4,\ldots,20$ and $\varepsilon = 10^{-5}$ and 10^{-7}. The calculations for q_F converge to approximately 1.02 W for $M \geqslant 3$(i.e., $\Delta x \leqslant 0.00533$ m) and $\varepsilon = 10^{-7}$. To see the effect of radiation, the problem is also run with the radiation terms eliminated. The calculations for q_F without radiation converge to about 0.575 W. Thus, the thermal radiation contributes 43.6% to the total rate of heat transfer.

Assuming that no errors have been made in our analysis and program inputs, we can be reasonably sure of an accurate solution, since the calculations for q_F converge for increasing M and decreasing ε. However, to further reinforce our confidence in the solution accuracy, calculations are made for q_F for longer fin lengths to enable us to compare the results with the analytical solution of Example 2-15. The comparison of numerical calculations for q_F for fins of various lengths with the analytical result for a long fin is made in Fig. E3-16c. Notice that the numerical calculations approach the analytical solution of Example 2-15 for lengths of the order of 0.3 m and greater.

The temperature distribution in the fin is shown in Table E3-16. The temperature drop along this short fin is seen to be very small, such that the fin efficiency is quite high.

TABLE E3-16   Numerical calculations for temperature distribution in fin

m	T_m (K)
1	353.0
2	352.7
3	352.5
4	352.3
5	352.1
6	351.9
7	351.8
8	351.73
9	351.67

EXAMPLE 3-17

(a)   Develop a FORTRAN program for the solution of the problem shown in Fig. E3-17a which utilizes the Gauss–Seidel iteration method and a grid network with $\Delta x = L/M$, where M can be any integer value.

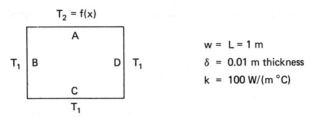

FIGURE E3-17a Two-dimensional heat transfer in rectangular plate.

(b) Utilize this program to obtain a more accurate numerical solution to Example 3-14.

(c) Utilize this program to obtain a solution for the case in which $T_1 = 0°C$ and $T_2 = 100°C$.

Solution

(a) *FORTRAN Program*: A finite-difference grid with $M = 6$ is shown in Fig. E3-17b for this conduction-heat-transfer problem. The interior nodal equations take the general form

$$4T_{m,n} = T_{m+1,n} + T_{m-1,n} + T_{m,n+1} + T_{m,n-1}$$

with $m = 2, 3, \ldots, M-1$, and $n = 2, 3, \ldots, N-1$. Our formulation consists of $Z[=(M-2)(N-2)]$ such equations plus the boundary conditions.

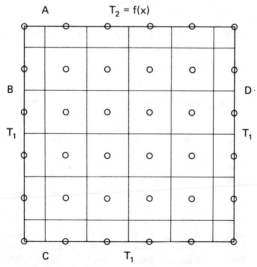

FIGURE E3-17b Finite-difference grid for rectangular plate.

Using FORTRAN computer language, this system of equations is put into the iterative form

$$T(J,I,2) = (T(J+1,I,1) + T(J-1,I,1) + T(J,I+1,1) + T(J,I-1,1))/4.$$

where $J = m = 2,3,4,...,M-1$, and $I = n = 2,3,4,...,N-1$. The third index represents the new (index 2) and old (index 1) iteration values for the nodal temperatures.

A flowchart and FORTRAN program which employs the Gauss–Seidel iteration technique are presented in Figs. A-E-1 and A-E-2 of the Appendix. Whereas the program developed in Example 3-15 only checks for iterative convergence of one nodal temperature, the program developed for this example requires that all nodal temperatures satisfy the convergence criterion. The inputs for T_1, T_2, L, w, δ, and k are specified in accordance with the problem shown in Fig. E3-17a.

(b) $f(x) = 100°C \sin(\pi x/L)$, $T_1 = 0°C$: Calculations for q_A obtained by running this program on a digital computer are shown in Fig. E3-17c for values of $M = 2$, 3, 4,...,20 and for $\varepsilon = 10^{-1}$ and 10^{-4}. Again we notice that the accuracy of the numerical solution depends upon both M and ε! The calculations for q_A are seen to converge to approximately -200.7 W for M equal to 9 with $\varepsilon = 10^{-4}$. The exact solution given by Eq. (3-67) is -200 W, such that the error is only 0.35%.

Calculations are also easily obtained for q_B, q_C, and q_D. For example, the calculations for q_C converge to approximately -17.5 W at a value of M equal to 13. The exact solution for q_C is -17.3 W, such that the error is about 1.5%.

Calculations are shown in Table E3-17 for the nodal temperatures obtained for $M = 9$. The accuracy of these predictions is extremely good, with a minimum error of about 0.1%.

TABLE E3-17 Numerical calculations for temperature distribution in rectangular plate; $T_2 = 100°C \sin(\pi x/L)$

0.0	34.20	64.28	86.60	98.48	98.48	86.6	64.28	34.20	0.0
0.0	24.16	45.41	61.18	69.57	69.57	61.1	45.41	24.16	0.0
0.0	17.03	32.01	43.13	49.04	49.04	43.1	32.01	17.03	0.0
0.0	11.96	22.48	30.28	34.44	34.44	30.2	22.48	11.96	0.0
0.0	8.33	15.65	21.09	23.98	23.98	21.0	15.65	8.33	0.0
0.0	5.70	10.72	14.44	16.42	16.42	14.4	10.72	5.70	0.0
0.0	3.77	7.08	9.54	10.84	10.84	9.5	7.08	3.77	0.0
0.0	2.28	4.29	5.78	6.58	6.58	5.7	4.29	2.28	0.0
0.0	1.08	2.02	2.73	3.10	3.10	2.7	2.02	1.08	0.0
0.0	0.0	0.0	0.0	0.0	0.0	0.0	0.0	0.0	0.0

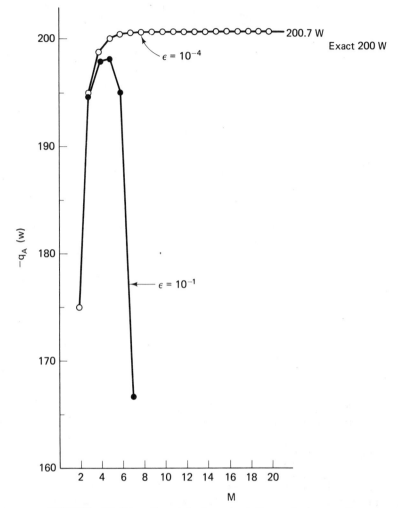

FIGURE E3-17c Numerical calculations for q_A; T_2 $100°C\sin(\pi x/L)$.

(c) $T_2 = f(x) = 100°C$, $T_1 = 0°C$: The program presented in Figs. A-E-1 and A-E-2 is easily run for $T_2 = 100°C$. Calculations produced for q_A and q_C for this condition are shown in Fig. E3-17d for $M = 2,3,4,...,20$ and $\varepsilon = 10^{-4}$. For this problem, q_C converges to approximately -22.2 W, but convergence is nowhere in sight for q_A. Referring back to Example 3-10, the exact solution for q_C is -22.1 W, but q_A is actually unbounded. Thus, our finite-difference solution is consistent with the exact analytical solution.

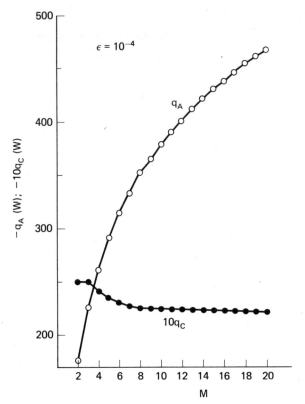

FIGURE E3-17d   Numerical calculations for q_A and q_C; $T_2 = 100°C$.

Direct methods

The most commonly used direct methods for solving systems of equations include Gaussian elimination and matrix methods.

Gaussian Elimination   In the Gaussian elimination method, the nodal equations, Eqs. (3-69-1) through (3-69-Z), are put into the form

$$a_{11}T_1 + a_{12}T_2 + \ldots + a_{1j}T_j + \ldots + a_{1Z}T_Z = C_1 \qquad (3\text{-}72\text{-}1)$$

$$a'_{22}T_2 + \ldots + a'_{2j}T_j + \ldots + a'_{2Z}T_Z = C'_2 \qquad (3\text{-}72\text{-}2)$$

$$\vdots$$

$$a'_{jj}T_j + \ldots + a'_{jZ}T_Z = C'_j \qquad (3\text{-}72\text{-}j)$$

$$\vdots$$

$$a'_{ZZ}T_Z = C'_Z \qquad (3\text{-}72\text{-}Z)$$

This arrangement is obtained by first dividing Eq. (3-69-1) by a_{11}. The resulting equation is used to algebraically eliminate T_1 from each of the remaining equations. For example, by combining Eqs. (3-69-1) and (3-69-2), we obtain

$$a'_{2j} = a_{2j} - a_{1j} a_{21}/a_{11} \qquad j = 1, 2, \ldots, Z$$
$$C'_2 = C_2 - C_1 a_{21}/a_{11} \tag{3-73}$$

The second equation is then used to eliminate T_2 from the remaining equations. This elimination procedure is continued until the final equation is obtained. This technique produces a solution for the unknown nodal temperature T_Z of the form

$$T_Z = \frac{C'_Z}{a'_{ZZ}} \tag{3-74}$$

The other unknowns can then be calculated in reverse order (i.e., $T_{Z-1}, T_{Z-2}, \ldots, T_j, \ldots, T_1$). This elimination procedure is simplified by the fact that a given nodal temperature T_j appears in no more than four equations for conduction problems involving two dimensions, such as x and y.

Matrix For more complex problems which necessitate computer solutions, matrix-solving programs are widely used which perform Gaussian elimination, matrix inversion, and other functions. Several standard subroutines are available for this purpose, such as the IBM SIMQ. For an introduction to matrix methods, which are restricted to linear equations, [4] can be consulted.

Unsteady analysis method

Because our next topic deals with the numerical solution of unsteady multidimensional conduction-heat-transfer problems, we merely mention here that steady-state problems are sometimes solved by means of unsteady analyses which are carried through to the steady-state limit.

3-6-3 Unsteady Systems

As indicated in Sec. 3-3-3, both explicit and implicit numerical finite-difference formulations can be developed for unsteady multidimensional problems. Because the explicit type of formulation which results from the use of a forward time difference for $\Delta T/\Delta t$ produces algebraic equations that can be solved by straightforward computational techniques, this method will be emphasized. Brief consideration will also be given to the alternative implicit finite-difference approach which is developed by the use of a backward time difference for $\Delta T/\Delta t$.

Explicit method

The unsteady three-dimensional (t,x,y) forward difference energy equation for interior nodes is given by Eq. (3-45). This equation can be rearranged such that the temperature at time $(\tau+1)\Delta t$ is expressed explicitly in terms of the temperature at the earlier instant of time $\tau\Delta t$ as follows (for $\dot{q}=0$):

$$T_{m,n}^{\tau+1} = \frac{\alpha\Delta t}{\Delta x^2}\left(T_{m+1,n}^{\tau} + T_{m-1,n}^{\tau} + T_{m,n+1}^{\tau} + T_{m,n-1}^{\tau}\right)$$
$$+ \left(1 - \frac{4\alpha\Delta t}{\Delta x^2}\right)T_{m,n}^{\tau} \tag{3-75}$$

Similar explicit equations can be written for the energy transfer at the exterior nodes. For example, the explicit energy equation for a regular exterior node at $m=M$ with surface heat flux q_s'' is given by

$$T_{M,n}^{\tau+1} = \frac{\alpha\Delta t}{\Delta x^2}\left(2T_{M-1,n}^{\tau} + T_{M,n+1}^{\tau} + T_{M,n-1}^{\tau}\right)$$
$$+ \left(1 - \frac{4\alpha\Delta t}{\Delta x^2}\right)T_{M,n}^{\tau} - \frac{2\alpha\Delta t}{\Delta x}\frac{q_s''}{k} \tag{3-76}$$

For convection heat transfer at the boundary,

$$q_s'' = q_c'' = h(T_{M,n}^{\tau} - T_F) \tag{3-77}$$

such that Eq. (3-76) takes the form

$$T_{M,n}^{\tau+1} = \frac{\alpha\Delta t}{\Delta x^2}\left(2\frac{h\Delta x}{k}T_F + 2T_{M-1,n}^{\tau} + T_{M,n+1}^{\tau} + T_{M,n-1}^{\tau}\right)$$
$$+ \left[1 - \frac{\alpha\Delta t}{\Delta x^2}\left(2\frac{h\Delta x}{k} + 4\right)\right]T_{M,n}^{\tau} \tag{3-78}$$

With the initial temperature distribution specified, the explicit interior and exterior nodal equations provide the means by which the temperature distribution can be obtained at the next increment of time, $1\Delta t$. The temperature distribution at $1\Delta t$ can then be used as an input to calculate the distribution at $2\Delta t$. This calculation procedure can be continued to obtain the temperature distribution over the number of time increments desired. Hence, future nodal temperatures can be calculated directly by the use of this explicit numerical finite-difference approach.

In regard to the mechanics of building up a solution for $T_{m,n}^{\tau+1}$, the increments Δt and Δx (and Δy, which has been set equal to Δx) must be selected such that the coefficients associated with the $T_{m,n}^{\tau}$ term for each nodal equation must be equal to or greater than zero. If $\alpha \Delta t / \Delta x^2$ does not satisfy this criterion, our finite-difference "solution" will blow up. The instability occurs because a mathematical condition is produced that violates the second law of thermodynamics. This point is illustrated in Example 3-18. (The mathematical basis for this instability is discussed in [9] and [10].) For unsteady three-dimensional (t, x, y) systems, this criterion takes the form

$$\frac{\alpha \Delta t}{\Delta x^2} \leqslant \frac{1}{X} \tag{3-79}$$

with X given in Table 3-2 for several boundary conditions. Notice that the minimum value of X for an unsteady three-dimensional analysis is 4.

With $\alpha \Delta t / \Delta x^2$ set equal to $\frac{1}{4}$, Eq. (3-75) reduces to the convenient form

$$T_{m,n}^{\tau+1} = \frac{1}{4} \left(T_{m+1,n}^{\tau} + T_{m-1,n}^{\tau} + T_{m,n+1}^{\tau} + T_{m,n-1}^{\tau} \right) \tag{3-80}$$

Utilizing this value of $\alpha \Delta t / \Delta x^2$, we see that $T_{m,n}^{\tau+1}$ is equal to the arithmetic average of the four surrounding nodal temperatures at $t = \tau \Delta t$.

Similar limits must be placed on the selection of Δx and Δt for unsteady two-dimensional (t, x) and four-dimensional (t, x, y, z) systems. For example, the following criterion must be satisfied for two-dimensional systems:

$$\frac{\alpha \Delta t}{\Delta x^2} \leqslant \frac{1}{X} \tag{3-81}$$

where X is given in Table 3-2. For this case, the minimum value of X is 2.

TABLE 3-2 Stability parameter for unsteady explicit formulation

	X	
Boundary condition	*Three-dimensional: (t, x, y)*	*Two-dimensional: (t, x)*
Isothermal	4	2
Convection	$4 + 2h \Delta x / k$	$2 + 2h \Delta x / k$
Blackbody thermal radiation	$4 + 2\sigma F_{s-R} \Delta x \, T_{m,n}^3 / k$	$2 + 2\sigma F_{s-R} \Delta x \, T_m^3 / k$
Specified heat flux	4	2

With $\alpha \Delta t / \Delta x^2$ set equal to $\frac{1}{2}$, the finite-difference formulation for interior nodes takes the form

$$T_m^{\tau+1} = \frac{1}{2} (T_{m+1}^{\tau} + T_{m-1}^{\tau}) \tag{3-82}$$

The fact that $T_m^{\tau+1}$ is equal to the average of the two adjacent nodal temperatures at the preceding increment of time provides the basis for the graphical approach to solving unsteady two-dimensional (t, x) conduction-heat-transfer problems which is presented in Sec. 3-8-2.

As a consequence of these stability criteria, the computational approach developed above generally requires the use of quite small increments in time. However, the fact that this explicit finite-difference procedure produces straightforward, direct calculations of future nodal temperatures sometimes compensates for the restrictions in Δt.

EXAMPLE 3-18

Demonstrate that Eq. (3-75) violates the second law of thermodynamics for values of $\alpha \Delta t / \Delta x^2 > \frac{1}{4}$.

Solution

To demonstrate this point, we consider the case in which the sides of a square plate $(L \times L \times \delta)$ initially at $110\,°F$ are suddenly brought to $100\,°F$ at time t. To simplify our test case, we utilize a grid with $\Delta x = L/2$, as shown in Fig. E3-18. At the instant $t = \tau \Delta t$, the nodal temperatures surrounding node (m, n) are given by

$$T_{m-1,n}^{\tau} = T_{m+1,n}^{\tau} = T_{m,n-1}^{\tau} = T_{m,n+1}^{\tau} = 100\,°F$$

but

$$T_{m,n}^{\tau} = 110\,°F$$

We now want to know the temperature at node (m, n) at the next

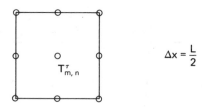

FIGURE E3-18   Finite-difference grid for square plate.

increment of time. Therefore, we utilize Eq. (3-75),

$$T_{m,n}^{\tau+1} = \frac{\alpha \Delta t}{\Delta x^2} 400 \, °\text{F} + \left(1 - 4\frac{\alpha \Delta t}{\Delta x^2}\right) 110 \, °\text{F} \qquad \text{(a)}$$

We select a value of Δt such that $\alpha \Delta t/\Delta x^2$ is greater than $\frac{1}{4}$, say

$$\frac{\alpha \Delta t}{\Delta x^2} = \frac{1}{3}$$

Substituting this value into Eq. (a), we obtain

$$T_{m,n}^{\tau+1} = \frac{400 \, °\text{F}}{3} + \left(1 - \frac{4}{3}\right) 110 \, °\text{F} = 96.7 \, °\text{F}$$

Because $T_{m,n}^{\tau+1}$ is less than $T_{m,n}^{\tau}$, we conclude that heat has been conducted out of node (m,n) to the surrounding nodes. However, the fact that $T_{m,n}^{\tau+1}$ is actually less than $100 \, °\text{F}$ indicates that heat has been transferred in directions of increasing temperature during part of the Δt time increment. This result is a violation of the second law of thermodynamics, which requires that heat cannot be transferred from a low-temperature system to a high-temperature system without the input of work.

By examining Eq. (a), it can be seen that the use of any time increment for which $\alpha \Delta t/\Delta x^2$ is greater than $\frac{1}{4}$ will bring about a violation of the second law. If, however, Δt is selected such that $\alpha \Delta t/\Delta x^2 \leqslant \frac{1}{4}$, $T_{m,n}^{\tau+1}$ does not fall below the surrounding nodal temperatures and no violation occurs.

EXAMPLE 3-19

A plate ($k = 50 \, \text{W}/(\text{m} \, °\text{C})$, $\alpha = 2 \times 10^{-5} \, \text{m}^2/\text{s}$) of 4 mm thickness is initially at $0 \, °\text{C}$. One side of the plate is then suddenly brought to a temperature of $100 \, °\text{C}$, with the other side maintained at $0 \, °\text{C}$. Develop a reasonably accurate explicit finite-difference solution for the temperature distribution and rate of heat transfer.

Solution

To get a feeling for the explicit solution technique, we first utilize the finite-difference grid with $\Delta x = L/4 = 1$ mm shown in Fig. E3-19a. Notice that this grid produces only three unknown nodal temperatures (T_2^{τ}, T_3^{τ}, and T_4^{τ}). An explicit finite-difference energy balance is developed

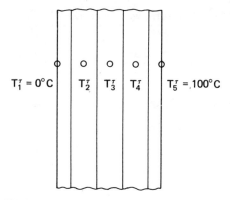

FIGURE E3-19a   Finite-difference grid for plane wall.

for one of these three interior nodes (m) as follows:

$$q_x = q_{x+\Delta x} + \frac{\Delta E_s}{\Delta t}$$

$$-kA\frac{T_m^\tau - T_{m-1}^\tau}{\Delta x} = -kA\frac{T_{m+1}^\tau - T_m^\tau}{\Delta x} + \rho A\,\Delta x\,c_v\frac{T_m^{\tau+1} - T_m^\tau}{\Delta t} \qquad \text{(a)}$$

The solution for $T_m^{\tau+1}$ is

$$T_m^{\tau+1} = \frac{\alpha\,\Delta t}{\Delta x^2}(T_{m-1}^\tau + T_{m+1}^\tau) + \left(1 - 2\frac{\alpha\,\Delta t}{\Delta x^2}\right)T_m^\tau \qquad \text{(b)}$$

To maintain stability, the coefficients associated with the T_m^τ term for each nodal equation must be equal to or greater than zero. Therefore, we require

$$1 - 2\alpha\frac{\Delta t}{\Delta x^2} \geqslant 0 \qquad \text{or} \qquad \frac{\alpha\,\Delta t}{\Delta x^2} \leqslant \frac{1}{2} \qquad \text{(c)}$$

Setting $\alpha\,\Delta t/\Delta x^2 = \frac{1}{2}$, Eq. (b) becomes

$$T_m^{\tau+1} = \frac{1}{2}(T_{m-1}^\tau + T_{m+1}^\tau) \qquad \text{(d)}$$

and Δt is given by

$$\Delta t = \frac{1}{2}\frac{\Delta x^2}{\alpha} = \frac{1}{2}\frac{(10^{-3}\text{ m})^2}{2\times10^{-5}\text{ m}^2/\text{s}} = 0.025\text{ s}$$

To summarize, our three nodal equations are

$$T_2^{\tau+1} = \frac{1}{2}(T_1^\tau + T_3^\tau) = \frac{1}{2}(0\,°C + T_3^\tau) \qquad\qquad \text{(e)}$$

$$T_3^{\tau+1} = \frac{1}{2}(T_2^\tau + T_4^\tau) \qquad\qquad \text{(f)}$$

$$T_4^{\tau+1} = \frac{1}{2}(T_3^\tau + T_5^\tau) = \frac{1}{2}(T_3^\tau + 100\,°C) \qquad\qquad \text{(g)}$$

The solution to these equations after the first increment of time ($t = \Delta t$) is

$$T_2^1 = \frac{1}{2}(0\,°C + 0\,°C) = 0\,°C$$

$$T_3^1 = \frac{1}{2}(0\,°C + 0\,°C) = 0\,°C$$

$$T_4^1 = \frac{1}{2}(0\,°C + 100\,°C) = 50\,°C$$

These results are then substituted back into Eqs. (e) through (g) to obtain predictions for T_2^2, T_3^2, and T_4^2. This procedure is continued as the solution is built up for increasing time steps. The predictions for T_m^τ are summarized in Table E3-19 for the number of time steps required to reach steady conditions. Note that the final steady-state profile is linear, which is consistent with the simple one-dimensional analysis developed in Chap. 2.

Observing that no change occurs in the nodal temperatures (to three significant figures) for $\tau \geqslant 19$, the length of time required to reach steady-state conditions is estimated as follows:

$$\begin{aligned} t_{ss} &= \tau_{ss}\Delta t = 19(0.025\ \text{s}) \\ &= 0.475\ \text{s} \end{aligned} \qquad\qquad \text{(h)}$$

It should be mentioned that the calculations are generally stopped when the percent change in all nodal temperatures falls below a specified steady-state convergence criterion, SSC. For instance, with SSC set equal to 0.005, the calculations would be terminated at $\tau = 15$.

To determine the rate of heat transfer from the surface $x = L$ at time t, we apply the first law of thermodynamics to node (5) as follows:

$$-kA\frac{T_5^\tau - T_4^\tau}{\Delta x} = q_L + \frac{\Delta E_s}{\Delta t}$$

TABLE E3-19   Numerical calculations for T_m^τ

τ	T_2^τ	T_3^τ	T_4^τ
0	0.	0.	50.
1	0.	25.	50.
2	12.5	25.	62.5
3	12.5	37.5	62.5
4	18.8	37.5	68.8
5	18.8	43.8	68.8
6	21.9	43.8	71.9
7	21.9	46.9	71.9
8	23.4	46.9	73.4
9	23.4	48.4	73.4
10	24.2	48.4	74.2
11	24.2	49.2	74.2
12	24.6	49.2	74.6
13	24.6	49.6	74.6
14	24.8	49.6	74.8
15	24.8	49.8	74.8
16	24.9	49.8	74.9
17	24.9	49.9	74.9
18	25.0	49.9	75.0
19	25.0	50.0	75.0
20	25.0	50.0	75.0

or

$$q_L = -\frac{kA}{\Delta x}(T_5^\tau - T_4^\tau) - \rho A\,\Delta x c_v\left(\frac{T_5^{\tau+1} - T_5^\tau}{\Delta t}\right) \tag{i}$$

where $T_5^{\tau+1} = T_5^\tau = 100\,°\text{C}$. Thus, to predict heat-transfer flux q_L'' at any instant t, the calculations for T_m^τ are substituted into Eq. (i). For example, at $t = 0.275$ s (i.e., $\tau = 11$), we obtain

$$q_L'' = \frac{k}{\Delta x}(74.2\,°\text{C} - 100\,°\text{C}) = -1.29\frac{\text{MW}}{\text{m}^2} \tag{j}$$

As we have seen in Examples 3-15 through 3-17, one way to assure a reasonably accurate numerical finite-difference solution is to compare the solutions for smaller and smaller subvolume sizes. Because of the obvious computational involvement in producing solutions for smaller values of Δx and Δt, a simple FORTRAN program is developed for calculating the nodal temperatures and heat-transfer rate at the surface. The general nodal equation given by Eq. (b) is written in FORTRAN

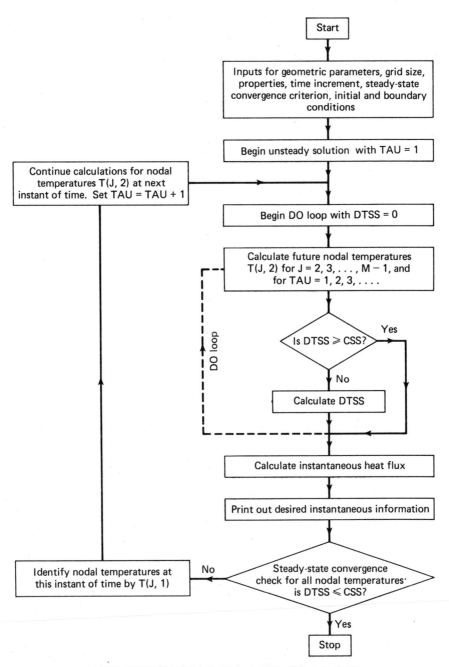

FIGURE E3-19b Program flowchart for Example 3-19.

```
              C        EXAMPLE 3-19
0001                   REAL K
0002                   REAL L
0003                   DIMENSION T(100,2)
              C
              C        INPUTS FOR GEOMETRIC PARAMETERS.
0004                   L=0.014
              C
              C        INPUTS FOR GRID SIZE.
0005                   M=10
0006                   M1=M-1
0007                   DX=L/M1
              C
              C        INPUTS FOR PROPERTIES.
0008                   K=50.
0009                   ALPHA=2.E-5
              C        INPUTS FOR TIME INCREMENTS.
0010                   S=0.4
0011                   DTIME=S*(DX)**2./ALPHA
              C
              C        INPUT FOR STEADY STATE CONVERGENCE CRITERION.
0012                   CSS=1.E-4
              C
              C        INPUTS FOR INITIAL CONDITIONS.
0013                   TI=1.E-5
0014                   DO 1 J=1,M
              C
              C        INPUTS FOR BOUNDARY CONDITIONS.
0015                 1 T(J,1)=TI
0016                   T1=1.E-5
0017                   T(1,1)=T1
0018                   T(1,2)=T(1,1)
0019                   TM=100.
0020                   T(M,1)=TM
0021                   T(M,2)=T(M,1)
              C
              C        BEGIN UNSTEADY SOLUTION WITH TAU = 1.
0022                   TAU=1
              C
              C        BEGIN DO LOOP WITH DTSS = 0.
0023                 2 DTSS=0.
              C
              C        CALCULATE FUTURE NODAL TEMPERATURES T(J,2).
0024                   DO 3 J=2,M1
0025                   T(J,2)=(T(J+1,1)+T(J-1,1))*S+(1.-2.*S)*T(J,1)
              C
              C        CHECK FOR CONVERGENCE TO STEADY STATE OF ALL NODAL TEMPERATURE
0026                   IF (DTSS .GE. CSS) GO TO 3
              C
              C        CALCULATE DTSS.
0027                   DTSS=ABS((T(J,2)-T(J,1))/T(J,2))
0028                 3 CONTINUE
              C
              C        CALCULATE INSTANTANEOUS HEAT FLUX.
0029                   QFL=-K/DX*(T(M,2)-T(M1,2))
              C
              C        PRINT OUT DESIRED INSTANTANEOUS INFORMATION.
0030                   TIME=TAU*DTIME
0031                   PRINT 90,M,TAU,TIME
0032                   PRINT 92,QFL
0033                90 FORMAT(5X,I5,5X,3(F12.5,5X))
0034                92 FORMAT (70X,'HEAT TRANSFER FLUX (QFL) = ',E12.5)
              C
              C        STEADY STATE CONVERGENCE CHECK FOR ALL NODAL TEMPERATURES.
0035                   IF (DTSS .LE. CSS) GO TO 50
              C
              C        IDENTIFY NODAL TEMPERATURES AT THIS INSTANT OF TIME BY T(J,1).
0036                   DO 4 J=2,M1
0037                 4 T(J,1)=T(J,2)
              C
              C        CONTINUE CALCULATION FOR NODAL TEMPERATURES T(J,2) AT NEXT
              C        INSTANT OF TIME SET TAU=TAU+1.
0038                   TAU=TAU+1
0039                   GO TO 2
0040                50 STOP
0041                   END
```

FIGURE E3-19c   FORTRAN program for Example 3-19.

computer language as

$$T(J,2) = (T(J+1,1) + T(J-1,1)*S + (1.-2.*S)*T(J,1)) \qquad (k)$$

where $J = m = 2,3,4,\ldots, M-1$, and $S = \alpha \Delta t / \Delta x^2$. The second index designates the time stations $\tau + 1$ (index 2) and τ (index 1). The flowchart and FORTRAN program are given in Figs. E3-19b and E3-19c. In addition to calculating the instantaneous nodal temperatures for any specified grid space Δx and time increment Δt, this program calculates the heat flux at $x = L$. The instantaneous heat flux q''_L is obtained from Eq. (i),

$$q''_L = -\frac{k}{\Delta x}(T_5^\tau - T_4^\tau)$$

or

$$QFL = -K/\dot{D}X*(T(M,2) - T(M-1,2)) \qquad (1)$$

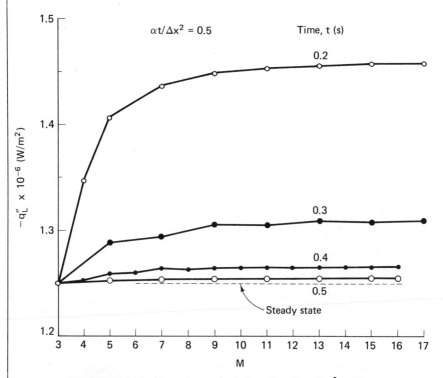

FIGURE E3-19d Numerical calculations for q''_L; $\alpha t / \Delta x^2 = 0.5$.

Calculations obtained for q_L'' by running this program are shown in Fig. E3-19d (to four significant figures) as a function of M and t with $\alpha \Delta t/\Delta x^2 = 0.5$. Note that as M increases, both Δx and Δt become smaller. Based on this result, we conclude that a reasonably accurate solution is obtained for $M \gtrsim 10$, but considerable error occurs for M much less than 10.

To see the effect of $\alpha \Delta t/\Delta x^2$ on our numerical solution, predictions are shown in Fig. E3-19e for the heat flux q_L'' at surface $x = L$ vs. τ for $\alpha \Delta t/\Delta x^2$ equal to 0.4, 0.5, and 0.6, with $M = 10$. For $\alpha \Delta t/\Delta x^2 \leqslant 0.5$, the solution is stable and converges to the proper steady-state value. However, for $\alpha \Delta t/\Delta x^2 = 0.6$, the numerical solution is seen to be unstable and divergent. This result underscores the importance of establishing the appropriate stability criteria for explicit finite-difference formulations of unsteady problems.

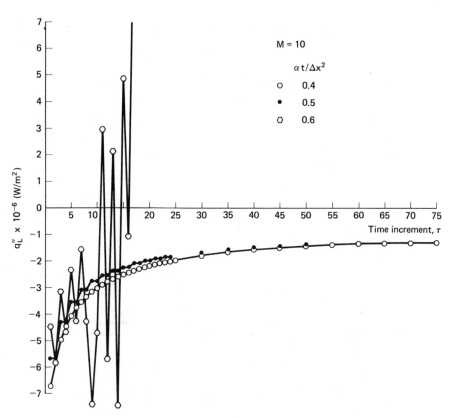

FIGURE E3-19e   Numerical calculations for q_L''; several values of $\alpha t/\Delta x^2$.

This computer program can be modified in order to print out the temperature distribution at any value of τ. This task is suggested as an assignment for the student (see Prob. 3-60).

Implicit method

For problems in which the computational time becomes a factor, finite difference formulations can be developed which involve the use of backward differences in the time increments. Referring to Eq. (3-47), we are reminded that this type of formulation does not produce explicit calculations for $T_{m,n}^{\tau+1}$ in terms of $T_{m,n}^{\tau}$. Instead, like the situation encountered in our solution of steady-state problems, the entire system of Z equations must be solved simultaneously to develop predictions for $T_{m,n}^{\tau+1}$. However, this method has the advantage of being stable for all values of Δt. Consequently, larger increments in t can be utilized in this approach, such that less computer time is sometimes required than is necessary in the explicit approach.

EXAMPLE 3-20

Develop an implicit finite-difference solution for the unsteady problem of Example 3-19.

Solution

The finite-difference grid for an implicit formulation is identical to the grid for an explicit formulation. Applying the first law of thermodynamics to interior node (m), we write

$$q_x = q_{x+\Delta x} + \frac{\Delta E_s}{\Delta t}$$

$$-kA\frac{T_m^{\tau} - T_{m-1}^{\tau}}{\Delta x} = -kA\frac{T_{m+1}^{\tau} - T_m^{\tau}}{\Delta x} + \rho\,\Delta V\,c_v\frac{T_m^{\tau} - T_m^{\tau-1}}{\Delta t}$$

where $\Delta E_s/\Delta t$ has been approximated by the backward difference, rather than the forward difference used in Example 3-19. Rearranging this equation, we have

$$T_{m-1}^{\tau} + T_{m+1}^{\tau} - \left(2 + \frac{\Delta x^2}{\alpha\,\Delta t}\right)T_m^{\tau} + \frac{\Delta x^2}{\alpha\,\Delta t}\,T_m^{\tau-1} = 0$$

or

$$T_{m-1}^{\tau+1} + T_{m+1}^{\tau+1} - \left(2 + \frac{\Delta x^2}{\alpha \Delta t}\right) T_m^{\tau+1} + \frac{\Delta x^2}{\alpha \Delta t} T_m^{\tau} = 0 \qquad \text{(a)}$$

For this problem, in which both surface temperatures are specified, m takes values of $2, 3, \ldots, M-1$, such that a total of $M-2$ unknown nodal temperatures at each time station $\tau = 1, 2, 3, \ldots$. However, unlike the explicit formulation of Example 3-19, these equations must be solved simultaneously for each value of τ. Therefore, the direct and indirect methods introduced in Sec. 3-6-2 can.be utilized.

Because of its simplicity, the Gauss–Seidel iterative method is utilized. Our general nodal equation is therefore put into the form

$$T_m^{\tau+1} = \frac{1}{2 + \dfrac{\Delta x^2}{\alpha \Delta t}} \left(T_{m-1}^{\tau+1} + T_{m+1}^{\tau+1} + \frac{\Delta x^2}{\alpha \Delta t} T_m^{\tau} \right) \qquad \text{(b)}$$

where $m = 2, 3, \ldots, M-1$, and $\tau = 1, 2, 3 \ldots$. A finite-difference FORTRAN program is given in Fig. A-E-3, which solves this equation iteratively for $T_2^{\tau+1}, T_3^{\tau+1}, T_4^{\tau+1}, \ldots, T_{M-1}^{\tau+1}$ at each increment of time $\tau = 1, 2, 3, \ldots$. This program is patterned after the FORTRAN program presented in Example 3-17.

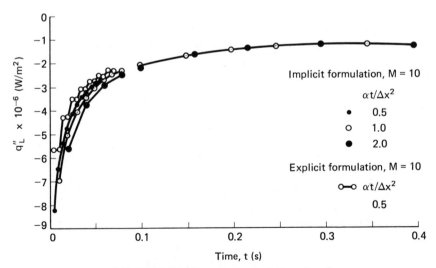

FIGURE E3-20   Numerical calculations for q_L''.

Calculations produced by this program for the rate of heat transfer per unit area from the surface at $x = L$ with $\alpha \Delta t / \Delta x^2$ equal to 0.5, 1, and 2, and $M = 10$ are compared in Fig. E3-20 with the explicit calculations of Example 3-19. Although much larger increments in Δt can be used in the implicit approach, a price is paid in accuracy of the solution for small values of t.

3-7 ANALOGY APPROACH

3-7-1 Introduction

Analogies have been found to exist between heat transfer and (1) the flow of electric current, (2) fluid flow, and (3) membrane behavior, which provide useful tools for developing predictions for the temperature distribution and rate of heat transfer in multidimensional systems. Because of its prominence in heat-transfer work, the electrical analogy will be presented in this section. (An introduction to the fluid flow and membrane analogies is given by Schneider [11].) Based on this analogy, the numerical finite-difference formulation concept will be put into the form of an R/C network format, and an experimental electrical analog method will be introduced.

3-7-2 Electrical Analogy

As shown in Chaps. 1 and 2, electrical circuits can be designed in which the flow of current simulates the rate of heat transfer in one-dimensional systems. The electrical/heat-transfer analogy developed to this point is summarized in Table 3-3. However, the analogy between current flow and heat transfer is more far reaching in that it extends to multidimensional electric fields and thermal systems. The differential and numerical finite-difference formulations provide the basis for this broader electrical analogy.

TABLE 3-3 Electrical analogy

Electrical		Thermal	
E_e	Voltage (V)	T	Temperature (°C)
I_e	Current (A)	q	Rate of heat transfer (W)
R_e	Resistance (Ω)	R	Thermal resistance (°C/W)
C_e	Capacitance (F)	$C\ (\equiv \rho V c_v)$	Thermal capacitance (J/°C)

To see this point, we first compare the Fourier equation (in Cartesian coordinates), Eq. (3-11),

$$\frac{\partial^2 T}{\partial x^2} + \frac{\partial^2 T}{\partial y^2} + \frac{\partial^2 T}{\partial z^2} = \frac{1}{\alpha}\frac{\partial T}{\partial t} \tag{3-83}$$

with the following differential equation that governs the distribution of voltage in an electrically conducting multidimensional system:

$$\frac{\partial^2 E_e}{\partial x^2} + \frac{\partial^2 E_e}{\partial y^2} + \frac{\partial^2 E_e}{\partial z^2} = R_e C_e \frac{\partial E_e}{\partial t} \tag{3-84}$$

Based on the similarity between these two equations, we conclude that an electrical field can be set up within an electrical conductor which corresponds to the thermal field in the modeled heat-transfer problem, with equipotential lines and orthogonal paths of electric current flow in the voltage field being representative of isotherms and paths of heat flow, respectively.

R/C network

We now want to transform the standard finite-difference representation of a heat-transfer problem into a network of equivalent resistances and capacitances. To do this, we develop an energy balance on the interior node shown in Fig. 3-7(a) [see Fig. 3-4(b) on page 134]; that is

$$\dot{q}\,\Delta V + \Delta q_x + \Delta q_y - \Delta q_{x+\Delta x} - \Delta q_{y+\Delta y} = \frac{\Delta E_s}{\Delta t} \tag{3-85}$$

Utilizing the Fourier law of conduction, the heat-transfer rates can be expressed in terms of thermal resistances.

$$\Delta q_x = -k\delta\,\Delta y\,\frac{T_{m,n}^\tau - T_{m-1,n}^\tau}{\Delta x} = \frac{T_{m-1,n}^\tau - T_{m,n}^\tau}{1/(\delta k)}$$

$$\Delta q_{x+\Delta x} = \frac{T_{m,n}^\tau - T_{m+1,n}^\tau}{1/(\delta k)} \tag{3-86}$$

$$\Delta q_y = \frac{T_{m,n-1}^\tau - T_{m,n}^\tau}{1/(\delta k)}$$

and

$$\Delta q_{y+\Delta y} = \frac{T_{m,n}^\tau - T_{m,n+1}^\tau}{1/(\delta k)}$$

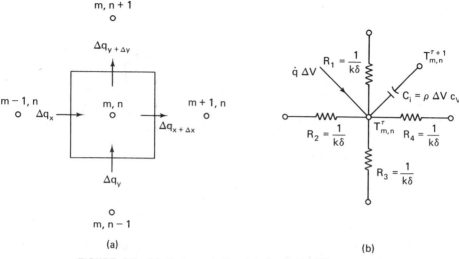

FIGURE 3-7 (a) Representative interior finite-difference subvolume. (b) Equivalent thermal network for interior subvolume.

Utilizing this result and a forward-difference approximation for $\partial T/\partial t$, Eq. (3-85) can be rewritten in the explicit form

$$\dot{q}\,\Delta V + \sum \frac{T_i^\tau - T_{m,n}^\tau}{R_i} = C_i \frac{T_{m,n}^{\tau+1} - T_{m,n}^\tau}{\Delta t} \qquad (3\text{-}87)$$

where $R_i = R_k = 1/(\delta k)$, $C_i = \rho\,\Delta V\,c_v$, and T_i^τ represents the temperatures of the nodes that surround the (m,n) node. Equation (3-87) is represented in terms of an analogous electrical circuit in Fig. 3-7(b).

Similarly, the energy balance for the exterior node shown in Fig. 3-8(a) can be written as

$$\dot{q}\,\Delta V + \Delta q_x + \Delta q_y - \Delta q_s - \Delta q_{y+\Delta y} = \frac{\Delta E_s}{\Delta t} \qquad (3\text{-}88)$$

where

$$\Delta q_x = \frac{T_{m-1,n}^\tau - T_{m,n}^\tau}{1/(\delta k)}$$

$$\Delta q_y = \frac{T_{m,n-1}^\tau - T_{m,n}^\tau}{1/(2\delta k)} \qquad (3\text{-}89)$$

$$\Delta q_{y+\Delta y} = \frac{T_{m,n}^\tau - T_{m,n+1}^\tau}{1/(2\delta k)}$$

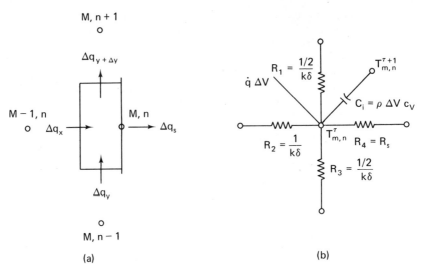

FIGURE 3-8 (a) Representative exterior finite-difference sub-volume. (b) Equivalent thermal network for exterior subvolume.

The rate of heat transfer Δq_s from the surface by conduction, convection, or thermal radiation, can also be written in terms of a thermal resistance, such that Eq. (3-88) can be put into the form of Eq. (3-87). The thermal circuit for this exterior node is shown in Fig. 3-8(b). Notice that R_i is dependent upon the geometry of the node and the boundary condition.

In summary, we find that Eq. (3-87) applies to both interior and exterior nodes and to both steady and unsteady multidimensional heat-transfer problems. In effect, this equation states that the summation of the currents flowing into the (m, n) node from the surroundings is equal to the flow of current into the capacitor. For steady-state conditions $T_{m,n}^{\tau+1} = T_{m,n}^{\tau}$, and no current flows to the capacitor. For this case, the summation of current flow into the (m, n) node is zero.

It should be mentioned that Eq. (3-87) must satisfy stability requirements which restrict Δt. For example, for $\dot q = 0$, Δt must be less than or equal to $C_i / \Sigma(1/R_i)$. If the restrictions imposed on Δt become too severe, an implicit formulation can be developed by merely utilizing a backward difference in $\partial T / \partial t$; that is,

$$\dot q\, \Delta V + \Sigma\, \frac{T_i^\tau - T_{m,n}^\tau}{R_i} = C_i \frac{T_{m,n}^\tau - T_{m,n}^{\tau-1}}{\Delta t} \qquad (3\text{-}90)$$

Equations (3-87) and (3-90) provide the basis for building up electrical networks which are analogous to steady and unsteady multidimensional conduction-heat-transfer problems. To illustrate, a finite-difference

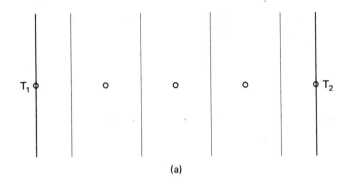

(a)

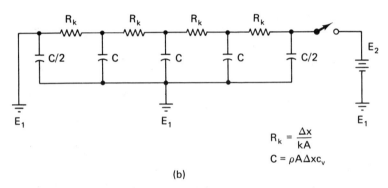

$$R_k = \frac{\Delta x}{kA}$$

$$C = \rho A \Delta x c_v$$

(b)

FIGURE 3-9 Unsteady conduction heat transfer in a plane wall. (a) Finite-difference grid. (b) R/C network.

grid and an electrical network are shown in Fig. 3-9 for unsteady conduction in a plane wall.

These electrical networks lend themselves to the development of numerical solutions. Comprehensive computer programs have been developed which determine the solution of R/C network representations of thermal (and other) systems. One such program, the Systems Improved Numerical Differencing Analyzer (SINDA) program, has been developed by Chrysler, TRW, and NASA. Typical of such programs, a manual is available [12] for the input options and operational features of SINDA, as well as a full explanation of the mechanics involved in these capabilities.

Analog field plotter

The analogy between electric and thermal fields provides the basis for an experimental electrical analogy method. A convenient experimental arrangement by which the electric field within steady multidimensional

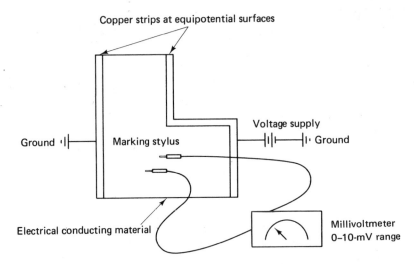

FIGURE 3-10   Representative analog field plotter arrangement.

systems can be measured is shown in Fig. 3-10 in the context of a rectangular plate. For many two-dimensional systems, electrical conducting paper, thin strips of certain metals such as Inconel, or a shallow saline bath can be patterned after the conduction system of interest. Such experimental arrangements are known as *analog field plotters*. In regard to boundary conditions, a uniform surface temperature is modeled by maintaining a uniform voltage at the surface, and insulated surfaces correspond to surfaces which are not connected to voltage sources. Lines of constant voltage which correspond to isotherms are then found by utilizing a millivoltmeter. Using the resulting measured system of equipotential lines, orthogonal current flow lines can be sketched in. (The flow lines can sometimes be determined by reversing the electrical boundary conditions.) The resulting network of equipotential lines and current flow lines represent the isotherms and heat-flow paths in the analogous heat-transfer system. The usefulness of this information will be seen in the next section when we consider the graphical solution technique.

EXAMPLE 3-21

In Chap. 2 it was concluded that one-dimensional heat transfer occurs in composite walls consisting of only two materials in parallel. It was also concluded that the heat transfer in composite walls consisting of combined series–parallel resistances is not one-dimensional. Demonstrate the validity of these conclusions by the use of an analog field plotter.

Solution

Electrical circuits which are analogous to conduction heat transfer in (a) a simple parallel composite and (b) a series–parallel composite were constructed of strips of 0.025 in. Inconel (No. 600 cold rolled), as shown in Figs. E3-21a and b. Voltages were then applied to the ends of these two composite strips and the equipotential lines shown in the two sketches were measured. These lines are clearly one-dimensional in Fig. E3-21a and two-dimensional in Fig. E3-21b. Therefore, we conclude that the heat transfer in simple parallel composites consisting of two thermal resistances is one-dimensional, but that the heat transfer in combined series–parallel composites is two-dimensional.

It follows that the heat transfer in the parallel composite can be represented by the one-dimensional thermal circuit shown in Fig. E3-21c. Representing the electrical resistivity of Inconel by ρ_e, the electrical

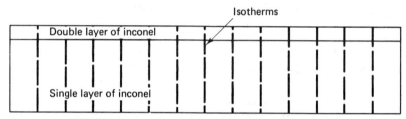

FIGURE E3-21a Composite wall; parallel arrangement.

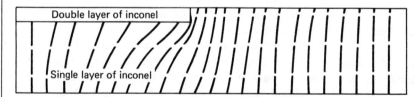

FIGURE E3-21b Composite wall; series–parallel arrangement.

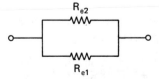

FIGURE E3-21c Thermal circuit for parallel arrangement.

resistance of the single layers and double layers are given by ($\delta = 0.025$ in.)

$$R_{e1} = \frac{\rho_e L}{\delta(0.027 \text{ m})} \qquad R_{e2} = \frac{\rho_e L}{2\delta(0.005 \text{ m})} = 2.7 R_{e1}$$

Because of the two dimensionality of the system shown in Fig. E3-21b, the heat transfer in composites of this type can only be approximated by a one-dimensional thermal circuit when the criterion introduced in Sec. 2-2-4 of Chap. 2 is satisfied.

3-8 GRAPHICAL APPROACH

3-8-1 Introduction

Graphical solution techniques have been developed for both steady and unsteady multidimensional conduction-heat-transfer problems. The use of these approaches in analyzing two-dimensional conduction heat transfer will be seen to provide further insight into these fairly complex processes.

3-8-2 Steady Two-Dimensional Systems

The key to the graphical approach to solving steady two-dimensional problems is the fact that isotherms and heat-flow lines are orthogonal (perpendicular). This point is reflected in the Fourier law of conduction itself and was touched upon in our brief study of the electrical analogy. To illustrate this point, isotherms and heat-flow lines are sketched in Fig. 3-11 for a fairly simple two-dimensional problem, such that a network of curvilinear squares is constructed.

Because heat is transferred along the M paths formed by adjacent heat-flow lines, the heat transfer in a single heat-flow lane is essentially one-dimensional (with respect to the curved coordinates, which follow the path ξ taken by individual heat-flow lanes) and is given by the following form of the Fourier law:

$$q_\xi = -kA_\xi \frac{dT}{d\xi} \tag{3-91}$$

The rate of heat transfer across each curvilinear square within the mth heat-flow path can be approximated by

$$q_m = -k\delta \frac{w}{\Delta\xi} \Delta T = -\frac{\Delta T}{R_n} \tag{3-92}$$

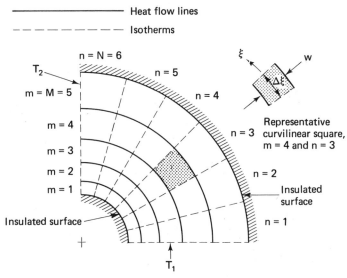

FIGURE 3-11 Cylindrical section with $r_1 = 2$ cm and $r_2 = 7.4$ cm.

where ΔT is the mean temperature drop, $\Delta \xi$ the mean length, w the mean width, and δ the mean thickness associated with the nth curvilinear square. Because $\Delta \xi = w$ for full curvilinear squares, the quantity R_n is equal to $1/(k\delta)$.

Assuming uniform thermal conductivity k and plate thickness δ, the rate of heat transfer in the mth heat-flow path can also be expressed in terms of the total temperature drop $T_1 - T_2$ by

$$q_m = \frac{T_1 - T_2}{\displaystyle\sum_{n=1}^{N_m} R_n} = \frac{k\delta}{N_m}(T_1 - T_2) \qquad (3\text{-}93)$$

where N_m is the number of curvilinear squares in the mth heat-flow lane. The total rate of heat transfer in the system can now be obtained by summing the rates for each heat-flow path; this is shown in the following equation.

$$q = k\delta(T_1 - T_2) \sum_{m=1}^{M} \frac{1}{N_m} \qquad (3\text{-}94)$$

for uniform k and δ. For simple problems in which the same number of full curvilinear squares N occur in each heat-flow path, Eq. (3-94) reduces to

$$q = k\delta \frac{M}{N}(T_1 - T_2) \tag{3-95}$$

where N and M take on integer values.

Suggestions for developing freehand plots of curvilinear networks have been developed by Bewley [13] and are summarized by Kreith [14]. Although the development of such sketches is an art, one can be guided by experimental electrical analogy measurements.

EXAMPLE 3-22

(a) Obtain an approximate solution for the rate of heat transfer in the hollow quarter-cylinder shown in Fig. 3-11 [$\delta = 1$ m, $r_1 = 2$ cm, $r_2 = 7.4$ cm, $k = 125$ W/(m °C), $T_1 = 150$°C, and $T_2 = 35$°C].
(b) Solve for the heat transfer in this problem for the case in which the surface at $\theta = 0$ rad is exposed to a fluid with $T_F = 35$°C and $\bar{h} = 100$ W/(m² °C).

Solution

Referring to Fig. 3-11, in which a network of curvilinear squares has already been developed, we see that $N = 6$ and $M = 5$.
(a) The rate of heat transfer for the case in which the surface temperatures are specified is given by Eq. (3-95); that is,

$$q = k\delta \frac{M}{N}(T_1 - T_2) = \left(125 \frac{W}{m\ °C}\right)(1\ m)\left(\frac{5}{6}\right)(150\ °C - 35\ °C)$$

$$= 12{,}000\ W$$

Note that the equivalent thermal resistance for this problem is

$$R_k = \frac{N}{Mk\delta} = \frac{1}{104} \frac{°C}{W}$$

(b) For the system with convection at one surface, we sketch the analogous electrical circuit in Fig. E3-22. The rate of heat transfer through this circuit is

$$q = \frac{150\ °C - 35\ °C}{\dfrac{1}{104} \dfrac{°C}{W} + \dfrac{m^2\ °C}{100\ W(0.054\ m)(1\ m)}} = 590\ W$$

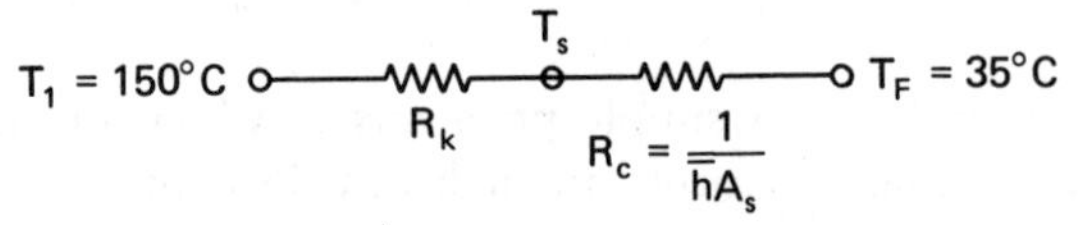

FIGURE E3-22   Thermal circuit.

The surface temperature is then expressed by

$$T_s = T_F + R_c q$$

$$= 35\,^{\circ}\text{C} + \frac{590\text{ W}}{5.4\text{ W}/^{\circ}\text{C}} = 144\,^{\circ}\text{C}$$

3-8-3   Unsteady Two-Dimensional Systems

Based on our finite-difference formulation for unsteady two-dimensional systems with $\alpha\,\Delta t/\Delta x^2$ set equal to $\frac{1}{2}$, we found that the temperature $T_m^{\tau+1}$ at any interior node is equal to the arithmetic average of the temperature of the adjacent nodes at the instant $t = \tau\,\Delta t$ [see Eq. (3-82) and/or Example 3-19]; that is,

$$T_m^{\tau+1} = \frac{1}{2}\left(T_{m+1}^{\tau} + T_{m-1}^{\tau}\right) \tag{3-96}$$

This fact provides the basis for the development of a simple graphical solution technique for unsteady two-dimensional conduction heat-transfer problems. With T_{m-1}^{τ} and T_{m+1}^{τ} known, the arithmetic average is easily constructed, as illustrated in Fig. 3-12. Notice that $T_m^{\tau+1}$ is represented by the intersection of the vertical line at x and the straight line between T_{m-1}^{τ} and T_{m+1}^{τ}.

The procedure followed in developing a graphical solution to such unsteady problems with specified surface temperatures is outlined as follows: (1) divide the body into M strips Δx in length, letting the direction x serve as the abscissa and utilizing the largest scale possible; (2) compute

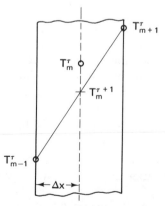

FIGURE 3-12   Construction of arithmetic average approximation for $T_m^{\tau+1}$.

the value of Δt in accordance with

$$\Delta t = \frac{\Delta x^2}{2\alpha} \tag{3-97}$$

(3) sketch the initial temperature within the body with the ordinate serving as the temperature coordinate; (4) draw straight lines between every other nodal temperature to obtain the temperature at the next increment of time; and (5) continue this process until the desired time span has been covered. These steps are reflected in the temperature–time Schmidt plot developed in Example 3-23.

The use of the graphical approach in analyzing problems involving convection boundary conditions is discussed in [15].

EXAMPLE 3-23

Develop a graphical solution for the temperature distribution in a plate $[k = 50 \text{ W}/(\text{m }°\text{C}), \alpha = 2 \times 10^{-5} \text{ m}^2/\text{s}]$ 3 mm thick with surface temperatures maintained at 750 °C and 0 °C. The plate is initially at 750 °C. Also determine the rate of heat transfer from the plate at the 0 °C surface after 0.1 s and estimate the time required to reach steady-state conditions.

Solution

As a first approximation, we utilize the 1-mm grid shown in Fig. E3-23. Our time increments are calculated as follows:

$$\Delta t = \frac{\Delta x^2}{2\alpha} = \frac{10^{-6} \text{ s}}{2(2 \times 10^{-5})} = 0.025 \text{ s}$$

Starting with a uniform temperature distribution, future nodal calculations are built up in the sketch in accordance with the steps outlined above. The time increment for each temperature is shown in Fig. E3-23. Notice that the profile is approximately linear for $\tau = 6$, such that the time required to reach steady-state conditions is given by

$$t_{ss} = 6 \, \Delta t = 6(0.025) \text{ s} = 0.150 \text{ s}$$

To obtain the rate of heat transfer at the cooler surface at $t = 0.1$ s, we measure the slope after the fourth time increment; that is,

$$\frac{\Delta T}{\Delta x} \simeq -\frac{280 °\text{C}}{1 \text{ mm}} \qquad \text{at } t = 0.1 \text{ s}$$

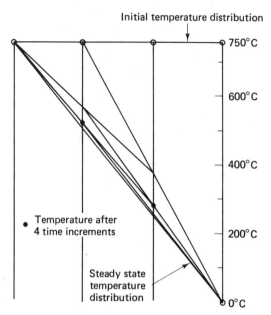

FIGURE E3-23 Schmidt plot for unsteady conduction heat transfer in plane wall.

Coupling this graphical measurement with the Fourier law of conduction, we obtain

$$q = -kA\left(-\frac{280\,°\mathrm{C}}{0.001\ \mathrm{m}}\right)$$

$$q'' = 50\,\frac{\mathrm{W}}{\mathrm{m}^2}\,\frac{280}{10^{-3}} = 140 \times 10^5\,\frac{\mathrm{W}}{\mathrm{m}^2}$$

3-9   SUMMARY: PRACTICAL SOLUTION RESULTS

In this chapter we have introduced the formulation and solution concepts that are required in the analyses of general multidimensional conduction-heat-transfer problems. These principles can be applied to a wide range of problems. As we have seen, the analytical, numerical, analogical, and graphical solution approaches all have their place in the science of heat transfer, with numerical methods providing the cornerstone for analysis of more complex problems.

Over the years these several approaches have been utilized to develop solutions to a great many multidimensional conduction-heat-transfer problems. As a result, design equations and charts have been developed for the that are commonly encountered in practice. These practical solution results are summarized in this section for standard systems. This important design information is cast in the form of (1) thermal resistance R tables for steady processes, (2) heat transfer q and Q charts for unsteady conditions, and (3) temperature charts for unsteady processes. This presentation is patterned after the practical solution results given in Chap. 2.

The thermal resistance R of several representative steady two- and three-dimensional conduction-heat-transfer systems are given in Tables 3-4 and 3-5. A comprehensive summary of conduction shape factors $S(kS = 1/R)$ is given by Hahne and Grigull [19].

Instantaneous rate and accumulative total heat transfer are given in Table 3-6 and in Fig. 3-13 for four standard unsteady two-dimensional processes with constant surface temperature T_1. In each of these systems, the initial temperature is T_i, with the surface suddenly brought to T_1 at $t = 0$.

Instantaneous rate and accumulative total heat transfer are given in Fig. 3-14 for unsteady two-dimensional heat transfer in a flat plate of thickness L which is suddenly exposed to a fluid on both sides. Similar heat-transfer charts are presented in Figs. A-F-1 and A-F-2 for unsteady convection from circular cylinders and spheres. Charts have also been developed which account for thermal radiation effects [22], [23].

Referring back to Chap. 2, we found that for approximate one-dimensional conditions associated with small values of the Biot number, the instantaneous rate q and total accumulative Q heat transfer for systems initially at T_i which are suddenly exposed to a fluid are given by [Eqs. (2-136) and (2-137)]

$$\frac{q}{\bar{h}A_s(T_i - T_F)} = \exp\left(-\frac{\bar{h}A_s t}{\rho V c_v}\right) \tag{3-98}$$

$$\frac{Q}{Q_{max}} = 1 - \exp\left(-\frac{\bar{h}A_s \tau}{\rho V c_v}\right) \tag{3-99}$$

where $Q_{max} = \rho V c_v (T_i - T_F)$. These equations are written in terms of the Biot number Bi $(= \bar{h}\ell/k)$ as

$$\frac{q}{\bar{h}A_s(T_i - T_F)} = \exp\left(-\text{Bi}\frac{\alpha t}{\ell^2}\right) \tag{3-100}$$

$$\frac{Q}{Q_{max}} = 1 - \exp\left(-\frac{1}{\text{Bi}}\frac{\bar{h}^2\alpha\tau}{k^2}\right) \tag{3-101}$$

where $\ell = V/A_s = L/2$ for this flat-plate geometry. These one-dimensional

TABLE 3-4 Thermal resistance: steady two-dimensional

System	Thermal resistance R
Hollow Circular Cylinder Section: surface at $\theta = 0$ and $\theta = \theta_1$ at T_1 and T_2; surfaces at $r = r_1$ and $r = r_2$ insulated.	$\dfrac{\theta_1/(Lk)}{\ln{(r_2/r_1)}}$
Circular Cylinder Buried Horizontally in Semi-infinite Medium: $L \gg r_1$.	$\dfrac{\cosh^{-1}{(Z/r_1)}}{2\pi L k}$
Circular Cylinder Buried Vertically in Semi-infinite Medium: $L \gg r_1$.	$\dfrac{\ln{(2L/r_1)}}{\pi L k}$
Two Circular Cylinders Buried Horizontally in Infinite Medium: $L \gg r_1, r_2$.	$\dfrac{\cosh^{-1}{\dfrac{Z - r_1 - r_2}{2 r_1 r_2}}}{2\pi L k}$
Sphere Buried in Semi-infinite Medium.	$\dfrac{1 - r_1/(2Z)}{4\pi r_1 k}$

Source: Summarized from [16]–[18].

TABLE 3-5 Thermal resistance: steady three-dimensional

System	*Thermal resistance R*
Circular Cylinder Buried Horizontally in Semi-infinite Medium: L short.	$\dfrac{\ln{(L/r_1)} - \ln{[L/(2Z)]}}{2\pi L k}$
Two Plane Walls with Edge Section: inside dimension greater than δ.	$\dfrac{1}{(\dfrac{aL}{\delta} + \dfrac{bL}{\delta} + 0.54L)k}$
Corner Section of Three Plane Walls: inside dimensions greater than δ.	$\dfrac{1}{0.15\delta k}$

Source: Summarized from [16]–[18] and Chap. 3.

TABLE 3-6 Heat transfer: Unsteady two-dimensional—surface suddenly brought to T_1

System	Heat Transfer	
	Instantaneous rate q	*Accumulative total Q*
Semi-infinite Solid Initially at T_i with Surface Suddenly Brought to T_1.	$\dfrac{kA\,(T_i - T_F)}{\sqrt{\pi \alpha t}}$	$\dfrac{2kA\,(T_i - T_F)}{\sqrt{\pi \alpha / \tau}}$
$T_1 \searrow$ $\vdash\!\!\longrightarrow \infty$	See Fig. 3–13	See Fig. 3–13
Plane Wall Initially at T_i with both Surfaces Suddenly Brought to T_1.	$\dfrac{kA}{L}\,4\displaystyle\sum_{n=1}^{\infty} \exp\left(-n^2\pi^2\alpha t/L^2\right)(T_i - T_F)$	
$T_1 \longrightarrow \quad \longleftarrow T_1$ $\vdash\!\!L\!\!\dashv$	See Fig. 3–13	See Fig. 3–13
Solid Circular Cylinder Initially at T_i with Surface Suddenly Brought to T_1. $L \gg r_1$. $T_1 \nearrow$	See Fig. 3–13	See Fig. 3–13
Solid Sphere Initially at T_i with Surface Brought to T_1.	$\dfrac{kA}{L}\,2\displaystyle\sum_{n=1}^{\infty} \exp\left(-n^2\pi^2\alpha t/r_1^2\right)(T_i - T_F)$	
$T_1 \nearrow$	See Fig. 3–13	See Fig. 3–13

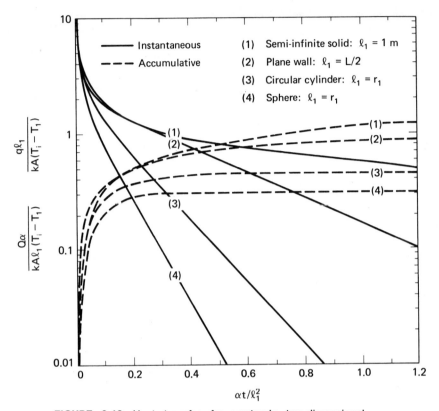

FIGURE 3-13 Heat transfer for unsteady two-dimensional processes—surfaces suddenly brought to T_J. (From *Conduction Heat Transfer* by P. J. Schneider; Fig. 10-6, p. 246. Copyright 1955 by Addison-Wesley Publishing Company, Inc. Reading, Mass. Used with permission.)

equations are compared with the exact two-dimensional solution for the flat plate in Fig. 3-14(a) and (b) for $Bi = 10$, 1, 0.1, and 0.01. (Charts for Q are given in Appendix F for cylindrical and spherical geometries.) As indicated in Chap. 2, the one-dimensional and two-dimensional analyses are in good agreement (within 5%) for Bi of the order of 0.1 and less. However, the appropriateness of these one-dimensional equations is seen to degenerate as the Biot number increases. These conclusions can also be shown to hold for the circular cylinder and spherical geometries.

The instantaneous heat-transfer rate is given in Fig. 3-15 for a semi-infinite solid initially at T_i which is suddenly exposed to a fluid. Because of its infinite extent ($Bi = \bar{h}\ell/k = \infty$), the simple one-dimensional analysis can never be used in analyzing the heat transfer in a semi-infinite solid.

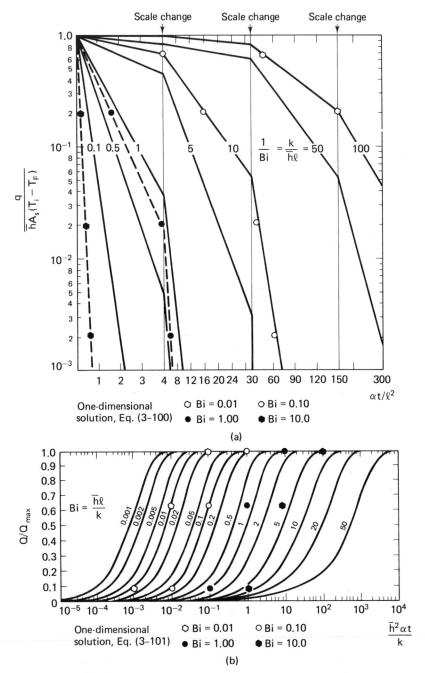

FIGURE 3-14 (a) Instantaneous heat-transfer rate for unsteady two-dimensional plane wall—both surfaces suddenly exposed to fluid. (b) Chart for accumulative heat transfer: unsteady two-dimensional plane-wall system—both surfaces suddenly exposed to fluid. (From *Fundamentals of Heat Transfer* by Grober et al. Copyright 1961 by McGraw-Hill Book Company. Used with permission of McGraw-Hill Book Company.)

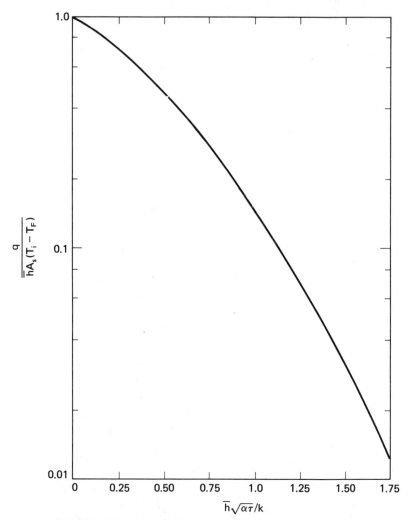

FIGURE 3-15 Unsteady heat transfer rate in semi-infinite solid with surface suddenly exposed to fluid.

Graphs known as Heisler temperature charts are shown in Fig. 3-16(a) and (b) for the unsteady temperature distribution in a flat plate initially at T_i which is suddenly exposed to a fluid. Figure 3-16(a) gives the centerline temperature T_0 as a function of time. The temperature at other positions within the wall is given in Fig. 3-16(b) in terms of the centerline temperature. (Similar charts are presented in Appendix F for circular cylinder and spherical geometries. A temperature chart is also given in Appendix F for a semi-infinite solid.)

One-dimensional solution, Eq. (3-102)

	Bi
○	0.01
●	0.10
○	1.00
●	10.0

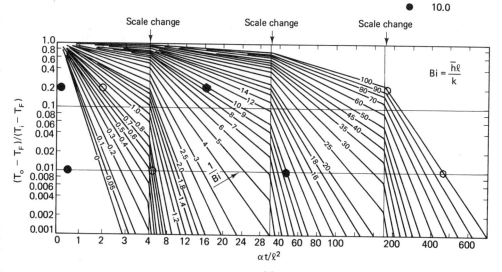

(a)

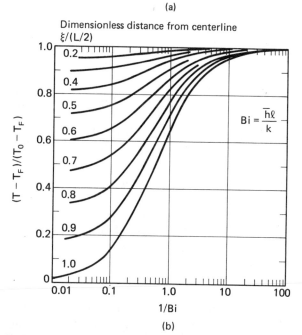

(b)

FIGURE 3-16 Heisler temperature charts for a flat plate of thickness L with surface suddenly exposed to a fluid. (a) Instantaneous midplane temperature T_0. (b) Instantaneous temperature distribution as a function of centerline temperature T_0. (From Heisler [20].)

Notice in Fig. 3-16(b) that for values of the Biot number Bi of the order of 0.1 and less (i.e., $1/\text{Bi} > 10$), the temperature T throughout the wall is essentially independent of location. That is, T is approximately equal to the centerline temperature T_0. To reinforce this point, the one-dimensional solution for T (or T_0) developed in Chap. 2 [Eq. (2-135)],

$$\frac{T_0 - T_F}{T_i - T_F} = \exp\left(-\frac{\bar{h} A_s t}{\rho V c_v}\right) = \exp\left(-\text{Bi}\frac{\alpha t}{\ell^2}\right) \tag{3-102}$$

is compared with the predictions shown in Fig. 3-16(a) for Biot numbers equal to 10, 1.0, 0.1, and 0.01; ℓ is equal to the half-width of the plate $L/2$. The simple one-dimensional analysis is seen to be in good agreement with the general two-dimensional solution for values of the Biot number of the order of 0.1 and less.

The relationship between these temperature charts and the q chart given in Fig. 3-14(a) is brought out in Prob. 3-69.

EXAMPLE 3-24

A sphere ($r_1 = 3$ cm) with surface temperature maintained at 150 °C is buried in the ground [$k = 0.06$ W/(m °C)] at a depth of 1 m. The surface temperature of the ground is 15 °C. Determine the rate of heat transfer from the sphere.

Solution

From Table 3-4, we find that the thermal resistance R is given by

$$R = \frac{1 - r_1/(2Z)}{4\pi r_1 k} = \frac{1 - 0.03/2}{4\pi(0.03 \text{ m})[0.06 \text{ W/(m °C)}]} = 43.5\frac{°\text{C}}{\text{W}}$$

Hence, the rate of heat transfer is given by

$$q = \frac{T_1 - T_2}{R} = \frac{(150 °\text{C} - 15 °\text{C})}{43.5 °\text{C/W}}$$
$$= 3.10 \text{ W}$$

EXAMPLE 3-25

A 20-in.-thick aluminum plate initially at 75 °F is suddenly exposed to a fluid with $T_F = 250$ °F and $\bar{h} = 3000$ Btu/(h ft^2 °F). Determine the length of time for the centerline temperature to rise to 230°F, assuming negligible radiation effects. Also determine the surface temperature at this instant.

Solution

We utilize the Heisler charts to obtain the solution. The dimensionless temperature and Biot number are calculated first. The properties of aluminum are obtained from Table A-C-2 in the Appendix.

$$\frac{T_0 - T_F}{T_i - T_F} = \frac{230°F - 250°F}{75°F - 250°F} = 0.114$$

$$\text{Bi} = \frac{\bar{h}}{k}\frac{L}{2} = \frac{3000\ \text{Btu}/(\text{h ft}^2\ °F)}{117\ \text{Btu}/(\text{h ft }°F)}\left(\frac{20}{12}\text{ft}\right)\frac{1}{2} = 21.4$$

$$\frac{1}{\text{Bi}} = 0.0468$$

Referring to Fig. 3-16(a), we find

$$\frac{\alpha t}{\ell^2} = \frac{4\alpha t}{L^2} = 1.1$$

$$t = \frac{1.1(20\ \text{ft}/12)^2}{4(9.09 \times 10^{-4}\ \text{ft}^2/\text{s})} = 840\ \text{s}$$

To find the surface temperature at this instant, we utilize Fig. 3-16(b) with $\xi/L/2$ set equal to unity. Based on this figure, we obtain

$$\frac{T_s - T_F}{T_i - T_F} = 0.07$$

$$T_s = 0.07(75°F - 250°F) + 250°F = 238°F$$

EXAMPLE 3-26

A brass sphere 50 cm in diameter initially at 80°C is placed in a cooling fluid with $T_F = 15°C$ and $\bar{h} = 500$ W/(m² °C). Determine the length of time required for the center of the sphere to cool to 30°C.

Solution

First we calculate the Biot number. The properties of brass are obtained from Table A-C-1 in the Appendix.

$$\text{Bi} = \frac{\bar{h}}{k}\frac{V}{A_s} = \frac{\bar{h}}{k}\frac{r_0}{3}$$

$$= \frac{500\ \text{W}/(\text{m}^2\ °C)}{111\ \text{W}/(\text{m }°C)}\frac{0.25\ \text{m}}{3} = 0.375 = \frac{1}{2.67}$$

Because $Bi > 0.1$, we will utilize the Heisler charts in Appendix F instead of the approximate one-dimensional solution developed in Chap. 2. The dimensionless temperature is

$$\frac{T_0 - T_F}{T_i - T_F} = \frac{30\,°C - 15\,°C}{80\,°C - 15\,°C} = 0.23$$

Referring to Fig. A-F-1, we obtain

$$\frac{\alpha t}{r_0^2} = 0.8$$

Hence, the time is

$$t = \frac{0.8(0.25\,\text{m})^2}{3.41 \times 10^{-5}\,\text{m}^2/\text{s}} = 1.46 \times 10^3\,\text{s}$$

EXAMPLE 3-27

The short circular cylinder shown in Fig. E3-27a, which is initially at a uniform temperature T_i, is suddenly exposed to a fluid with h and T_F specified. Demonstrate how the instantaneous temperature at any location within the cylinder can be determined by the use of Heisler charts.

Solution

The differential formulation for this three-dimensional (t, z, r) problem takes the form [for $\psi = (T - T_F)/(T_i - T_F)$]

$$\frac{1}{r}\frac{\partial}{\partial r}\left(r\frac{\partial \psi}{\partial r}\right) + \frac{\partial^2 \psi}{\partial z^2} = \frac{1}{\alpha}\frac{\partial \psi}{\partial t} \tag{a}$$

$$\psi = 1 \qquad \text{at } t = 0 \tag{b}$$

$$\frac{\partial \psi}{\partial r} = 0 \qquad \text{at } r = 0 \tag{c}$$

$$k\frac{\partial \psi}{\partial z} = h\psi \qquad \text{at } z = 0 \tag{d}$$

$$-k\frac{\partial \psi}{\partial r} = h\psi \qquad \text{at } r = r_0 \tag{e}$$

$$-k\frac{\partial \psi}{\partial z} = h\psi \qquad \text{at } z = L \tag{f}$$

These equations can be reduced to two simpler problems by assuming

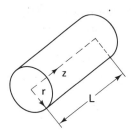

Fluid at T_F with h known

$T(0, r, z) = T_i$

$L = 3$ cm
$D = 2$ cm

FIGURE E3-27a Short circular cylinder.

the product solution

$$\psi(t,r,z) = \psi_1(t,r)\psi_2(t,z) \tag{g}$$

Using this substitution, we obtain

$$\frac{1}{r}\frac{\partial}{\partial r}\left(r\frac{\partial\psi_1}{\partial r}\right) = \frac{1}{\alpha}\frac{\partial\psi_1}{\partial t} \qquad \frac{\partial^2\psi_2}{\partial z^2} = \frac{1}{\alpha}\frac{\partial\psi_2}{\partial t} \tag{h,i}$$

$$\psi_1 = 1 \qquad \text{at } t = 0 \qquad\qquad \psi_2 = 1 \qquad \text{at } t = 0 \tag{j,k}$$

$$-k\frac{\partial\psi_1}{\partial r} = h\psi_1 \quad \text{at } r = r_0 \qquad -k\frac{\partial\psi_2}{\partial z} = h\psi_2 \quad \text{at } z = L \tag{l,m}$$

$$\frac{\partial\psi_1}{\partial r} = 0 \qquad \text{at } r = 0 \qquad\qquad k\frac{\partial\psi_2}{\partial z} = h\psi_2 \quad \text{at } z = 0 \tag{n,o}$$

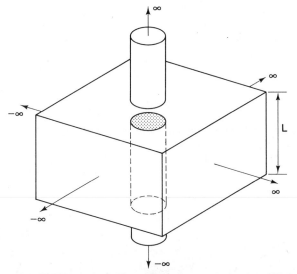

FIGURE E3-27b Intersection of flat plate and circular cylinder.

The solutions to both of these problems are represented by the Heisler charts given in Figs. 3-16 and A-F-1. Therefore, we see that the solution to our multidimensional problem is equal to the product of the solutions to two simpler unsteady problems. This concept is illustrated in Fig. E3-27b. (Similar results can be obtained for other geometries, such as the square rod in Prob. 3-71.)

PROBLEMS

3-1. (a) Show that the Fourier law of conduction can be written in the form

$$q'' = -k\nabla T = -k\left(\frac{\partial T}{\partial x}\mathbf{i} + \frac{\partial T}{\partial y}\mathbf{j} + \frac{\partial T}{\partial z}\mathbf{k}\right)$$

(b) Write the Fourier equation, Eq. (3-11), in vector form.

3-2. Develop differential formulations for the convecting fin shown in Fig. P3-2 for the conditions listed below.
(a) $Bi < 0.1$ tip insulated   (c) $Bi > 0.1$ tip at T_1
(b) $Bi > 0.1$ tip insulated   (d) $Bi > 0.1$ convection from tip

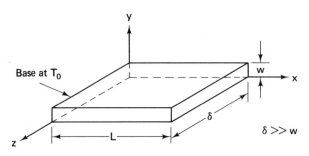

FIGURE P3-2

3-3. Develop differential formulations for the convecting fin shown in Fig. P3-3 for the conditions listed below.
(a) $Bi < 0.1$ tip insulated   (c) $Bi > 0.1$ tip at T_1
(b) $Bi > 0.1$ tip insulated   (d) $Bi > 0.1$ convection from tip

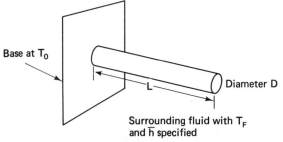

FIGURE P3-3

3-4. Develop differential formulations for the fin shown in Fig. P3-3 for the conditions listed below, if it is exposed to blackbody radiation with an enclosure at T_R. Assume that convection is negligible.
(a) $Bi < 0.1$ tip insulated
(b) $Bi > 0.1$ tip insulated
(c) $Bi > 0.1$ thermal radiation from tip

3-5. Write the differential formulation for the systems shown in Fig. P3-5.

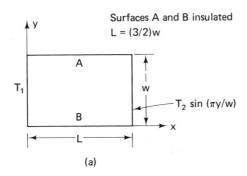

(a)

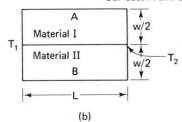

(b)

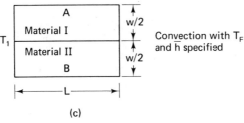

(c)

FIGURE P3-5

3-6. A fin with a 1-cm-diameter circular cross section initially at T_0 is suddenly exposed to a fluid at T_F with h known. Develop the differential formulation for small-Biot-number conditions.

3-7. A sphere initially at T_i is suddenly exposed to blackbody thermal radiation with its surroundings at T_R. F_{s-R} is equal to unity. Develop the differential formulation.

3-8. Develop the differential formulation for the composite system shown in Fig. P3-8.

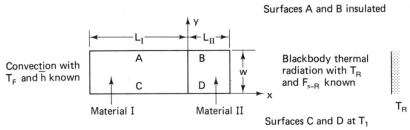

FIGURE P3-8

3-9. (a) Show $dq_{\theta+d\theta} = dq_\theta + [\partial(dq_\theta)/\partial\theta]d\theta$ for the cylindrical coordinate system. (b) Write the following equations in cylindrical coordinates: (i) Fourier equation; (ii) Laplace equation; (iii) Poisson equation.

3-10. Transform Eq. (3-13a) into cylindrical coordinates by utilizing the coordinate transformation given by Eqs. (3-15) and (3-16).

3-11. The spherical coordinate system is shown in Appendix B. Notice that ϕ varies from 0 to π rad and θ ranges from 0 to 2π. Utilizing this coordinate system, show that the differential energy equation for unsteady conduction heat transfer in a sphere takes the form

$$\frac{1}{r^2}\frac{\partial}{\partial r}\left(r^2\frac{\partial T}{\partial r}\right) + \frac{1}{r^2\sin\phi}\frac{\partial}{\partial\phi}\left(\sin\phi\frac{\partial T}{\partial\phi}\right) + \frac{1}{r^2\sin^2\phi}\frac{\partial^2 T}{\partial\theta^2} + \frac{\dot{q}}{k} = \frac{1}{\alpha}\frac{\partial T}{\partial t}$$

3-12. A fin is attached to the surface of a flat plate as shown in Fig. P3-12. The plate and fin are made of different materials. Develop the differential formulation for this problem for the case in which the plate thickness is significant and the fin Biot number is of the order of unity.

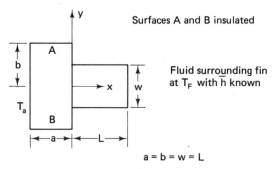

$$a = b = w = L$$

FIGURE P3-12

3-13. A plate is initially at temperature T_i. One side of the plate is then suddenly exposed to a fluid with temperature T_F and coefficient of heat transfer specified. The other side of the plate is maintained at T_1. Write the differential formulation for this problem.

3-14. Develop the differential formulation for the cylindrical system shown in Fig. P3-14.

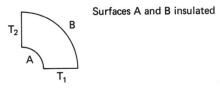

FIGURE P3-14

3-15. Write the differential formulation for the system shown in Fig. P3-15.

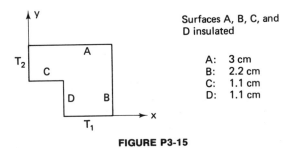

Surfaces A, B, C, and
D insulated

A: 3 cm
B: 2.2 cm
C: 1.1 cm
D: 1.1 cm

FIGURE P3-15

3-16. Develop the differential formulation for heat transfer in the tapered rod shown in Fig. P3-16.

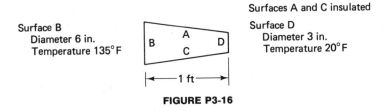

Surfaces A and C insulated

Surface B
  Diameter 6 in.
  Temperature 135°F

Surface D
  Diameter 3 in.
  Temperature 20°F

FIGURE P3-16

3-17. Show that the central difference approximation for $\partial T/\partial x$ at the interface between two subvolumes has an error of the order of Δx^2.

3-18. With $\Delta x^2/(\alpha\,\Delta t)$ set equal to 6 in an explicit finite-difference formulation for unsteady four-dimensional conduction heat transfer with $\dot{q}=0$, demonstrate that the temperature $T_{m,n,p}^{\tau+1}$ (where $z=p\Delta z$) is equal to the arithmetic average of the 6 surrounding nodal temperatures at the instant $t=\tau\,\Delta t$.

3-19. Set up the finite-difference grid for the radiating fin with rectangular cross section shown in Fig. P3-19 for the conditions listed below. Identify the unknown nodal temperatures and write their nodal equations. Also write an expression for the rate of heat transfer at the base.
  (a) $\text{Bi}<0.1$, $\Delta x=L$ tip insulated
  (b) $\text{Bi}<0.1$, $\Delta x=L/2$ radiation from tip
  (c) $\text{Bi}<0.1$, $\Delta x=L/2$ radiation from tip
  (d) $\text{Bi}>0.1$, $\Delta x=L$ tip insulated
  (e) $\text{Bi}>0.1$, $\Delta x=L/2$ radiation from tip
  (f) $\text{Bi}>0.1$, $\Delta x=L/3$ radiation from tip

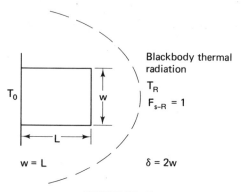

FIGURE P3-19

3-20. Set up the finite-difference grid for the convecting fin shown in Fig. P3-20 for the conditions listed below. Identify the unknown nodal temperatures and write their nodal equations. Write an expression for the rate of heat transfer at the base.
  (a) $Bi < 0.1$, $\Delta x = L$ tip insulated
  (b) $Bi < 0.1$, $\Delta x = L$ tip not insulated
  (c) $Bi < 0.1$, $\Delta x = L/2$ tip insulated
  (d) $Bi < 0.1$, $\Delta x = L/3$ tip insulated
  (e) $Bi < 0.1$, $\Delta x = L/4$ perimeter insulated with convection from tip
  (f) $Bi > 0.1$, $\Delta x = L/2$ tip insulated
  (g) $Bi > 0.1$, $\Delta x = L/4$ tip not insulated
  (h) $Bi > 0.1$, $\Delta x = L/4$ tip insulated

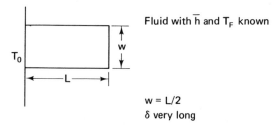

FIGURE P3-20

3-21. Write the solution for temperature distribution in a rectangular plate with the surface at $x = 0$ maintained at T_2 and the other three surfaces at T_1.

3-22. Evaluate c_n in Eq. (3-60) for the case in which $f(x) = T_1 + T_2 \sin(\pi x / L)$.

3-23. Develop expressions for q_x and q_D for steady-state heat transfer in the rectangular solid of Fig. 3-5 for the case in which $F(x) = T_1 + T_2 \sin(\pi x / L)$.

3-24. A solid 10-cm-diameter aluminum sphere is initially at 100°C. Its surface is then suddenly changed to 0°C. (a) Develop the differential formulation for this heat transfer problem. (b) The solution for the temperature distribution in this problem is given by [11]

$$\frac{T - T_1}{T_i - T_1} = \frac{2}{\pi} \sum_{n=1}^{\infty} \frac{(-1)^{n+1}}{n} \exp\left(-n^2\pi^2\alpha t / r_0^2\right) \frac{\sin\left(n\pi r / r_0\right)}{r / r_0}$$

(i) Show that this equation satisfies the differential equation, and (ii) develop expressions for the instantaneous rate q and accumulative total Q heat transfer from the surface of the sphere. (This series is uniformly convergent for $t > 0$ and $0 \ll r \ll r_0$.)

3-25. Develop a numerical finite-difference formulation with $\Delta x = L/2$ for the square plate shown in Fig. P3-25. Then solve for the unknown nodal temperature and develop predictions for the rates of heat transfer from each surface. (Note that the rate of heat transfer through the surfaces of each corner node is indeterminant.) (More accurate numerical solutions for this system are the subject of Examples 3-14 and 3-17.)

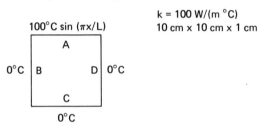

k = 100 W/(m °C)
10 cm x 10 cm x 1 cm

100°C sin $(\pi x / L)$

FIGURE P3-25

3-26. Develop a numerical finite-difference formulation with $\Delta x = L/2$ for the square plate shown in Fig. P3-26. Then solve for the unknown nodal temperatures and develop predictions for the rates of heat transfer from surfaces A and B. (Note that the heat transfer in corner nodes involving convection is determinant.)

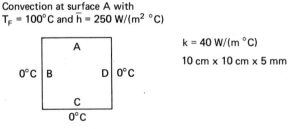

Convection at surface A with
$T_F = 100°C$ and $\bar{h} = 250$ W/(m^2 °C)

k = 40 W/(m °C)
10 cm x 10 cm x 5 mm

FIGURE P3-26

3-27. Develop a numerical finite-difference formulation with $\Delta x = L = 1$ cm for the rectangular plate shown in Fig. P3-27. Then solve for the unknown nodal temperature and develop predictions for the rates of heat transfer from each surface. (Note that the heat transfer through the surfaces of corner nodes involving radiation is determinant. The differential formulation for this problem is given in Example 3-1.)

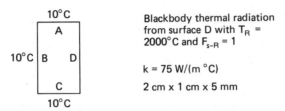

Blackbody thermal radiation
from surface D with T_R =
2000°C and F_{s-R} = 1

k = 75 W/(m °C)

2 cm x 1 cm x 5 mm

FIGURE P3-27

3-28. (a) Develop a numerical finite-difference formulation with $\Delta x = L/2$ for the square plate shown in Fig. P3-28. Then solve for the unknown nodal temperatures and develop predictions for the rates of heat transfer from each surface. (b) Solve this problem with the limiting grid spacing $\Delta x = L$.

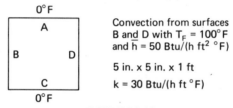

Convection from surfaces
B and D with T_F = 100°F
and $\bar{h}$ = 50 Btu/(h ft² °F)

5 in. x 5 in. x 1 ft

k = 30 Btu/(h ft °F)

FIGURE P3-28

3-29. Develop a numerical finite-difference formulation with $\Delta x = 1$ cm for the fin unit shown in Fig. P3-29. Also develop predictions for the rate of heat transfer from surface A.

Surface A at 100°C

Surfaces B insulated

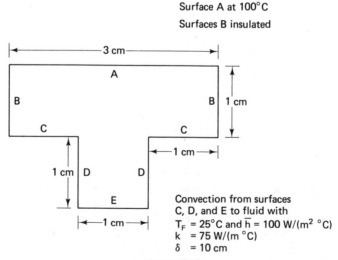

Convection from surfaces
C, D, and E to fluid with
T_F = 25°C and $\bar{h}$ = 100 W/(m² °C)
k = 75 W/(m °C)
δ = 10 cm

FIGURE P3-29

3-30. (a) Develop a numerical finite-difference formulation with $\Delta x = L/2$ for the square plate shown in Fig. P3-30. Then solve for the unknown nodal temperatures and develop predictions for the rates of heat transfer from each surface. (b) Solve this problem with $\Delta x = L$.

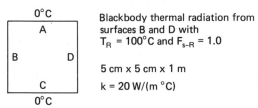

FIGURE P3-30

3-31. (a) Develop a numerical finite-difference formulation with $\Delta x = L/2$ for the square plate shown in Fig. P3-31. Then solve for the unknown nodal temperatures and develop predictions for the rates of heat transfer from each surface. (b) Solve this problem with $\Delta x = L$.

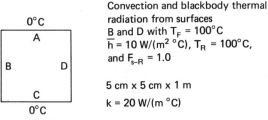

FIGURE P3-31

3-32. Solve Prob. 3-26 for the case in which 0.5 MW/m^3 is uniformly generated within the body.

3-33. Develop a numerical finite-difference formulation with $\Delta x = L$ for the square plate shown in Fig. P3-33. Then solve for the unknown nodal temperatures and develop predictions for the rates of heat transfer from each surface.

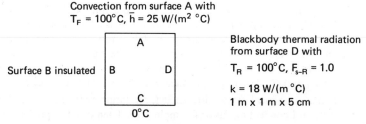

FIGURE P3-33

3-34. Develop a numerical finite-difference formulation with $\Delta x = 10$ cm for the composite plate shown in Fig. P3-34. Then solve for the unknown nodal temperatures and develop predictions for the rates of heat transfer from the two surfaces.

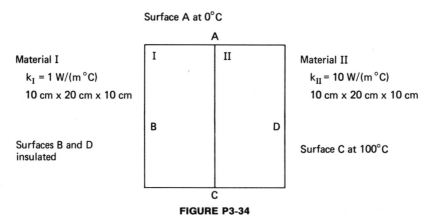

Surface A at 0°C

Material I
$k_I = 1$ W/(m °C)
10 cm x 20 cm x 10 cm

Material II
$k_{II} = 10$ W/(m °C)
10 cm x 20 cm x 10 cm

Surfaces B and D
insulated

Surface C at 100°C

FIGURE P3-34

3-35. Develop a numerical finite-difference formulation with $\Delta x = 10$ cm for the composite plate shown in Fig. P3-35. Then solve for the unknown nodal temperatures and develop predictions for the rates of heat transfer from the two surfaces.

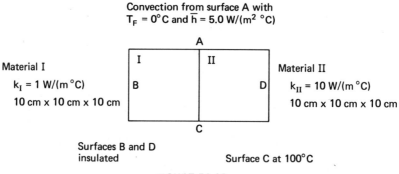

Convection from surface A with
$T_F = 0°$C and $\bar{h} = 5.0$ W/(m² °C)

Material I
$k_I = 1$ W/(m °C)
10 cm x 10 cm x 10 cm

Material II
$k_{II} = 10$ W/(m °C)
10 cm x 10 cm x 10 cm

Surfaces B and D
insulated

Surface C at 100°C

FIGURE P3-35

3-36. (a) A 1-cm-diameter 3-cm-long steel fin [$k = 43$ W/(m °C)] transfers heat from the wall of a heat exchanger at 200°C to a fluid at 25°C with $\bar{h} = 120$ W/(m² °C). Develop a numerical finite-difference formulation with $\Delta x = L/2$. Then solve for the unknown nodal temperatures and develop a prediction for the rate of heat transfer from the fin. (The development of more accurate numerical solutions for this system is the subject of Examples 3-13 and 3-15.) (b) Solve this problem with $\Delta x = L$ and $\Delta x = L/3$.

3-37. Develop a numerical finite-difference formulation with $\Delta x = L/2$ for the radiating and convecting fin of Example 3-16. Then develop predictions for the rate of heat transfer from the fin for the first few increments of time.

3-38. Develop a numerical finite-difference formulation with $\Delta y = w$ for the plate of Example 3-2, where $T_i = 0°C$, $T_F = 100°C$, $\bar{h} = 50$ W/(m² °C), $k = 35$ W/(m °C), $\alpha = 2 \times 10^{-5}$ m²/s, and $w = 1$ cm. Also calculate the instantaneous heat flux at the surface at $t = 1, 2,$ and 3 s.

3-39. Repeat Prob. 3-38 for $\Delta y = w/2$.

3-40. The fin of Prob. 3-36 is initially at 25°C. The base is then suddenly brought to 200°C. Develop a numerical finite-difference formulation with $\Delta x = L/2$. Then solve for the unknown nodal temperature and develop predictions for the instantaneous rate of heat transfer from the fin.

3-41. The fin of Example 2-13 is initially at 25°C. The base is then suddenly brought to a temperature of 200°C. Develop a numerical finite-difference formulation with $\Delta x = L/2$. Then solve for the unknown nodal temperatures and develop predictions for the instantaneous rate of heat transfer from the fin.

3-42. The square plate of Prob. 3-26 is initially at 0°C. Surface A is then suddenly exposed to convection with $T_F = 100°C$ and $\bar{h} = 25$ W/(m² °C). Develop a numerical finite-difference formulation with $\Delta x = L/2$. Then solve for the unknown nodal temperatures and the instantaneous rate of heat transfer at surface A.

3-43. The rectangular plate of Prob. 3-27 is initially at 10°C. Surface D is suddenly exposed to blackbody radiation. Develop a numerical finite-difference formulation with $\Delta x = 1$ cm. Then determine the instantaneous rate of heat transfer across surface D.

3-44. Utilize the FORTRAN program of Example 3-15 to obtain the temperature distribution in the fin for 5-mm increments in x.

3-45. Modify the FORTRAN program of Example 3-15 to account for convection heat transfer through the tip.

3-46. Modify the FORTRAN program presented in Example 3-17 to determine the temperature distribution and rate of heat transfer at face A for the case in which face A is exposed to the blackbody thermal radiation with a surface at 1000°C with $F_{s-R} = 0.5$.

3-47. Utilize the FORTRAN program presented in Example 3-17 to obtain the temperature distribution and the rate of heat transfer at face A for the case in which the boundary temperatures are given as follows: face A: 100°C $\cos(\pi x / L)$; face B; 100°C $\sin[\pi x/(2w)]$; face C: 0°C; face D: 0°C.

3-48. Develop an accurate numerical finite-difference solution for Prob. 3-26.

3-49. Develop an accurate numerical finite-difference solution for Prob. 3-27.

3-50. Develop an accurate numerical finite-difference solution for Prob. 3-28.

3-51. Develop an accurate numerical finite-difference solution for Prob. 3-29.

3-52. Develop an accurate numerical finite-difference solution for Prob. 3-30.

3-53. Develop an accurate numerical finite-difference solution for Prob. 3-31.

3-54. Develop an accurate numerical finite-difference solution for Prob. 3-32.

3-55. Develop an accurate numerical finite-difference solution for Prob. 3-33.

3-56. Develop an accurate numerical finite-difference solution for Prob. 3-35.

3-57. Develop an accurate numerical finite-difference solution for the rate of heat transfer in the hollow square steel rod shown in Fig. P3-57.

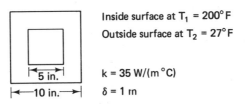

Inside surface at $T_1 = 200°F$
Outside surface at $T_2 = 27°F$

$k = 35$ W/(m °C)

$\delta = 1$ m

FIGURE P3-57

3-58. A 5-in.-long 1-in.-diameter blackbody fin [$k = 100$ Btu/(h ft^2 °F)] extends from a wall at 1000°F. Develop a numerical finite-difference formulation and FORTRAN program for the temperature distribution and rate of heat transfer at the base for the case in which the surroundings are at $T_R = 250°F$ with $F_{s-R} = 0.4$, radiation occurs at tip, and convection is negligible.

3-59. Develop a numerical finite-difference formulation for the spine shown in Fig. P3-59. Also develop an accurate numerical solution for the rate of heat transfer from the base.

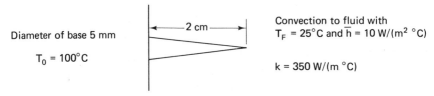

Diameter of base 5 mm

$T_0 = 100°C$

2 cm

Convection to fluid with
$T_F = 25°C$ and $\bar{h} = 10$ W/(m^2 °C)

$k = 350$ W/(m °C)

FIGURE P3-59

3-60. Modify the FORTRAN program of Example 3-19 to produce predictions for the temperature at 0.2 s, 0.3 s, 0.4 s, and 0.5 s for increments of 0.5 mm.

3-61. Develop a computer solution for the temperature distribution and instantaneous rate of heat transfer in the plate of Example 3-19 for the case in which one side is exposed to a fluid with $T_F = 100°C$ and $\bar{h} = 150$ W/(m^2 °C) and the other side is kept at 0°C.

3-62. Develop an accurate numerical finite-difference solution for Prob. 3-38. Also determine the length of time required to reach steady state.

3-63. Develop an accurate numerical finite-difference solution for Prob. 3-40. Also obtain the steady-state heat-transfer rate and the length of time required to reach steady state.

3-64. Develop an accurate numerical finite-difference solution for Prob. 3-41. Also obtain the steady-state heat-transfer rate and the length of time required to reach steady state.

3-65. Develop an accurate numerical finite-difference solution for Prob. 3-42. Also obtain the steady-state heat-transfer rate at surface A and the length of time required to reach steady state.

3-66. Develop an accurate numerical finite-difference solution for Prob. 3-43. Also obtain the steady-state heat-transfer rate at surface D and the length of time required to reach steady state.

3-67. Develop an approximate graphical solution to Prob. 3-57. Simplify your analysis by utilizing system symmetry.

3-68. Utilize the graphical approach to determine the rate of heat transfer in a cylindrical quarter section such as the one shown in Fig. 3-11 for $r_1 = 1$ cm, $r_2 = 10$ cm, $T_1 = -50°C$, $T_2 = 250°C$, and $k = 25$ W/(m °C).

3-69. Show how Fig. 3-14(a) can be developed from the Heisler charts given in Fig. 3-16(a) and (b).

3-70. A Heisler chart for accumulated heat transfer for unsteady heat transfer in a long solid cylinder which is suddenly exposed to a fluid is given in Fig. A-F-1(a) of the Appendix. Show that the simple one-dimensional solution for Q is in close agreement with the Heisler curves for Biot number less than 0.1.

3-71. A long hot steel bar (10 cm × 10 cm) initially at 500°C is suddenly placed in a cooling fluid at 15°C with $\bar{h} = 1000$ W/(m² °C), $k = 35$ W/(m°C), and $\alpha = 1.5 \times 10^{-5}$ m²/s. Utilize the Heisler charts and the principle of superposition to determine the length of time for the centerline temperature to reach 50°C.

3-72. A 2-cm-thick copper plate initially at 100°C is suddenly exposed to a cooling fluid at 0°C with a very high coefficient of heat transfer. Develop a graphical solution for the temperature distribution and rate of heat transfer from the surface. Utilize a grid spacing of $L/10$.

3-73. A large copper plate of 1 in. thickness initially at 98°F is suddenly immersed in a well-stirred fluid kept at 5°F with $\bar{h} = 75$ Btu/(h ft² °F). Determine the time required for the surface of the plate to reach 30°F. Also determine the accumulative heat transfer over this period of time.

3-74. A concrete wall of 2 ft thickness is initially at a uniform temperature of 1100°F. The wall is suddenly exposed on both sides to a convecting fluid with $T_F = 100°F$ and $\bar{h} = 5$ Btu/(h ft² °F). Determine the temperature at the center of the slab after 5 h and 20 h. Assume $k = 0.694$ Btu/(h ft² °F), and $\alpha = 8.6 \times 10^{-6}$ ft²/s.

3-75. The surface of a very thick slab of iron initially at 200°C is suddenly exposed to a fluid with $T_F = 10°C$ and $\bar{h} = 75$ W/(m² °C). Determine the surface temperature and rate of heat transfer at increments of 60 s for a period of 10 min.

3-76. Solve Prob. 3-75 for the case in which the slab is 5 cm thick, with both surfaces exposed to the fluid. Also determine the total accumulated heat transfer over the 10-min interval and the centerline temperature.

3-77. A long 5-in.-diameter copper cylinder initially at $-100°F$ is suddenly exposed to a fluid with $T_F = 27°F$ and $\bar{h} = 100$ Btu/(h ft² °F). Determine the centerline and surface temperatures at increments of 60 s for a 10-min period of time. Determine the length of time for the cylinder to essentially reach steady-state conditions.

3-78. A semi-infinite 5-cm-diameter copper cylinder initially at 200°C is suddenly exposed to a fluid with $T_F = 50°C$ and $\bar{h} = 250$ W/(m² °C). Calculate the temperatures at the axis and the surface of the cylinder 5 cm from the end 1 min after the process has begun.

3-79. Develop an analytical solution for the temperature distribution in the rectangular solid shown in Fig. 3-5 for the case in which both surfaces A and D are maintained at T_2, with surfaces B and C at T_1. Also develop an expression for the rate of heat transfer across the base at $x = 0$.

3-80. The semi-infinite rectangular solid shown in Fig. P3-80 has a base temperature of T_2, with $T = T_1$ at $x = 0$ and $x = L$. (a) Utilize the product solution approach to show that the temperature distribution takes the form

$$\frac{T - T_1}{T_2 - T_1} = \frac{4}{\pi} \sum_{n=0}^{\infty} \frac{1}{2n+1} \exp[-(2n+1)\pi y/L] \sin\frac{(2n+1)\pi x}{L}$$

(b) Develop an expression for the rate of heat transfer q_y and show that $q_y \to 0$ as $y \to \infty$, and $q_y \to \infty$ as $y \to 0$.

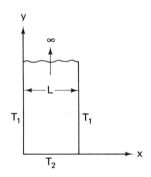

FIGURE P3-80

3-81. The very long plate of thickness w shown in Fig. P3-81 is initially at uniform temperature T_i. Its surfaces are then suddenly brought to a temperature T_1. (a) Utilize the product solution approach to show that the unsteady temper-

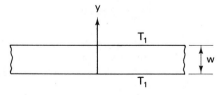

FIGURE P3-81

ature distribution in the plate takes the form

$$\frac{T - T_1}{T_i - T_1} = \frac{4}{\pi} \sum_{n=0}^{\infty} \frac{1}{2n+1} \exp\left[-(2n+1)^2 \pi^2 \alpha t / w^2 \right] \sin \frac{(2n+1)\pi y}{w}$$

(This series is uniformly convergent for $t > 0$ and $0 \leqslant y \leqslant w$.) (b) Develop expressions for q and Q. (To check your solution for q as $\tau \to \infty$, note that $\sum_{n=0}^{\infty} 1/(2n+1)^2 = \pi^2/8$.)

REFERENCES

[1] MYERS, G. E., *Analytical Methods in Conduction Heat Transfer*. New York: McGraw-Hill Book Company, 1971.

[2] KREITH, F., *Principles of Heat Transfer*, 3rd ed. New York: Intext Press, Inc., 1973.

[3] HOLMAN, J. P., *Heat Transfer*, 4th ed. New York: McGraw-Hill Book Company, 1976.

[4] KREYSZIG, E., *Advanced Engineering Mathematics*, 3rd ed. New York: John Wiley & Sons, Inc., 1972.

[5] OZISIK, M. N., *Boundary Value Problems of Heat Conduction*. Scranton, Pa.:, International Textbook Company, 1968.

[6] ARPACI, V. S., *Conduction Heat Transfer*. Reading, Mass.: Addison-Wesley Publishing Company, Inc., 1966.

[7] CARSLAW, H. S., and J. C. JAEGER, *Conduction of Heat in Solids*. London: Oxford University Press, 1947.

[8] KARLEKAR, B. V., and R. M. DESMOND, *Engineering Heat Transfer*. St. Paul, Minn.: West Publishing Co., 1977.

[9] FOX, L., *Numerical Solution of Ordinary and Partial Differential Equations*. Reading, Mass.: Addison-Wesley Publishing Company, Inc., 1962.

[10] SMITH, G. D., *Numerical Solution of Partial Differential Equations with Exercises and Worked Solutions*. London: Oxford University Press, 1965.

[11] SCHNEIDER, P. J., *Conduction Heat Transfer*. Reading, Mass.: Addison-Wesley Publishing Company, Inc., 1974.

[12] SMITH, J. P., *SINDA User's Manual*, NASA Contract 9-10435, 1971.

[13] BEWLEY, L. V., *Two-Dimensional Fields in Electrical Engineering*. New York: Macmillan Publishing Co., Inc., 1948.

[14] KREITH, F., *Principles of Heat Transfer*, 2nd ed. London: International Textbook Company Ltd., 1965.

[15] JAKOB, M., *Heat Transfer*, vol. 1. New York: John Wiley & Sons, Inc., 1949.

[16] LANGMUIR, I., E. O. ADAMS, and F. A. MEIKLE, "Flow of Heat through Furnace Walls," *Trans. Am. Electrochem. Soc.*, **24**, 1913, 53.

[17] RUDENBERG, R., "Die Ausbreitung der Luft- und Erdfelder um Hochapannungaleitungen besonders bei Erd- und Kurzschterssen," *Electrotech. Z.*, **46**, 1945, 1342.

[18] ANDREWS, R. V., "Solving Conductive Heat Transfer Problems with Electrical-Analogue Shape Factors," *Chem. Eng. Progr.*, **5**, 1955, 67.

[19] HAHNE, E., and U. GRIGULL, "Formfaktor und Formwiderstand der stationärem mehrdimensionalen Wärmeleitung," *Int. J. Heat Mass Transfer*, **18**, 1975, 75.

[20] HEISLER, M. P., "Temperature Charts for Induction and Constant Temperature Heating," *Trans. ASME*, **69**, 1947, 227–236.

[21] OZISIK, M. N., *Basic Heat Transfer*. New York: McGraw-Hill Book Company, 1977.

[22] ZERKLE, R. D., and J. E. SUNDERLAND, "Transient Temperature Distribution in Slab Subject to Thermal Radiation," *J. Heat Transfer*, **87**, 1965, 117.

[23] CHUNG, B. T. F., and L. T. YEH, "Unsteady Heat Conduction in a Slab with Uniform Heat Source and Radiative Boundary Condition," *J. Spacecraft Rocket*, **10**, 1973, 483.

CHAPTER 4

RADIATION HEAT TRANSFER

4-1 INTRODUCTION

Radiation heat transfer is defined as the transfer of energy across a system boundary by means of an electromagnetic mechanism which is caused solely by a temperature difference. Some of the more basic and practical aspects of thermal radiation were introduced in Chap. 1. We now want to take a closer look at the physical mechanism, properties, and geometric factors associated with thermal radiation. In addition, modeling concepts are developed and design information is presented in this chapter which provide the basis for practical thermal analysis of radiation heat transfer. Practical thermal analyses will be developed for blackbody and diffuse nonblackbody surfaces, with special consideration given to unsteady systems, participating medium, combined mode processes, and solar radiation.

4-2 PHYSICAL MECHANISM

Whenever a charged particle undergoes acceleration, energy possessed by the particle is converted into a form of energy known as electromagnetic radiation. Electromagnetic radiation includes cosmic rays, gamma rays, X rays, ultraviolet radiation, visible light, infrared radiation, microwaves, radar, radio waves, and ultrasonic electrical waves. Electromagnetic radiation can be produced by various means, depending on the type of charged particles that are involved in the process. To illustrate, γ rays are produced by the fission of nuclei or by radioactive disintegration, X rays by the bombardment of metals with high-energy electrons, and radio waves by the excitation of certain crystals or by the flow of alternating current through electric conductors. Of particular interest to us is electromagnetic radiation which is produced by rotational and vibrational movements of the subatomic particles, atoms, and molecules of a substance. Because the level of energy associated with the fluctuating motion of these small oscillators is indicated by temperature, the resulting electromagnetic radiation is referred to as *thermal radiation*. In other words, thermal radiation represents the conversion of internal thermal energy of a substance into electromagnetic energy.

All the various types of electromagnetic waves are characterized by a frequency ν and a propagation velocity in free space (a vacuum or transparent medium) equal to the speed of light c. The speed of light c in a gas, liquid, or solid is related to the speed of light in a vacuum c_0 $(=3\times10^8$ m/s$)$ by the index of refraction n $(\equiv c_0/c)$. (The index of refraction of air and most gases is essentially unity, but for liquids and solids such as water and glass is of the order of 1.5.) The wavelength λ is defined in terms of ν and c by

$$\lambda = \frac{c}{\nu} \tag{4-1}$$

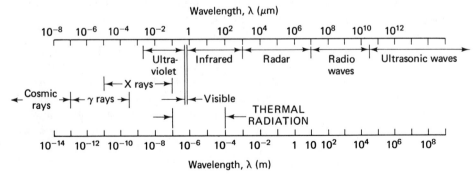

FIGURE 4-1 Electromagnetic wave spectrum.

The standard unit for λ is the micrometer μm ($\equiv 10^{-6}$ m). Whereas the propagation velocity and the wavelength of a radiant beam depend on the medium, the frequency depends only on the radiating source and is independent of the substance through which it is transmitted.

The various types of electromagnetic radiation are characterized according to wavelength or frequency by the *electromagnetic spectrum*, which is shown in Fig. 4-1. Notice that the spectrum of electromagnetic radiation ranges from the very short wavelength cosmic-ray, γ-ray, and X-ray phenomena, through the intermediate-wavelength thermal radiation, to long-wavelength radio waves and ultrasonic waves. The thermal radiation wavelength band extends essentially from 0.1 to 100 μm, which includes ultraviolet (0.1 to 0.38 μm), visible (0.38 to 0.76 μm), and infrared (0.76 to 100 μm) regions. The thermal radiation portion of the electromagnetic spectrum is focused upon in Fig. 4-2. Note that *solar radiation* lies within the heart of the thermal radiation wavelength band.

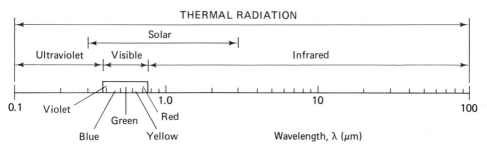

FIGURE 4-2 Electromagnetic spectrum for thermal radiation wavelength band.

4-3   THERMAL RADIATION PROPERTIES

4-3-1   Introduction

The exchange of thermal radiation between surfaces is a function of (1) surface emission properties; (2) surface absorption, reflection, and transmission properties; and (3) properties of the medium that lies within the path of the thermal radiation. Each of these issues is considered in this section.

4-3-2   Surface Emission Properties

The rate of thermal radiation emitted by a body is dependent upon the surface temperature T_s, the nature of the surface, and the electromagnetic radiation wavelength λ or frequency ν. The effect of each of these factors on the emission of thermal radiation must be considered.

Total emissive power

The effect of emitter surface temperature T_s on the rate of thermal radiation emitted by a body is seen by examining the *total emissive power* E, which was introduced in Chap. 1. Because the total emissive power represents the total rate of thermal radiation emitted per unit surface area (i.e., thermal radiation flux) over all wavelengths, it is sometimes designated by $E_{0 \to \infty}$ instead of E. However, for convenience, we will retain the symbol E.

As indicated in Chap. 1, a surface that emits the maximum possible thermal radiation at any given temperature is called a *blackbody*. The total emissive power for thermal radiation in a vacuum from such ideal emitting blackbody surfaces is given by the *Stefan–Boltzmann law*,

$$E_b = \sigma T_s^4 \tag{4-2}$$

where the Stefan–Boltzmann constant σ is equal to 5.67×10^{-8} W/(m^2 K^4). Referring back to Fig. 1-6, we are reminded that blackbody thermal radiation flux ranges from very significant levels for source temperatures of the order of 1000 K and above to quite small and often negligible quantities for normal environmental temperatures.

TABLE 4-1 Emissivities of representative surfaces

Surface	Emissivity, ε	Temperature, T(K)
Aluminum:		
Polished	0.04	500
Commercial sheet	0.09	370
Brass:		
Polished	0.07	320
Dull	0.22	320
Copper:		
Polished	0.041	340
Slightly polished	0.12	320
Polished, lightly tarnished	0.05	320
Dull	0.15	320
Oxidized at 1030 K	0.50	590
Nickel, polished	0.09	270
Silver, polished	0.02	300
Stainless steel 18-8, polished	0.25	310
Tungsten, polished	0.33	3400
Asphalt	0.93	310
Glass, Pyrex	0.88	420
Parsons black paint	0.98	240
Lampblack paint	0.96	310

Source: Based on data primarily from [1] and [2].

Although some surfaces and geometrical configurations approach ideal emitting conditions, perfect blackbody surfaces do not exist. (The nature of ideal blackbody thermal radiating surfaces will be explored further in Sec. 4-3-3.) The total emissive power of real nonblackbody surfaces is expressed in terms of E_b by

$$E = \varepsilon E_b \qquad (4\text{-}3)$$

where the *emissivity* ε ranges from zero to unity. (Because ε accounts for the thermal radiation emitted over all wavelengths into the entire hemispherical space above a surface, it is also commonly referred to as the total hemispherical emissivity.) It is important to note that ε is a property which is dependent only on the nature of the surface and its temperature T_s. The emissivity is given for common surfaces in Table 4-1, Fig. 4-3, and in Table A-C-7 of the Appendix; very comprehensive tabulations of radiation properties are available in *Thermophysical Properties of Matter* by Touloukian et al. [1]–[3] and in *Thermal Radiation Properties Survey* by Gubareff et al. [4]. Referring to Table 4-1, we observe that blackbody

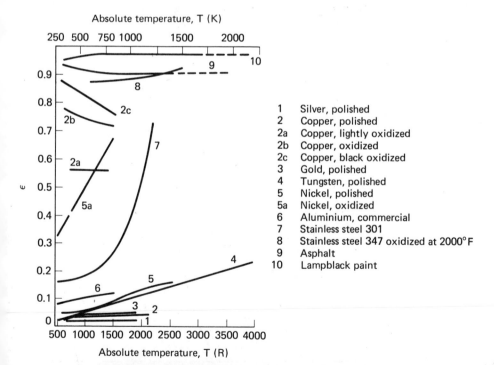

FIGURE 4-3   Dependence of emissivity ε of various surfaces on surface temperature. (Based on data from Touloukian and De Witt [2] and Gubareff et al. [4].)

conditions are approached by surfaces coated with lampblack paint. On the other hand, metals have emissivities that range from very low values for polished surfaces to fairly high values for surfaces that have been oxidized or anodized. However, modifying terms such as polished, commercial finish, oxidized, anodized, and so on, which are used to describe the nature of a surface are very subjective. As pointed out by Sparrow and Cess [5], because of the ambiguity of such terms, it is unwise to assume that radiative property values reported in literature apply with high precision to other similarly described materials.

Although the total emissive power E of real surfaces is less than E_b, E always increases with emitter surface temperature. This point is reinforced by Fig. 4-3, which shows the variation of ε with temperature. By multiplying ε by E_b $(= \sigma T_s^4)$, we find that E increases with increasing temperature, even for the materials for which ε itself decreases.

Subtotal emissive power

To see the effect of wavelength on thermal radiation, we consider the energy flux emitted from a surface over wavelength from zero to λ. This *subtotal emissive power* $E_{0 \to \lambda}$ is shown in Fig. 4-4 as a function of λ for a

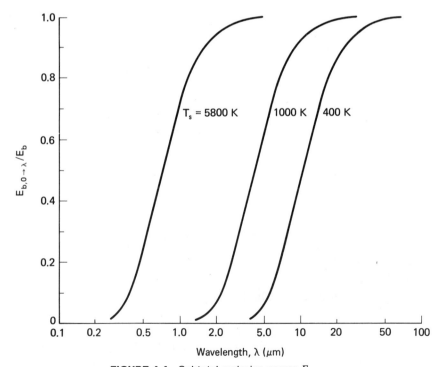

FIGURE 4-4   Subtotal emissive power $E_{b,o \to \lambda}$.

blackbody at several temperatures. $E_{b,0\to\lambda}$ increases from zero at small values of λ and approaches the total emissive power E_b as λ becomes large. Consistent with the electromagnetic spectrum shown in Figs. 4-1 and 4-2, we find that the significant contribution to the thermal radiation for these temperatures occurs within wavelengths of about 0.1 to 100 μm. For solar radiation which has an effective blackbody source temperature of roughly 5800 K, the wavelength band essentially lies between 0.30 and 3.0 μm, with about 98% of the energy associated with $\lambda < 3.0$ μm. On the other hand, the wavelength range for a surface temperature of 400 K is mainly between 3.0 and 40 μm. As a matter of fact, less than 1% of the thermal radiation emitted at environmental temperatures below 400 K is contained in the part of the electromagnetic spectrum for which $\lambda < 3.0$ μm.

A more general representation of the subtotal emissive power for a blackbody which applies to the entire temperature range is given in Fig. 4-5 and Table A-G-1 of the Appendix in terms of $E_{b,0\to\lambda}/E_b$ vs. λT_s.

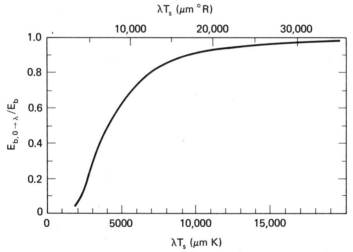

FIGURE 4-5   General representation of subtotal emissive power $E_{b,o\to\lambda}$.

Monochromatic emissive power

To complete the picture concerning the effect of wavelength on thermal radiation emitted from a surface, we introduce the *monochromatic emissive power* e_λ, which is defined as the thermal radiation flux emitted per unit wavelength $d\lambda$. This important thermal radiation emission property is related to the subtotal emissive power $E_{0\to\lambda}$ by

$$e_\lambda = \frac{dE_{0\to\lambda}}{d\lambda} \tag{4-4}$$

or

$$E_{0 \to \lambda} = \int_0^\lambda e_\lambda \, d\lambda \qquad (4\text{-}5)$$

As λ becomes large, it follows that

$$E = \int_0^\infty e_\lambda \, d\lambda \qquad (4\text{-}6)$$

Blackbody Thermal Radiation   Theoretical predictions based on quantum theory were developed for the monochromatic emissive power for blackbody thermal radiation by M. Planck in 1901. The famous *Planck law* [6] takes the form

$$e_{b\lambda} = \frac{C_1}{\lambda^5 \{ \exp[\, C_2 / (\lambda T_s)] - 1 \}} \qquad (4\text{-}7)$$

where $C_1 = 3.74 \times 10^8$ W $\mu\text{m}^4/\text{m}^2$ and $C_2 = 1.44 \times 10^4$ μm K. This equation is shown in Fig. 4-6 for several temperatures, with λ taken as the independent variable. For each temperature, $e_{b\lambda}$ is seen to increase from zero at low wavelengths to a peak, and to then gradually fall back toward zero.

The peak in $e_{b\lambda}$ increases and shifts to shorter wavelengths as the temperature increases. The wavelength at which the peak occurs is given as a function of temperature by *Wien's displacement law*,

$$(\lambda T_s)_{\max} = 2990 \ \mu\text{m K} \qquad (4\text{-}8)$$

This shift in $\lambda_{\max}$ and increase in $e_{b\lambda}$ with increasing temperature is responsible for the familiar change in color of heat-treated steel, which goes from a dull red at around 700 °C to bright red, then to bright yellow, and finally becomes glowing white at approximately 1300 °C. Referring to Fig. 4-6, we observe that little thermal radiation emitted by low-temperature blackbodies lies in the portion of the electromagnetic spectrum from 0.38 to 0.76 μm that is visible to the eye. However, as the temperature increases, more and more of the thermal radiation falls within the visible range, thus producing this array of color.

EXAMPLE 4-1

For practical purposes, the sun is generally considered to be a blackbody radiator with an effective temperature of about 5800 K. Estimate the fraction of energy emitted by the sun which falls in the visible region of the electromagnetic spectrum.

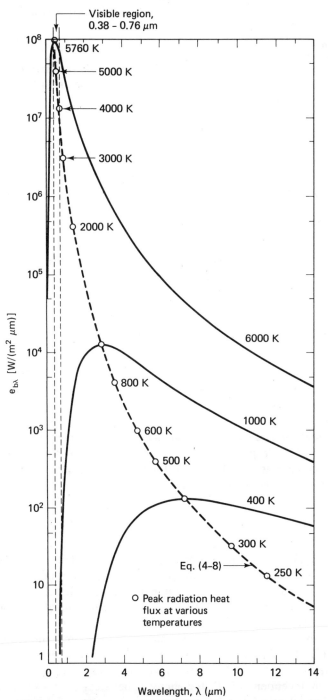

FIGURE 4-6 Monochromatic emissive power of a blackbody for several temperatures.

Solution

We utilize Table A-G-1 to determine the fraction of solar radiation that falls in the wavelength bands from 0 to 0.38 μm and from 0 to 0.76 μm. Referring to Table A-G-1, we find that

$$\frac{E_{b,0\to\lambda_1}}{E_b} = 0.102$$

for $\lambda_1 T_s = (0.38\ \mu\text{m})\ (5800\ \text{K}) = 2200\ \mu\text{m K}$, and

$$\frac{E_{b,0\to\lambda_2}}{E_b} = 0.55$$

for $\lambda_2 T_s = (0.76\ \mu\text{m})\ (5800\ \text{K}) = 4410\ \mu\text{m K}$. The difference between these two values gives the fraction of solar radiation falling in the visible spectrum.

$$\frac{E_{b,\lambda_1\to\lambda_2}}{E_b} = 0.55 - 0.102 = 0.448$$

Therefore, approximately 44.8% of solar radiation is visible to the human eye, with about 10.2% lying in the ultraviolet zone and 45% in the infrared region.

Thermal Radiation Emitted from Real Surfaces   The monochromatic emissive power e_λ of polished copper, anodized aluminum, and a blackbody are compared in Fig. 4-7. The ratio $e_\lambda / e_{b\lambda}$ at any given wavelength is known as the *monochromatic emissivity* ε_λ:

$$\varepsilon_\lambda = \frac{e_\lambda}{e_{b\lambda}} \tag{4-9}$$

Like the emissivity ε, ε_λ is a property that is dependent on the surface alone. The monochromatic emissivities of these two real surfaces are shown in Fig. 4-8. Note that ε_λ of these surfaces are less than unity and vary rather irregularly with λ. To at least some extent, all real surfaces exhibit these same characteristics. To reinforce this point, the monochromatic emissivities of several other materials at room temperature are shown in Fig. 4-9. An impressive listing of data for ε_λ is given in [1–4] for many types of surfaces. Data in these references indicate that ε_λ is essentially independent of T_s for many substances. For example, ε_λ is nearly independent of surface temperature for metals such as polished copper, polished iron, and tungsten and for nonmetals such as carbon,

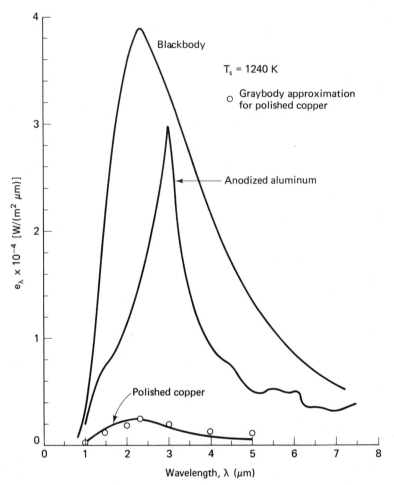

FIGURE 4-7 Monochromatic emissive power for blackbody and representative real surfaces at 1240 K.

Pyrex, and certain carbides. However, large changes occur in ε_λ with T_s for many nonmetallic substances, such as aluminum oxide (Al_2O_3).

The relationship between the emissivity ε and the monochromatic emissivity ε_λ is obtained by writing

$$\varepsilon = \frac{E}{E_b}$$

$$= \frac{\displaystyle\int_0^\infty e_\lambda \, d\lambda}{\displaystyle\int_0^\infty e_{b\lambda} \, d\lambda} = \frac{\displaystyle\int_0^\infty \varepsilon_\lambda e_{b\lambda} \, d\lambda}{\displaystyle\int_0^\infty e_{b\lambda} \, d\lambda} \tag{4-10}$$

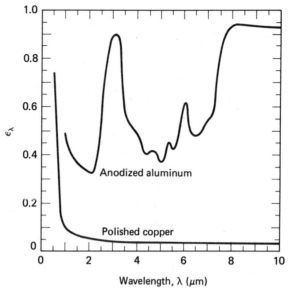

FIGURE 4-8 Monochromatic emissivity of polished copper and anodized aluminum. (From Dunkle et al. [7] and Seban [8].)

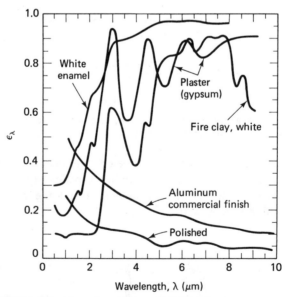

FIGURE 4-9 Monochromatic emissivity of several materials. (From Dunkle et al. [7] and Sieber [9].)

Note that for the case in which ε_λ is independent of λ, Eq. (4-10) reduces to

$$\varepsilon = \varepsilon_\lambda \qquad (4\text{-}11)$$

Surfaces that satisfy this equation are known as *graybodies*. Referring to Fig. 4-8, we see that ε_λ of the polished copper surface exhibits approximate graybody behavior for wavelengths greater than 1 to 2. A graybody approximation for the monochromatic emissive power $e_{g\lambda}$ for real surfaces such as this is given by

$$e_{g\lambda} = \varepsilon e_{b\lambda} \qquad (4\text{-}12)$$

This graybody approximation for the monochromatic emissive power of polished copper is shown in Fig. 4-7. Notice that $e_{g\lambda}$ follows the same wavelength-dependence pattern as $e_{b\lambda}$. The anodized aluminum and the other substances shown in Fig. 4-9 are observed to exhibit distinct non-graybody characteristics. Furthermore, the polished copper surface is non-gray for wavelengths less than about unity.

Thermal radiation intensity

Referring to Fig. 4-10, the *intensity* I is defined as the total rate of thermal radiation emitted per unit solid angle $d\omega$ and per unit area normal to the direction ϕ, θ. (The intensity is sometimes defined in the literature in terms of the total radiant energy emitted, reflected, and transmitted from a surface.) For a surface area dA_s, the projected area is simply $dA_s \cos \phi$. The solid angle $d\omega$ is defined by

$$d\omega = \frac{dA_r}{r^2} \qquad (4\text{-}13)$$

where the area dA_r of the hemispherical surface element is shown in Fig. 4-10. The intensity I is written in terms of the total emissive power E,

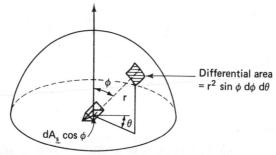

FIGURE 4-10 Geometric perspective for concept of intensity.

which is the rate of energy emitted per unit surface area dA_s, as

$$I = \frac{dE}{\cos \phi \, d\omega} \tag{4-14}$$

Rearranging this equation and assuming no variations in the emissive properties with the azimuthal angle θ, we have

$$\frac{dE}{d\omega} = e_\phi = I \cos \phi \tag{4-15}$$

where e_ϕ is the directional emissive power.

An important aspect of ideal blackbody emitting surfaces is that the intensity I is the same in all directions. Surfaces that exhibit this characteristic are said to be *diffuse* emitters. Many real surfaces such as industrially rough surfaces approach diffuse conditions. Utilizing Eq. (4-15), I is equal to the directional emissive power normal to the surface e_0 for diffuse surfaces. For this case, Eq. (4-15) reduces to the form

$$e_\phi = e_0 \cos \phi \tag{4-16}$$

which is known as the *Lambert cosine law*.

The *directional emissivity* ε_ϕ,

$$\varepsilon_\phi = \frac{e_\phi}{e_{b\phi}} = \frac{I}{I_b} \tag{4-17}$$

is shown in Fig. 4-11 for several real surfaces. Notice that the nonconductors are essentially diffuse for ϕ less than about 40 degrees but violate the Lambert cosine law for larger angles, with ε_ϕ falling to very small values. The metallic surfaces obey the Lambert cosine law over about the same range in ϕ, but ε_ϕ increases quite sharply before falling to zero at 90 degrees. Consistent with these findings, a hot metallic sphere will appear brighter near the base ($\phi \simeq 80$ degrees) than at the center ($\phi \simeq 0$ degrees). The opposite holds true for a nonmetallic sphere. These directional effects are generally accounted for in design by utilizing the emissivity ε which accounts for the radiant energy emitted into the entire hemispherical space above the surface. The emissivity ε is usually set equal to some fraction of ε_0. For example, $\varepsilon/\varepsilon_0 \simeq 1.2$ for bright metallic surfaces and $\varepsilon/\varepsilon_0 \simeq 0.96$ for nonconductors.

EXAMPLE 4-2

Develop an expression for the total emissive power E of a surface in terms of the intensity I.

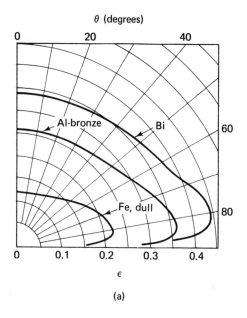

(a)

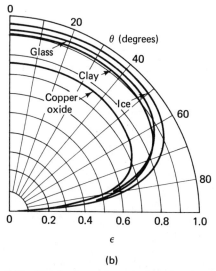

(b)

FIGURE 4-11 Distribution of total directional emissivity for several surfaces. (a) Metal surfaces. (b) Electric nonconducting surfaces. (From Schmidt and Eckert [10].)

Solution

Based on Eqs. (4-13) and (4-14), we write

$$dE = I \frac{\cos \phi}{r^2} dA_r$$

Referring to Fig. 4-10, dA_r is equal to $(r\, d\phi)\, (r \sin \phi\, d\theta)$, such that dE becomes

$$dE = I \cos \phi \sin \phi\, d\theta\, d\phi$$

To obtain E, we integrate over the hemisphere (i.e., $0 \leqslant \theta \leqslant 2\pi, 0 \leqslant \phi \leqslant \pi/2$).

$$E = \int_0^{\pi/2} \int_0^{2\pi} I \cos \phi \sin \phi\, d\theta\, d\phi$$

For a diffuse surface, I is uniform and is brought outside the integrals,

$$E = I \int_0^{\pi/2} \int_0^{2\pi} \cos \phi \sin \phi\, d\theta\, d\phi \qquad (a)$$

By employing the double-angle trigonometric formula (see Prob. 4-12), we obtain

$$E = \pi I$$

Thus, for a diffuse surface, E is equal to I times one-half the solid angle of a sphere.

4-3-3   Surface Irradiation Properties

Total Irradiation properties

As illustrated in Fig. 4-12, thermal radiation incident upon a surface is absorbed, reflected, and transmitted through the body. The *absorptivity* α, *reflectivity* ρ, and *transmissivity* τ were defined in Chap. 1. These total hemispherical surface irradiation properties account for the fractions of incident thermal radiation flux G at *all* wavelengths over the entire hemisphere above a surface that are absorbed, reflected, and transmitted. The incoming thermal radiation flux G is called the *irradiation*. Referring to Fig. 4-12, the thermal irradiation received by the surface is distributed as follows:

Thermal radiation flux absorbed	αG
Thermal radiation flux reflected	ρG
Thermal radiation flux transmitted	τG
Total irradiation	G

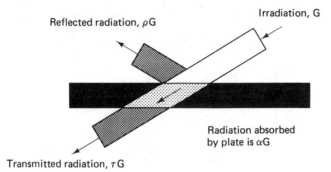

Reflected radiation, ρG

Irradiation, G

Radiation absorbed by plate is αG

Transmitted radiation, τG

FIGURE 4-12 Absorption, reflection, and transmission of thermal radiation incident on a surface.

The relationship between these surface irradiation properties is given by Eq. (1-19),

$$\alpha + \rho + \tau = 1 \tag{4-18}$$

Except for a few materials such as glass, rock salt, and other inorganic crystals, most solids are essentially opaque, with τ equal to zero.

Although all real surfaces reflect and/or transmit at least some thermal radiation, the concept of an ideal blackbody surface which absorbs all incident irradiation (i.e., $\alpha = 1$, $\rho = 0$, $\tau = 0$) is extremely important. As we have already seen, it is for such ideal absorbing and emitting surfaces that the pioneering theoretical studies by Stefan, Boltzmann, Planck, and others were developed. Moreover, the radiative performance of ideal blackbody surfaces provides a standard against which the performance of real surfaces can be compared. The term "blackbody" stems from the observation that surfaces which absorb nearly all of the thermal radiation in the visible part of the electromagnetic spectrum are black in color as a result of the absence of reflected light. The eye is a very good indicator of reflected visible thermal radiation, but is totally insensitive to the reflection of thermal radiation outside this narrow wavelength spectrum. It just so happens that surfaces that appear black in color generally are also good absorbers of thermal radiation outside the visible range.

The total hemispherical absorption, reflection, and transmission properties are dependent upon the nature and temperature T_R of the emitting source and upon the character of the receiving surface. The importance of T_R and the type of surface is shown in Fig. 4-13. In this figure, the absorptivity α is shown as a function of emitter source temperature T_R for several common materials which are at room temperature. Notice that the white fire clay, which would be judged by the eye to be a

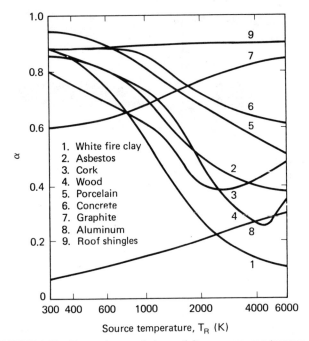

FIGURE 4-13 Dependence of absorptivity α on source temperature of incident thermal radiation. (From Sieber [9].)

poor absorber of thermal radiation, is actually a very good absorber ($\alpha \gtrsim 0.8$) of thermal radiation which is emitted from sources with temperatures below 500 K. However, white fire clay reflects most of the incoming solar radiation ($\alpha \simeq 0.1$) which is associated with an effective blackbody source temperature of approximately 5800 K.

Data for α, ρ, and τ are available in [1]–[4] and elsewhere for many surfaces. However, these data are generally restricted to situations involving irradiation from ideal emitting sources at one or two values of T_R. For example, considerable data are available for surfaces that receive solar radiation and radiation from blackbody sources at 300 K. But aside from the calculations by Sieber [9] which are shown in Fig. 4-13, relatively little information is available in the literature on the general effect of T_R on the surface irradiation properties. To circumvent this problem, α can be evaluated on the basis of tabulated data for ε for certain situations. This important practical point is developed in the next section.

EXAMPLE 4-3

No perfect blackbody surface has been found to exist. Referring to Table A-C-7 in the Appendix, we see that emissivities of the order of 0.95 to 0.98 are common for near-blackbodies such as surfaces coated with flat black lacquer and asbestos board. However, geometrical blackbodies can be constructed which perform even closer to the ideal. Show that blackbody conditions are approached by a small hole in the wall of a large cavity with a highly absorbing surface.

Solution

A spherical cavity with a small opening in its wall is shown in Fig. E4-3. If we trace the path of an incident ray of thermal radiation entering the cavity, we find that the ray is reflected within the interior of the cavity many times, with a large part of the energy being absorbed each time. When the reflected ray eventually reaches the opening and escapes, its energy content is extremely small. Because nearly all of the thermal radiation entering the cavity is absorbed, blackbody conditions are approached in this system.

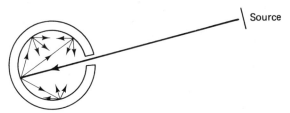

FIGURE E4-3 Approximate geometric blackbody.

EXAMPLE 4-4

The filament of an incandescent lamp operates at 2500 K. Assuming approximate graybody characteristics, estimate the fraction of radiant energy emitted by the filament that falls in the visible spectrum.

Solution

Because the spectral distribution of radiant energy emitted by a graybody is the same as for a blackbody, we utilize Table A-G-1 to solve this problem. Following the pattern established in Example 4-1, we have

$$\frac{E_{b,0\to\lambda_1}}{E_b} = \frac{E_{g,0\to\lambda_1}}{E_g} = 0.00021$$

for $\lambda_1 T_s = (0.38 \ \mu m)(2500 \ K) = 950 \ \mu m \ K$, and

$$\frac{E_{b,0\to\lambda_2}}{E_b} = \frac{E_{g,0\to\lambda_2}}{E_g} = 0.0522$$

for $\lambda_2 T_s = (0.76 \ \mu m)(2500 \ K) = 1900 \ \mu m \ K$. Therefore, the fraction of energy that falls in the visible part of the spectrum is

$$\frac{E_{g,\lambda_1 \to \lambda_2}}{E_g} = 0.0522 - 0.00021 = 0.052$$

For this operating temperature, only 5.2% of the radiant energy dissipated by the filament produces light, with most of the remaining energy producing infrared heating.

Monochromatic Irradiation properties

Spectral surface irradiation properties are now defined which account for the effect of wavelength λ. The *monochromatic absorptivity* α_λ is the fraction of incident thermal radiation with wavelength λ that is absorbed by a surface. Similarly, the *monochromatic reflectivity* ρ_λ and the *monochromatic transmissivity* τ_λ represent the fractions of incoming thermal radiation with wavelength λ that are reflected and transmitted, respectively. These important spectral surface irradiation properties are related by

$$\alpha_\lambda + \rho_\lambda + \tau_\lambda = 1 \qquad (4\text{-}19)$$

The relationship between the monochromatic absorptivity α_λ of a receiving surface and its monochromatic emissivity ε_λ is given by the *Kirchhoff law* for thermal radiation, which states essentially that α_λ and ε_λ are equal for a system in thermodynamic equilibrium. Although Kirchhoff's law strictly only applies to systems in which the temperature T_R of the source of the thermal radiation is equal to T_s, this law has been found to apply to nonisothermal conditions for many practical situations. This extension of the Kirchhoff law takes the form

$$\alpha_\lambda = \varepsilon_\lambda \qquad (4\text{-}20)$$

(The limitations of this equation are discussed in [1], [5], [11], and [18].) In this regard, the monochromatic emissivities shown in Figs. 4-8 and 4-9 for several metallic and nonmetallic materials are also equal to the monochromatic absorptivities of these surfaces. In addition to the considerable amount of data available for ε_λ and α_λ, extensive tabulations of data have also been developed for ρ_λ and τ_λ. Again, [1]–[4] are very good sources of data for these spectral surface irradiation properties.

The monochromatic transmissivity τ_λ is shown in Fig. 4-14(a) and (b) for ordinary window glass and type 2A diamond. It is observed that glass transmits thermal radiation quite well in the low-wavelength visible range of the electromagnetic spectrum. However, in the longer-wavelength in-

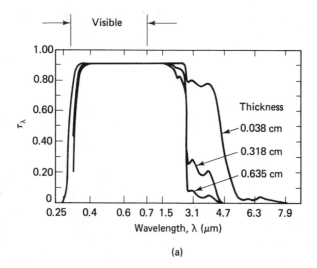

(a)

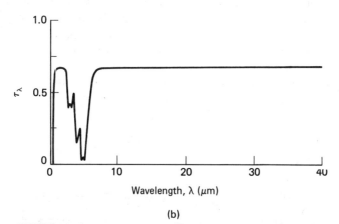

(b)

FIGURE 4-14 (a) Monochromatic transmissivity τ_λ of glass (with 0.02 Fe_2O_3) at room temperature. (From Dietz [12].) (b) Monochromatic transmissivity τ_λ of type 2A diamond at room temperature. (From Seal [13].)

frared part of the spectrum ($\lambda \gtrsim 2.6$ μm), the glass is nearly opaque to thermal radiation, with most of the energy being absorbed and reflected. As we have already mentioned, this nongraybody characteristic of glass is responsible for the greenhouse effect, in which glass is transparent to short-wavelength irradiation from the sun but is essentially opaque to longer-wavelength thermal radiation emitted by the low-temperature interior of an enclosure. Some plastic films such as polyethylene have similar characteristics.

On the other hand, the type 2A diamond is transparent throughout much of the infrared region (6 μm $\gtrsim\lambda\gtrsim$40 μm), as well as in visible range. Because of its transparency across such a broad range of wavelengths and because of its great strength, type 2A diamond was utilized in the development of two 18.2-mm-diameter windows for the 1978 Pioneer Venus Space Mission. The function of the windows was to protect the infrared radiometer equipment from the extremely hostile Venusian environment. The conditions withstood by the diamond window included "a 4-month journey through the cold and vacuum of space, entry decelerations to 565g, searing heat (the Venusian surface is red hot), crushing pressure (100 times that of the earth's atmosphere), and a highly corrosive atmosphere containing sulfuric acid and other aggressive gases" [13].

Representative measurements are shown in Fig. 4-15 for the monochromatic reflectivity ρ_λ for several surfaces. These surfaces are essentially opaque, such that α_λ is equal to $1-\rho_\lambda$. Notice that ρ_λ of the pure aluminum surface is quite high for all wavelengths. But ρ_λ for aluminum surfaces with certain coatings such as lead sulfide fall to much smaller values for wavelengths below 3 μm. This selective characteristic of certain types of metallic surfaces is very important in solar applications.

In order to express the absorptivity α in terms of α_λ (or ϵ_λ) we first introduce the *monochromatic irradiation* g_λ, which is defined as the irradiation flux per unit wavelength. The total irradiation G and the monochromatic irradiation g_λ are themselves related by

$$G = \int_0^\infty g_\lambda \, d\lambda \qquad (4-21)$$

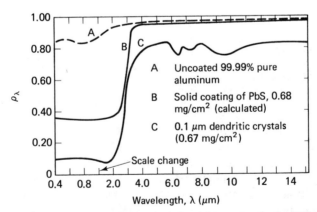

FIGURE 4-15 Monochromatic reflectivity ρ_λ for lead sulfide coatings on aluminum substrates. (From Williams et al. [14].)

The absorptivity is now formally defined by the equation

$$\alpha = \frac{\int_0^\infty \alpha_\lambda g_\lambda \, d\lambda}{\int_0^\infty g_\lambda \, d\lambda} \tag{4-22}$$

For the many practical situations in which $\alpha_\lambda = \varepsilon_\lambda$, this expression becomes

$$\alpha = \frac{\int_0^\infty \varepsilon_\lambda g_\lambda \, d\lambda}{\int_0^\infty g_\lambda \, d\lambda} \tag{4-23}$$

This equation provides a basis for the evaluation of α in terms of ε for several situations. First, for approximate graybody conditions, ε_λ is essentially independent of wavelength and equal to ε, such that Eq. (4-23) reduces to the *graybody* α *approximation*

$$\alpha = \varepsilon \tag{4-24}$$

where both α and ε are evaluated at T_s. Although very few materials exist for which ε_λ is constant over the entire wavelength, some substances exhibit approximate graybody characteristics over significant parts of the spectrum. The key issue when using this graybody α approximation is that ε_λ (and α_λ) be essentially uniform in the wavelength range where there are appreciable amounts of both emitted and incident radiation. For example, this graybody α approximation could be used for a polished copper surface with T_s and T_R both less than approximately 1000 K, but should not be used for situations in which T_R is much greater than 1000 K.

For nongraybody surfaces, the use of Eq. (4-23) in evaluating α requires that the monochromatic irradiation g_λ be specified. For situations involving blackbody or graybody radiation sources, g_λ in Eq. (4-23) can be replaced by $e_{b\lambda}$; that is,

$$\alpha = \frac{\int_0^\infty \varepsilon_\lambda(T_s) e_{b\lambda}(T_R) \, d\lambda}{\int_0^\infty e_{b\lambda}(T_R) \, d\lambda} \tag{4-25}$$

By comparing Eqs. (4-10) and (4-25), we see that the absorptivity is approximately equal to the emissivity for an approximate blackbody or graybody source with T_R equal to T_s. However, it should be noted that the rate of radiation heat transfer is zero for such isothermal conditions. It follows that the most practical consequence of Eq. (4-25) is its application to cases in which T_s and T_R are *not* equal. The first and most obvious extension of this law applies to systems in which the difference between T_s and T_R is small. For such cases, we simply utilize the *isothermal* α *approximation*

$$\alpha = \varepsilon(T_s) \qquad (4\text{-}26)$$

which is equivalent to the graybody α approximation given by Eq. (4-24). However, because of the strong dependence of the surface irradiation properties on the characteristics of the incident thermal radiation for nongraybody surfaces, the following *nonisothermal* α *approximation* is generally preferred:

$$\alpha = \varepsilon(T_R) \qquad (4\text{-}27)$$

This equation is obtained from Eq. (4-25) by merely setting $\varepsilon_\lambda(T_s)$ equal to $\varepsilon_\lambda(T_R)$. As we have already noted, this restriction on ε_λ is essentially satisfied by many metals and certain nonmetallic substances. [For thermal radiation from a blackbody or graybody source incident on a metallic surface with T_R low enough to exclude significant radiation in the near-infrared, visible, or ultraviolet ranges, some investigators prefer an α approximation of the form $\alpha = \varepsilon(\sqrt{T_s T_R}\,)$.]

For nongraybody surfaces for which $\varepsilon_\lambda(T_s)$ is not approximately equal to $\varepsilon_\lambda(T_R)$, Eq. (4-25) can be integrated numerically, provided that $\varepsilon_\lambda(T_s)$ is known. This approach was used by Sieber [9] to obtain the absorptivities shown in Fig. 4-13.

To evaluate the graybody (or isothermal) and nonisothermal α approximations, we consider data for commercial-finish aluminum surfaces. Referring back to Fig. 4-8, we see that the aluminum surface exhibits nongraybody characteristics. A comparison of the data for ε shown in Fig. 4-3 with the values of α given in Fig. 4-13 reveals a very close agreement between the two. This result reinforces our confidence in the usefulness of the nonisothermal α approximation. On the other hand, the very fact that α for aluminum shown in Fig. 4-13 ranges from a value less than 0.1 for a source temperature T_R of 300 K to 0.3 for T_R equal to 6000 K alerts us to the limitation of the graybody or isothermal α approximation for this material. For that matter, except for applications involving small temperature differences, the use of the graybody α approximation would lead to serious error in the analysis of radiation from surfaces 1 through 8 listed in Fig. 4-13.

For situations in which most of the irradiation upon a surface originates from a single approximate blackbody or graybody source, the nonisothermal approximation is fairly easy to employ. For example, in the heating of a small metal ball in a furnace, α can readily be estimated by using Eq. (4-27). However, in applications involving multiple nongraybody surfaces and complex reflection patterns, the nonisothermal α approximation is less reliable and is not so easily administered, since the irradiation upon a surface originates from various sources. Consequently, the simple graybody α approximation is often utilized, at least as a first cut.

Directional effects

Finally, mention should be made of the directional characteristics of irradiation surface properties. Two limiting types of reflection are illustrated in Fig. 4-16. The reflection is *diffuse* if the intensity of the reflected thermal radiation is constant for all angles of irradiation and reflection. On the other hand, if the angle of reflection is equal to the angle of incidence, the reflection is *specular*. Although real surfaces have neither diffuse nor specular irradiation surface properties, many common surfaces can be placed in one or the other of these categories for purpose of design. For example, industrially rough surfaces essentially possess diffuse properties, with polished and smooth surfaces exhibiting near-specular characteristics.

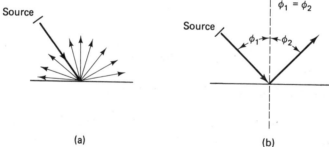

FIGURE 4-16 Types of reflection. (a) Diffuse reflection. (b) Spectral reflection.

EXAMPLE 4-5

Assuming that a glass plate transmits 90% of the incident thermal radiation in the wavelength range from 0.29 to 2.70 μm (see Fig. 4-14), is opaque outside this range, and reflects 5% for all wavelengths, estimate the total transmissivity τ of the glass for solar radiation.

Solution

Assuming that the sun behaves as a blackbody radiator at 5800 K, we utilize Table A-G-1 to calculate the fraction of solar radiation that falls in the wavelength band from 0.29 to 2.70 μm. It follows that

$$\frac{E_{b,0 \to \lambda_1}}{E_b} = 0.0254$$

for $\lambda_1 T_R = (0.29\ \mu\text{m})(5800\ \text{K}) = 1680\ \mu\text{m K}$, and

$$\frac{E_{b,0 \to \lambda_2}}{E_b} = 0.972$$

for $\lambda_2 T_R = (2.70\ \mu\text{m})(5800\ \text{K}) = 15{,}700\ \mu\text{m K}$. Thus, the fraction of the solar radiation falling in the wavelength band for which τ_λ is 0.9 is

$$\frac{E_{b,\lambda_1 \to \lambda_2}}{E_b} = 0.972 - 0.0254 = 0.947$$

It follows that the total transmissivity τ is

$$\tau = (0.947)(0.9) = 0.852$$

To calculate the total absorptivity, we write

$$\begin{aligned} \alpha &= 1 - \rho - \tau \\ &= 1 - 0.05 - 0.852 \\ &= 0.098 \end{aligned}$$

4-3-4   Thermal Radiation Properties of Gases

We are reminded that a vacuum provides the ideal medium for the transfer of thermal radiation from one surface to another. Thermal radiation passes through such evacuated spaces at the speed of light. Of course, in most practical situations at least some fluid, often in the form of a gas, lies in the path of the thermal radiation. Elementary gases such as oxygen (O_2), hydrogen (H_2), nitrogen (N_2), and dry air, which have symmetrical molecular structures, are essentially transparent to thermal radiation at low to moderate temperatures. Hence, the presence of such *nonparticipating gases* can generally be ignored. But, other polyatomic gases, such as water vapor (H_2O), carbon dioxide (CO_2), sulfur dioxide (SO_2), and various hydrocarbons absorb and emit significant amounts of thermal radiation. Gases such as these are known as *participating gases*. For example, water vapor contained at 1 atm pressure and 100°C between two plates 1 m apart would emit as much as 55% of the energy that would be emitted by a blackbody at 100°C with the same surface area as the plates. Carbon dioxide at the same temperature and pressure would emit about 20%. But carbon monoxide, which is diatomic, would only emit about 3% of the energy that would be emitted by an ideal radiator. Hence, the presence of participating gases between thermal radiating surfaces can have very important effects.

Thermal radiation emission and absorption characteristics of participating gases are generally much more complicated than for opaque solids. Whereas the emission and absorption of thermal radiation for opaque materials are surface phenomena, the thickness, shape, surface area, pressure, and temperature distribution can all affect thermal radiation in gases. Representative measurements for α_λ are shown in Fig. 4-17 for carbon dioxide in terms of the wave number $1/\lambda$. The absorbing spectrum of this important gas is seen to consist of distinct narrow bands. These absorbing and emitting patterns are typical of participating gases.

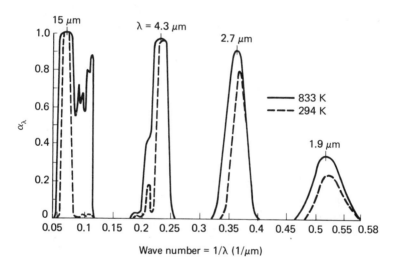

FIGURE 4-17 Monochromatic absorptivity α_λ for carbon dioxide. (From Edwards [15].)

Concerning the more practical information pertaining to the total properties which account for all wavelengths and all directions in which thermal radiation passes, Hottel and Egbert [16] have developed the charts shown in Fig. 4-18(a) and (b) for water vapor and carbon dioxide total emissivities ε. Note that ε increases with increasing *mean beam length* L_e and partial pressure. The mean beam length, which accounts for all possible directions the thermal radiation may take, is listed in Table 4-2 for several standard systems. For situations in which L_e has not been evaluated, it is generally approximated by $L_e = 3.6V/A_s$. The data in Fig. 4-18(a) and (b) were obtained for a total pressure of 1 atm. Correction

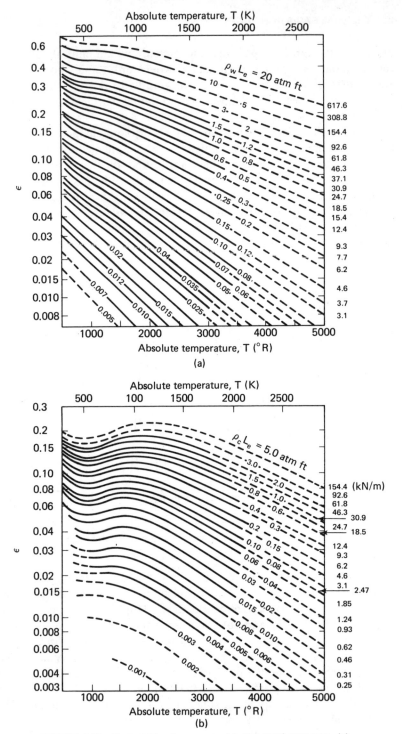

FIGURE 4-18 Emissivities for gases at 1 atm total pressure. (a) Water vapor. (b) Carbon dioxide. (From Hottel and Egbert [16].)

TABLE 4-2 Mean equivalent beam length L_e for radiation from entire gas volume

Gas volume	Characteristic dimension	L_e
Volume between two infinite planes	Separation distance L	$1.8L$
Cube; radiation to any face	Edge L	$0.60L$
Sphere	Diameter D	$0.65D$
Circular cylinder with $L = D$; radiation to entire surface	Diameter D	$0.60D$
Circular cylinder, with semi-infinite length; radiation to entire base	Diameter D	$0.65D$

Source: From [17] and [18].

charts are available in [17] for systems under other total pressures. For systems involving combustion, both water vapor and carbon dioxide are present. For such situations, the total gas emissivity is simply equal to the sum of the emissivities for each component, minus a small correction factor that accounts for the mutual emission that takes place between the two gases. Corrections that account for this mutual emission factor are presented in [17]. This small difference can generally be neglected as a first approximation.

In regard to the total absorptivity α of participating gases, Hottel [17] has developed approximate graybody correlations for water vapor and carbon dioxide which take the forms

$$\alpha_w = \varepsilon_w(T_s)\left(\frac{T_m}{T_s}\right)^{0.45} \tag{4-28a}$$

$$\alpha_c = \varepsilon_c(T_s)\left(\frac{T_m}{T_s}\right)^{0.65} \tag{4-28b}$$

where T_m is the mean temperature of the gas and T_s the surface temperature. ε_w and ε_c are evaluated from Fig. 4-18 with the parameters $P_w L_e$ and $P_c L_e$ replaced by $P_w L_e T_s/T_m$ and $P_c L_e T_s/T_m$. For a mixture of water vapor and carbon dioxide, the absorptivity of the mixture is simply equal to the sum of α_w and α_c, minus a small correction.

It should be mentioned that in the high-temperature process of combustion which produces nonluminous products such as H_2O and CO_2, clouds of carbon particles radiate intensely at short wavelengths in the visible-light region of the electromagnetic spectrum. This visible thermal radiation which is emitted during combustion is what is referred to as the *flame*. The total emissivities of luminous flames range from values of the order of 0.2 for gaseous hydrocarbon fuels to almost unity for fuels such as oil which are burned under conditions of large carbon/hydrogen ratios. In engineering applications, the emission from luminous flames generally must be determined experimentally.

EXAMPLE 4-6

The walls of a cubical furnace 1 m on a side are maintained at 500 K. The products of combustion at 1 atm and 1500 K consist of 20% CO_2, 15% H_2O, and 65% N_2 by volume. Assuming that the walls are essentially black, estimate the gas emissivity ε_m and absorptivity α_m.

Solution

According to Dalton's law of partial pressure, the partial pressure of an ideal gas in a mixture is equal to the product of the total pressure of the mixture and the volume fraction of the gas. It follows that

$$P_w = (0.15)(1 \text{ atm}) = 0.15 \text{ atm}$$
$$P_c = (0.2)(1 \text{ atm}) = 0.2 \text{ atm}$$

According to Table 4-2, the mean equivalent beam length L_e is given by

$$L_e = 0.6L = 0.6(1 \text{ m}) = 0.6 \text{ m}$$

Thus, the pressure/length parameters are

$$P_w L_e = 0.15 \text{ atm}(0.6 \text{ m}) = 0.09 \text{ atm m} = 0.295 \text{ atm ft}$$
$$P_c L_e = 0.2 \text{ atm}(0.6 \text{ m}) = 0.12 \text{ atm m} = 0.393 \text{ atm ft}$$

Referring to Fig. 4-18, the emissivities ε_w and ε_c are evaluated for a mean gas temperature T_m of 1500 K (1230°C).

$$\varepsilon_w = 0.08$$
$$\varepsilon_c = 0.094$$

Thus, the total emissivity of the gas is approximated by

$$\varepsilon_m = \varepsilon_w + \varepsilon_c = 0.08 + 0.094$$
$$= 0.174$$

By reference to [17], we find that the correction for ε_m which accounts for the mutual emission of the two participating gases is of the order of -10%. Therefore, as a first approximation, we are safe in assuming $\varepsilon_m \simeq 0.174$.

To estimate the absorptivities, we calculate the modified pressure/length parameters.

$$P_w L_e \frac{T_s}{T_m} = (0.295 \text{ atm ft}) \frac{500 \text{ K}}{1500 \text{ K}} = 0.0983 \text{ atm ft}$$

$$P_c L_e \frac{T_s}{T_m} = (0.393 \text{ atm ft}) \frac{500 \text{ K}}{1500 \text{ K}} = 0.131 \text{ atm ft}$$

Evaluating the emissivities at T_s, we have

$$\varepsilon_w(T_s) = 0.097$$
$$\varepsilon_c(T_s) = 0.075$$

The absorptivities are now calculated by utilizing Eqs. (4-28a) and (4-28b).

$$\alpha_w = \varepsilon_w(T_s) \left(\frac{T_m}{T_s} \right)^{0.45}$$

$$= 0.097 \left(\frac{1500 \text{ K}}{500 \text{ K}} \right)^{0.45} = 0.159$$

$$\alpha_c = \varepsilon_c(T_s) \left(\frac{T_m}{T_s} \right)^{0.65}$$

$$= 0.075 \left(\frac{1500 \text{ K}}{500 \text{ K}} \right)^{0.65} = 0.153$$

Referring to [17], we find the correction for mutual absorption is less than 1%. It follows that the total absorptivity of the gas is about

$$\alpha_m = \alpha_w + \alpha_c$$
$$= 0.159 + 0.153 = 0.312$$

4-4   RADIATION SHAPE FACTOR

4-4-1   Introduction

As a final step toward our objective of developing practical predictions for radiation heat transfer, attention is turned to important geometric aspects of thermal radiation exchange between surfaces.

The *radiation shape factor* F_{s-R} introduced in Chap. 1 is defined as the fraction of thermal radiation leaving a diffuse surface A_s that passes through a nonparticipating medium to surface A_R. F_{s-R} is also commonly referred to as *view factor, configuration factor,* and *angle factor*. Because we will be dealing with systems involving more than one source, it is convenient to replace the subscript R by the source surface identification index j, where $j = 1, 2, 3, \ldots, N$. Thus, the radiation shape factor will generally be denoted by F_{s-j}, except for systems involving only two surfaces, for which case we will write F_{1-2}.

To illustrate, we consider the two simple systems shown in Fig. 4-19(a) and (b). The radiation shape factors F_{1-2} for the infinitely long parallel-plate system of Fig. 4-19(a) are both unity, because all the energy leaving either surface reaches the other surface. For the concentric-sphere arrangement shown in Fig. 4-19(b), F_{1-2} is also equal to unity. But, because only a fraction of the energy leaving the outer spherical surface A_2 reaches the inner surface A_1, F_{2-1} is less than unity.

We now turn our attention to fundamental principles that apply to radiation shape factors and design curves for several standard geometries.

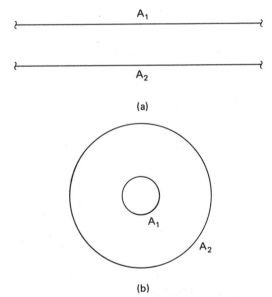

(a)

(b)

FIGURE 4-19   (a) Parallel-plate system for which $F_{1-2} = 1$ and $F_{2-1} = 1$. (b) Concentric sphere system for which $F_{1-2} = 1$, but $F_{2-1} < 1$.

4-4-2 · Law of Reciprocity

The general relationship between F_{s-j} and F_{j-s} is given by the *reciprocity law*,

$$A_s F_{s-j} = A_j F_{j-s} \qquad (4\text{-}29)$$

This relationship is developed in [5] and [11] on the basis of geometric considerations (see Prob. 4-59). Based on this principle, we see that F_{2-1} for the spherical system shown in Fig. 4-19(b) is

$$F_{2-1} = \frac{A_1}{A_2} F_{1-2} = \left(\frac{r_1}{r_2}\right)^2 \qquad (4\text{-}30)$$

Note that F_{2-1} approaches zero as r_1/r_2 decreases, and approaches unity as r_1 approaches r_2.

4-4-3   Summation Principles

Another important geometric concept pertains to the radiation shape factors from surface s to N surfaces forming an enclosure. This first summation principle takes the form

$$\sum_{j=1}^{N} F_{s-j} = 1 \qquad (4\text{-}31)$$

For example, for the three-surface enclosure shown in Fig. 4-20, we have

$$F_{1-2} + F_{1-3} = 1 \qquad (4\text{-}32)$$

This principle also requires that

$$F_{2-1} + F_{2-3} = 1 \qquad (4\text{-}33)$$

and

$$F_{3-1} + F_{3-2} = 1 \qquad (4\text{-}34)$$

For systems involving concave curved surfaces, we must include the term F_{s-s}, which accounts for the fraction of thermal radiation leaving surface s that is directly incident upon itself. Thus, for the spherical system of Fig. 4-19(b), we have

$$F_{2-1} + F_{2-2} = 1 \qquad (4\text{-}35)$$

A second summation principle states that the total radiation shape factor is equal to the sum of its parts. To illustrate, the radiation shape

FIGURE 4-20   Trisurface enclosure.

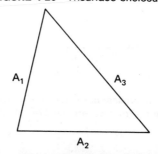

factor from surface A_1 to the combined surfaces of A_2 and A_3 in Fig. 4-20 is

$$F_{1-(2,3)} = F_{1-2} + F_{1-3} \qquad (4\text{-}36)$$

By employing the law of reciprocity, this equation is put into the useful form

$$(A_2 + A_3)F_{(2,3)-1} = A_2 F_{2-1} + A_3 F_{3-1} \qquad (4\text{-}37)$$

4-4-4   Design Curves

Relationships have been developed for radiation shape factors for a great many geometries. The theoretical approach to this task involves the use of the intensity concept for diffuse surfaces, as illustrated in Example 4-8.

Because the theoretical evaluation of radiation shape factors is generally quite involved, standard design curves are heavily relied upon in practice. Design equations developed by Mackey et al. [19] for radiation shape factors are given in Figs. 4-21 through 4-23 for several arrangements that are commonly encountered in practice. More comprehensive listings of radiation shape factors are given in [11], [17], [20]–[22]. As shown in Example 4-7, information on radiation shape factors for standard geometries such as those given in Figs. 4-21 and 4-22 can sometimes be extended

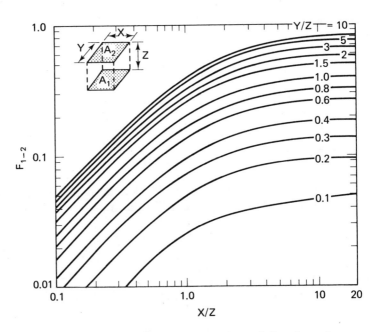

FIGURE 4-21 Theoretical calculations for radiation shape factor: parallel plates. (From Eckert and Drake [18].)

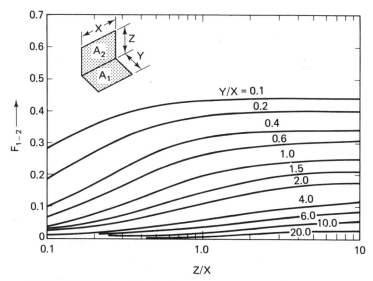

FIGURE 4-22 Theoretical calculations for radiation shape factor: perpendicular rectangles with a common edge. (From Eckert and Drake [18].)

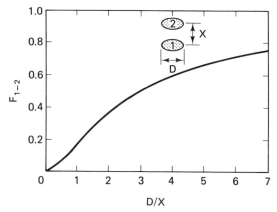

FIGURE 4-23 Theoretical calculations for radiation shape factor: parallel circular disks. (From Eckert and Drake [18].)

to other geometrical arrangements of practical interest by utilizing the reciprocity and summation principles.

EXAMPLE 4-7

Determine the radiation shape factor for surfaces A_1 and A_2 shown in Fig. E4-7.

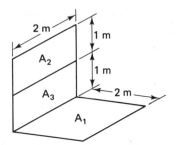

FIGURE E4-7   Trisurface system.

Solution

According to the second summation principle, we write

$$F_{1-(2,3)} = F_{1-2} + F_{1-3}$$

or

$$F_{1-2} = F_{1-(2,3)} - F_{1-3}$$

According to Fig. 4-22, $F_{1-3} = 0.15$ and $F_{1-(2,3)} = 0.2$, such that

$$F_{1-2} = 0.2 - 0.15 = 0.05$$

To obtain F_{2-1}, we utilize the principle of reciprocity.

$$F_{2-1} = \frac{A_1}{A_2} F_{1-2} = \left(\frac{2}{1}\right)(0.05) = 0.1$$

EXAMPLE 4-8

Develop an expression for the radiation shape factor from the differential element dA_1 to the finite disk A_2 shown in Fig. E4-8.

Solution

Referring to Fig. E4-8, the total rate of thermal radiation diffusely emitted by surface element dA_1 is $E_1 \, dA_1$. To determine the portion of

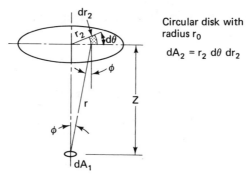

FIGURE E4-8 Parallel circular disks.

this energy that reaches surface A_2, we return to the definition of intensity given by Eq. (4-14),

$$I_1 = \frac{dE_1}{\cos \phi \, d\omega} \tag{4-14}$$

or

$$dE_1 = I_1 \cos \phi \, d\omega \tag{a}$$

where ϕ is the angle between the normal to surface dA_1 and the line r drawn between the area elements dA_1 and dA_2. The solid angle $d\omega$ is given by Eq. (4-13),

$$d\omega = \frac{dA_r}{r^2} \tag{4-13}$$

where dA_r is the differential area normal to the line r; that is,

$$dA_r = dA_2 \cos \phi \tag{b}$$

Thus, the thermal radiation flux emitted from dA_1 that reaches dA_2 is

$$dE_1 = I_1 \cos^2 \phi \, \frac{dA_2}{r^2} \tag{c}$$

where $I_1 = E_1/\pi$ for diffuse conditions (see Example 4-2).

The radiation shape factor $F_{dA_1-dA_2}$ is simply equal to the ratio dE_1/E_1; that is,

$$F_{dA_1-dA_2} = \frac{1}{\pi} \cos^2 \phi \frac{dA_2}{r^2} \tag{d}$$

To obtain $F_{dA_1-A_2}$, we must integrate over the entire surface A_2.

$$F_{dA_1-A_2} = \frac{1}{\pi} \int_{A_2} \cos^2 \phi \frac{dA_2}{r^2} \tag{e}$$

Noting that

$$dA_2 = r_2 \, d\theta \, dr_2 \tag{f}$$

we obtain

$$F_{dA_1-A_2} = \frac{1}{\pi} \int_0^{r_0} \int_0^{2\pi} \cos^2 \phi \frac{r_2 d\theta \, dr_2}{r^2} \tag{g}$$

The quantities r and ϕ are expressed in terms of r_2 by

$$r^2 = Z^2 + r_2^2 \tag{h}$$

$$\cos \phi = \frac{Z}{(Z^2 + r_2^2)^{1/2}} \tag{i}$$

Thus, Eq. (g) takes the form

$$F_{dA_1-A_2} = \frac{1}{\pi} \int_0^{r_0} \int_0^{2\pi} \frac{Z^2 r_2 \, d\theta \, dr_2}{(Z^2 + r_2^2)^2}$$

$$= 2Z^2 \int_0^{r_0} \frac{r_2 \, dr_2}{(Z^2 + r_2^2)^2} \tag{j}$$

Setting $Z^2 + r_2^2 = \xi$, we continue the integration process,

$$F_{dA_1-A_2} = Z^2 \int_{Z^2}^{Z^2+r_0^2} \frac{d\xi}{\xi^2}$$

$$= -\frac{Z^2}{\xi} \bigg|_{Z^2}^{Z^2+r_0^2} = Z^2 \left(\frac{1}{Z^2} - \frac{1}{Z^2 + r_0^2} \right)$$

$$= \frac{r_0^2}{Z^2 + r_0^2} \tag{k}$$

Notice that $F_{dA_1-A_2}$ appropriately approaches unity as the radius r_0 of the disk becomes large.

4-5 PRACTICAL THERMAL ANALYSIS OF RADIATION HEAT TRANSFER

4-5-1 Introduction

Now that we have a basic understanding of the physical mechanism, properties, and geometric factors pertaining to the thermal radiation phenomenon, we are in a position to develop predictions for the rate of heat transfer (i.e., *net* rate of exchange of thermal radiation) that occurs in diffuse thermal radiating systems. Therefore, we turn our attention to the practical thermal analysis of radiation heat transfer.

To illustrate the perspective that we will be taking, consider the radiating body in Fig. 4-24. The thermal radiation heat transfer q_R from surface A_s to other surrounding surfaces will be analyzed in sections to follow for cases in which the surface temperature T_s and thermal radiation properties are uniform over A_s. But we must recognize that the rate of thermal radiation heat transfer q_R from surface A_s must be in balance with changes in the rate of storage and the rates of energy transfer into and out of the body through other surfaces and by other means. That is, the energy transfer associated with this body must satisfy the first law of thermodynamics,

$$\sum \dot{E}_i = q_R + \frac{\Delta E_s}{\Delta t} \qquad (4\text{-}38)$$

where $\sum \dot{E}_i$ is the net rate of energy transfer into the body, not including q_R. For steady-state conditions, $\Delta E_s / \Delta t$ is zero with T_s being independent of time, such that Eq. (4-38) reduces to

$$\sum \dot{E}_i = q_R \qquad (4\text{-}39)$$

This equation simply indicates that the rate of thermal radiation heat transfer from surface A_s with steady surface temperature T_s must be

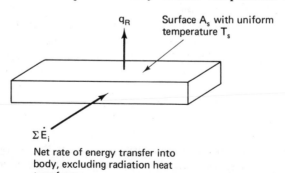

FIGURE 4-24 Thermal radiating body.

replaced from an outside source, such as by the electrical generation of power within the body. If, on the other hand, a net rate of thermal radiation is received by surface A_s such that q_R is negative, this same rate of energy must be transferred from the body into a heat sink if steady-state conditions are to be maintained. For example, energy can be taken out of the back surface of the body by radiation or convection.

For unsteady conditions, T_s is time-dependent and q_R must satisfy Eq. (4-38). That is, q_R must be in balance with the rate of energy brought into a body $\sum \dot{E}_i$ and the rate of change in energy stored within the body $\Delta E_s / \Delta t$. In this regard, for bodies with negligible thermal resistance to conduction (i.e., small thermal radiation Biot number Bi_R), the temperature throughout the body is essentially uniform. For such bodies, a lumped approximation can be made for $\Delta E_s / \Delta t$, with Eq. (4-38) taking the simpler form (for constant properties)

$$\sum \dot{E}_i = q_R + mc_v \frac{dT_s}{dt} \tag{4-40}$$

Keeping in mind that the energy transfer associated with the body of each radiating surface must satisfy the first law of thermodynamics, we now move on to the development of practical thermal analyses for the radiation heat transfer between two or more diffuse thermal radiating surfaces with uniform properties and uniform heating. To begin with, we focus on radiating surfaces in which conduction and convection are not significant. In addition to analyzing ideal blackbody and diffuse non-blackbody thermal radiating systems under steady-state conditions with a nonparticipating medium, consideration is given to systems involving unsteady conditions, participating medium, combined modes, and solar radiation. Approaches to analyzing more complex systems involving surfaces with nondiffuse characteristics, nonuniform properties, and nonuniform heating are introduced in references [5] and [11].

To simplify our notation, the subscript R for radiation heat transfer will be omitted throughout the remainder of this chapter, except for cases involving combined modes of heat transfer. Accordingly, the rate of radiation heat transfer from surface A_s to a second surface A_j will be designated by q_{s-j}. In addition, the total radiation-heat-transfer rate from surface A_s to N surfaces of an enclosure will be represented by q_s.

4-5-2 Blackbody Thermal Radiation

The analysis of radiation heat transfer in systems involving ideal blackbody surfaces is quite straightforward because the energy transfer is direct, with no reflections. In this section we will consider basic two-surface and multisurface blackbody systems. The analysis of these ideal systems provides a foundation for the practical analysis of real non-blackbody surfaces which is considered in Sec. 4-5-3. In addition, the results of this analysis provide us with a new viewpoint concerning the analogy between the flow of electric current and radiation heat transfer.

Bisurface systems

Representative blackbody systems consisting of two surfaces are shown in Fig. 4-25(a) and (b). Whereas no thermal radiation enters or leaves the closed system of Fig. 4-25(a), thermal radiation is propagated through the open spaces or "windows" of the radiatively open system of Fig. 4-25(b).

To determine the radiation heat transfer rate between two blackbody surfaces A_1 and A_2 in radiatively open or closed systems, we must recognize that the rate of thermal radiation leaving A_1 that reaches and is absorbed by A_2 is $A_1 F_{1-2} E_{b1}$. Similarly, the rate of thermal radiation

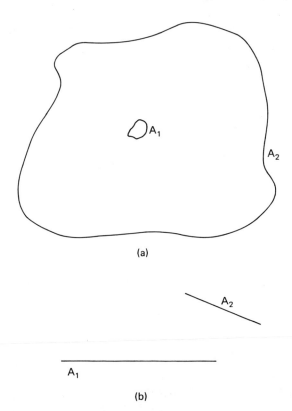

(a)

(b)

FIGURE 4-25   Bisurface thermal radiation systems. (a) Closed system. (b) Open system.

emitted by A_2 that is absorbed by A_1 is $A_2 F_{2-1} E_{b2}$. It follows that the rate of radiation heat transfer q_{1-2} from A_1 to A_2 in bisurface systems is equal to the net exchange of thermal radiation between A_1 and A_2; that is,

$$q_{1-2} = A_1 F_{1-2} E_{b1} - A_2 F_{2-1} E_{b2} \qquad (4\text{-}41)$$

Utilizing the principle of reciprocity, which states that $A_1 F_{1-2}$ and $A_2 F_{2-1}$ are equal, this equation can be written as

$$q_{1-2} = A_1 F_{1-2} (E_{b1} - E_{b2}) \qquad (4\text{-}42)$$

Finally, introducing the Stefan–Boltzmann law, Eq. (4-2), we have

$$q_{1-2} = A_1 F_{1-2} \sigma (T_1^4 - T_2^4) \qquad (4\text{-}43)$$

Multisurface systems

To determine the rate of radiation heat transfer from blackbody surface area A_s to N blackbody surfaces in radiatively closed or open systems such as those shown in Fig. 4-26(a) and (b), we must account for

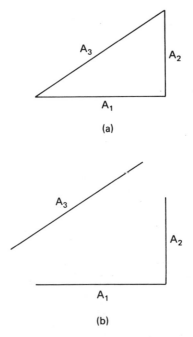

(a)

(b)

FIGURE 4-26 Trisurface thermal radiation systems. (a) Closed system. (b) Open system.

the exchange of thermal radiation between surface A_s and each of the N surfaces. The net rate of thermal radiation exchange between A_s and one of the surfaces A_j is simply

$$q_{s-j} = A_s F_{s-j} E_{bs} - A_j F_{j-s} E_{bj}$$
$$= A_s F_{s-j}(E_{bs} - E_{bj}) \tag{4-44}$$

It follows that the total rate of radiation heat transfer q_s from A_s to all N blackbody surfaces is given by

$$q_s = \sum_{j=1}^{N} A_s F_{s-j}(E_{bs} - E_{bj}) \tag{4-45}$$

With E_{bs} and E_{bj} specified by the Stefan–Boltzmann law, we have

$$q_s = \sum_{j=1}^{N} A_s F_{s-j} \sigma(T_s^4 - T_j^4) \tag{4-46}$$

EXAMPLE 4-9

Demonstrate the validity of the principle of reciprocity by analyzing the radiation heat transfer between two blackbodies A_1 and A_2 which are at the same temperature.

Solution

The rate of radiation heat transfer between two blackbodies is given by Eq. (4-41),

$$q_{1-2} = A_1 F_{1-2} E_{b1} - A_2 F_{2-1} E_{b2}$$

Of course, q_{1-2} must be zero and E_{b1} is equal to E_{b2}, since both surfaces are at the same temperature. It follows that

$$(A_1 F_{1-2} - A_2 F_{2-1}) E_{b1} = 0$$

Because

$$E_{b1} = E_{b2} = \sigma T_1^4$$

the factor $(A_1 F_{1-2} - A_2 F_{2-1})$ must be zero. Thus, the reciprocity law

$$A_1 F_{1-2} = A_2 F_{2-1}$$

must be satisfied.

This same reasoning can be applied to any two blackbody surfaces A_s and A_j, with the result that

$$A_s F_{s-j} = A_j F_{j-s}$$

A more rigorous derivation of the principle of reciprocity is developed in [5] and [11] which is based solely on geometric considerations. This approach is considered in Prob. 4-59.

Thermal radiation networks

As we have seen in Chaps. 1 through 3, the practical thermal analysis of many basic heat-transfer processes can be facilitated by use of the standard electrical analogy concept in which temperature T is taken to be analogous to electrical potential E_e. Based on this electrical analogy of the first kind, the radiation-heat-transfer resistance R_R between two blackbody surfaces is given by

$$R_R = \frac{1}{A_s F_{s-j} \sigma (T_s + T_j)(T_s^2 + T_j^2)} \tag{4-47}$$

This analogy concept is indeed useful for systems such as the ones shown in Fig. 4-25(a) and (b), which involve only two thermal radiating surfaces with known temperatures. However, because R_R is a function of temperature, the utility of this approach decreases as the number of thermal radiating surfaces with unknown temperature increases.

An alternative electrical analogy has been developed which more readily lends itself to the analysis of the more complex thermal radiation problems involving multiple surfaces and unknown surface temperatures. In this electrical analogy approach the total emissive power E is taken to be analogous to electrical voltage E_e. Equations (4-42) and (4-45) provide the basis for this powerful radiation-heat-transfer analysis tool. Simply put, the radiation-heat-transfer resistance R_{s-j} between surfaces A_s and A_j based on this electrical analogy of the second kind is given by

$$R_{s-j} = \frac{1}{A_s F_{s-j}} \tag{4-48}$$

To illustrate, analogous electrical circuits of the second kind (which we will refer to as *thermal radiation networks*) are utilized in Examples 4-10 through 4-12 for blackbody systems involving two, three, and four surfaces.

EXAMPLE 4-10

Referring to Fig. E4-10a, the blackbody surfaces A_1 and A_2 are at 27°C and 500°C, respectively. Determine the rate of radiation heat transfer between A_1 and A_2. Also sketch the thermal circuits of the first and second kinds.

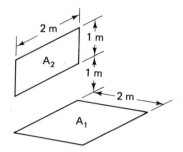

FIGURE E4-10a Open thermal radiating system.

Solution

The rate of radiation heat transfer is given by Eq. (4-42),

$$q_{1-2} = A_1 F_{1-2}(E_{b1} - E_{b2})$$

Referring back to Example 4-7 on page 269, we see that $F_{1-2} = 0.05$. Calculating q_{1-2}, we obtain

$$q_{1-2} = (4 \text{ m}^2)(0.05)\left(5.67 \times 10^{-8}\frac{W}{m^2\,K^4}\right)\left[(300\text{ K})^4 - (773\text{ K})^4\right] \times$$

$$(4)(0.05)(5.67)(3^4 - 7.73^4)\text{ W} = -3960\text{ W}$$

Thermal circuits of the first and second kinds are shown in Figs. E4-10b and E4-10c. The thermal resistances R_R and R_{1-2} are given by

$$R_R = \frac{1}{A_1 F_{1-2}\sigma(T_1 + T_2)(T_1^2 + T_2^2)}$$

$$= 0.12\frac{K}{W}$$

and

$$R_{1-2} = \frac{1}{A_1 F_{1-2}} = \frac{5}{m^2}$$

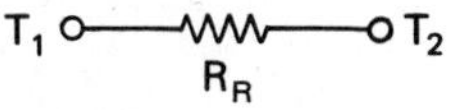

$$T_1 \circ\!\!-\!\!\text{\Large ww}\!\!-\!\!\circ T_2$$
$$R_R$$

FIGURE E4-10b Thermal circuit of first kind.

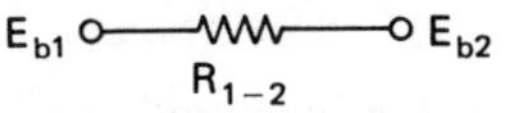

$$E_{b1} \text{ o——\!\!\!\bigvee\!\!\!\bigvee\!\!\!\bigvee\!\!\!—— o } E_{b2}$$
$$R_{1-2}$$

FIGURE E4-10c   Thermal radiation network.

EXAMPLE 4-11

Utilize the thermal-radiation-network approach to evaluate the rate of radiation heat transfer between each of the three blackbody surfaces of the cylindrical enclosure shown in Fig. E4-11a.

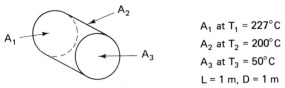

A_1 at $T_1 = 227°C$
A_2 at $T_2 = 200°C$
A_3 at $T_3 = 50°C$
$L = 1$ m, $D = 1$ m

FIGURE E4-11a   Cylindrical enclosure.

Solution

The thermal radiation network is shown in Fig. E4-11b, where

$$E_{b1} = \sigma T_1^4 = \left(5.67 \times 10^{-8} \frac{W}{m^2\,K^4}\right)(773\ K)^4 = 3540 \frac{W}{m^2}$$

$$E_{b2} = \sigma T_2^4 = \sigma(473\ K)^4 = 2840 \frac{W}{m^2}$$

$$E_{b3} = \sigma T_3^4 = \sigma(323\ K)^4 = 617 \frac{W}{m^2}$$

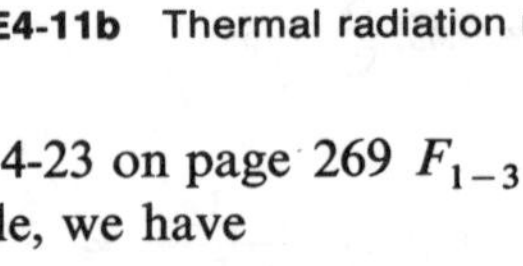

E_{b3}

R_{1-3} R_{2-3}

$E_{b1} \text{ o——\!\!\!\bigvee\!\!\!\bigvee\!\!\!\bigvee\!\!\!—— o } E_{b2}$
R_{1-2}

FIGURE E4-11b   Thermal radiation network.

Referring to Fig. 4-23 on page 269 F_{1-3} is equal to 0.17. Utilizing the summation principle, we have

$$F_{1-2} = 1 - F_{1-3} = 0.83$$

and

$$F_{3-2} = 0.83$$

The thermal resistances are

$$R_{1-2} = \frac{1}{A_1 F_{1-2}}$$

$$= \frac{1}{[\pi(1 \text{ m})^2/4](0.83)} = \frac{1.53}{\text{m}^2}$$

$$R_{1-3} = \frac{1}{A_1 F_{1-3}} = \frac{7.47}{\text{m}^2}$$

$$R_{2-3} = \frac{1}{A_2 F_{2-3}} = \frac{1}{A_3 F_{3-2}} = \frac{1.53}{\text{m}^2}$$

The rates of radiation heat transfer flowing in this network are now easily calculated.

$$q_{1-2} = \frac{E_{b1} - E_{b2}}{R_{1-2}} = \frac{(3540 - 2840)\text{W}/\text{m}^2}{1.53/\text{m}^2} = 458 \text{ W}$$

$$q_{1-3} = \frac{E_{b1} - E_{b3}}{R_{1-3}} = \frac{(3540 - 617)\text{W}/\text{m}^2}{7.47/\text{m}^2} = 391 \text{ W}$$

$$q_{2-3} = \frac{E_{b2} - E_{b3}}{R_{2-3}} = \frac{(2840 - 617)\text{W}/\text{m}^2}{1.53/\text{m}^2} = 1450 \text{ W}$$

The total rate of radiation heat transfer from each surface is

$$q_1 = q_{1-2} + q_{1-3} = 849 \text{ W}$$
$$q_2 = q_{2-3} + q_{2-1} = q_{2-3} - q_{1-2} = 992 \text{ W}$$
$$q_3 = q_{3-1} + q_{3-2} = -q_{1-3} - q_{2-3} = -1841 \text{ W}$$

Note that

$$q_1 + q_2 + q_3 = 0$$

EXAMPLE 4-12

The blackbody plates shown in Fig. E4-12a are located in a very large blackbody enclosure A_4 which is at 27°C. The back side of each plate is insulated, and 1 kW is electrically generated in plate 3. Sketch the thermal radiation network for this problem, and determine the temperature of surface A_3 and the rate of radiation heat transfer between A_1 and A_3.

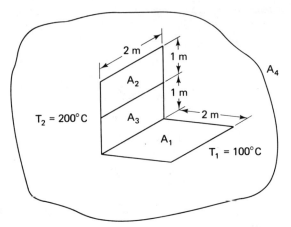

FIGURE E4-12a Closed thermal radiating system.

Solution

Referring back to Example 4-7, we have

$$F_{1-2}=0.05 \qquad F_{1-3}=0.15$$

By reciprocity,

$$F_{3-1}=\frac{A_1}{A_3}F_{1-3}=2(0.15)=0.3$$

$$F_{2-1}=\frac{A_1}{A_2}F_{1-2}=2(0.05)=0.1$$

It is clear that $F_{2-3}=0$. By employing the summation principle, we obtain

$$F_{1-4}=1-F_{1-2}-F_{1-3}=1-0.05-0.15=0.8$$
$$F_{2-4}=1-F_{2-1}-F_{2-3}=1-0.1-0=0.9$$
$$F_{3-4}=1-F_{3-1}-F_{3-2}=1-0.3-0=0.7$$

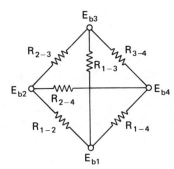

FIGURE E4-12b Thermal radiation network.

The thermal radiation network for this four-surface system is shown in Fig. E4-12b, where

$$E_{b1} = \sigma(373 \text{ K})^4 = 1100 \frac{W}{m^2}$$

$$E_{b2} = \sigma(473 \text{ K})^4 = 2840 \frac{W}{m^2}$$

$$E_{b4} = \sigma(300 \text{ K})^4 = 459 \frac{W}{m^2}$$

$$R_{1-3} = \frac{1}{A_1 F_{1-3}} = \frac{1}{(4 \text{ m}^2)(0.15)} = \frac{1.67}{m^2}$$

$$R_{1-2} = \frac{1}{A_1 F_{1-2}} = \frac{1}{(4 \text{ m}^2)(0.05)} = \frac{5}{m^2}$$

$$R_{1-4} = \frac{1}{A_1 F_{1-4}} = \frac{1}{(4 \text{ m}^2)(0.8)} = \frac{0.313}{m^2}$$

$$R_{2-3} = \frac{1}{A_2 F_{2-3}} = \frac{1}{(2 \text{ m}^2)(0)} = \frac{\infty}{m^2}$$

$$R_{3-4} = \frac{1}{A_3 F_{3-4}} = \frac{1}{(2 \text{ m}^2)(0.7)} = \frac{0.714}{m^2}$$

$$R_{2-4} = \frac{1}{A_2 F_{2-4}} = \frac{1}{(2 \text{ m}^2)(0.9)} = \frac{0.556}{m^2}$$

To determine the unknown potential E_{b3}, we perform an energy balance at surface A_3.

$$q_3 = 1 \text{ kW} = \frac{E_{b3} - E_{b1}}{R_{1-3}} + \frac{E_{b3} - E_{b2}}{R_{2-3}} + \frac{E_{b3} - E_{b4}}{R_{3-4}}$$

Making substitutions and solving for E_{b3}, we have

$$E_{b3} = \frac{1 \text{ kW} + \dfrac{E_{b1}}{R_{1-3}} + \dfrac{E_{b2}}{R_{2-3}} + \dfrac{E_{b4}}{R_{3-4}}}{\dfrac{1}{R_{1-3}} + \dfrac{1}{R_{2-3}} + \dfrac{1}{R_{3-4}}}$$

$$= \frac{(1000 + 1100/1.67 + 2840/\infty + 459/0.714) \text{ W}}{(1/1.67 + 1/\infty + 1/0.714) \text{ m}^2}$$

$$= 1150 \frac{W}{m^2}$$

To calculate T_3, we write

$$E_{b3} = \sigma T_3^4$$

$$T_3 = \left[\frac{1150 \text{ W/m}^2}{5.67 \times 10^{-8} \text{ W/(m}^2 \text{ K}^4)} \right]^{1/4} = 377 \text{ K}$$

$$= 103 °C$$

The rate of radiation heat transfer between A_1 and A_3 is

$$q_{1-3} = \frac{E_{b1} - E_{b3}}{R_{1-3}}$$

$$= \frac{(1100 - 1150)\text{W/m}^2}{1.67/\text{m}^2}$$

$$= -29.9 \text{ W}$$

To check our solution, the summation $q_1 + q_2 + q_3 + q_4$ can be shown to be equal to zero (see Prob. 4-19).

4-5-3   Nonblackbody Thermal Radiation from Diffuse Opaque Surfaces

As we have seen, some real surfaces exhibit approximate blackbody characteristics, but most thermal radiation systems encountered in practice involve distinctly nonblackbody surfaces. Radiation heat transfer in common opaque nonblackbody systems is complicated by the occurrence of reflections. For systems involving more than one nonblackbody surface, the multiple reflection patterns of thermal radiation often become extremely complex. Fortunately, these complexities can be rather easily overcome for systems involving diffuse surfaces with uniform properties and heating conditions by the use of a practical thermal network approach. Numerous real surface systems found in practice approximate these conditions. To develop this practical analysis approach for diffuse nonblackbody opaque systems, we first introduce the concept of radiosity.

Radiosity

Referring to Fig. 4-27, we define the *radiosity* J_s as the rate of thermal radiation emitted and reflected per unit surface area from A_s.

$$J_s = E_s + \rho_s G_s \tag{4-49}$$

Recall that the irradiation G_s is the thermal radiation flux incident on the surface A_s.

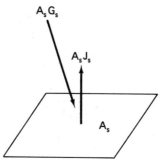

FIGURE 4-27 Thermal radiating surface A_s: concept of radiosity.

The total rate of radiation heat transfer from an opaque surface A_s is simply equal to the difference between the rate of thermal radiation leaving, $A_s J_s$, and the rate of thermal radiation coming in, $A_s G_s$; that is,

$$q_s = A_s(J_s - G_s) \qquad (4\text{-}50)$$

To eliminate the irradiation G_s, we combine Eqs. (4-49) and (4-50).

$$q_s = A_s\left[J_s - \frac{1}{\rho_s}(J_s - E_s)\right]$$

$$= \frac{A_s}{\rho_s}\left[E_s - (1-\rho_s)J_s\right] \qquad (4\text{-}51)$$

Since the body is opaque,

$$\rho_s = 1 - \alpha_s \qquad (4\text{-}52)$$

and we have

$$q_s = \frac{A_s \alpha_s}{1-\alpha_s}\left(\frac{\varepsilon_s}{\alpha_s}E_{bs} - J_s\right) \qquad (4\text{-}53)$$

Based on this equation, the total rate of radiation heat transfer q_s can be represented by a thermal network element as shown in Fig. 4-28; the node potentials are $(\varepsilon_s/\alpha_s)E_{bs}$ and J_s, and the thermal resistance R_s is

$$R_s = \frac{1-\alpha_s}{A_s \alpha_s} = \frac{\rho_s}{A_s \alpha_s} \qquad (4\text{-}54)$$

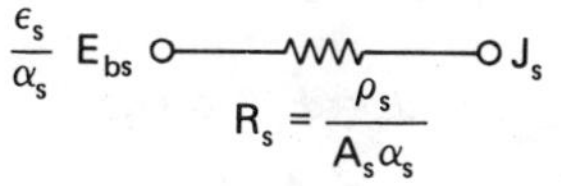

$$\frac{\varepsilon_s}{\alpha_s}E_{bs} \circ\!\!-\!\!\mathsf{WW}\!\!-\!\!\circ J_s$$
$$R_s = \frac{\rho_s}{A_s \alpha_s}$$

FIGURE 4-28 Thermal radiation network element for q_s.

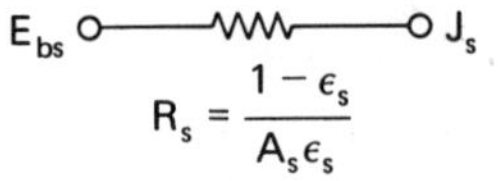

$$E_{bs} \circ \!\!-\!\!\!-\!\!\!\!\wedge\!\!\wedge\!\!\wedge\!\!-\!\!\!-\!\!\circ J_s$$

$$R_s = \frac{1 - \epsilon_s}{A_s \epsilon_s}$$

FIGURE 4-29   Thermal radiation network for q_s: graybody conditions.

As indicated in Sec. 4-3, the total absorptivity α_s of a surface is dependent upon the sources from which the irradiation G_s originates. Because of the reflection patterns associated with nonblackbody surface systems, special care must be taken in the evaluation of α_s for nongraybody conditions, as illustrated in several examples in the next section. For systems with surfaces that exhibit approximate graybody characteristics, α_s is independent of the source of the various components of incoming irradiation, and α_s can be set equal to ϵ_s. For this limiting condition, we have

$$q_s = \frac{A_s \epsilon_s}{1 - \epsilon_s} (E_{bs} - J_s) \tag{4-55}$$

The thermal radiation network element for this simple situation is shown in Fig. 4-29. The node potentials for this case are simply E_{bs} and J_s, and R_s is given by

$$R_s = \frac{1 - \epsilon_s}{A_s \epsilon_s} = \frac{\rho_s}{A_s \epsilon_s} \tag{4-56}$$

Equations (4-53) and its limiting form Eq. (4-55), together with the corresponding thermal network elements given in Figs. 4-28 and 4-29, provide the key building block for the development of a simple practical analysis of radiation heat transfer. However, before we turn our attention to the completion of our practical analysis approach for bisurface and multisurface systems, it should be mentioned that another approach to handling Eqs. (4-49) and (4-50) that is favored by some is to eliminate the radiosity J_s. This alternative method is introduced in Example 4-13.

EXAMPLE 4-13

Develop an alternative formulation for radiation heat transfer from a surface by eliminating J_s in Eqs. (4-49) and (4-50) instead of G_s.

Solution

Starting with Eqs. (4-49) and (4-50),

$$J_s = E_s + \rho_s G_s \tag{4-49}$$

$$q_s = A_s (J_s - G_s) \tag{4-50}$$

we eliminate J_s, to obtain

$$q_s = A_s(E_s + \rho_s G_s - G_s)$$
$$= A_s(E_s - \alpha_s G_s) \tag{a}$$

The irradiation G_s falling on surface A_s is expressed in terms of the radiosities of the surrounding surfaces by

$$A_s G_s = \sum_{j=1}^{N} A_j F_{j-s} J_j \tag{b}$$

Utilizing the principle of reciprocity, this equation takes the form

$$A_s G_s = \sum_{j=1}^{N} A_s F_{s-j} J_j \tag{c}$$

or

$$G_s = \sum_{j=1}^{N} F_{s-j} J_j \tag{d}$$

The total rate of thermal radiation absorbed by surface A_s is

$$\alpha_s G_s = \sum_{j=1}^{N} \alpha_{sj} F_{s-j} J_j \tag{e}$$

where α_{sj} is the absorptivity component associated with the J_j source. Substituting this result into Eq. (a), we have

$$q_s = A_s \left(E_s - \sum_{j=1}^{N} \alpha_{sj} F_{s-j} J_j \right) \tag{f}$$

Although this equation does not conveniently lend itself to electrical analogy representation, as does Eq. (4-53), it does provide a base for the analytical or numerical solution of more complex thermal radiation systems. For the simple limiting case of graybody conditions, Eq. (f) reduces to

$$q_s = A_s \varepsilon_s \left(E_{bs} - \sum_{j=1}^{N} F_{s-j} J_j \right) \tag{g}$$

As in the case of Eq. (4-53), Eqs. (f) and (g) apply to systems with uniform emission, reflection, and irradiation over each surface.

As a point of interest, it is noted that the total absorptivity α_s at surface A_s can be obtained by combining Eqs. (d) and (e); that is,

$$\alpha_s = \frac{\displaystyle\sum_{j=1}^{N} \alpha_{sj} F_{s-j} J_j}{\displaystyle\sum_{j=1}^{N} F_{s-j} J_j} \tag{h}$$

Bisurface systems

We consider radiation heat transfer between two opaque diffuse surfaces A_1 and A_2. For uniform radiosities, the rate of thermal radiation leaving A_1 that reaches A_2 is $A_1 F_{1-2} J_1$, and from A_2 to A_1 is $A_2 F_{2-1} J_2$. Therefore, the thermal radiation-heat-transfer rate q_{1-2} from surface A_1 to A_2 is

$$q_{1-2} = A_1 F_{1-2} J_1 - A_2 F_{2-1} J_2 \tag{4-57}$$

or, since $A_1 F_{1-2} = A_2 F_{2-1}$,

$$q_{1-2} = A_1 F_{1-2}(J_1 - J_2) \tag{4-58}$$

This equation provides the basis for the thermal network element shown in Fig. 4-30. (For surfaces with nonuniform radiosities, q_{1-2} is given by a more general expression which is developed in Prob. 4-27.)

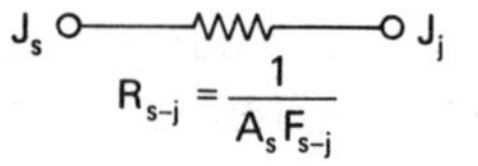

$$R_{s-j} = \frac{1}{A_s F_{s-j}}$$

FIGURE 4-30 Thermal radiation network element between potentials J_s and J_j.

To close our two-surface system analysis, we couple the nodes J_1 and J_2 to the surface potentials $(\varepsilon_1/\alpha_1)E_{b1}$ and $(\varepsilon_2/\alpha_2)E_{b2}$ by using the thermal network element given in Fig. 4-28. The resulting thermal radiation network is shown in Fig. 4-31. Because Eq. (4-58) was developed for surfaces with uniform radiosities, this thermal radiation network only applies to systems with uniform emission, reflection, and irradiation over each surface. The requirement that the reflected radiation flux be uniform is strictly satisfied only for symmetrical systems, such as infinite parallel plates or concentric spheres, with each surface having a uniform temperature and uniform thermal radiation properties. However, this convenient thermal radiation network approach is often applied to nonsymmetrical systems as a first approximation.

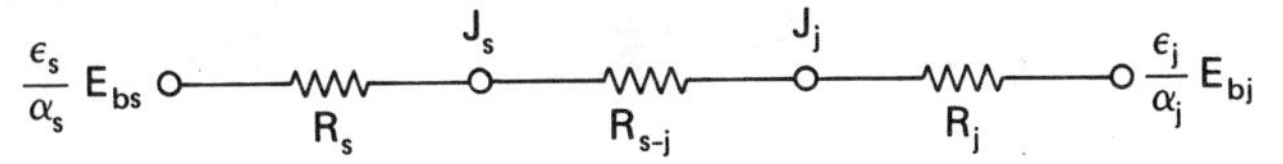

$$\frac{\epsilon_s}{\alpha_s} \; E_{bs} \; \circ \!\!-\!\!\!\mathsf{\Lambda\Lambda\Lambda}\!\!\!\!\overset{J_s}{-\!\!\circ\!\!-}\!\!\!\mathsf{\Lambda\Lambda\Lambda}\!\!\!\!\overset{J_j}{-\!\!\circ\!\!-}\!\!\!\mathsf{\Lambda\Lambda\Lambda}\!\!-\!\!\circ \; \frac{\epsilon_j}{\alpha_j} \; E_{bj}$$
$$ R_s R_{s\text{-}j} R_j$$

FIGURE 4-31   Thermal radiation network for bisurface system.

The rate of radiation heat transfer q_{1-2} associated with this thermal radiation network is given by

$$q_{1-2} = \frac{(\varepsilon_1/\alpha_1)E_{b1} - (\varepsilon_2/\alpha_2)E_{b2}}{R_1 + R_{1-2} + R_2}$$

$$= \frac{(\varepsilon_1/\alpha_1)E_{b1} - (\varepsilon_2/\alpha_2)E_{b2}}{\dfrac{1-\alpha_1}{A_1\alpha_1} + \dfrac{1}{A_1 F_{1-2}} + \dfrac{1-\alpha_2}{A_2\alpha_2}} \qquad (4\text{-}59)$$

For blackbody conditions, R_1 and R_2 are both zero, such that the thermal radiation network reduces to the one shown in Fig. E4-10c on page 280, and Eq. (4-59) reduces to Eq. (4-42). For approximate graybody conditions, Eq. (4-59) takes the form

$$q_{1-2} = \frac{E_{b1} - E_{b2}}{\dfrac{1-\varepsilon_1}{A_1\varepsilon_1} + \dfrac{1}{A_1 F_{1-2}} + \dfrac{1-\varepsilon_2}{A_2\varepsilon_2}} \qquad (4\text{-}60)$$

EXAMPLE 4-14

Consider exchange of radiation heat transfer between the small surface A_1 and an enclosure A_2 shown in Fig. E4-14a. Demonstrate that the enclosure approximates blackbody conditions as A_2 becomes large.

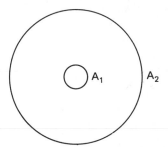

FIGURE E4-14a   Closed thermal radiating system.

Solution

The thermal radiation network for this system is shown in Fig. E4-14b. For very large values of A_2, most of the energy incident upon A_2

$$R_1 = \frac{\rho_1}{A_1\alpha_1} \qquad R_{1-2} = \frac{1}{A_1 F_{1-2}} \qquad R_2 = \frac{\rho_2}{A_2\alpha_2}$$

$$\frac{\epsilon_1}{\alpha_1} E_{b1} \; \text{O} \!-\!\!\!\text{ww}\!-\!\!\!\text{o}\!-\!\!\!\text{ww}\!-\!\!\!\text{o}\!-\!\!\!\text{ww}\!-\!\!\text{O} \; \frac{\epsilon_2}{\alpha_2} E_{b2}$$

FIGURE E4-14b Thermal radiation network.

originates from itself, unless the body is a perfect reflector. Consequently, the nonisothermal α approximation reduces to the graybody α approximation with ϵ_2/α_2 equal to unity. Further, the resistance R_2 approaches zero as A_2 increases. For these conditions, the radiosity J_2 is essentially equal to E_{b2}, such that blackbody conditions are approached at the surface of the enclosure.

Incidently, the thermal radiation network for this system implicitly accounts for thermal radiation from surface A_2 to itself. This point is expanded upon in Prob. 4-29.

EXAMPLE 4-15

Develop an expression for the net rate of exchange of thermal radiation between two diffuse nonblackbody infinite parallel plates A_1 and A_2 without the use of radiosity.

Solution

The reflection and absorption pattern of thermal radiation originally emitted from surface A_1 is schematically shown in Fig. E4-15. The energy rate content of this thermal radiation before absorption and reflection is $E_1 A_1$. Of this amount of radiation power, the quantity $\alpha_2 E_1 A_1$ is absorbed by A_2 and $\rho_2 E_1 A_1$ is reflected back toward A_1. Of this reflected energy rate, the quantity $\rho_1(\rho_2 E_1 A_1)$ is rereflected toward surface A_2. Surface A_2 absorbs $\alpha_2(\rho_1\rho_2 E_1 A_1)$ of this rate of energy and

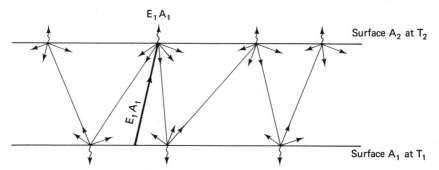

FIGURE E4-15 Nonblackbody thermal radiation between infinite parallel plates.

reflects the remainder, $\rho_2(\rho_1\rho_2 E_1 A_1)$. The total rate of thermal radiation originating from surface A_1 that is absorbed by A_2 is

$$\alpha_2 E_1 A_1 \left(1 + \rho_1 \rho_2 + (\rho_1 \rho_2)^2 + \ldots + (\rho_1 \rho_2)^n + \ldots \right)$$

or

$$\alpha_1 E_1 A_1 \sum_{n=0}^{\infty} (\rho_1 \rho_2)^n$$

In a similar fashion, we can show that the total rate of thermal radiation originating from surface A_2 that is absorbed by A_1 is

$$\alpha_1 E_2 A_2 \sum_{n=0}^{\infty} (\rho_1 \rho_2)^n$$

It follows that the net rate of thermal radiation exchange between A_1 and A_2 is

$$q_{1-2} = (\alpha_2 E_1 A_1 - \alpha_1 E_2 A_2) \left[\sum_{n=0}^{\infty} (\rho_1 \rho_2)^n \right] \qquad \text{(a)}$$

where $A_1 = A_2$.

Referring to [23], the series $\sum_{n=0}^{\infty} x^n$ converges to $1/(1-x)$ for x less than unity. Thus, the rate of radiation heat transfer is

$$q_{1-2} = \frac{A_1(\alpha_2 E_1 - \alpha_1 E_2)}{1 - \rho_1 \rho_2} \qquad \text{(b)}$$

Introducing the identities $E_1 = \varepsilon_1 E_{b1}$, $E_2 = \varepsilon_2 E_{b2}$, $\rho_1 = 1 - \alpha_1$, and $\rho_2 = 1 - \alpha_2$, we obtain

$$q_{1-2} = \frac{A_1 \left[\left(\dfrac{\varepsilon_1}{\alpha_1} \right) E_{b1} - \left(\dfrac{\varepsilon_2}{\alpha_2} \right) E_{b2} \right]}{\dfrac{1}{\alpha_1} + \dfrac{1}{\alpha_2} - 1} \qquad \text{(c)}$$

which is equivalent to Eq. (4-59) for $A_1 = A_2$ and $F_{1-2} = 1$. For graybody conditions, we have

$$q_{1-2} = \frac{A_1(E_{b1} - E_{b2})}{\dfrac{1}{\varepsilon_1} + \dfrac{1}{\varepsilon_2} - 1} \qquad \text{(d)}$$

This equation is equivalent to Eq. (4-60) for this infinite parallel-plate geometry.

The fact that this direct approach is consistent with the radiosity-based network approach should reinforce our confidence in the more efficient network method. Further, because this direct method becomes extremely unwieldly for multisurface systems, the concept of radiosity is the key to developing practical thermal analyses for radiation processes.

EXAMPLE 4-16

Surface A_2 of the system shown in Fig. E4-16a is a graybody with emissivity of 0.56 and surface A_1 is a blackbody. Determine the rate of radiation heat transfer between A_1 and A_2 if $T_1 = 27°C$ and $T_2 = 500°C$.

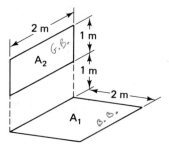

FIGURE E4-16a   Open thermal radiating system.

Solution

Because surface A_2 is a graybody, we have

$$\alpha_2 = \varepsilon_2 = 0.56$$

The radiation shape factor is given in Example 4-10 on page 279 as

$$F_{1-2} = 0.05$$

Assuming diffuse radiation, the approximate thermal radiation network for this nonsymmetrical bisurface system is shown in Fig. E4-16b, where $E_{b1} = 459$ W/m^2, $E_{b2} = 20,200$ W/m^2, and

$$R_{1-2} = \frac{1}{A_1 F_{1-2}} = \frac{5}{m^2}$$

$$R_2 = \frac{\rho_2}{A_2 \alpha_2} = \frac{1 - 0.56}{(2 \text{ m}^2)(0.56)} = \frac{0.393}{m^2}$$

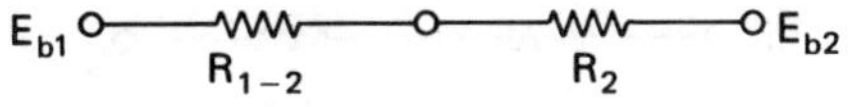

$$E_{b1} \circ\!\!-\!\!\bigwedge\!\!\bigwedge\!\!-\!\!\circ\!\!-\!\!\bigwedge\!\!\bigwedge\!\!-\!\!\circ E_{b2}$$
$$\qquad R_{1-2} \qquad\qquad R_2$$

FIGURE E4-16b Thermal radiation network.

It follows that the rate of radiation heat transfer q_{2-1} is approximately

$$q_{2-1} = -q_{1-2} = \frac{(20,200 - 459)\,\mathrm{W/m^2}}{(5 + 0.393)/\mathrm{m^2}}$$

$$= 3660\,\mathrm{W}$$

This result lies about 7.5% below the value of 3960 W obtained in Example 4-10 for blackbody radiation from both surfaces.

To improve the accuracy of our analysis for this nonsymmetrical system, surface A_2 can be subdivided into smaller areas (see Prob. 4-28).

EXAMPLE 4-17

Surface A_2 of the system shown in Fig. E4-16a is made of oxidized nickel and surface A_1 is a blackbody. Estimate the rate of radiation heat transfer between A_1 and A_2 if $T_1 = 27°C$ and $T_2 = 500°C$.

Solution

The emissivity of the oxidized nickel surface A_2 is found in Fig. 4-3 on page 239 to be approximately 0.63 for a temperature of 500°C and 0.32 at 27°C. Therefore, we take

$$\varepsilon_2 = 0.63$$

Noting that all the thermal radiation from A_1 to A_2 is emitted by blackbody A_1, which is at a temperature of 27°C, we employ the nonisothermal α approximation to obtain

$$\alpha_2 = 0.32$$

As we have already seen, $F_{1-2} = 0.05$.

The thermal radiation network for this system is shown in Fig. E4-17 with $(\varepsilon_2/\alpha_2)E_{b2}$ and R_2 given by

$$\frac{\varepsilon_2}{\alpha_2}E_{b2} = \frac{0.63}{0.32}\left(20,200\,\frac{\mathrm{W}}{\mathrm{m^2}}\right) = 39,800\,\frac{\mathrm{W}}{\mathrm{m^2}}$$

$$R_2 = \frac{\rho_2}{A_2\alpha_2} = \frac{1 - 0.32}{(2\,\mathrm{m^2})(0.32)} = \frac{1.06}{\mathrm{m^2}}$$

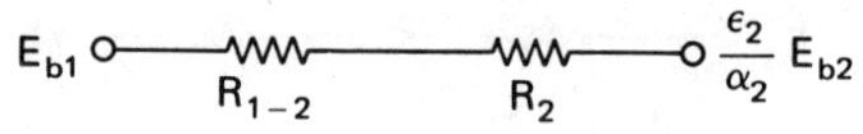

FIGURE E4-17   Thermal radiation network.

The rate of radiation heat transfer is given by

$$q_{2-1} = \frac{(\varepsilon_2/\alpha_2)E_{b2} - E_{b1}}{R_{1-2} + R_2}$$

$$= \frac{(39,800 - 459)\,\text{W}/\text{m}^2}{(5 + 1.06)/\text{m}^2}$$

$$= 6490\,\text{W}$$

As a point of interest, the graybody α approximation gives

$$\alpha_2 = 0.63$$

and

$$R_2 = \frac{\rho_2}{A_2\alpha_2} = \frac{1 - 0.63}{(2\,\text{m}^2)(0.63)} = \frac{0.294}{\text{m}^2}$$

The resulting prediction for the rate of radiation heat transfer is

$$q_{2-1} = \frac{E_{b2} - E_{b1}}{R_{1-2} + R_2}$$

$$= \frac{(20,200 - 459)\,\text{W}/\text{m}^2}{(5 + 0.294)/\text{m}^2}$$

$$= 3730\,\text{W}$$

which is a substantial 40% below the value obtained by utilizing the preferred nonisothermal α approximation.

    As in Example 4-16, the accuracy of our analysis for this nonsymmetrical system can be improved by breaking surfaces A_1 and A_2 into smaller areas.

Multisurface systems

Following the pattern of our analysis of bisurface systems, the net rate of thermal radiation between opaque diffuse surfaces A_s and A_j of a multisurface system with uniform radiosities is

$$q_{s-j} = A_s F_{s-j} J_s - A_j F_{j-s} J_j \tag{4-61}$$

or, based on the principle of reciprocity,

$$q_{s-j} = \frac{J_s - J_j}{R_{s-j}}$$ (4-62)

where

$$R_{s-j} = \frac{1}{A_s F_{s-j}}$$ (4-63)

Thus, we have a general thermal radiation network element such as the one shown in Fig. 4-30 for each surface combination.

The radiosity J_s associated with each surface is linked to its surface potential $(\varepsilon_s/\alpha_s)E_{bs}$ by means of the element shown in Fig. 4-28. To illustrate, thermal radiation networks are shown in Figs. 4-32 and 4-33 for representative three-surface and four-surface diffuse systems. For blackbody surfaces, R_s is zero, and these thermal radiation networks reduce to the networks shown in Examples 4-11 and 4-12. And, for graybody conditions, α_s is set equal to ε_s. As in the case of bisurface systems, these multisurface thermal radiation networks strictly only apply to symmetrical systems, but are commonly used as a first estimate for nonsymmetrical systems.

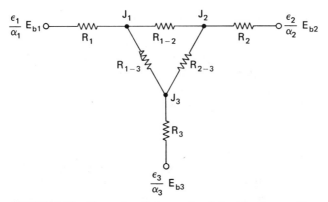

FIGURE 4-32 Thermal radiation network for trisurface system.

Referring back to the development of Eq. (4-53),

$$q_s = \frac{A_s \alpha_s}{1 - \alpha_s}\left(\frac{\varepsilon_s}{\alpha_s}E_{bs} - J_s\right)$$ (4-53)

which is the basis for the network element of Fig. 4-28, we are again reminded that the rate of radiation heat transfer q_s under steady-state

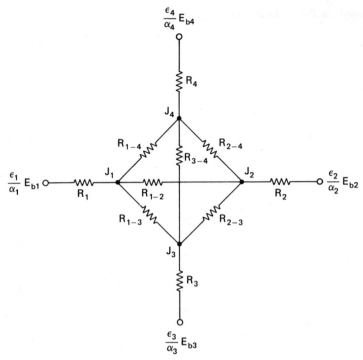

FIGURE 4-33 Thermal radiation network for four-surface system.

conditions must be in balance with the net rate of energy entering the body, $\Sigma \dot{E}_i$. Certain situations occur in practice in which the only significant energy transfer to or from a body is caused by thermal radiation. Two rather special cases of practical importance that fall into this category are thermal radiation shields and reradiating surfaces.

Thermal Radiation Shields As suggested in Chap. 1, the heat transfer through a wall can be greatly reduced by utilizing a composite construction that consists of layers of highly reflective materials separated by evacuated spaces. The thin plates or shells utilized in such superinsulative composite walls are known as *thermal radiation shields*. Referring to Fig. 4-34(a), we see that under steady-state conditions the rate of radiation heat transfer q_{1-s} from enclosure surface A_1 to the single thermal radiation shield is equal to the rate of radiation heat transfer q_{s-2} from the shield to enclosure surface A_2. Although such highly reflective surfaces have specular characteristics, the total radiation in this symmetrical system and in an infinite parallel-plate system can be treated as diffuse. A thermal radiation network is shown for this arrangement in Fig. 4-34(b). Note that the

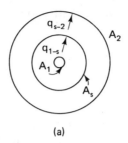

(a)

$$\frac{\epsilon_1}{\alpha_1} \, E_{b1} \, \circ\!\!-\!\!\mathsf{WW}\!\!-\!\!\circ\overset{J_1}{}\!\!-\!\!\mathsf{WW}\!\!-\!\!\circ\overset{J_{s1}}{}\!\!-\!\!\mathsf{WW}\!\!-\!\!\circ\!\!-\!\!\mathsf{WW}\!\!-\!\!\circ\overset{J_{s2}}{}\!\!-\!\!\mathsf{WW}\!\!-\!\!\circ\overset{J_2}{}\!\!-\!\!\mathsf{WW}\!\!-\!\!\circ \, \frac{\epsilon_2}{\alpha_2} \, E_{b2}$$

$$\qquad R_1 \qquad R_{1-s} \qquad R_{s1} \qquad R_{s2} \qquad R_{s-2} \qquad R_2$$

(b)

FIGURE 4-34   Thermal radiation shield. (a) Concentric sphere system. (b) Thermal radiation network.

thermal radiation shield is assumed to be very thin with negligible resistance to conduction. Based on this thermal radiation network, the rate of radiation heat transfer through the shield from surface A_1 to A_2 is

$$q_{1-2} = \frac{(\epsilon_1/\alpha_1)E_{b1} - (\epsilon_2/\alpha_2)E_{b2}}{\dfrac{\rho_1}{A_1\alpha_1} + \dfrac{1}{A_1 F_{1-s}} + \dfrac{\rho_{s1}}{A_s\alpha_{s1}} + \dfrac{\rho_{s2}}{A_s\alpha_{s2}} + \dfrac{1}{A_s F_{s-2}} + \dfrac{\rho_2}{A_2\alpha_2}} \tag{4-64}$$

with F_{1-s} and F_{s-2} both equal to unity. Because the shield surfaces are highly reflective, the thermal radiation reaching surface A_1 primarily originates from surface A_1 itself. Therefore, the nonisothermal α approximation essentially reduces to the graybody α approximation (i.e., $\epsilon_1/\alpha_1 \simeq 1$). The same is true for surface A_2, with $\epsilon_2/\alpha_2 \simeq 1$. As a final simplification, the radiation properties of the shield can generally be taken as the average of its two surfaces; that is,

$$\alpha_s = \frac{\alpha_{s1} + \alpha_{s2}}{2}$$

$$\rho_s = \frac{\rho_{s1} + \rho_{s2}}{2} = 1 - \alpha_s \tag{4-65}$$

Introducing these simplifications, q_{1-2} is given by

$$q_{1-2} = \frac{E_{b1} - E_{b2}}{\dfrac{1-\epsilon_1}{A_1\epsilon_1} + \dfrac{1}{A_1} + \dfrac{1-\epsilon_2}{A_2\epsilon_2} + \dfrac{2-\epsilon_s}{A_s\epsilon_s}} \tag{4-66}$$

For N shields, it is a simple matter to show that q_{1-2} becomes

$$q_{1-2} = \frac{E_{b1} - E_{b2}}{\dfrac{1-\varepsilon_1}{A_1 \varepsilon_1} + \dfrac{1}{A_1} + \dfrac{1-\varepsilon_2}{A_2 \varepsilon_2} + \displaystyle\sum_{j=1}^{N} \dfrac{1}{A_{sj}} \dfrac{2-\varepsilon_{sj}}{\varepsilon_{sj}}} \qquad (4\text{-}67)$$

For a parallel-plate arrangement with each reflective shield having approximately the same radiation properties, this equation reduces to

$$
\begin{aligned}
q_{1-2} &= \frac{A_1(E_{b1} - E_{b2})}{\dfrac{1-\varepsilon_1}{\varepsilon_1} + 1 + \dfrac{1-\varepsilon_2}{\varepsilon_2} + \dfrac{2-\varepsilon_s}{\varepsilon_s} N} \\[2ex]
&= \frac{A_1(E_{b1} - E_{b2})}{\dfrac{1}{\varepsilon_1} + \dfrac{1}{\varepsilon_2} - 1 + \dfrac{2-\varepsilon_s}{\varepsilon_s} N}
\end{aligned}
\qquad (4\text{-}68)
$$

Taking one step further, for the case in which the radiation properties of all the surfaces are approximately equal, we have

$$q_{1-2} = \frac{\varepsilon A_1 (E_{b1} - E_{b2})}{(2-\varepsilon)(1+N)} \qquad (4\text{-}69)$$

An examination of Eqs. (4-67) through (4-69) reveals that the rate of radiation heat transfer becomes small as (1) the reflectivity of the shield increases toward unity (i.e., as α_s or ε_s falls toward zero, and (2) the number of shields increases.

EXAMPLE 4-18

A blackbody plate at $0°F$ exchanges radiation with a parallel stainless steel 301 plate at $1500°F$. Determine the percent reduction in heat transfer if a thin polished aluminum plate with emissivity and absorptivity approximately equal to 0.08 is placed between these two large plates.

Solution

Referring to Fig. 4-3, ε_2 is approximately 0.5 and, based on the nonisothermal α approximation,

$$\alpha_2 = \varepsilon_2(T_1) \simeq 0.16 \qquad (a)$$

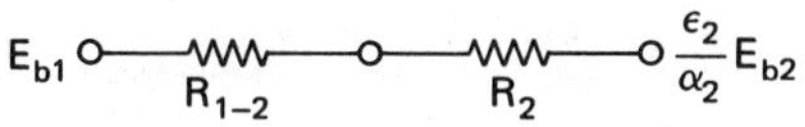

FIGURE E4-18a Thermal radiation network for parallel plates.

The thermal radiation network for the parallel plates without the shield is shown in Fig. E4-18a, where

$$E_{b1} = \sigma T_1^4 = 0.171 \times 10^{-8} \frac{\text{Btu}}{\text{h ft}^2 \, ^\circ\text{R}^4} (460^\circ\text{R})^4$$

$$= 76.6 \frac{\text{Btu}}{\text{h ft}^2}$$

$$E_{b2} = \sigma T_2^4 = \sigma(1960^\circ\text{R})^4 = 2.52 \times 10^4 \frac{\text{Btu}}{\text{h ft}^2}$$

$$\frac{\varepsilon_2}{\alpha_2} E_{b2} = \frac{0.5}{0.16} \left(2.52 \times 10^4 \frac{\text{Btu}}{\text{h ft}^2} \right) = 7.88 \times 10^4 \frac{\text{Btu}}{\text{h ft}^2}$$

$$R_{1-2} = \frac{1}{A_1 F_{1-2}} = \frac{1}{A_1}$$

$$R_2 = \frac{\rho_2}{A_2 \alpha_2} = \frac{1 - 0.16}{A_1(0.16)} = \frac{5.25}{A_1}$$

Calculating the rate of radiation heat transfer q_{2-1}, we obtain

$$q_{2-1} = \frac{(\varepsilon_2/\alpha_2)E_{b2} - E_{b1}}{R_{1-2} + R_2}$$

$$= \frac{(7.88 \times 10^4 - 76.6) \, \text{Btu}/(\text{h ft}^2)}{(1 + 5.25)/A_1}$$

$$q_{2-1}'' = 1.26 \times 10^4 \frac{\text{Btu}}{\text{h ft}^2} \tag{b}$$

The thermal radiation network for the situation in which a thin plate lies between A_1 and A_2 is shown in Fig. E4-18b, where

$$R_{1-s} = \frac{1}{A_1 F_{1-s}} = \frac{1}{A_1}$$

$$R_{s1} = R_{s2} = \frac{\rho_s}{A_s \alpha_s} = \frac{1 - 0.08}{A_s(0.08)} = \frac{11.5}{A_1}$$

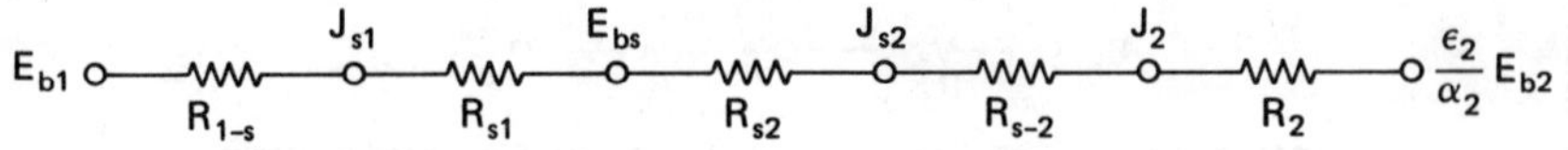

E_{b1} o—WW—o—WW—o—WW—o—WW—o—WW—o $\frac{\varepsilon_2}{\alpha_2}$ E_{b2}

R_{1-s} R_{s1} R_{s2} R_{s-2} R_2

J_{s1} E_{bs} J_{s2} J_2

FIGURE E4-18b Thermal radiation network for parallel plates with shield.

As we have already noted, the emissivity of the stainless steel plate at 1500°F is approximately 0.5. Because most of the radiation incident upon A_2 actually originates at surface A_2 itself, we evaluate α_2 at a source temperature of 1500°F, as a first estimate; that is,

$$\alpha_2 = \varepsilon_2(T_2) = 0.5 \tag{c}$$

It follows that

$$R_2 = \frac{\rho_2}{A_2 \alpha_2} = \frac{1 - 0.5}{A_1(0.5)} = \frac{1}{A_1}$$

The rate of radiation heat transfer is then

$$q_{2-1} = \frac{E_{b2} - E_{b1}}{R_{1-s} + R_{s1} + R_{s2} + R_{s-2} + R_2}$$

$$= \frac{(2.52 \times 10^4 - 76.6) \text{Btu}/(\text{h ft}^2)}{(1 + 11.5 + 11.5 + 1 + 1)/A_1}$$

$$q''_{2-1} = 966 \frac{\text{Btu}}{\text{h ft}^2} \tag{d}$$

Thus, based on this approximate analysis, we have a reduction in the radiation heat transfer of about 92%.

To determine the approximate temperature of the radiation shield, we write

$$q_{2-1} = \frac{E_{bs} - E_{b1}}{R_{1-s} + R_{s1}}$$

$$E_{bs} = E_{b1} + (R_{1-s} + R_{s1})q_{2-1}$$

$$= \left[76.6 + (1 + 11.5)966\right] \frac{\text{Btu}}{\text{h ft}^2}$$

$$\sigma T_s^4 = 12{,}200 \frac{\text{Btu}}{\text{h ft}^2}$$

$$T_s = 1630°\text{R} = 1170°\text{F} \tag{e}$$

Our analysis can be refined by evaluating the absorptivity at surface A_2 at a temperature which better represents the actual source of radiation. For example, the radiosity J_{s2} represents the radiant energy incident upon A_2. Therefore, we could evaluate α_2 at the temperature of a blackbody with total emissive power equal to J_{s2}. This point is considered further in Prob. 4-35.

Reradiating Surfaces In the furnace arrangement of Fig. 4-35(a), the plate on the insulated floor is heated to a steady-state temperature which is totally governed by the thermal radiation from the walls at T_2 and ceiling at T_1. Under steady-state conditions, the total rate of radiation heat transfer q_s from the plate to its enclosure is clearly zero. Bodies such as this with q_s equal to zero are said to have *reradiating surfaces*.

For the usual case in which the reflectivity ρ_s of an irradiating surface is not equal to unity, the thermal resistance R_s given by Eq. (4-54) or (4-56)

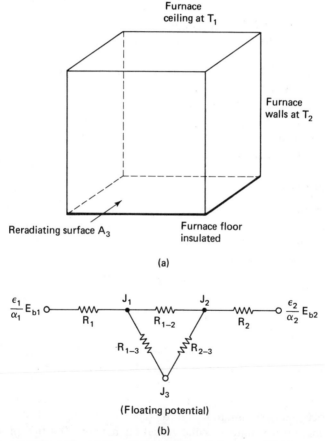

FIGURE 4-35 (a) Reradiating surface in furnace. (b) Thermal radiation network.

is finite. Thus, with q_s set equal to zero in Eq. (4-55), we conclude that for irradiating surfaces with diffuse characteristics, the floating potential E_{bs} must be equal to the radiosity J_s. This conclusion gives rise to the thermal radiation network shown in Fig. 4-35(b) for the system in Fig. 4-35(a). Solving this network for q_{1-2}, we obtain

$$q_{1-2} = \frac{(\varepsilon_1/\alpha_1)E_{b1} - (\varepsilon_2/\alpha_2)E_{b2}}{\dfrac{\rho_1}{A_1\alpha_1} + \dfrac{\rho_2}{A_2\alpha_2} + R_{1-2}} \qquad (4\text{-}70)$$

where

$$R_{1-2} = \left[A_1F_{1-2} + \frac{1}{1/(A_1F_{1-3}) + 1/(A_2F_{2-3})} \right]^{-1}$$

Incidently, for polished metallic surfaces such as copper (refer back to Fig. 4-8), the absorptivity α_s and emissivity ε_s are very small, with ρ_s being almost equal to unity. The thermal resistance R_s for such highly reflective surfaces is very large. If we neglect directional aspects, the thermal network for a near-perfect reflector with specified temperature will be exactly the same as for a reradiating surface with floating potential. But, in reality, important nondiffuse specular characteristics of polished surfaces enter into the picture for many geometric arrangements that compromise the accuracy of Eq. (4-62).

EXAMPLE 4-19

Outline the steps required for determining the temperature of the reradiating plate surface A_3 shown in Fig. 4-35(a) if the temperature T_1 of the ceiling is 1000°C and the temperature T_2 of the walls is 500°C. Radiation from the interior of the furnace exhibits graybody conditions with emissivity equal to 0.8 and the plate is made of commercial aluminum.

Solution

First, the radiation properties ε_3 and α_3 of surfaces A_3 can be obtained from Fig. 4-3 and by utilizing either the nonisothermal or graybody α approximation. According to the graybody α approximation, $\alpha_1 = \varepsilon_1 = 0.8$ and $\alpha_2 = \varepsilon_2 = 0.8$. Second, the radiation shape factors can be evaluated by utilizing Fig. 4-21 on page 268 and the principles of reciprocity and summation. Third, referring to Fig. 4-35 which represents the approximate thermal radiation network for this problem, q_1 can be calculated from Eq. (4-70), J_1 and J_2 can be calculated by writing

$$q_1 = \frac{(\varepsilon_1/\alpha_1)E_{b1} - J_1}{R_1}$$

$$q_1 = -q_2 = \frac{J_2 - (\varepsilon_2/\alpha_2)E_{b2}}{R_2}$$

q_{1-3} can be calculated from

$$q_{1-3} = q_1 - \frac{J_1 - J_2}{R_{1-2}}$$

and $E_{b3}(=J_3)$ can be obtained from

$$q_{1-3} = \frac{J_1 - E_{b3}}{R_{1-3}}$$

Radiation factor

In looking back over the analyses developed for diffuse thermal radiation between opaque surfaces, we observe that the rate of radiation heat transfer q_{s-j} between two surfaces A_s and A_j can be expressed in the compact form

$$q_{s-j} = A_s \mathscr{F}_{s-j}\left(\frac{\varepsilon_s}{\alpha_s}E_{bs} - \frac{\varepsilon_j}{\alpha_j}E_{bj}\right) \tag{4-71}$$

We will refer to $\mathscr{F}_{s-j}$ as the *radiation factor*. For approximate graybody conditions, Eq. (4-71) reduces to

$$q_{s-j} = A_s \mathscr{F}_{s-j}(E_{bs} - E_{bj})$$
$$= A_s \mathscr{F}_{s-j}\sigma(T_s^4 - T_j^4) \tag{4-72}$$

For example, for thermal radiation exchange in a bisurface system, $\mathscr{F}_{1-2}$ is given by

$$\mathscr{F}_{1-2} = \frac{1}{\dfrac{\rho_1}{\alpha_1} + \dfrac{1}{F_{1-2}} + \dfrac{A_1}{A_2}\dfrac{\rho_2}{\alpha_2}} \tag{4-73}$$

This equation reduces to

$$\mathscr{F}_{1-2} = \frac{1}{\dfrac{1-\varepsilon_1}{\varepsilon_1} + \dfrac{1}{F_{1-2}} + \dfrac{A_1}{A_2}\dfrac{1-\varepsilon_2}{\varepsilon_2}} \tag{4-74}$$

TABLE 4-3 Radiation factor $\mathscr{F}_{1-2}$

System	$\mathscr{F}_{1-2}$ General	$\mathscr{F}_{1-2}$ Graybody conditions
Surfaces A_1 and A_2	$\left(\dfrac{\rho_1}{\alpha_1}+\dfrac{1}{F_{1-2}}+\dfrac{A_1}{A_2}\dfrac{\rho_2}{\alpha_2}\right)^{-1}$	$\left(\dfrac{\epsilon_1-1}{\epsilon_1}+\dfrac{1}{F_{1-2}}+\dfrac{A_1}{A_2}\dfrac{\epsilon_2-1}{\epsilon_2}\right)^{-1}$
Infinite parallel plates	$\left(\dfrac{\rho_1}{\alpha_1}+1+\dfrac{\rho_2}{\alpha_2}\right)^{-1}$	$\left(\dfrac{1}{\epsilon_1}+\dfrac{1}{\epsilon_2}-1\right)^{-1}$
Concentric cylinders or spheres	$\left(\dfrac{\rho_1}{\alpha_1}+1+\dfrac{D_1}{D_2}\dfrac{\rho_2}{\alpha_2}\right)^{-1}$	$\left(\dfrac{1-\epsilon_1}{\epsilon_1}+1+\dfrac{A_1}{A_2}\dfrac{1-\epsilon_2}{\epsilon_2}\right)^{-1}$
Small body A_1 inside large body A_2	$\left(\dfrac{\rho_1}{\alpha_1}+1\right)^{-1}$	ϵ_1
Blackbody surfaces A_1 and A_2	F_{1-2}	F_{1-2}
Surfaces A_1 and A_2 with one radiation shield		$\left(\dfrac{1-\epsilon_1}{\epsilon_1}+1+\dfrac{A_1}{A_2}\dfrac{1-\epsilon_2}{\epsilon_2}+\dfrac{A_1}{A_s}\dfrac{2-\epsilon_s}{\epsilon_s}\right)^{-1}$
Parallel plates with N radiation shields; $\epsilon_{sj}=\epsilon_s$		$\left(\dfrac{1-\epsilon_1}{\epsilon_1}+1+\dfrac{1-\epsilon_2}{\epsilon_2}+\dfrac{2-\epsilon_s}{\epsilon_s}N\right)^{-1}$

for approximate graybody conditions, and to

$$\mathscr{F}_{1-2}=F_{1-2} \qquad (4\text{-}75)$$

for the limiting blackbody case.

For purposes of design, $\mathscr{F}_{s-j}$ is listed in Table 4-3 for several practical arrangements. For the more complex geometries for which $\mathscr{F}_{s-j}$ is not tabulated, the thermal radiation network should be solved systematically by numerical or analytical techniques. The systematic solution of thermal radiation problems is discussed in the following section.

Systematic solution approach

To develop a systematic solution for the radiation heat transfer in a thermal network, an energy balance can be made on each J node. For the moment, we focus our attention on systems with opaque diffuse surfaces. Referring to the J_s node shown in Fig. 4-36, we of course have our building-block equation,

$$q_s=\frac{A_s\alpha_s}{\rho_s}\left(\frac{\epsilon_s}{\alpha_s}E_{bs}-J_s\right) \qquad (4\text{-}53)$$

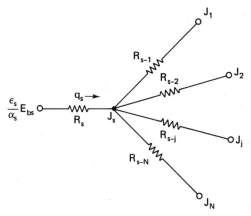

FIGURE 4-36 Segment of thermal radiation network involving J_s node.

Based on the first law of thermodynamics, q_s must also satisfy

$$q_s = \sum_{j=1}^{N} A_s F_{s-j}(J_s - J_j) \tag{4-76}$$

To obtain a nodal equation for J_s for the case in which the surface temperature T_s is known, we combine Eqs. (4-53) and (4-76), with the result

$$\frac{\alpha_s}{\rho_s} J_s + \sum_{j=1}^{N} F_{s-j}(J_s - J_j) = \frac{\epsilon_s}{\rho_s} E_{bs} \tag{4-77}$$

On the other hand, if the rate of radiation heat transfer q_s from a surface is specified with T_s being unknown, then Eq. (4-76) serves as the nodal equation for J_s. A nodal equation of one of these types can be written for each of the N surfaces. For example, for the three-surface system of Fig. 4-37, we write

$$\left(\frac{\alpha_1}{\rho_1} + F_{1-2} + F_{1-3} \right) J_1 - F_{1-2}J_2 - F_{1-3}J_3 = \frac{\epsilon_1}{\rho_1} E_{b1} \tag{4-78-1}$$

$$- F_{2-1}J_1 + \left(\frac{\alpha_2}{\rho_2} + F_{2-1} + F_{2-3} \right) J_2 - F_{2-3}J_3 = \frac{\epsilon_2}{\rho_2} E_{b2} \tag{4-78-2}$$

$$- F_{3-1}J_1 - F_{3-2}J_2 + (F_{3-1} + F_{3-2})J_3 = \frac{q_3}{A_3} \tag{4-78-3}$$

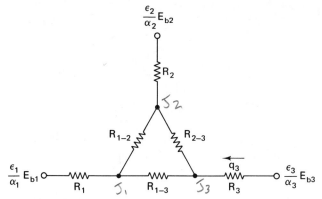

FIGURE 4-37 Thermal radiation network for trisurface system with T_1, T_2, and q_3 specified.

In general, for N surfaces we write

$$a_{11}J_1 + a_{12}J_2 + \cdots + a_{1j}J_j + \cdots + a_{1N}J_N = C_1 \qquad (4\text{-}79\text{-}1)$$

$$a_{21}J_1 + a_{22}J_2 + \cdots + a_{2j}J_j + \cdots + a_{2N}J_N = C_2 \qquad (4\text{-}79\text{-}2)$$

$$\vdots$$

$$a_{j1}J_1 + a_{j2}J_2 + \cdots + a_{jj}J_j + \cdots + a_{jN}J_N = C_j \qquad (4\text{-}79\text{-}j)$$

$$\vdots$$

$$a_{N1}J_1 + a_{N2}J_2 + \cdots + a_{Nj}J_j + \cdots + a_{NN}J_N = C_N \qquad (4\text{-}79\text{-}N)$$

The N unknowns $J_1, J_2, \ldots, J_j, \ldots, J_s, \ldots, J_N$ in these N equations can be solved by the analytical or numerical techniques which were introduced in Chap. 3. For example, the Gauss–Seidel approach can be used to obtain a hand calculation or computer solution. Or, hand calculation and computer solutions can be affected by powerful matrix methods. For situations in which the graybody α approximation can be utilized, the solutions are quite straightforward. However, for nongraybody conditions, α_s is expressed in terms of the unknown radiosities J_j, such that this approach requires iteration. For this situation, α_s can be specified by Eq. (h) in Example 4-13.

Once the radiosities have been determined, the unknown radiation-heat-transfer rate q_s or temperature T_s can be determined for each surface by employing Eq. (4-53).

As an alternative approach, the equation developed in Example 4-13 can be used. But here again, because of the dependence of the various components of absorptivity α_{sj} on the radiosities J_j, this approach also involves iteration, unless graybody conditions are assumed.

EXAMPLE 4-20

The following information is available for the very long three-surface graybody enclosure shown in Fig. E4-20:

$$T_1 = 200°C \qquad \varepsilon_1 = 0.2$$
$$T_2 = 27°C \qquad \varepsilon_2 = 0.7$$
$$q_3'' = 1 \text{ kW/m}^2 \qquad \varepsilon_3 = 0.5$$

Determine the radiation-heat-transfer fluxes q_1'' and q_2'' from surfaces A_1 and A_2 and the temperature T_3 of surface A_3.

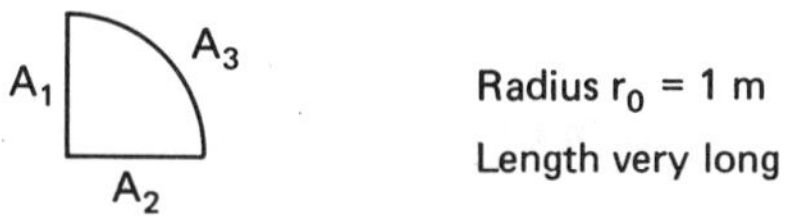

A_1 A_3
A_2

Radius r_0 = 1 m

Length very long

FIGURE E4-20   Trisurface enclosure.

Solution

The approximate thermal radiation network for this nonsymmetrical system is given in Fig. 4-37.

We first consider the geometric aspects of the problem. As seen in Prob. 4-10, F_{1-2} is equal to 0.293. Based on the reciprocity and summation principles, we have

$$F_{2-1} = \frac{A_1}{A_2} F_{1-2} = (1)(0.293) = 0.293$$

$$F_{1-3} = 1 - F_{1-2} = 1 - 0.293 = 0.707$$

$$F_{3-1} = \frac{A_1}{A_3} F_{1-3} = \frac{4}{2\pi} (0.707) = 0.45$$

$$F_{3-2} = \frac{A_2}{A_3} F_{2-3} = 0.45$$

Because A_3 is concave, we also have

$$F_{3-3} = 1 - F_{3-1} - F_{3-2} = 0.1$$

The total emissive powers E_{b1} and E_{b2} are

$$E_{b1} = \sigma T_1^4 = \left(5.67 \times 10^{-8} \frac{W}{m^2 \, K^4} \right)(473 \text{ K})^4 = 2.84 \frac{kW}{m^2}$$

$$E_{b2} = \sigma T_2^4 = \sigma (300 \text{ K})^4 = 0.459 \frac{kW}{m^2}$$

And, because we are dealing with graybody surfaces, the properties are

$$\alpha_1 = \varepsilon_1 = 0.2 \qquad \rho_1 = 1 - \alpha_1 = 0.8$$
$$\alpha_2 = \varepsilon_2 = 0.7 \qquad \rho_2 = 1 - \alpha_2 = 0.3$$
$$\alpha_3 = \varepsilon_3 = 0.5 \qquad \rho_3 = 1 - \alpha_3 = 0.5$$

The nodal equations at the three radiosity nodes are given by Eqs. (4-78-1), (4-78-2), and (4-78-3). Substituting for the various potentials, radiation shape factors, and radiation properties, we obtain

$$\left(\frac{0.2}{0.8} + 0.293 + 0.707\right)J_1 - 0.293J_2 - 0.707J_3 = \frac{0.2}{0.8}\left(2.84\frac{kW}{m^2}\right)$$

$$-0.293J_1 + \left(\frac{0.7}{0.3} + 0.293 + 0.707\right)J_2 - 0.707J_3 = \frac{0.7}{0.3}\left(0.459\frac{kW}{m^2}\right)$$

$$-0.45J_1 - 0.45J_2 + (0.45 + 0.45)J_3 = 1\frac{kW}{m^2}$$

or

$$1.25J_1 - 0.293J_2 - 0.707J_3 = 0.71\frac{kW}{m^2}$$

$$-0.293J_1 + 3.33J_2 - 0.707J_3 = 1.07\frac{kW}{m^2}$$

$$-0.45J_1 - 0.45J_2 + 0.9J_3 = 1\frac{kW}{m^2}$$

These three equations are easily solved for J_1, J_2, and J_3 by elimination, iteration, or other means. The radiosity values are found to be

$$J_1 = 2.5\frac{kW}{m^2} \qquad J_2 = 1.17\frac{kW}{m^2} \qquad J_3 = 2.94\frac{kW}{m^2}$$

With the radiosity values known, predictions are obtained for q_1'' and q_2'' by utilizing Eq. (4-53).

$$q_s = \frac{A_s\alpha_s}{\rho_s}\left(\frac{\varepsilon_s}{\alpha_s}E_{bs} - J_s\right)$$

$$q_1'' = \frac{2840 - 2500}{0.8/0.2}\frac{W}{m^2} = 85\frac{W}{m^2}$$

$$q_2'' = \frac{459 - 1170}{0.3/0.7}\frac{W}{m^2} = -1660\frac{W}{m^2}$$

Because q_3'' is known, Eq. (4-53) is rearranged to obtain the surface temperature T_3.

$$E_{b3} = \frac{\alpha_3}{\varepsilon_3}\left(\frac{\rho_3}{A_3\alpha_3}q_3 + J_3\right)$$

$$= \left(\frac{0.5}{0.5}1000 + 2940\right)\frac{W}{m^2} = 3940\frac{W}{m^2}$$

$$T_3 = \left(\frac{E_{b3}}{\sigma}\right)^{\frac{1}{4}} = \left[\frac{3940\ W/m^2}{5.67\times10^{-8}\ W/(m^2\ K^4)}\right]^{\frac{1}{4}}$$

$$= 513\ K = 240°C$$

To check our solution, we calculate the rates of heat transfer per unit length from each surface.

$$\frac{q_1}{L} = q_1''r_0 = 85\frac{W}{m^2}(1\,m) = 85\frac{W}{m}$$

$$\frac{q_2}{L} = -1660\frac{W}{m^2}(1\,m) = -1660\frac{W}{m}$$

$$\frac{q_3}{L} = 1000\frac{W}{m^2}\left(\frac{2\pi}{4}\,m\right) = 1570\frac{W}{m}$$

Summing these three values, we find that the system is in balance to within three significant figures.

4-5-4   Unsteady Thermal Radiation Systems

The principles pertaining to thermal radiation presented in this chapter apply for unsteady as well as steady conditions. However, for unsteady systems, we must account for the rate of change in stored energy within bodies with unsteady temperature and we must be aware of changes in the thermal radiation properties which may be brought about by the change in temperature with time.

As suggested earlier in this chapter, practical lumped analyses can sometimes be developed for unsteady thermal radiating systems in which the radiation Biot numbers $Bi_R(\equiv \overline{h}_R \ell / k)$ of the bodies with unsteady temperature are of the order of 0.1 and less. The analysis of such systems follows the pattern established in Sec. 2-7 for unsteady one-dimensional convection heat transfer.

To illustrate, we consider the situation in which a small metallic ball initially at T_i with surface area A_1 is placed in an oven with steady surface temperature T_2. Because the unsteady temperature T_1 within the ball is essentially uniform at any instant t for small values of Bi_R, the lumped differential form of the first law is written as (for constant properties)

$$q_{2-1} = \frac{dU}{dt} = \rho V c_v \frac{dT_1}{dt} \tag{4-80}$$

For approximate graybody conditions, q_{2-1} is given by Eq. (4-72),

$$q_{2-1} = A_1 \mathscr{F}_{1-2} \sigma \left(T_2^4 - T_1^4 \right) \tag{4-81}$$

where $\mathscr{F}_{1-2}$ is given by Eq. (4-73),

$$\mathscr{F}_{1-2} = \frac{1}{\dfrac{\rho_1}{\alpha_1} + \dfrac{1}{F_{1-2}} + \dfrac{A_1}{A_2} \dfrac{\rho_2}{\alpha_2}} \tag{4-73}$$

Setting F_{1-2} equal to unity and assuming that the variation in radiation properties α_1 and α_2 is small as the temperature T_1 changes from T_i to T_2, $\mathscr{F}_{1-2}$ is taken as constant. Hence, our lumped differential formulation becomes

$$\frac{dT_1}{dt} + \frac{\sigma A_1}{\rho V c_v} \mathscr{F}_{1-2} \left(T_1^4 - T_2^4 \right) = 0 \tag{4-82}$$

$$T_1 = T_i \quad \text{at } t = 0 \tag{4-83}$$

The solution of this nonlinear ordinary differential equation is developed in Example 4-21.

For unsteady systems in which the thermal radiation Biot number is not small or for nonsymmetrical conditions, a multidimensional analysis must be developed which accounts for the effects of conduction heat transfer. The topic of mixed-mode thermal radiation is discussed in Sec. 4-5-6.

EXAMPLE 4-21

The walls of a furnace are kept at 1230°C. A thin copper plate (1 m by 1 m by 1 mm) initially at 50°C is then placed in the oven with all surfaces exposed to radiation from the furnace walls. All surfaces can be considered to be black. Determine the temperature of the plate and the rate of heat transfer as a function of time. How long does it take for the plate to reach steady state?

Solution

Assuming that the thermal radiation Biot number is less than 0.1 (see Prob. 4-46), the differential formulation for this problem is given by Eqs. (4-82) and (4-83).

$$q_{1-2} + \frac{dU}{dt} = 0 \tag{a}$$

$$\frac{dT_1}{dt} + \frac{\sigma A_1 \mathcal{F}_{1-2}}{\rho V c_v}(T_1^4 - T_2^4) = 0 \tag{b}$$

$$T_1 = T_i = 323 \text{ K} \tag{c}$$

where $\mathcal{F}_{1-2} = F_{1-2} = 1$ for blackbody radiation.

Separating the variables in this nonlinear first-order differential equation, we obtain

$$\frac{dT_1}{T_1^4 - T_2^4} = \frac{dT_1}{(T_1^2 - T_2^2)(T_1^2 + T_2^2)} = -\frac{\sigma A_1 \mathcal{F}_{1-2}}{\rho V c_v} dt \tag{d}$$

The left-hand side of this relationship can be broken into partial fractions as follows:

$$\frac{dT_1}{(T_1^2 - T_2^2)(T_1^2 + T_2^2)} = \left(\frac{C_1}{T_1^2 - T_2^2} + \frac{C_2}{T_1^2 + T_2^2}\right)dT_1 \tag{e}$$

or

$$1 = C_1 T_1^2 + C_1 T_2^2 + C_2 T_1^2 - C_2 T_2^2 \tag{f}$$

where C_1 and C_2 must be evaluated. By equating the coefficients associated with T_1, we obtain

$$0 = C_1 + C_2 \tag{g}$$

We also have

$$1 = (C_1 - C_2)T_2^2 \tag{h}$$

Hence, C_1 and C_2 are given by

$$C_1 = -C_2 \qquad C_1 = \frac{1}{2T_2^2} \tag{i, j}$$

Therefore, our differential equation takes the simpler form

$$\frac{dT_1}{2T_2^2(T_1^2 - T_2^2)} - \frac{dT_1}{2T_2^2(T_1^2 + T_2^2)} = -\frac{\sigma A_1 \mathcal{F}_{1-2}}{\rho V c_v} dt \tag{k}$$

This equation can be integrated with the help of integration tables. The result is

$$\frac{1}{4T_2^3}\ln\left|\frac{T_1-T_2}{T_1+T_2}\right| - \frac{1}{2T_2^3}\tan^{-1}\frac{T_1}{T_2} = -\frac{\sigma A_1 \mathscr{F}_{1-2}}{\rho V c_v}t + C_3 \qquad (1)$$

Applying the initial condition, we have

$$C_3 = \frac{1}{2T_2^3}\left(\frac{1}{2}\ln\left|\frac{T_i-T_2}{T_i+T_2}\right| - \tan^{-1}\frac{T_i}{T_2}\right) \qquad (m)$$

The solution now can be written as

$$\frac{1}{2}\left(\ln\left|\frac{T_1-T_2}{T_1+T_2}\right| - \ln\left|\frac{T_i-T_2}{T_i+T_2}\right|\right) - \left(\tan^{-1}\frac{T_1}{T_2} - \tan^{-1}\frac{T_i}{T_2}\right)$$
$$= -2T_2^3\frac{\sigma A_1 \mathscr{F}_{1-2}}{\rho V c_v}t \qquad (n)$$

Setting $\mathscr{F}_{1-2}=1$, $A_1=2$ m^2, $V=10^{-3}$ m^3, $T_2=1500$ K, $\rho=8950$ kg/m^3, and $c_v=0.383$ kJ/(kg °C), Eq. (n) reduces to the form

$$t = \frac{s}{0.223}\left[\left(\tan^{-1}\frac{T_1}{1500\text{ K}} - 0.212\right) - \frac{1}{2}\left(\ln\left|\frac{T_1-1500\text{ K}}{T_1+1500\text{ K}}\right| + 0.438\right)\right] \qquad (o)$$

This equation is solved for t by taking T_1 as the independent variable. The resulting predictions for T_1 are shown in Fig. E4-21.

To estimate the time required for steady state to occur, we set T_1 equal to 99% of T_R (i.e., $T_1 = 1490$ K). For this value of T_1, Eq. (o) gives

$$t_{ss} = 14.4 \text{ s}$$

The rate of radiation heat transfer at any instant of time t is obtained by writing

$$q_R = \sigma A_1 F_{1-2}\left(T_1^4 - T_2^4\right)$$

where T_1 is taken from Fig. E4-21. Calculations for q_R'' are shown in Fig. E4-21.

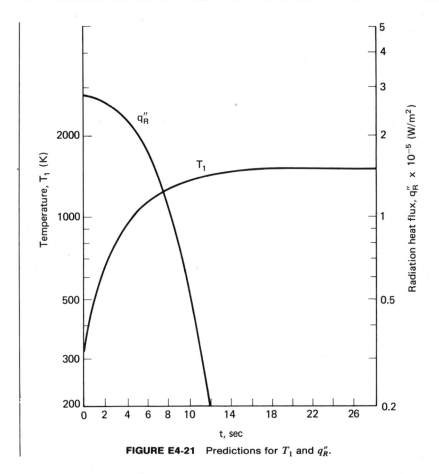

FIGURE E4-21 Predictions for T_1 and q_R''.

4-5-5 Transparent Medium

Because of the significance of thermal radiation in transparent gases and glass panes, we now develop practical analyses for radiation heat transfer through participating gaseous and solid mediums. Attention is first given to gases which emit, absorb, and transmit thermal radiation, but which exhibit negligible reflection and scattering characteristics. Then consideration is turned to thermal radiation through transparent solids such as glass.

Participating gases

To introduce some of the basic concepts involved in the analysis of thermal radiation in participating gases, we restrict ourselves to simplistic but very practical approximate gray gas systems. The basic approach in which participating gases are treated as graybodies was introduced and developed by Hottel in the 1930s in the context of high-gas-temperature systems with highly oxidized or contaminated walls with high emissivity and absorptivity. The gas emissivity given in Fig. 4-18(a) and (b) on page 262 for water vapor and carbon dioxide were obtained by Hottel and Egbert [16],[17] for such systems. Consequently, the gray gas techniques are most reliable for enclosures with blackbody or near-blackbody surfaces. For this reason, we will focus our attention on blackbody and gray near-blackbody enclosures.

Blackbody Enclosures   We first consider gray gas/blackbody enclosure systems.

Gray Gas/Single-Surface Enclosure.   For thermal radiation between an isothermal blackbody enclosure to the gas medium, q_{s-m} is simply equal to the difference between (1) the thermal radiation emitted by the enclosure which is absorbed by the gas, $\alpha_m(A_s F_{s-m} E_{bs})$, and (2) the thermal radiation emitted by the gas, $A_m F_{m-s} \varepsilon_m E_{bm}$; that is,

$$q_{s-m} = \alpha_m(A_s F_{s-m} E_{bs}) - A_m F_{m-s} \varepsilon_m E_{bm} \tag{4-84}$$

Utilizing reciprocity, it follows that

$$q_{s-m} = A_s F_{s-m}(\alpha_m E_{bs} - \varepsilon_m E_{bm}) \tag{4-85}$$

where $F_{s-m} = 1$. Referring back to Sec. 4-3, the emissivity ε_m of the gas is evaluated at the mean gas temperature T_m and the influence of the source temperature T_s on the gas absorptivity α_m can be accounted for by utilizing approximate graybody correlations of the form of Eqs. (4-28a) and (4-28b).

EXAMPLE 4-22

The walls of a cubical furnace 1 m on a side are maintained at 227°C. The products of combustion at 1 atm and 1230°C consist of 20% CO_2, 15% H_2O, and 65% N_2. Assuming that the walls are essentially black, estimate the radiation heat transfer between the gas and the walls.

Solution

The gas emissivity and absorptivity are shown in Example 4-6 on page 264 to be approximately

$$\varepsilon_m = 0.174$$
$$\alpha_m = 0.312$$

The total emissive powers E_{bs} and E_{bm} are given by

$$E_{bs} = \sigma T_s^4 = 5.67 \times 10^{-8} \frac{\text{W}}{\text{m}^2 \text{ K}^4} (500 \text{ K})^4$$

$$= 3.54 \times 10^3 \frac{\text{W}}{\text{m}^2}$$

$$E_{bm} = \sigma T_m^4 = \sigma (1500 \text{ K})^4$$

$$= 2.87 \times 10^5 \frac{\text{W}}{\text{m}^2}$$

The radiation heat transfer between the burning gases and the surface of the furnace are calculated by utilizing Eq. (4-85).

$$q_{m-s} = -q_{s-m} = A_s F_{s-m} (\varepsilon_m E_{bm} - \alpha_m E_{bs})$$

$$= (6 \text{ m}^2)(1) [0.174(2.87 \times 10^5) - 0.312(3.54 \times 10^3)] \frac{\text{W}}{\text{m}^2}$$

$$= 293 \text{ kW}$$

Gray Gas/Multisurface Enclosure. For multisurface blackbody enclosures, we have equations similar to Eq. (4-84) for each surface. For example, for two surfaces A_s and A_j we have

$$q_{s-m} = A_s F_{s-m} (\alpha_{ms} E_{bs} - \varepsilon_m E_{bm}) \tag{4-86}$$

and

$$q_{j-m} = A_j F_{j-m} (\alpha_{mj} E_{bj} - \varepsilon_m E_{bm}) \tag{4-87}$$

where the second subscript on the absorptivity of the gas medium designates the source of the irradiation.

In addition, an expression must be developed for the rate of thermal radiation q_{s-j} transmitted through the medium from surface s to surface j. The rate of thermal radiation leaving surface A_s that reaches A_j is $A_s F_{s-j} E_{bs} \tau_{ms}$ and the rate from surface A_j to A_s is $A_j F_{j-s} E_{bj} \tau_{mj}$. Therefore, the rate of thermal radiation from A_s to A_j is

$$q_{s-j} = A_s F_{s-j} (E_{bs} \tau_{ms} - E_{bj} \tau_{mj}) \tag{4-88}$$

If the surface temperatures T_s and T_j are not too different, τ_{ms} and τ_{mj} can be represented by an average value τ_m, such that Eq. (4-88) becomes

$$q_{s-j} = A_s F_{s-j} \tau_m (E_{bs} - E_{bj}) \tag{4-89}$$

Following through, α_{ms} and α_{mj} in Eqs. (4-86) and (4-87) can be set equal to α_m. Otherwise, Eq. (4-88) can be left in its present form and the distinction between α_{ms} and α_{mj} in Eqs. (4-86) and (4-87) can be retained.

Equations (4-86) through (4-89) provide the basis for the thermal radiation network representation of multisurface blackbody systems involving a participating gray gas. For example, the parallel-plate system in Fig. 4-38(a) is represented by the network shown in Fig. 4-38(b) for the case in which $\tau_{m1} \simeq \tau_{m2} \simeq \tau_m$ and $\alpha_{m1} \simeq \alpha_{m2} \simeq \alpha_m$. Based on this thermal network, we see that the total radiation-heat-transfer rate q_m from the medium is

$$
\begin{aligned}
q_m &= \frac{(\varepsilon_m/\alpha_m)E_{bm} - E_{b1}}{R_{1-m}} + \frac{(\varepsilon_m/\alpha_m)E_{bm} - E_{b2}}{R_{2-m}} \\
&= A_1 F_{1-m}\alpha_m\left(\frac{\varepsilon_m}{\alpha_m}E_{bm} - E_{b1}\right) + A_2 F_{2-m}\alpha_m\left(\frac{\varepsilon_m}{\alpha_m}E_{bm} - E_{b2}\right)
\end{aligned}
\tag{4-90}
$$

The rate q_m must also satisfy the first law of thermodynamics,

$$
\sum \dot{E}_i = q_m + \frac{\Delta E_s}{\Delta t}
\tag{4-91}
$$

Surface A_1 at T_1

Participating gas with average temperature T_m

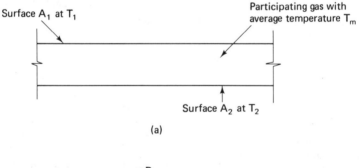

Surface A_2 at T_2

(a)

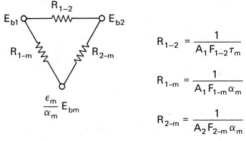

$$R_{1-2} = \frac{1}{A_1 F_{1-2}\tau_m}$$

$$R_{1-m} = \frac{1}{A_1 F_{1-m}\alpha_m}$$

$$R_{2-m} = \frac{1}{A_2 F_{2-m}\alpha_m}$$

(b)

FIGURE 4-38   Radiation heat transfer in a bisurface blackbody system with participating gas. (a) Parallel-plate system. (b) Thermal radiation network.

which, for steady-state conditions, reduces to

$$q_m = \sum \dot{E}_i \tag{4-92}$$

If no energy is transferred to the medium from external sources (i.e., $\sum \dot{E}_i = 0$), then q_m is zero under steady-state conditions and the node E_{bm} becomes a simple floating point. Under these passive equilibrium conditions, the same rate of energy is emitted by the medium as is absorbed. The solution to the thermal network for this simple case gives rise to

$$q_{1-2} = A_1 \left[F_{1-2}\tau_m + \cfrac{\alpha_m}{\cfrac{1}{F_{1-m}} + \cfrac{A_1}{A_2 F_{2-m}}} \right] (E_{b1} - E_{b2}) \tag{4-93}$$

Notice that for a nonparticipating gas with $\alpha_m = 0$, this equation reduces to Eq. (4-42),

$$q_{1-2} = A_1 F_{1-2}(E_{b1} - E_{b2}) \tag{4-42}$$

Near-Blackbody Enclosures   For situations in which the emissivity of the wall of a single enclosure is of the order of 0.8 and larger, Hottel [17] has shown that the net rate of radiation heat transfer can be approximated by multiplying Eq. (4-85) by the factor $(\varepsilon_s + 1)/2$.

$$q_{s-m} = A_s F_{s-m}(\alpha_m E_{bs} - \varepsilon_m E_{bm}) \frac{\varepsilon_s + 1}{2} \tag{4-94}$$

But it should be emphasized that this approximation is only valid for near-blackbody surfaces.

For enclosures with low emittance surfaces, the gray gas method is not appropriate. For these more complex-type problems, more comprehensive analyses must be developed which account for the band absorption characteristics of the gas. Such higher-order analyses are discussed in [5], [11], and [24].

Transparent solids

As indicated in the preceding sections, the dependence of medium irradition properties on the source is sometimes important in dealing with participating gases. This factor is very critical in important applications involving the transfer of solar radiation through a glass medium into an enclosure. Therefore, we now focus attention on the dependence of the medium properties on the irradiation source.

Blackbody Bisurface System   The analysis of thermal radiation between two blackbody surfaces A_1 and A_2 which are separated by a
nonreflecting transparent solid medium is identical to the analysis developed in the previous section for a gray gas/multisurface blackbody enclosure. The basic equations for a parallel-plate bisurface system are taken
from Eqs. (4-86) through (4-88); that is,

$$q_{1-m} = A_1 F_{1-m}(\alpha_{m1} E_{b1} - \varepsilon_m E_{bm}) \tag{4-95}$$

$$q_{2-m} = A_2 F_{2-m}(\alpha_{m2} E_{b2} - \varepsilon_m E_{bm}) \tag{4-96}$$

$$q_{1-2} = A_1 F_{1-2}(E_{b1}\tau_{m1} - E_{b2}\tau_{m2}) \tag{4-97}$$

where

$$\alpha_{m1} + \tau_{m1} = 1 \tag{4-98}$$

and

$$\alpha_{m2} + \tau_{m2} = 1 \tag{4-99}$$

In order to develop a thermal radiation network for the important
case in which τ_{m1} and τ_{m2} are significantly different, we rearrange Eqs.
(4-95) through (4-97) as follows:

$$q_{1-m} = A_1 F_{1-m} \frac{\alpha_{m1} E_{b1} - \varepsilon_m E_{bm}}{E_{b1} - E_{bm}} (E_{b1} - E_{bm}) \tag{4-100}$$

$$q_{2-m} = A_2 F_{2-m} \frac{\alpha_{m2} E_{b2} - \varepsilon_m E_{bm}}{E_{b2} - E_{bm}} (E_{b2} - E_{bm}) \tag{4-101}$$

$$q_{1-2} = A_1 F_{1-2} \frac{E_{b1}\tau_{m1} - E_{b2}\tau_{m2}}{E_{b1} - E_{b2}} (E_{b1} - E_{b2}) \tag{4-102}$$

These equations provide the basis for the thermal radiation network shown
in Fig. 4-39.

The absorptivities and transmissivities can be approximated by the
method developed in Example 4-13. Of course, for situations in which the
surface temperatures T_1 and T_2 are of the same order of magnitude,
$\tau_{m1} \simeq \tau_{m2}$ (and $\alpha_{m1} \simeq \alpha_{m2}$) such that this network reduces to the thermal
radiation network shown in Fig. 4-38(b). However, for important applica-
tions involving the transfer of solar or high-temperature radiation through
glass into enclosures, say from A_1 to A_2, we have a maximum difference
between τ_{m1} and τ_{m2}, with the glass being essentially transparent to the
high-temperature radiation ($\tau_{m1} \simeq 1, \alpha_{m1} \simeq 0$) and nearly opaque to the energy emitted within the enclosure ($\tau_{m2} \simeq 0, \alpha_{m2} \simeq 1$). By referring to Fig. 4-39

$$R_{1-2} = \frac{E_{b1} - E_{b2}}{A_1 F_{1-2}(E_{b1} \tau_{m1} - E_{b2} \tau_{m2})}$$

$$R_{1-m} = \frac{E_{b1} - E_{bm}}{A_1 F_{1-m}(\alpha_{m1} E_{b1} - \epsilon_m E_{bm})}$$

$$R_{2-m} = \frac{E_{b2} - E_{bm}}{A_2 F_{2-m}(\alpha_{m2} E_{b2} - \epsilon_m E_{bm})}$$

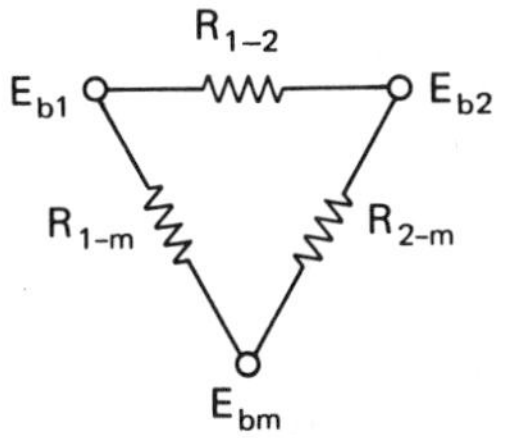

FIGURE 4-39 Thermal radiation network for blackbody bi-surface system with transparent solid medium.

and by reexamining Eqs. (4-95) through (4-97), we see that this combination of glass properties allows a large net rate of thermal radiation to be transmitted through the glass.

By performing an energy balance on the E_{bm} node, we have (neglecting convection)

$$q_{1-m} = q_{m-2} \tag{4-103}$$

That is, the rate of radiation heat transfer between A_1 and the glass is equal to the net rate of radiant exchange between the glass and A_2. It follows that the total rate of radiation heat transfer from A_1 is given by

$$q_1 = q_{1-2} + q_{1-m} \tag{4-104}$$

The rate of radiation heat transfer q_{1-2} transmitted directly through the glass can be evaluated by using Eq. (4-102). To evaluate q_{1-m}, we must solve for E_{bm}. This is done by utilizing Eq. (4-103). Substituting for q_{1-m} and q_{m-2} in this equation, we obtain

$$A_1 F_{1-m}(\alpha_{m1} E_{b1} - E_{bm}) = A_2 F_{2-m}(E_{bm} - \alpha_{m2} E_{b2})$$

$$E_{bm} = \frac{A_1 F_{1-m} \alpha_{m1} E_{b1} + A_2 F_{2-m} \alpha_{m2} E_{b2}}{A_1 F_{1-m} + A_2 F_{2-m}} \tag{4-105}$$

Employing the principle of reciprocity and recognizing that $F_{m-1} = 1$ and

$F_{m-2} = 1$, we have

$$E_{bm} = \frac{\alpha_{m1}E_{b1} + \alpha_{m2}E_{b2}}{2}$$

(4-106)

The substitution of this expression for E_{bm} into Eq. (4-100) or (4-101) gives

$$q_{1-m} = q_{m-2} = A_m F_{m-1}\left(\alpha_{m1}E_{b1} - \frac{\alpha_{m1}E_{b1} + \alpha_{m2}E_{b2}}{2}\right)$$

$$= A_m\left(\frac{\alpha_{m1}E_{b1} - \alpha_{m2}E_{b2}}{2}\right)$$

(4-107)

Returning to Eq. (4-104), q_1 becomes

$$q_1 = A_1 F_{1-2}(E_{b1}\tau_{m1} - E_{b2}\tau_{m2}) + A_m\left(\frac{\alpha_{m1}E_{b1} - \alpha_{m2}E_{b2}}{2}\right)$$

(4-108)

Based on geometric considerations, F_{1-2} can be set equal to F_{1-m}, such that $A_1 F_{1-2} = A_1 F_{1-m} = A_m F_{m-1} = A_m$. Thus, our final expression for q_1 takes the form

$$q_1 = A_m\left(E_{b1}\tau_{m1} - E_{b2}\tau_{m2} + \frac{\alpha_{m1}E_{b1} - \alpha_{m2}E_{b2}}{2}\right)$$

$$= \frac{A_m}{2}\left[E_{b1}(1 + \tau_{m1}) - E_{b2}(1 + \tau_{m2})\right]$$

(4-109)

This rate of thermal radiation heat transfer from the high-temperature source is equal to the total rate of thermal radiation received by A_2, $-q_2$.

EXAMPLE 4-23

Determine the rate of radiation heat transfer from the interior of a furnace with surface area of 1 m² at 2000°C through a 0.1-m² glass plate to the interior of a large room with a 10-m² surface area at 27°C. The glass properties are specified as $\tau_\lambda = 0.90$ for 0.29 μm $< \lambda$ < 2.7 μm, $\tau_\lambda = 0$ outside this range, and $\rho = 0$ for all wavelengths. Assume that the walls of the furnace and room can be approximated as blackbodies.

Solution

Following the pattern established in Example 4-5 on page 259, we obtain $E_{b,\lambda_1 \to \lambda_2}/E_b$ for both source temperatures in the wavelength range 0.29 μm $< \lambda < 2.7$ μm. For the 2270 K temperature source, the corresponding λT range is 659 μm K $< \lambda T < 6140$ μm K. Referring to

Table A-G-1, we find

$$\frac{E_{b,\lambda_1 \to \lambda_2}}{E_b} = 0.748 - 0.17 \times 10^{-7} \simeq 0.748$$

It follows that

$$\tau_{m1} = (0.9)(0.748) = 0.673$$

For the 300 K source, the λT range is from 87 to 810 μm K and

$$\frac{E_{b,\lambda_1 \to \lambda_2}}{E_b} \simeq 0.738 \times 10^{-4}$$

such that τ_{m2} is essentially zero.

To obtain the total rate of radiation heat transfer we merely employ Eq. (4-107).

$$q_{1-2} = A_m \left(E_{b1}\tau_{m1} - E_{b2}\tau_{m2} + \frac{\alpha_{m1}E_{b1} - \alpha_{m2}E_{b2}}{2} \right)$$

where $\tau_{m1} = 0.673$, $\tau_{m2} = 0$, $\alpha_{m1} = 0.327$, $\alpha_{m2} = 1$, and

$$E_{b1} = \sigma T_1^4 = 5.67 \times 10^{-8} \frac{W}{m^2 \, K^4} (2270 \, K)^4$$

$$= 1.51 \times 10^6 \frac{W}{m^2}$$

$$E_{b2} = \sigma T_2^4 = \sigma(300 \, K)^4 = 459 \frac{W}{m^2}$$

Substituting into Eq. (4-109), we have

$$q_{1-2} = 0.1 \, m^2 \left[(1.51 \times 10^6)(0.673) - 0 + \frac{(0.327)(1.51 \times 10^6) - (1)(459)}{2} \right] \frac{W}{m^2}$$

$$= 1.26 \times 10^5 \, W$$

For purpose of comparison, the radiation-heat-transfer rate is calculated for an opening with no glass plate.

$$q_{1-2} = A_m(E_{b1} - E_{b2})$$

$$= 0.1 \, m^2 (1.51 \times 10^6 - 459) \frac{W}{m^2}$$

$$= 1.51 \times 10^5 \, W$$

Thus, we find that the glass plate reduces the rate of radiation heat transfer by about 16%.

Effects of Diffuse Nonblackbody Surfaces and Reflecting Medium   Of course, the transfer of thermal radiation through a transparent solid medium generally occurs in the context of nonblackbody surfaces. In addition, reflection by the surfaces of the medium is sometimes significant. For example, the reflectivity of glass is usually of the order of 0.1. These complexities are approximately accounted for in Examples 4-24 and 4-25 by utilizing the radiosity concept.

EXAMPLE 4-24

Develop a thermal radiation network for diffuse nonblackbody parallel plates which are separated by a nonreflecting transparent solid medium.

Solution

To analyze radiation heat transfer between two diffuse surfaces A_1 and A_2 which are separated by a transparent solid medium, we utilize the concepts that have been introduced in the previous two sections. First, the net rate of radiation heat transfer from each surface is expressed in terms of radiosity by relationships of the form of Eq. (4-53); that is,

$$q_1 = \frac{A_1 \alpha_1}{\rho_1} \left(\frac{\epsilon_1}{\alpha_1} E_{b_1} - J_1 \right) \tag{a}$$

$$q_2 = \frac{A_2 \alpha_2}{\rho_2} \left(\frac{\epsilon_2}{\alpha_2} E_{b_2} - J_2 \right) \tag{b}$$

A second set of equations is written for the net rate of radiation heat transfer between each surface and the medium A_m. By recognizing that q_{1-m} must be equal to the difference between (1) the energy leaving A_1 that is absorbed by the medium, $\alpha_{m1}(A_1 F_{1-m} J_1)$, and (2) the thermal radiation which is emitted by the medium and reaches A_1, $A_m F_{m-1} \epsilon_m E_{bm}$, we have

$$q_{1-m} = \alpha_{m1}(A_1 F_{1-m} J_1) - A_m F_{m-1} \epsilon_m E_{bm}$$
$$= A_1 F_{1-m}(\alpha_{m1} J_1 - \epsilon_m E_{bm}) \tag{c}$$

Similarly, q_{2-m} takes the form

$$q_{2-m} = A_2 F_{2-m}(\alpha_{m2} J_2 - \epsilon_m E_{bm}) \tag{d}$$

Finally, the rate of radiation heat transfer between A_1 and A_2 is equal to the difference between (1) the rate of thermal radiation leaving A_1 that reaches A_2, $A_1 F_{1-2} J_1 \tau_{m1}$, and (2) the rate from A_2 to A_1, $A_2 F_{2-1} J_2 \tau_{m2}$.

$$q_{1-2} = A_1 F_{1-2}(J_1 \tau_{m1} - J_2 \tau_{m2}) \tag{e}$$

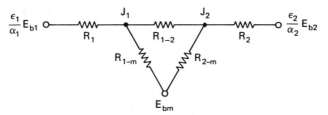

FIGURE E4-24   Thermal radiation network.

Equations (a) through (e) provide the basis for the thermal radiation network shown in Fig. E4-24, where

$$R_{1-2} = \frac{J_1 - J_2}{A_1 F_{1-2}(J_1 \tau_{m1} - J_2 \tau_{m2})}$$

$$R_{1-m} = \frac{J_1 - E_{bm}}{A_1 F_{1-m}(\alpha_{m1} J_1 - \epsilon_m E_{bm})}$$

$$R_{2-m} = \frac{J_2 - E_{bm}}{A_2 F_{2-m}(\alpha_{m2} J_2 - \epsilon_m E_{bm})}$$

Note that this network reduces to the one shown in Fig. 4-39 for blackbody surfaces A_1 and A_2.

EXAMPLE 4-25

Develop a thermal radiation network for two diffuse nonblackbody parallel plates which are separated by a reflecting transparent solid medium.

Solution

To account for reflection from either side of the solid medium in a bisurface system, we utilize the radiosity concept. The radiosity J_{m1} from one side of the solid is written in terms of the thermal radiation emitted and reflected as

$$J_{m1} = \varepsilon_m E_{bm} + \rho_{m1} G_{m1} \tag{a}$$

where G_{m1} is the irradiation from surface A_1. It is important to note that the thermal radiation transmitted through the solid medium is *not* included in our defining equation for radiosity, but rather is treated separately. The net rate of radiation heat transfer from one side of the solid medium q_{m1} (not including energy transmitted through the solid) is expressed by

$$q_{m1} = A_m(\varepsilon_m E_{bm} - \alpha_{m1} G_{m1}) \tag{b}$$

Combining Eqs. (4-103) and (4-105) to eliminate G_{m1}, we have

$$q_{m1} = A_m \left[\varepsilon_m E_{bm} - \frac{\alpha_{m1}}{\rho_{m1}} (J_{m1} - \varepsilon_m E_{bm}) \right]$$

$$= \frac{A_m}{\rho_{m1}} \left[E_{bm} \varepsilon_m (\rho_{m1} + \alpha_{m1}) - \alpha_{m1} J_{m1} \right]$$

$$= \frac{A_m}{\rho_{m1}} (1 - \tau_{m1}) \left(\varepsilon_m E_{bm} - \frac{\alpha_{m1} J_{m1}}{1 - \tau_{m1}} \right) \qquad (c)$$

Similarly, an equation can be written for q_{m2} of the form

$$q_{m2} = \frac{A_m}{\rho_{m2}} (1 - \tau_{m2}) \left(\varepsilon_m E_{bm} - \frac{\alpha_{m2} J_{m2}}{1 - \tau_{m2}} \right) \qquad (d)$$

To obtain an expression for the net rate of thermal radiation from surface A_1 to the solid medium excluding the energy transmitted through the medium, we take the difference between (1) the rate of nontransmitted thermal radiation leaving A_1 that reaches the medium, $A_1 F_{1-m} J_1 (1 - \tau_{m1})$, and (2) the rate of nontransmitted thermal radiation leaving the medium that reaches A_1, $A_m F_{m-1} J_m$. That is,

$$q_{1-m} = A_1 F_{1-m} J_1 (1 - \tau_{m1}) - A_m F_{m-1} J_{m1}$$

$$= A_1 F_{1-m} \left[J_1 (1 - \tau_{m1}) - J_{m1} \right]$$

$$= A_1 F_{1-m} (1 - \tau_{m1}) \left(J_1 - \frac{J_{m1}}{1 - \tau_{m1}} \right) \qquad (e)$$

Similarly, we can write

$$q_{2-m} = A_2 F_{2-m} (1 - \tau_{m2}) \left(J_2 - \frac{J_{m2}}{1 - \tau_{m2}} \right) \qquad (f)$$

To account for the thermal radiation transmitted through the medium, we utilize Eq. (e) in Example 4-24,

$$q_{1-2} = A_1 F_{1-2} (J_1 \tau_{m1} - J_2 \tau_{m2}) \qquad (g)$$

Finally, the rate of radiation heat transfer from each surface is given by Eq. (4-53).

$$q_1 = \frac{A_1 \alpha_1}{\rho_1} \left(\frac{\varepsilon_1}{\alpha_1} E_{b1} - J_1 \right) \qquad (h)$$

$$q_2 = \frac{A_2 \alpha_2}{\rho_2} \left(\frac{\varepsilon_2}{\alpha_2} E_{b2} - J_2 \right) \qquad (i)$$

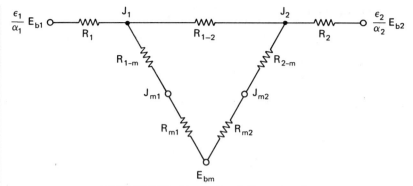

FIGURE E4-25   Thermal radiation network.

Equations (c) through (i) permit us to construct the thermal radiation network shown in Fig. E4-25, where

$$R_{1-2} = \frac{J_1 - J_2}{A_1 F_{1-2}(J_1 \tau_{m1} - J_2 \tau_{m2})}$$

$$R_{1-m} = \frac{J_1 - J_{m1}}{A_1 F_{1-m}(1 - \tau_{m1})\left(J_1 - \dfrac{J_{m1}}{1 - \tau_{m1}}\right)}$$

$$R_{m1} = \frac{J_{m1} - E_{bm}}{\dfrac{A_m}{\rho_m}(1 - \tau_{m1})\left(\varepsilon_m E_{bm} - \dfrac{\alpha_{m1} J_{m1}}{1 - \tau_{m1}}\right)}$$

4-5-6   Thermal Radiation Systems with Combined Modes

Most thermal radiation problems encountered in practice involve at least one of the other two modes of heat transfer. For example, radiation heat transfer in a furnace includes conduction heat transfer through the walls and within bodies being heated as well as convection heat transfer within and without the furnace. Another example is the transfer of heat through superinsulative walls, which involves all three modes.

Practical analyses were developed in Chaps. 1 through 3 for mixed-mode blackbody radiation systems by utilizing the coefficient of radiation heat transfer $\bar{h}_R$ or the thermal resistance R_R. This practical approach is extended to more realistic one-dimensional nonblackbody combined-mode systems in Examples 4-26 and 4-27. For more complex multidimensional mixed-mode systems which involve nonuniform surface temperatures, numerical solution techniques are required.

EXAMPLE 4-26

A lightly oxidized thin copper plate with a 1-m^2 surface area is mounted horizontally outdoors with the lower surface insulated. Determine the temperature of the plate on a clear winter night for which the air temperature is 1 °C and $\bar{h}$ is 10 W/(m^2 °C). (*Note:* According to [25], the sky can be considered as a blackbody radiator with temperature given by

$$T_{sky} = T_{air} - A \tag{a}$$

where $A = 20$ °C in the winter and $A = 6$ °C in the summer.)

Solution

Focusing attention on the radiation heat transfer, the thermal radiation network is given in Fig. E4-26a where

$$T_R = T_{air} - 20\,°C = 254\ K$$

$$E_{bR} = \sigma T_R^4$$

$$R_{s-R} = \frac{1}{A_s F_{s-R}} = \frac{1}{A_s}$$

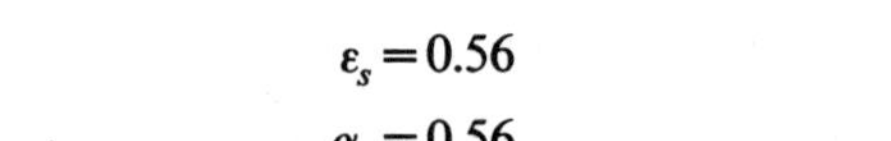

FIGURE E4-26a Thermal radiation network.

Utilizing Fig. 4-3 and the nonisothermal or graybody α approximation, we have

$$\varepsilon_s = 0.56$$

$$\alpha_s = 0.56$$

Therefore, R_s is given by

$$R_s = \frac{\rho_s}{A_s \alpha_s} = \frac{1 - 0.56}{A_s(0.56)} = \frac{0.786}{A_s}$$

The rate of radiation heat transfer is now expressed in terms of the unknown surface potential T_s by

$$q_R = \frac{(\varepsilon_s/\alpha_s)E_{bs} - E_{bR}}{R_s + R_{s-R}}$$

$$= \frac{E_{bs} - E_{bR}}{(0.786 + 1)/A_s} = \frac{\sigma A_s(T_s^4 - T_R^4)}{1.79} \tag{b}$$

where $T_R = 254$ K.

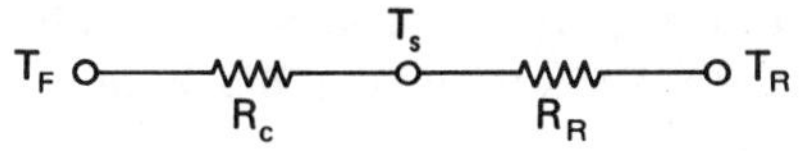

$$T_F \circ \!\!\!-\!\!\!-\!\!\!\!\!\bigwedge\!\!\!\bigwedge\!\!\!-\!\!\!\!\!\overset{T_s}{\circ}\!\!\!-\!\!\!\bigwedge\!\!\!\bigwedge\!\!\!-\!\!\!-\!\!\!\circ T_R$$
$$\qquad R_c \qquad\qquad R_R$$

FIGURE E4-26b   Thermal circuit.

Turning to the overall problem, a mixed-mode thermal circuit is shown in Fig. E4-26b. We obtain R_R by rearranging Eq. (b) as follows:

$$q_R = \left[\frac{\sigma A_s}{1.79} (T_s + T_R)(T_s^2 + T_R^2) \right] (T_s - T_R)$$

$$= \frac{T_s - T_R}{R_R}$$

$$R_R = \frac{1.79}{\sigma A_s (T_s + T_R)(T_s^2 + T_R^2)}$$

Of course, R_c is simply equal to $1/(\bar{h} A_s)$.

To determine the unknown surface temperature T_s, an energy balance is performed on the T_s node.

$$q_c + q_R = 0$$

$$\bar{h} A_s (T_s - T_F) + \frac{\sigma A_s}{1.79} (T_s^4 - T_R^4) = 0$$

$$T_s = T_F - \frac{\sigma}{1.79 \bar{h}} (T_s^4 - T_R^4)$$

$$= 274 \text{ K} - \frac{3.17 \times 10^{-9}}{\text{K}^3} \left[T_s^4 - (254 \text{ K})^4 \right]$$

$$= 287 \text{ K} - \frac{3.17 \times 10^{-9}}{\text{K}^3} T_s^4$$

This equation is solved by iteration, with the result that

$$T_s \simeq 270 \text{ K} = -3 \,^{\circ}\text{C}$$

Thus, we find that although the air temperature is $1\,^{\circ}\text{C}$, the surface temperature is below freezing because of radiation to the sky.

EXAMPLE 4-27

Determine the rate of heat transfer through a superinsulative spherical wall which consists of an inner surface at $-40\,^{\circ}\text{C}$ with a radius 1.01 m, a 1-mm evacuated space, and a 10-cm-thick layer of insulation [$k = 0.60$ W/(m $^{\circ}$C)]. The surrounding fluid is at $35\,^{\circ}$C with $\bar{h} = 150$ W/(m^2 $^{\circ}$C).

The inner radiative surface is constructed of a highly reflective material with $\rho = 0.9$ while the other surface is a blackbody.

Solution

The thermal circuit for this mixed-mode radiation, conduction, and convection system is shown in Fig. E4-27a. The thermal resistances for conduction and convection are given by

$$R_k = \frac{r_3 - r_2}{4\pi r_3 r_2 k}$$

$$= \frac{(1.111 - 1.011) \text{ m}}{4\pi (1.111 \text{ m})(1.011 \text{ m})0.6 \text{ W}/(\text{m } °\text{C})}$$

$$= 0.0118 \frac{°\text{C}}{\text{W}}$$

$$R_c = \frac{1}{\bar{h} A_s} = \frac{1}{[150 \text{ W}/(\text{m}^2 °\text{C})]4\pi (1.111 \text{ m})^2}$$

$$= 4.30 \times 10^{-4} \frac{°\text{C}}{\text{W}}$$

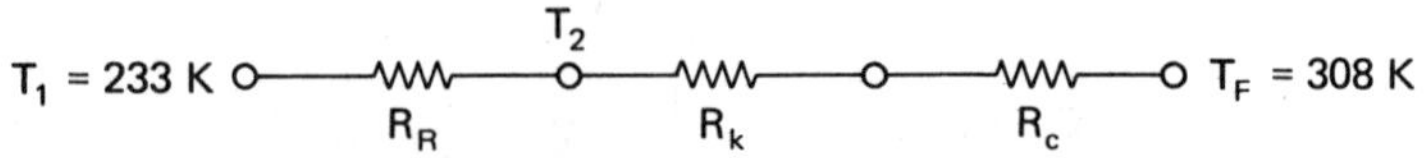

$T_1 = 233 \text{ K} \circ\!\!-\!\!\text{WW}\!\!-\!\!\overset{T_2}{\underset{R_R}{\circ}}\!\!-\!\!\text{WW}\!\!-\!\!\underset{R_k}{\circ}\!\!-\!\!\text{WW}\!\!-\!\!\circ \; T_F = 308 \text{ K}$

FIGURE E4-27a   Thermal circuit.

To determine the resistance R_R for thermal radiation, we utilize the thermal radiation network shown in Fig. E4-27b, where

$$R_1 = \frac{\rho_1}{A_1 \alpha_1} = \frac{0.9}{A_1 (0.1)} = \frac{9}{A_1}$$

$$R_{1-2} = \frac{1}{A_1 F_{1-2}} = \frac{1}{A_1}$$

$$E_{b2} = \sigma T_2^4$$

and, assuming that $\varepsilon_1 \simeq \alpha_1$,

$$\frac{\varepsilon_1}{\alpha_1} E_{b1} \simeq E_{b1} = \sigma T_1^4$$

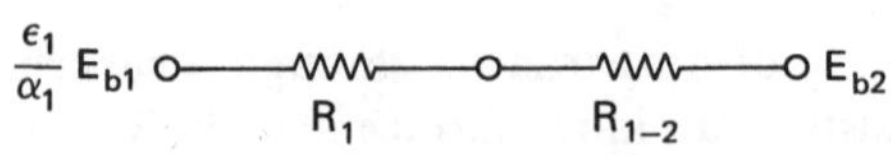

$\frac{\varepsilon_1}{\alpha_1} E_{b1} \circ\!\!-\!\!\text{WW}\!\!-\!\!\underset{R_1}{\circ}\!\!-\!\!\text{WW}\!\!-\!\!\circ E_{b2}$

FIGURE E4-27b   Thermal radiation network.

It follows that the rate of radiation heat transfer q_{1-2} can be written as

$$q_{1-2} = \frac{E_{b1} - E_{b2}}{9/A_1 + 1/A_1} = \frac{A_1(E_{b1} - E_{b2})}{10}$$

Rearranging this expression, we have

$$q_{1-2} = \frac{\left[A_1\sigma(T_1 + T_2)(T_1^2 + T_2^2)\right](T_1 - T_2)}{10}$$

or

$$q_{1-2} = \frac{T_1 - T_2}{R_R} \tag{a}$$

where $A_1 = 4\pi(1.01 \text{ m})^2$, $T_1 = 233$ K, and

$$R_R = \frac{10}{A_1\sigma(T_1 + T_2)(T_1^2 + T_2^2)}$$

Returning to the circuit in Fig. E4-27a, we perform an energy balance on the T_2 node, with the result that

$$\frac{233 \text{ K} - T_2}{R_R} = \frac{T_2 - 308 \text{ K}}{(1.18 \times 10^{-2} + 4.3 \times 10^{-4}) \text{ K/W}}$$

Rearranging, we obtain

$$T_2 = \frac{233 \text{ K}/R_R + 2.52 \times 10^4/\text{W}}{81.8 \text{ W/K} + 1/R_R}$$

This equation is solved for T_2 by iteration. Starting with $T_2 = 290$ K,

$$R_R = 0.191 \frac{\text{K}}{\text{W}}$$

$$T_2 = 304 \text{ K}$$

A second iteration gives

$$R_R = 0.174 \frac{\text{K}}{\text{W}}$$

$$T_2 \simeq 303 \text{ K}$$

Substituting this result into Eq. (a), we obtain

$$q_{1-2} = \frac{233 \text{ K} - 303 \text{ K}}{0.174 \text{ K/W}}$$

$$= -402 \text{ W}$$

As a point of interest, we note that for blackbody radiation at surface A_1, the resistance R_R reduces to

$$R_R = \frac{1}{A_1 \sigma (T_1 + T_2)(T_1^2 + T_2^2)}$$

which is a factor of 10 less than for the case in which a highly reflective surface is used. For the blackbody situation, we find

$$T_2 = 270 \text{ K}$$

and

$$q_R = -2.22 \times 10^4 \text{ W}$$

This heat loss is greater by a factor of 55 than for the case in which one reflective surface is employed. This result gives a good indication of why highly reflective surfaces are used in superinsulative walls.

4-6   SOLAR RADIATION

4-6-1   The Solar Resource

The sun is an essentially spherical body ($r_0 \simeq 0.695 \times 10^6$ km) of extremely high-temperature matter. Within its inner core ($r \gtrsim 0.23 r_0$), temperatures of the order of 8×10^6 K to 40×10^6 K are maintained by a continuous fusion process in which mass is converted into energy. This fusion process produces X-ray and γ-ray electromagnetic radiations which emanate from the high-density core. In the low-density gaseous region between $0.7 r_0$ and r_0, energy is also transported by convection. The temperature is believed to drop from about 130,000 K to 5000 K across this convective zone. The *photosphere*, which makes up the outer layer of the convection zone, is essentially opaque and well defined. Three gaseous layers lie outside the photosphere, with temperatures ranging from about 5000 K within the inner layer to as high as 10^6 K in the outer layer.

Solar radiation consists of energy that is emitted by the various layers, with the major contribution being provided by the photosphere.

Although the radiation emitted by the sun originates from the various temperature zones, for practical purposes the sun can be considered as a blackbody radiator at an effective temperature of about 5800 K. A more accurate value for the effective temperature of the sun according to Thekaekara [26] is 5762 K. Based on Thekaekara's estimate, the effective total emissive power of the sun is calculated to be about

$$E_{sun} = \sigma T_{sun}^4$$

$$= 5.67 \times 10^{-8} \frac{W}{m^2 \, K^4} (5762 \text{ K})^4$$

$$= 6.25 \times 10^7 \frac{W}{m^2}$$

The approximate monochromatic emissive power $e_{b\lambda}$ of the sun calculated by the use of the Planck law, Eq. (4-7), is shown in Fig. 4-6.

Of course, only a small fraction of this enormous solar radiation flux reaches the outer fringes of the earth's atmosphere, and an even smaller portion reaches the surface of the earth itself. Based on direct measurements of solar irradiation in the outer reaches of earth's atmosphere which have been obtained by high-altitude aircraft, balloons, and spacecraft, the monochromatic extraterrestrial solar irradiation $G_{s\lambda}$ reaching the earth's atmosphere for the mean earth–sun distance of about 1.50×10^8 km is given by the standard NASA curve shown in Fig. 4-40. This solar irradiation consists of very short wavelength γ rays, X rays, and ultraviolet rays, as well as thermal radiation. The total extraterrestrial solar irradiation G_s reaching the earth's atmosphere for the mean earth–sun distance is approximately 1350 W/m². Strictly speaking, G_s varies by about $\pm 3\%$ as a result of changes in the earth–sun distance and the conditions on the sun. However, for practical purposes G_s can be taken as a constant. For this reason, G_s is generally referred to as the *solar constant*.

The solar irradiation flux is further attenuated by the atmosphere before it reaches the surface of the earth. The γ-ray and X-ray components of the solar irradiation are absorbed in the ionosphere by nitrogen, oxygen, and other materials, and much of the ultraviolet radiation is absorbed by ozone. Most of the small amount of thermal radiation with wavelength greater that about 2.5 μm is absorbed by water vapor and carbon dioxide. Hence, the solar radiation finally reaching the surface of the earth primarily consists of thermal radiation with wavelengths ranging from 0.30 to 2.5 μm. This thermal solar radiation is itself attenuated by scattering and absorbing within the atmosphere. The geometric factors can be handled with the aid of celestial geometry, but the lack of meteorological information generally makes it impractical to base predictions for the terrestrial

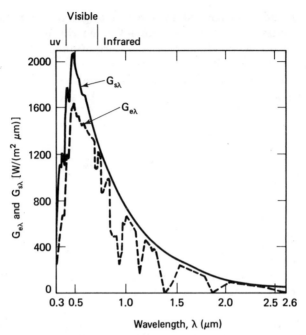

FIGURE 4-40 Representative monochromatic extraterrestrial solar irradiation $G_{s\lambda}$ and monochromatic solar irradiation G_e. (From Thekaekara [26].)

solar resource at any location on extraterrestrial solar irradiation. Rather, the terrestrial solar irradiation is usually measured over a reasonable length of time at the location of interest. Data of this type have been collected at a number of stations across the United States and Europe for many years. Numerous other stations are being set up to measure the solar resource throughout the world. Such data are provided on hourly, daily, and annual bases, with the annual solar irradiation data being of the most practical value in the design of solar systems. Representative experimental measurements for the monochromatic solar irradiation $G_{e\lambda}$ reaching the earth's surface for a very clear atmosphere are shown in Fig. 4-40. The irregularities in $G_{e\lambda}$ are caused by absorption of water vapor, carbon dioxide, and oxygen.

To illustrate the importance of weather conditions and time of day, the total terrestrial solar irradiation G_e measured in the springtime at Madison, Wisconsin, is shown in Fig. 4-41 for both clear and cloudy days.

Concerning the location of a receiving surface such as the one illustrated in Fig. 4-41, the solar irradiation is a function of the geometrical relation between the surface and the sun, which is continuously changing on a daily and annual basis. One of the most important practical aspects

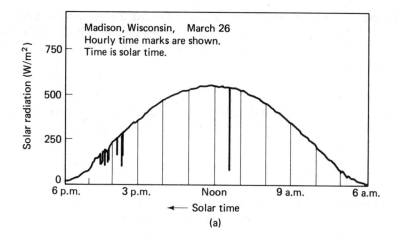

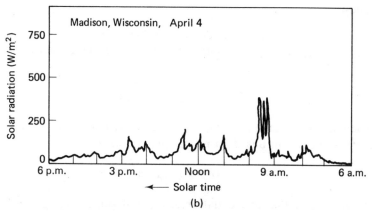

FIGURE 4-41 Total solar irradiation on a horizontal surface vs. time. (a) Clear day. (b) Cloudy day. (From *Solar Energy Thermal Processes* by J. A. Duffie and W. A. Beckman. Copyright 1974 by John Wiley & Sons, Inc. Reprinted by permission of John Wiley & Sons, Inc.)

involving location is the total number of hours of daytime per year. Another important factor pertains to the orientation of the surface itself. Of course, the maximum irradiation flux on a receiving surface at a given location on the earth is obtained by continual adjustment of the surface orientation to maintain normal solar incidence throughout the daylight hours. But such adjustment requires rotation about two axes which is generally not practical, except for high-temperature power generation stations. Based on the analysis of much experimental data, researchers in the area of solar radiation have recommended that for maximum annual collection, a receiver should be oriented toward the equator with a slope approximately equal to the latitude ϕ. For best winter collection the slope should be about $\phi + 10$ degrees, and for optional summer irradiation the slope should be approximately $\phi - 10$ degrees.

The solar radiation reaching the surface of the earth is absorbed and reflected. Some of the energy that is absorbed is reemitted, with the remainder being stored thermally in the land masses and oceans, chemically in vegetation through the process of photosynthesis, and mechanically in form of wind.

EXAMPLE 4-28

Assuming that the sun radiates as a blackbody at 5760 K, develop a prediction for the extraterrestrial irradiation G_s on the atmosphere of the earth.

Solution

Referring to the geometric relationship between the earth and the sun shown in Fig. E4-28, the solar irradiation G_s reaching the earth's outer atmosphere is given by

$$G_s \, dA_e = A_{sun} F_{A_{sun} - dA_e} E_{sun} \tag{a}$$

where the differential area dA_e is normal to the line r drawn between the earth and the sun. Based on reciprocity, we write

$$G_s = F_{dA_e - A_{sun}} E_{sun} \tag{b}$$

By visualizing the differential area dA_e and sun as two parallel disks, $F_{dA_e - A_{sun}}$ can be approximated by Eq. (k) of Example 4-8 on page 272,

$$F_{dA_e - A_{sun}} = \frac{r_0^2}{Z^2 + r_0^2} \tag{c}$$

where the distance Z between the earth and sun is about 1.5×10^8 km

$$D_{sun} = 1.39 \times 10^6 \text{ km } (8.64 \times 10^4 \text{ miles})$$
$$D_{earth} = 1.27 \times 10^4 \text{ km } (7.9 \times 10^3 \text{ miles})$$

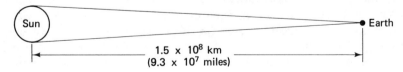

FIGURE E4-28   Earth–sun geometric relationship (not to scale).

and the radius of the sun r_0 is about 0.695×10^6 km. Combining Eqs. (b) and (c) and employing the Stefan–Boltzmann law, we have

$$
\begin{aligned}
G_s &= \frac{\sigma T_{\text{sun}}^4}{(Z/r_0)^2 + 1} \\
&= \frac{\left[5.67 \times 10^{-8} \text{ W}/(\text{m}^2 \text{ K}^4) \right] (5760 \text{ K})^4}{(1.5 \times 10^8 / 0.695 \times 10^6)^2 + 1} \\
&= 1340 \frac{\text{W}}{\text{m}^2}
\end{aligned}
$$

This value of the radiation constant G_s lies between the standard value of 1353 W/m^2 proposed in 1971 by Thekaekara and Drummond [27] and an earlier standard of 1322 W/m^2.

4-6-2 Solar Energy Systems

A variety of solar energy systems have been devised to satisfy applications ranging from the heating of water, the heating and cooling of residential and commerical buildings, and the operation of desalination plants, to the production of power on a commercial basis and the energizing of electronic equipment in remote locations on land, sea, and in space. Several basic types of solar energy systems include solar thermal systems, ocean thermal systems, wind systems, photovoltaic devices, as well as others. We will focus our attention on solar thermal and ocean thermal systems.

Solar thermal systems

All solar systems that receive, collect, store, and utilize thermal radiation directly from the sun fall under the heading of solar thermal. These systems consist of two basic components: a collector and a storage unit. Of course, these solar components must be interfaced with conversion devices (i.e., air-conditioning units and engines), loads, auxiliary energy supplies, and controls to obtain a total energy system.

The solar collector is actually a thermal radiation heat exchanger in that such devices transfer solar energy to a fluid. Flat-plate and focusing collectors are shown in Figs. 4-42 and 4-43. The simple flat-plate collector consists of a near-black solar energy absorbing surface, which collects and transfers the solar energy to a fluid; a transparent cover, which minimizes radiation and convection losses to the atmosphere; and back insulation, which minimizes conduction losses. Flat-plate collectors are utilized in applications requiring moderate temperatures up to about 100 °C above ambient. Applications involving the use of flat-plate collectors include water and environmental heating, refrigeration, and seawater desalination.

Collector is double glazed. Lo-Iron ASG Tru-Temp tempered glass is used for the first glazing, DuPont Teflon for the second glazing.

Absorber plate is an assembly of copper tubes recessed in an extruded aluminum plate which is primed and finished with Nextel "Black Velvet."

Collector housing is constructed of 20-gauge galvanized steel. The enclosure is completely lined with high-temperature fiberglas insulation around and below the absorber plate.

FIGURE 4-42   Flat-plate collector. (Courtesy of Northrup, Inc.)

The concentrating collector shown in Fig. 4-43 is capable of operating at much higher temperatures than flat-plate collectors. This single-axis system can operate at temperatures up to 315 °C. Systems with double-axis focusing are able to attain temperatures as high as 3600 °C.

Because of the intermittent nature of solar radiation, the storage unit is an essential part of most solar systems. Simply put, the procedure for storing energy in thermal solar energy systems consists of heating a fluid or solid which is confined in a well-insulated space. For example, hot water is often circulated from the collector into a heavily insulated tank. For systems that utilize air as the working fluid, the hot air is sometimes passed from the collector into a porous bed of stone which receives and stores the energy. However, simple methods such as these are generally restricted to storage periods of only a few days, at best, because of imperfect thermal insulation and size limitations. Other more sophisticated methods which involve materials that store energy by change of phase from solid to liquid have been found to reduce the storage volume and extend the length of time energy can be stored.

Ocean thermal systems

The oceans represent a tremendous natural collector of solar energy. As a consequence of the absorption of solar radiation, an approximate 20 °C temperature difference exists between the warm surface layers of the

(a)

Sun-following is accomplished by a small electric motor attached to a cable and pulley mechanism. The motor is controlled by two silicon cells.

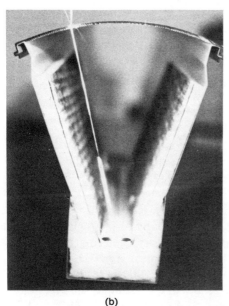

(b)

Curved Fresnel-type lens made of extruded acrylic.

Collector housing is a galvanized steel trough, insulated with fiberglas.

The tube is coated with a selective black surface that has a high absorbtivity for solar radiation but a low emissivity for long-wave infrared radiation.

FIGURE 4-43 (a) Concentrating solar collector. (b) Photograph of laser beam reflected by lens on concentrating solar collector. (Courtesy of Northrup, Inc.)

tropical oceans and the cold bottom layer. This 50-m-deep surface layer has an essentially uniform temperature of 25 °C, with 5 °C water lying at a depth of about 900 m [28]. Serious consideration is presently being given to harnessing this vast source of work potential. A model of a 10-MW (electric) self-contained OTEC power module is shown in Fig. 4-44. This module is designed for attachment to a surface platform at sea. A standard

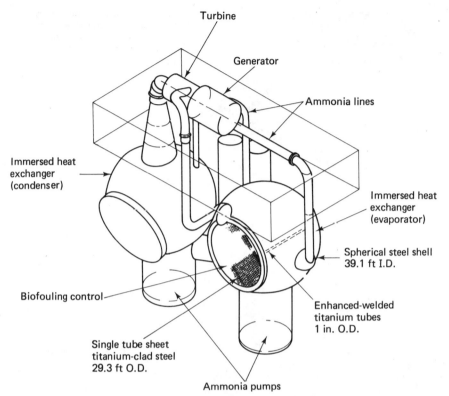

FIGURE 4-44   Model of a 10-MW(e) self-contained OTEC power module. (Courtesy of Lockheed Missiles and Space Company, Inc.)

power cycle is utilized in this closed-cycle system, with the energy being provided to the working fluid (ammonia) by the circulation of warm surface waters through the evaporator, and energy being removed in the condenser by cold water from the ocean depths. However, because of the relatively small 20°C temperature difference that is available, very large and efficient heat exchangers are required. For example, the evaporator for this system has a shell I.D. of 39 ft, 51,400 tubes of 1-in. O.D. and 26.2-ft length, with a total weight of 410 tons. The condenser is even larger. Because of size and effectiveness requirements, the heat exchangers dominate the overall ocean thermal plant cost.

4-7 CLOSURE

Practical thermal analyses have been developed in this chapter for thermal radiation heat transfer in ideal or near-ideal systems involving diffuse surfaces with uniform and known thermal radiation properties ε, α, ρ, and

τ, uniform heating conditions (including uniform emission, reflection, and emission), and nonparticipating or gray gases. The use of this basic approach in analyzing problems involving solar radiation, combined modes, and unsteady conditions has also been introduced. As we have seen, the concept of *radiosity* is the key to the development of practical analysis of radiation heat transfer. Either the thermal radiation network approach or the alternative approach of Example 3-13 can be used.

The radiosity concept and other approaches, such as the Monte Carlo method, are used in the analysis of nonideal systems involving surfaces with nondiffuse characteristics, complex spectral properties, and nonuniform heating. Introductions to the analysis of these types of non-ideal systems, as well as systems involving multidimensional combined modes and participating gases, are presented by Siegel and Howell [11] and Sparrow and Cess [5]. With regard to solar radiation which was briefly touched upon in Sec. 4-6, this timely topic is more thoroughly introduced by Duffie and Beckman [25].

PROBLEMS

4-1. Calculate the total emissive power of a blackbody at (a) 25 K; (b) 100 °F; (c) 1000 °C; (d) 10,000 °R.

4-2. Referring to Example 4-1, estimate the fraction of energy emitted by a blackbody at 1000 °C that falls in the visible region.

4-3. Estimate the absorptivity α of a stainless steel surface A_s at 500 °C which is exposed to radiation from a blackbody surface at (a) 0 °C; (b) 750 °C. (Assume graybody conditions.)

4-4. Estimate the total emissive power for a stainless steel surface at (a) 0 °C; (b) 500 °F; (c) 500 °C.

4-5. Referring to Prob. 1-12, determine the percent decrease in q_R which results from the use of a blackbody radiation shield.

4-6. Estimate the emissivity ε and absorptivity α of a gas at 1 atm and 1500 °C which consists of 40% CO_2 (by volume) and 60% N_2. The parallel-plate blackbody walls are at 500 °C, with L_e equal to 1 m.

4-7. Demonstrate that water vapor contained between parallel plates 1 m apart at 1 atm pressure and 100 °C emits as much as 55% of the energy that would be emitted by a blackbody at 100 °C with the same surface area as the plates.

4-8. Determine the radiation shape factor from the walls of a cubical furnace with sides of 1 m length to a 4-cm^2 window located in the center of one of the walls.

4-9. Determine the radiation shape factor from the walls of a 10-cm-long 5-cm-diameter pipe to one end which is open.

4-10. Demonstrate that the radiation shape factor F_{1-2} for the duct shown in Fig. E4-20 is equal to 0.293.

4-11. Determine the radiation shape factors F_{1-3} and F_{1-4} for the system shown in Fig. P4-11.

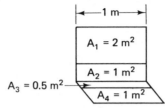

FIGURE P4-11

4-12. Utilize the following double-angle formula to perform the integration in Eq. (a) of Example 4-2:

$$\sin(a \pm b) = \sin a \cos b \pm \cos a \sin b$$

4-13. Determine the radiation-heat-transfer rate between two concentric blackbody spheres with radii $r_1 = 10$ cm and $r_2 = 25$ cm, and surface temperatures $T_1 = 1000\,°C$ and $T_2 = 30\,°C$.

4-14. Determine the rate of radiation heat transfer between two long concentric blackbody cylinders with radii $r_1 = 2$ in. and $r_2 = 10$ in., and surface temperatures $T_1 = -40\,°F$ and $T_2 = 100\,°F$. Sketch the thermal radiation network.

4-15. Determine the rate of radiation heat transfer between each surface of a long equilateral triangular duct for blackbody conditions with each side being 10 cm in length, $T_1 = 500°C$, $T_2 = 25°C$, and $T_3 = 200°C$. Also determine the total radiation heat flux from each surface. Sketch the thermal radiation network.

4-16. If the trisurface enclosure of Fig. 4-26(a) represents a very long triangular duct with $A_1 = A_2 = 100$ ft^2, determine the total radiation-heat-transfer rate from surfaces A_1 and A_2 and the temperature of surface A_3 for the case in which each surface is black and $T_1 = 70\,°F$, $T_2 = 300\,°F$, and $q_3'' = 100$ Btu/ (h ft^2).

4-17. Referring to the open radiation system of Fig. 4-26(b), determine the rate of radiation heat transfer between each surface for blackbody conditions with $A_1 = 4$ m^2, $A_2 = 2$ m^2, $A_3 = 5$ m^2, $F_{1-2} = 0.2$, $F_{1-3} = 0.4$, $F_{2-3} = 0.3$, $T_1 = 500\,°C$, $T_2 = 25\,°C$, and $T_3 = 200\,°C$. Explain why the total rate of heat transfer from each surface cannot be determined.

4-18. Assuming that the open trisurface system of Prob. 4-17 is surrounded by a blackbody at 0 K, determine radiation-heat-transfer rates between each surface and the total rate of radiation heat transfer from each surface. Sketch the thermal radiation network for this system.

4-19. Check the solution to Example 4-12 by showing that $q_1 + q_2 + q_3 + q_4 = 0$.

4-20. Sketch the approximate thermal radiation network for Example 4-10 for the case in which surface A_1 is a graybody with $\varepsilon_1 = 0.6$. Then determine q_{1-2}.

4-21. Sketch the approximate thermal radiation network for Example 4-10 for the case in which A_1 and A_2 are both graybodies with $\varepsilon_1 = 0.6$ and $\varepsilon_2 = 0.1$. Determine q_{1-2}.

4-22. Sketch the approximate thermal radiation network for Example 4-11 for the case in which A_2 is a graybody with $\varepsilon_2 = 0.8$. Determine the total rate of radiation heat transfer from A_2.

4-23. Sketch the thermal radiation network for two infinite parallel plates at temperatures of 100°C and 500°C; A_1 is a blackbody and A_2 is a graybody with $\varepsilon_2 = 0.8$. Also determine the radiation heat transfer flux q''_{1-2}.

4-24. Estimate the total rate of radiation heat transfer from A_2 in Example 4-11 for the case in which surfaces A_1 and A_2 are graybodies with $\varepsilon_1 = 0.2$ and $\varepsilon_2 = 0.5$.

4-25. Referring back to Prob. 4-3, determine the rate of radiation heat transfer for the case in which $F_{s-R} = 1$.

4-26. Develop an expression for the rate of radiation heat transfer in a blackbody system involving a differential area dA_1 and a finite area A_2. Then demonstrate that Eqs. (4-43) and (4-46) apply to nonsymmetrical as well as symmetrical systems.

4-27. Equation (4-58) strictly applies to symmetrical systems. To demonstrate this point, develop a more general expression for q_{1-2} that applies to nonsymmetrical systems. (Follow the approach used in Prob. 4-26.)

4-28. (a) Refine the approximate solution for the nonsymmetrical system of Example 4-16 by breaking A_2 into two 1-ft^2 areas. (b) Show that this refinement is unnecessary if both surfaces A_1 and A_2 are blackbodies.

4-29. (a) Develop a thermal radiation network for a concentric spherical system which accounts for the thermal radiation which leaves the outer surface and reaches itself. (It is suggested that this concave surface be broken into a number of smaller surfaces.) Then show that this more general network reduces to the simple thermal radiation network given in Example 4-14. (b) Demonstrate that the thermal radiation network in Example 4-14 is in error for an eccentric system.

4-30. Determine the rate of radiation heat transfer between the concentric spheres of Prob. 4-13 for the case in which a thin radiation shield with diameter of 15 cm is utilized. The emissivity and absorptivity of the shield material are both equal to 0.2.

4-31. Solve Prob. 4-30 for the case in which the shield is made of polished copper.

4-32. A thin radiation shield of 5 in. diameter is placed between the two concentric cylinders of Prob. 4-14. Determine the rate of radiation heat transfer and temperature of the shield for the case in which the emissivity and absorptivity of the shield are equal to 0.1.

4-33. A blackbody plate at 1000 °F exchanges radiation with an oxidized nickel plate at 100 °F. Determine the percent decrease in the rate of radiation heat transfer for the case in which a thin polished aluminum plate is placed between these two bodies.

4-34. Referring to Prob. 4-33, determine the number of polished aluminum shields required to reduce the rate of heat transfer by 500%.

4-35. Refine the solution for Example 4-18 by evaluating α_2 at the temperature of a blackbody with total emissive power equal to J_{s2}.

4-36. Both surfaces A_1 and A_2 of the system shown in Fig. E4-16 are constructed of oxidized nickel. Determine the rate of thermal radiation heat transfer between A_1 and A_2.

4-37. A blackbody 1-m-diameter spherical surface at $0°F$ exchanges radiation with a concentric 2-m-diameter sphere at $1500°F$. Determine the percent reduction in heat transfer if a polished aluminum shield with $\varepsilon \simeq \alpha \simeq 0.8$ is placed between these two bodies.

4-38. Referring to Example 4-19, calculate the rate of heat transfer q_1 in this reradiating system if $A_1 = 1$ m^2 and $A_2 = 4$ m^2.

4-39. Referring to Example 4-20, determine the rates of radiation heat transfer q_1, q_2, and q_3 if surface A_3 is a blackbody at $500°C$, $A_1 = 1$ m^2 and $A_2 = 4$ m^2.

4-40. An anodized aluminum surface can be taken as a graybody with $\varepsilon = 0.92$ for long wavelength thermal radiation associated with low temperatures (see Fig. 4-8). Reevaluate q_R in Example 2-15 by utilizing this information.

4-41. Solve Example 2-15 for the case in which $\alpha \simeq \varepsilon \simeq 0.1$.

4-42. Develop a FORTRAN program to solve Example 2-15 for the temperature distribution in radiating fins. Also compute the rate of heat transfer for comparison with the analytical solution.

4-43. Solve Example 3-16 for the case in which α and ε of the anodized aluminum are equal to 0.92.

4-44. Solve Prob. 3-41 for the case of a nonblackbody radiating fin with emissivity and absorptivity approximately equal to 0.7.

4-45. The floor A_1, ceiling A_2, and one wall A_3 of a 1-m^3 cubical enclosure are made of lightly oxidized copper and are at temperatures $T_1 = 100°C$, $T_2 = 300°C$, and $T_3 = 500°C$. The other three walls are reradiating surfaces. Determine the temperature of the reradiating surfaces.

4-46. Calculate the radiation Biot number for the thin plate of Example 4-21 for the case in which the thermal conductivity is equal to 250 W (m °C).

4-47. Solve Example 4-21 for the case in which a thin polished 1-cm-diameter copper plate initially at $50°C$ is placed in the oven.

4-48. On a cold clear night, air at $0°F$ passes over the surface of an electrically heated graybody radiator which is exposed to radiation from the sky. The mean convection coefficient $\bar{h}$ is equal to 7.5 Btu/(h ft^2 °F). Determine the rate of heat transfer per unit area for a radiator surface temperature of $100°F$.

4-49. The walls of a 5-ft-diameter spherical blackbody furnace are maintained at $200°F$. The gas within the furnace consists of 25% CO_2 by volume and 75% nitrogen at 1 atm pressure and $1000°F$. Evaluate the radiation properties of the gas and calculate the rate of heat transfer to the walls.

4-50. A gas turbine combustion chamber is 1 ft in diameter. The products of combustion are at $2000°F$ and 1 atm, and have a volumetric composition of 10% CO_2 and 20% H_2O. Determine the rate of radiation heat transfer assuming a very long combustion chamber with a wall temperature of $1000°F$.

4-51. A glass plate receives a solar radiation flux of 500 W/m^2. The glass properties are given as $\tau_\lambda = 0.92$ for $0.3 \ \mu m < \lambda < 3 \ \mu m$, $\tau_\lambda = 0$ outside this range, and $\rho_\lambda = 0$ for all wavelengths. Determine the rate of solar radiation transmitted through the glass.

4-52. Determine the rate of radiation heat transfer to the interior of a room with surface area of 20 m² at 27 °C. A solar radiation flux of 500 W/m² enters through a glass plate with a 1-m² area. The glass properties are given in Prob. 4-51. Assume that the walls of the room are black.

4-53. Solve Prob. 4-52 for the case in which the interior wall is a graybody with emissivity of 0.1.

4-54. Solve Prob. 4-52 for the case in which the reflectivity of the glass at all wavelengths is approximately 0.1.

4-55. A very thin stainless steel 301 plate with uniform temperature of 81°F exchanges radiant energy with a blackbody plate at 1000 °F. A cool fluid passes over the other side of the stainless steel plate with $\bar{h}$ equal to 250 Btu/(h ft² °F). Determine the temperature T_F of the fluid.

4-56. Show that the resistance to conduction within the radiation shield of Example 4-18 is negligible.

4-57. Solve Example 4-18 for the case in which a 1-cm-thick plate with $k = 30$ W/(m °C), and emissivity and absorptivity equal to 0.8.

4-58. Two parallel blackbody plates at $T_1 = 600$ K and $T_2 = 1000$ K are separated by a plate with $k \simeq 35$ W/(m °C), $\varepsilon \equiv 0.8$, and $\alpha = 0.8$. Determine the rate of radiation heat transfer for plate thicknesses of (a) 1 mm; (b) 10 cm.

4-59. To develop an expression for the radiation shape factor between the two bodies shown in Fig. P4-59, attention is first focused on the exchange of

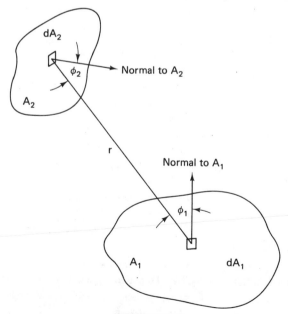

FIGURE P4-59

blackbody thermal radiation between the differential areas dA_1 and dA_2. Following the pattern established in the solution of Example 4-8, show that

$$F_{dA_1-A_2} = \int_{A_1} \frac{\cos \phi_1 \cos \phi_2}{\pi r^2} \, dA_2$$

$$F_{1-2} = \frac{1}{A_1} \int_{A_1} F_{dA_1-A_2} \, dA_1$$

and

$$F_{dA_2-A_1} = \int_{A_1} \frac{\cos \phi_1 \cos \phi_2}{\pi r^2} \, dA_1$$

$$F_{1-2} = \frac{1}{A_2} \int_{A_2} F_{dA_1-A_2} \, dA_2$$

Then demonstrate the validity of the principle of reciprocity.

REFERENCES

[1] TOULOUKIAN, Y. S., and D. P. DE WITT, *Thermophysical Properties of Matter*, Vol. 7, *Thermal Radiative Properties—Metallic Elements and Alloys*. New York: IFI/Plenum Data Corporation, 1970.

[2] TOULOUKIAN, Y. S., and D. P. DE WITT, *Thermophysical Properties of Matter*, Vol. 8, *Thermal Radiative Properties—Nonmetallic Solids*. New York: IFI/Plenum Data Corporation, 1970.

[3] TOULOUKIAN, Y. S., D. P. DE WITT, and R. S. HERNICZ, *Thermophysical Properties of Matter*, Vol. 9, *Thermal Radiative Properties—Coatings*. New York: IFI/Plenum Data Corporation, 1970.

[4] GUBAREFF, G. G., J. E. JANSEN, and R. H. TORBORG, "Thermal Radiation Properties Survey," Honeywell Research Center, Honeywell Regulator Company, Minneapolis, Minn., 1960.

[5] SPARROW, E. M., and R. D. CESS, *Radiation Heat Transfer*. Washington, D. C.: Hemisphere Publishing Corporation, 1978.

[6] PLANCK, M., *The Theory of Heat Radiation*. New York: Dover Publications, Inc., 1959.

[7] DUNKLE, R. V., J. T. GIER, and coworkers, "Snow Characteristics Project, Progress Report," University of California, Berkeley, 1953.

[8] SEBAN, R. A., "The Emissivity of Transition Metals in the Infrared," *J. Heat Transfer*, C87, 1965, 173–176.

[9] SIEBER, W., "Zusammensetzung der von Werk- und Baustoffen Zurückgeworfenen Wärmestrahlung," *Z. Tech. Physik*, 22, 1941, 130–135.

[10] SCHMIDT, E., and E. ECKERT, "Über die Richtungsverteilung der Wärmestrahlung," *Forsch. Geb. Ingenieurw.*, 6, 1935, 175–183.

[11] SIEGEL, R., and J. R. HOWELL, *Thermal Radiation Heat Transfer*. New York: McGraw-Hill, Book Company, 1972.

[12] DIETZ, A. G. H., "Diathermanous Materials and Properties of Surfaces," in *Space Heating with Solar Energy* by R. W. Hamilton. Cambridge, Mass.: The MIT Press, 1954.

[13] SEAL, M., "The Increasing Applications of Diamond as an Optical Material and in the Electronics Industry," *Ind. Diamond Rev.*, April 1978, 130–134.

[14] WILLIAMS, D. A., T. A. LAPPIN, and J. A. DUFFIE, "Selective Radiation Properties of Particular Coatings," *J. Engr. Power*, **95A**, 1963, 213.

[15] EDWARDS, D. K., "Radiation Interchange in a Nongray Enclosure Containing an Isothermal Carbon-Dioxide–Nitrogen Gas Mixture," *J. Heat Transfer*, **c84**, 1962, 1–11.

[16] HOTTEL, H. C., and R. S. EGBERT, "Radiant Heat Transmission from Water Vapor," *AIChE Trans.*, **38**, 1942.

[17] HOTTEL, H. C., "Radiant Heat Transmission," Chap. 4 in *Heat Transmission*, 3rd ed., by W. H. McAdams. New York: McGraw-Hill Book Company, 1954.

[18] ECKERT, E. R. G., and R.M. DRAKE, *Analysis of Heat and Mass Transfer*. New York: McGraw-Hill Book Company, 1972.

[19] MACKEY, C. O., L. T. WRIGHT, JR., R. E. CLARK, and N. R. GAY, "Radiant Heating and Cooling, Part *I*," *Cornell Univ. Eng. Expt. Sta. Bull.*, **32**, 1943.

[20] HOTTEL, H. C., "Radiant Heat Transmission," *Mech. Eng.*, **52**, 1930, 699–704.

[21] HOTTEL, H. C., "Radiant Heat Transmission between Surfaces Separated by Nonabsorbing Media," *ASME Trans.*, **53**, 1931, 265–271.

[22] HAMILTON, D. C., and W. R. MORGAN, "Radiant Interchange Configuration Factors," *NACA TN 2836*, 1952.

[23] KREYSZIG, E., *Advanced Engineering Mathematics*, 3rd ed. New York: John Wiley & Sons, Inc., 1972.

[24] GEBHART, B., *Heat Transfer*, 2nd ed. New York: McGraw-Hill Book Company, 1971.

[25] DUFFIE, J. A., and W. A. BECKMAN, *Solar Energy Thermal Processes*. New York: John Wiley & Sons, Inc., 1974.

[26] THEKAEKARA, M. P., "Data on Incident Solar Radiation," *Suppl. Proc. 20th Ann. Meeting Inst. Environ. Sci.*, **21**, 1974.

[27] THEKAEKARA, M. P., and A. J. DRUMMOND, "Standard Values for the Solar Constant and Its Spectral Components," *Nat. Phys. Sci.*, **229**, 1971, 6.

[28] ZENDER, C., "Solar Sea Power, *Mech. Eng.*, **99**, 1977, 26–29.

CHAPTER 5

CONVECTION HEAT TRANSFER: INTRODUCTION

To introduce the subject of convection heat transfer, we consider the main features that distinguish the various convection systems from one another, and the approaches to the analysis of convection heat transfer that are available to us.

5-1 CHARACTERIZING FACTORS ASSOCIATED WITH CONVECTION SYSTEMS

Convection flow systems can be categorized according to several basic factors. These several categories are now introduced.

5-1-1 Forced and Natural Convection

Forced convection and natural convection are two basic mechanisms by which fluid motion can be produced. *Forced convection* represents fluid flow that is caused by fans, pumps, compressors, and so on. Examples of

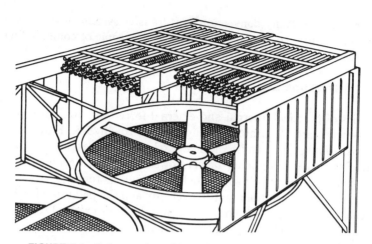

FIGURE 5-1 Cutaway view of forced-draft air cooler. (Courtesy of Yuba Heat Transfer Corporation.)

forced convection systems include the forced-draft air cooler shown pictorially in Fig. 5-1, as well as gas turbines, condensers, and evaporators in steam power plants and in refrigeration units, and oil and gas pipelines, to name only a few. *Natural convection* refers to fluid motion that is caused by temperature-induced density gradients within the fluid. Natural convection (or free convection) flow of air over a steam pipe is represented in Fig. 5-2. Notice that the less-dense air near the steam pipe rises while the heavier cool air falls. Other familiar examples of natural convection flow include circulation through fireplaces and the cooling of electronic devices.

Warm (lighter) air rises.

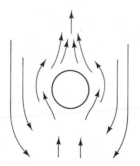

Cool (more dense) air falls to replace warm rising air.

FIGURE 5-2 Natural convection flow of air over a heated steam pipe.

In practice, many convection-heat-transfer systems involve both the forced and natural convection mechanisms. The topic of combined natural and forced convection is considered in Chap. 7.

5-1-2 Internal and External Flow

Examples of practical internal and external forced and natural convection flow systems are illustrated in Figs. 5-3 through 5-5. Flow in tubes, channels, annuli, tube banks, and heat exchangers are examples of internal flows. External (or boundary layer) flows involve such geometries as flat plates, finned surfaces, wing foils, cylinders, and so forth.

It is important to note that the velocity and temperature distributions are generally functions of axial location x in both external and internal flow fields. Such fields are said to be hydrodynamically and thermally

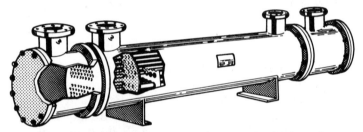

FIGURE 5-3 Typical industrial shell-and-tube heat exchanger—single pass construction on shell side and tube side. (Courtesy of Conseco.)

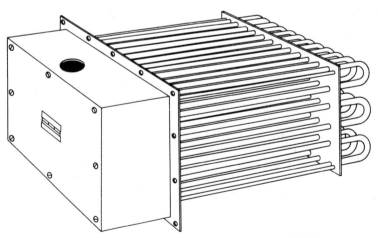

FIGURE 5-4 Tube bank: forced-air duct heater. (Courtesy of McGraw-Edison Co.)

FIGURE 5-5 External flow over wing foil. (Courtesy of H. Werle, ONERA, Paris.)

developing. However, under certain conditions associated with internal flow systems with constant cross-sectional area, the form of the velocity and temperature distributions become independent of axial location x. These flow fields are known as hydrodynamically and thermally fully developed. To illustrate this point, Fig. 5-6 shows representative distributions in axial velocity u for tube flow. The velocity distributions are seen to be developing in the entrance region and are fully developed (unchanging) in the region downstream. Similar developing and fully developed distributions in temperature T also occur.

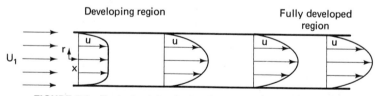

FIGURE 5-6 Typical axial velocity distributions for laminar tube flow.

5-1-3 Time and Space Dimensions

As in the analysis of conduction-heat-transfer systems, convection-heat-transfer processes are categorized according to the time and space dimensions that the temperature and velocity distributions are dependent upon. In our study, attention will be focused upon steady two-dimensional (x,r

or x,y) processes such as occur in a circular tube with uniform wall temperature or uniform wall heat flux. However, one should be aware of the importance of more complex multidimensional processes involving unsteady operation and nonuniform heating. For example, unsteady effects are important in the start-up of a boiler and in pulsating flows, and variations in the surface temperature or heat flux around the perimeter of a tube could be important if the tube is radiantly heated on only one side.

5-1-4 Laminar and Turbulent Flow

Laminar forced or natural convection flows exist when individual elements of fluid follow smooth streamline paths, whereas the flow is considered turbulent when the movement of elements of fluid is unsteady and random in nature. This important distinction between laminar and turbulent flow is demonstrated in Fig. 5-7, in which the path followed by fluid is marked by dye. In both cases, the main flow is from left to right. It should be noted that the random fluctuations associated with turbulent flow are superimposed upon the main flow.

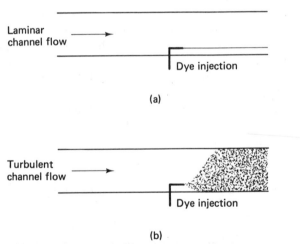

FIGURE 5-7 Typical dye streak patterns for channel or tube flow. (a) Laminar. (b) Turbulent.

One of the simplest ways of experimentally determining whether the flow is laminar or turbulent is to utilize very small electrical heating probes, such as those shown in Fig. 5-8, which can be mounted within a flow stream or flush with the surface of a wall. These *anemometer* probes are maintained at an essentially constant temperature by controlling the instantaneous electrical current flow. The instantaneous bridge voltage for

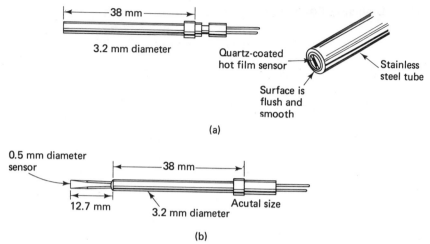

FIGURE 5-8 Anemometer probes. (a) Flush surface probe. (b) Standard probe with cylindrical sensor. (Courtesy of TSI Inc.)

a flush-mounted probe is shown in Fig. 5-9 for laminar and turbulent channel flow of liquid mercury. The unsteady character of the turbulent condition is clearly indicated by this signal. Similar signals are obtained from regular probes mounted in the flow stream. When properly calibrated, the signal from a regular anemometer probe can be used to determine the approximate velocity at the location of the probe.

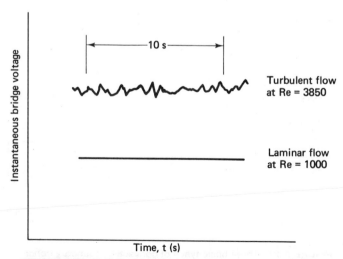

FIGURE 5-9 Signals from flush-mounted anemometer probe for channel flow of liquid mercury.

5-1-5 Boundary Conditions

The two simplest thermal boundary conditions encountered in convection heat transfer are the uniform wall heat flux and the uniform wall temperature conditions. A more general situation is also frequently encountered in heat exchangers which involves convection between two fluids which are separated by a wall. For situations such as this, the temperatures or heat fluxes at the surfaces are not known *a priori*. These three thermal boundary conditions are illustrated in Fig. 5-10. In this connection, a uniform wall

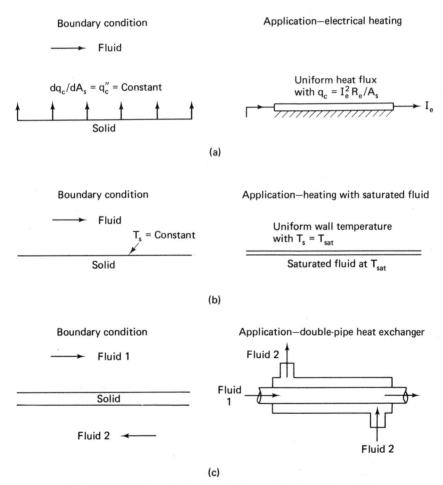

FIGURE 5-10 Three basic types of convection boundary conditions and their applications. (a) Uniform wall-heat flux. (b) Uniform wall-temperature. (c) Two fluids separated by a wall.

heat flux condition can be experimentally achieved by the generation of energy in the wall itself by the flow of electric current. A uniform wall temperature boundary condition is approximated for the case in which heat is convected from a saturated fluid at constant temperature and very high coefficient of heat transfer through a thin metallic wall into the fluid of interest. As mentioned in Sec. 5-1-3, more complex boundary conditions are sometimes encountered in practice which involve variations in wall temperature or heat flux around the system perimeter.

For momentum transfer, we can use the nonslip condition $u=0$ at a stationary wall for most fluids. Of course, if the wall itself possesses an axial velocity u_0, then $u=u_0$ at the wall. For nonporous walls, the transverse velocity v is zero at the wall. But for situations in which fluid actually passes through the wall, $v\neq0$.

5-1-6   Type of Fluid

The performance of convection-heat-transfer processes is a strong function of the properties of the fluid(s) involved. Four of the more important properties associated with both forced and natural convection heat transfer include the density ρ, viscosity μ, specific heat c_p, and thermal conductivity k.

In our treatment of convection heat transfer, attention will be focused upon incompressible processes; that is, on processes in which the density is essentially constant. Liquid flow systems can usually be treated as incompressible. In addition, gas flow systems that operate in the low-Mach-number subsonic range can generally be considered to be incompressible. However, compressibility effects must be accounted for in supersonic gas flow processes. In such systems, the pressure and density change dramatically with axial location.

In regard to the properties μ and k, many of the fluids encountered in practice can be classified as being Newtonian and isotropic. Such fluids satisfy the following fluid stress and conduction heat transfer laws for two-dimensional $(x,r$ or $x,y)$ conditions:

$$\tau = -\mu\frac{\partial u}{\partial r} \quad \text{or} \quad \tau = \mu\frac{\partial u}{\partial y} \qquad \text{Newtonian} \qquad (5\text{-}1a,b)$$

$$q_r'' = k\frac{\partial T}{\partial r} \quad \text{or} \quad q_y'' = -k\frac{\partial T}{\partial y} \qquad \text{isotropic} \qquad (5\text{-}2a,b)$$

Whereas most fluids are essentially isotropic, some important fluids such as blood exhibit distinct non-Newtonian characteristics. In our study of the fundamentals of convection heat transfer, we will concentrate entirely on Newtonian, isotropic-type fluids.

The dimensionless *Prandtl number*

$$Pr = \frac{\mu c_p}{k} \qquad (5\text{-}3)$$

is a key parameter in characterizing convection in single-phase fluids. The Prandtl numbers of several common fluids are given in Tables A-C-3 through A-C-5 in the Appendix and are summarized in Fig. 5-11. The Prandtl number is seen to range from very small values for liquid metals to very large values for highly viscous liquids such as oil. The Prandtl number for gases is generally of the order of unity. The physical significance of this property will be discussed further in Chap. 10.

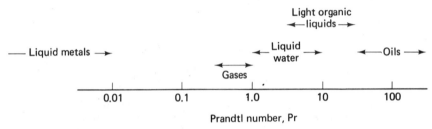

FIGURE 5-11   Prandtl-number spectrum of various fluids.

The performance of convection processes is also dependent upon whether the fluid exists as a single-phase gas or liquid, or as a multiple-phase substance. Of course, boiling and condensation are the most common types of multiple-phase convection processes. Both of these two-phase convection-heat-transfer processes are of great technological importance.

5-1-7   Other Factors

Numerous other factors are encountered in convection heat transfer which characterize the particular problem being considered. For example, for situations in which the temperature variations throughout the systems are not large, the properties are generally assumed to be independent of the temperature. However, for cases in which the properties vary noticeably with temperature, this temperature dependence must be accounted for. Further, factors such as kinetic energy, potential energy, energy dissipation, axial conduction, and three-dimensional effects can be neglected under certain conditions, but are sometimes quite important. In analyzing any convection-heat-transfer problem, care must be taken to ensure that all the significant factors are accounted for.

5-2 APPROACHES TO THE ANALYSIS OF CONVECTION HEAT TRANSFER

Convection heat transfer is generally more complex than conduction in a solid or in a stationary fluid because of the superimposed effects of fluid motion. Aside from this complicating factor, the basic concepts involved in the treatment of heat transfer from a solid/fluid interface to a fluid are the same as the treatment of the heat transfer within a solid. Similar to the analysis of conduction heat transfer, the complete theoretical solution of convection-heat-transfer problems requires the development of a mathematical formulation (differential, finite difference, finite element, integral) that represents the energy transport within the fluid itself. In addition, because energy is transported by fluid movement, a mathematical formulation must be developed for the fluid motion. Strictly speaking, it is the formulation and solution of such equations of motion and energy that provide the means of obtaining predictions for the velocity and temperature distributions within a fluid and the wall shear stress and convection heat transfer. The formulation and solution of the equations of motion and energy will be presented for several convection processes in Chap. 10, which pertains to the theoretical analysis of convection.

The engineer is often primarily interested in knowing the rate of heat transfer (or the surface temperature), the wall shear stress, and the pressure drop, and is not always concerned with the details of the temperature and velocity distributions within the fluid. Therefore, a much simpler lumped-analysis approach has been developed that provides a means for obtaining practical heat-transfer calculations. The practical thermal analysis approach involves the use of coefficients of friction or drag and heat transfer.

The local *Fanning friction factor* f is defined in terms of the local wall shear stress τ_s by

$$\tau_s = \rho U_F^2 \frac{f}{2} \tag{5-4}$$

where U_F is a reference velocity which depends on the geometry. (τ_s is expressed in terms of the axial velocity u by Eq. (5-1) with $r = r_0$ or $y = 0$.) The mean Fanning friction factor over an area of surface A_s is defined as

$$\bar{f} = \frac{1}{A_s} \int_{A_s} f \, dA_s \tag{5-5}$$

For two dimensional flow, Eq. (5-5) reduces to

$$\bar{f} = \frac{1}{L} \int_0^L f \, dx \tag{5-6}$$

It is important to note that $\bar{f}$ is defined in terms of f and *not* in terms of the mean wall shear stress $\bar{\tau}_s$.

To obtain an expression for $\bar{\tau}_s$ in terms of the friction factor for a two-dimensional flow field, we proceed as follows:

$$\bar{\tau}_s = \frac{1}{L} \int_0^L \tau_s \, dx$$

$$= \frac{1}{L} \int_0^L \rho U_F^2 \frac{f}{2} \, dx \tag{5-7}$$

For the case in which ρU_F^2 is constant, this equation reduces to the convenient form

$$\bar{\tau}_s = \frac{\rho U_F^2}{2} \frac{1}{L} \int_0^L f \, dx$$

$$= \rho U_F^2 \frac{\bar{f}}{2} \tag{5-8}$$

But, for situations in which ρU_F^2 is a function of x, Eq. (5-7) must be used, not Eq. (5-8).

The local coefficient of heat transfer h is defined by the *general Newton law of cooling*, Eq. (2-96),

$$q_c'' = \frac{dq_c}{dA_s} = h(T_s - T_F) \tag{5-9}$$

(The heat flux from the wall is given in terms of the distribution in temperature T by Eq. (5-2) with $r = r_0$ or $y = 0$.) The mean coefficient of heat transfer $\bar{h}$ is generally defined in terms of h by

$$\bar{h} = \frac{1}{A_s} \int_{A_s} h \, dA_s \tag{5-10}$$

or, for a two-dimensional system,

$$\bar{h} = \frac{1}{L} \int_0^L h \, dx \tag{5-11}$$

An expression can be written for the total rate of heat transfer q_c in terms of the coefficient of heat transfer by simply integrating Eq. (5-9).

$$q_c = \int_{q_c} dq_c = \int_{A_s} h(T_s - T_F) \, dA_s \tag{5-12}$$

For cases in which $T_s - T_F$ is uniform, this equation reduces to the simple form of the Newton law of cooling,

$$q_c = \bar{h} A_s (T_s - T_F) \tag{5-13}$$

As we shall see in the next several chapters, relationships are available for coefficients of friction and heat transfer for many standard convection processes. For situations in which these coefficients are known, the practical lumped analysis approach will be seen to produce predictions for the wall shear stress and the rate of heat transfer (or the surface temperature) for prescribed mass flow rate, and fluid-flow/heating conditions. However, the velocity and temperature distributions within the fluid cannot be obtained by the lumped approach. If the velocity and temperature profiles are required or if information pertaining to the coefficients of friction and heat transfer for the specific problem of interest is not available, the full theoretical analysis referred to above must be developed or experimental measurements must be made.

Practical thermal analyses are developed for standard single fluid convection flow systems in Chaps. 6 through 8. This material provides the basis for the evaluation and thermal design of heat exchangers, which is the topic of Chap. 9.

Finally, representative laminar and turbulent convection heat-transfer processes will be analyzed in Chap. 10, which deals with the theory of convection. This treatment will provide a theoretical basis for some of the friction factor and coefficient of heat-transfer correlations which are used in evaluation and thermal design, and will provide the reader with at least some background for dealing with more complex problems.

CHAPTER 6

CONVECTION HEAT TRANSFER:
Practical Thermal Analysis—
FORCED CONVECTION

6-1 INTRODUCTION

Forced-convection heat transfer is generally the most effective means of transporting energy from a solid to a fluid. Forced-convection processes are categorized according to the geometry of the heat-transfer surface. Three basic types of forced-convection systems include (1) internal flow in tubes, annuli, channels, and so on; (2) external flow over surfaces such as flat plates, cylinders, and spheres; and (3) flow across tube banks. The practical thermal analysis of each of these types of forced-convection processes will be considered in this chapter, with emphasis given to single-fluid/single-phase systems.

6-2 INTERNAL FLOW

6-2-1 Introduction

In our study of internal forced-convection flows, attention will be focused on systems with constant cross-sectional area, such as the circular tube shown in Fig. 6-1. For these systems, flow enters at a uniform velocity U_1 and heating is initiated at some distance x_0 from the entrance.

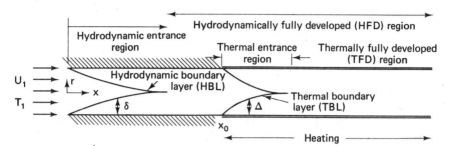

FIGURE 6-1 Hydrodynamic and thermal entrance regions and fully developed regions associated with flow in a circular tube.

6-2-2 Hydrodynamic Entrance and Fully Developed Regions

As fluid passes through such internal passages, the velocity is brought to zero at the wall. As the fluid proceeds downstream from the entrance, the effect of the wall on the velocity reaches farther and farther into the fluid because of viscous shearing forces. The region in which the velocity differs by at least 1% from the centerline velocity is known as the *hydrodynamic boundary layer* (HBL). The thickness of the HBL, designated by δ, grows as x increases, until it reaches the centerline. Beyond the axial location at which this occurs, the growth of the boundary layer is constrained and the velocity profile is independent of x; that is,

$$\frac{\partial u}{\partial x} = 0 \tag{6-1}$$

In this region, the flow is *hydrodynamically fully developed* (HFD). The region upstream of the HFD region is known as the *hydrodynamic entrance region*.

The *bulk stream velocity* U_b is defined in terms of the mass flow rate $\dot{m}$ at any axial location x by

$$\dot{m} = \rho A U_b = \int_{\dot{m}} d\dot{m} = \int_A \rho u \, dA \tag{6-2}$$

For incompressible flow ($\rho = $ constant) in systems with constant cross-sectional area A and mass flow rate $\dot{m}$, U_b is constant and is equal to the entering velocity U_1. The bulk flow rate is often expressed in the dimensionless format

$$\text{Re} = \frac{U_b D_H}{\nu} \tag{6-3}$$

where Re is the *Reynolds number*, ν ($\equiv \mu/\rho$) is the kinematic viscosity, and

D_H is the *hydraulic diameter*. D_H is defined by

$$D_H = \frac{4A}{p} \tag{6-4}$$

where p is the wetted perimeter; D_H is equal to the diameter D for flow in a circular tube and $2w$ for parallel-plate flow (w is the distance between the plates). The Reynolds number actually represents the ratio between the inertial and viscous forces which act upon the fluid. In addition to providing an efficient means of prescribing the significant hydrodynamic properties, the Reynolds number provides useful information pertaining to the nature of the flow field. For example, the flow is generally laminar if Re is less than about 2000, and the flow is turbulent if Re is much greater than this value. In the turbulent region, the flow is usually fully turbulent for $Re \gtrsim 10^4$ and transitional for $2000 \gtrsim Re \gtrsim 10^4$.

The Fanning friction factor for internal flow is defined by Eq. (5-4) with $U_F = U_b$; that is,

$$\tau_s = \rho U_b^2 \frac{f}{2} \tag{6-5}$$

The mean coefficient of friction over the length 0 to x for two-dimensional flow is given by

$$\bar{f} = \frac{1}{x} \int_0^x f \, dx \tag{6-6}$$

It should be noted that the hydrodynamic entrance region is characterized by a local coefficient of friction f which is a function of x. However, f is independent of x in the HFD region. To see this point, we couple the Newtonian shear law, Eq. (5-1), with the defining equation for f, Eq. (6-5), and the defining equation for HFD flow, Eq. (6-1), as follows:

$$\tau_s = \rho U_b^2 \frac{f}{2} = \mu \frac{\partial u}{\partial y} \bigg|_0 \tag{6-7}$$

$$\frac{d\tau_s}{dx} = \frac{\rho U_b^2}{2} \frac{df}{dx} = \mu \frac{\partial}{\partial y} \left(\frac{\partial u}{\partial x} \right) \bigg|_0 = 0 \tag{6-8}$$

Hence,

$$\tau_s = \text{constant} \tag{6-9}$$

and

$$f = \text{constant} \tag{6-10}$$

in the HFD region.

6-2-3 Thermal Entrance and Fully Developed Regions

Similar to the development of the hydrodynamic boundary layer, as fluid flows over a body, the temperature of the fluid in contact with the wall is brought to the surface temperature T_s. As the fluid moves downstream from the point at which heating or cooling is initiated, the effect of the wall on the temperature of the fluid penetrates deeper into the fluid, causing the buildup of a *thermal boundary layer* (TBL) such as is illustrated in Fig. 6-1.

Because the fluid is heated or cooled as it flows downstream, its enthalpy or energy content changes with x. To characterize this change, we define the *bulk stream temperature* T_b at any location x by

$$\dot{H}_b = \dot{m} c_p T_b = \int_m c_p T \, d\dot{m} = \int_A c_p T \rho u \, dA \qquad (6\text{-}11)$$

where $\dot{H}_b$ is the bulk stream enthalpy rate. Note that whereas T_b is a function of x for heating or cooling, except for saturated fluids, its counterpart U_b is constant for incompressible flow in systems with uniform cross-sectional area and mass flow rate.

The thickness Δ of the TBL grows with respect to x until it becomes constrained by the system geometry (see Fig. 6-1). At the axial location at which Δ becomes a constant, the dimensionless temperature profile becomes independent of x for uniform wall flux heating; that is,

$$\frac{\partial}{\partial x}\left(\frac{T - T_s}{T_b - T_s} \right) = 0 \qquad (6\text{-}12)$$

This is the defining equation for *thermal developed flow* (TFD). The thermal fully developed conditions represented by Eq. (6-12) also occur for uniform wall temperature heating and for certain nonuniform wall flux heating cases, but this region lies well downstream of the location at which the TBL ceases to grow.

The coefficient of heat transfer h for internal flow is defined by Eq. (5-9) with $T_F = T_b$.

$$q_c'' = \frac{dq_c}{dA_s} = h(T_s - T_b) \qquad (6\text{-}13)$$

The mean coefficient of heat transfer over the length 0 to x is defined by

$$\bar{h} = \frac{1}{x} \int_0^x h \, dx \qquad (6\text{-}14)$$

As will be seen, $\bar{h}$ is primarily used in the analysis of problems involving uniform wall-temperature conditions and in heat-exchanger applications.

Whereas the coefficients h and $\bar{h}$ are functions of x in the thermal entrance region, the TFD region is characterized by a coefficient h that is independent of x. To demonstrate this point, we combine the Fourier law of conduction, Eq. (5-2), with the general Newton law of cooling, Eq. (6-13), and the defining equation for TFD flow, Eq. (6-12), to obtain

$$q_c'' = h(T_s - T_b) = -k \frac{\partial T}{\partial y}\bigg|_0 \tag{6-15}$$

where k is the thermal conductivity of the fluid and $\partial T/\partial y|_0$ is the temperature gradient within the fluid at $y = 0$. Rearranging this equation, we obtain

$$h = -k \frac{\partial}{\partial y}\left(\frac{T}{T_s - T_b}\right)\bigg|_0 = k \frac{\partial}{\partial y}\left(\frac{T - T_s}{T_b - T_s}\right)\bigg|_0 \tag{6-16}$$

$$\frac{dh}{dx} = k \frac{\partial}{\partial y}\left[\frac{\partial}{\partial x}\left(\frac{T - T_s}{T_b - T_s}\right)\right]\bigg|_0 = 0 \tag{6-17}$$

Hence, we see that

$$h = \text{constant} \tag{6-18}$$

for TFD conditions.

The coefficient h is usually expressed in terms of the *Nusselt number* Nu,

$$\text{Nu} = \frac{hD_H}{k} \tag{6-19}$$

or the *Stanton number* St,

$$\text{St} = \frac{h}{\rho c_p U_b} \tag{6-20}$$

where k, ρ and c_p are properties of the fluid. These two dimensionless groups are related by

$$\text{St} = \frac{\text{Nu}}{\text{Re Pr}} \tag{6-21}$$

(The product Re Pr is known as the *Peclet number* Pe.) Likewise, the mean coefficient of heat transfer is generally expressed in terms of the mean Nusselt number $(\overline{\text{Nu}} = \bar{h}D_H/k)$ or mean Stanton number $[\overline{\text{St}} = \bar{h}/(\rho c_p U_b)]$.

Dimensional analysis is one approach to determining the pertinent dimensionless parameters in convection-heat-transfer processes. Dimensional analysis is also often utilized to develop empirical correlations for the coefficient of heat transfer on the basis of experimental data. This approach is based on the principle of dimensional continuity, which requires that all terms in every equation be dimensionally consistent. To properly apply the dimensional-analysis approach to convection, we must know what dimensional parameters the coefficient of heat transfer depends upon. These parameters can generally be established on the basis of a thorough physical understanding of the problem.

The mechanics involved in the general dimensional analysis approach are discussed by Buckingham [1], Bridgeman [2], and Langhaar [3], and introductions to the use of this tool in analyzing convection heat transfer are given by Kreith [4] and Karlekar and Desmond [5]. The use of this approach in analyzing forced convection heat transfer is illustrated in Example 6-1.

The dimensionless parameters that characterize a convection-heat-transfer process can also often be obtained by developing a theoretical analysis. This more comprehensive approach is introduced in Chap. 10.

EXAMPLE 6-1

Assuming that the coefficient of heat transfer for forced convection in a short tube is dependent upon the geometric, hydrodynamic, and thermal parameters listed in Table E6-1, utilize the dimensional-analysis approach to determine the key dimensionless groups for this problem.

TABLE E6-1 Forced convection parameters

Dimensional parameter	Symbol	Dimensions
Hydrodynamic:		
Bulk stream velocity	U_b	L/Θ
Viscosity	μ	$M/(L\Theta)$
Density	ρ	M/L^3
Thermal:		
Thermal conductivity	k	$ML/(\Theta^3 T)$
Specific heat	c_p	$L^2/(\Theta^2 T)$
Coefficient of heat transfer	$\bar{h}$	$M/(\Theta^3 T)$
Geometric:		
Diameter	D	L
Axial location	x	L

Solution

Based on the *Buckingham π theorem* [1], the number of independent dimensionless groups N which are associated with a physical phenomenon is equal to the total number of significant dimensional parameters I minus the number of fundamental dimensions J which are required to define the dimensions of all the I parameters. The relationship among the various dimensionless groups π_1, π_2, and so on, takes the general form

$$\pi_1 = F(\pi_2, \pi_3, \ldots, \pi_i, \ldots, \pi_N) \tag{a}$$

Referring to the table, we see that this problem involves eight dimensional parameters ($I=8$) and four fundamental dimensions ($J=4$). Therefore, according to the Buckingham π theorem, we have four dimensionless groups:

$$N = I - J = 8 - 4 = 4 \tag{b}$$

Following the procedure outlined by various authors [1]–[5], to determine the dimensionless groups π_i (where $i = 1,2,3,4$), we write

$$\pi_i = U_b^a \, \mu^b \, \rho^c \, k^d \, c_p^e \, \bar{h}^f \, D^g \, x^h$$

Substituting for the dimensions of each parameter, we have

$$\pi_i = \left[\frac{L}{\Theta}\right]^a \left[\frac{M}{L\Theta}\right]^b \left[\frac{M}{L^3}\right]^c \left[\frac{ML}{\Theta^3 T}\right]^d \times$$
$$\left[\frac{L^2}{\Theta^2 T}\right]^e \left[\frac{M}{\Theta^3 T}\right]^f [L]^g [L]^h$$

Because the π_i groups are dimensionless, the summation of the exponents of each fundamental dimension must be equal to zero. It follows that

$$\begin{align} b + c + d + f &= 0 \quad \text{for mass } M \tag{c}\\ a - b - 3c + d + 2e + g + h &= 0 \quad \text{for length } L \tag{d}\\ -a - b - 3d - 2e - 3f &= 0 \quad \text{for time } \Theta \tag{e}\\ -d - e - f &= 0 \quad \text{for temperature } T \tag{f} \end{align}$$

These four equations must be satisfied for each dimensionless group. But, because we have eight unknowns, the values of four of the exponents must be specified for each of the four dimensionless groups

π_1, π_2, π_3, and π_4. To help in selecting these four exponents for each dimensionless group, we will purposefully seek one thermal group (π_1) involving $\bar{h}$, one thermal group (π_2) excluding $\bar{h}$, one hydrodynamic group (π_3), and one geometric group (π_4).

To obtain dimensionless thermal group π_1, we set f equal to unity, and the hydrodynamic and geometric exponents a, b, and h equal to zero. Solving Eqs. (c) through (f) with these inputs, we obtain $c=0$, $d=-1$, $e=0$, and $g=1$. Therefore, this dimensionless thermal group is

$$\pi_1 = \frac{\bar{h}D}{k} \tag{g}$$

which is the mean *Nusselt number* $\overline{\mathrm{Nu}}$.

To formulate a second thermal group, b is set equal to unity, and a, f, and h are set equal to zero. Equations (c) through (f) then give rise to $c=0$, $d=-1$, $e=1$, and $g=0$, such that

$$\pi_2 = \frac{\mu c_p}{k} \tag{h}$$

which is recognized as the *Prandtl number* Pr.

To obtain a dimensionless hydrodynamic group, the thermal exponents d, e, and f and the geometric exponent h are set equal to zero. It follows that $a=1$, $b=-1$, $c=1$, and $g=1$, such that

$$\pi_3 = \frac{\rho D U_b}{\mu} \tag{i}$$

which is the *Reynolds number* Re.

Finally, to develop a dimensionless geometric group, we leave g unspecified with $h=1$, $b=0$, $e=0$, and $f=0$. Substituting these values into Eqs. (c) through (f), we obtain $a=0$, $c=0$, $d=0$, and $g=-1$. Hence, we have the natural dimensionless geometric grouping

$$\pi_4 = \frac{x}{D} \tag{j}$$

Based on these results, we can expect correlations for the coefficient of heat transfer associated with forced convection in smooth tubes to take the general form

$$\frac{\bar{h}D}{k} = \overline{\mathrm{Nu}} = F(\mathrm{Re}, \mathrm{Pr}, x/D) \tag{k}$$

And this is indeed the case, as we shall see in the next section, which presents design equations for the coefficient of heat transfer.

It should be noted that the actual functional relationship between $\overline{Nu}$ and Re, Pr, and x/D, which is dependent upon the specific operating conditions, can often be determined on the basis of experimental data. In this connection, Eq. (k) reduces the data correlation problem from one involving the eight variables U_b, μ, ρ, k, c_p, $\overline{h}$, D, and x to one with only the four dimensionless groups, $\overline{Nu}$, Re, Pr, and x/D. To obtain an empirical correlation for $\overline{h}$ in terms of U_b, μ, ρ, k, c_p, D, and x, it would be necessary to vary each of these seven parameters at least three or four times. Assuming that we obtained four data points for each of these seven variables, we would require 4^7, or about 16,400 individual measurements! On the other hand, to obtain a correlation for $\overline{Nu}$ in terms of Re, Pr, and x/D with as much information would only require 4^3, or 64, data points. Thus, although dimensional analysis cannot be used to determine the functional relationship between a parameter such as $\overline{h}$ and the other system parameters, this approach can be utilized to greatly simplify the correlation of experimental data to produce empirical correlations for design. (As demonstrated in Chap. 10, the relationship between the various system parameters can often be obtained on the basis of the theoretical approach.)

6-2-4   Coefficients of Friction and Heat Transfer

Information pertaining to the coefficients of friction and heat transfer is presented in this section for several common internal forced convection flows systems with uniform cross-sectional area. The underlying theoretical basis for some of the equations presented in this section is developed in Chap. 10. A more complete survey of correlations for friction and heat transfer is available in *Engineering Sciences Data* [6] and in the *Handbook of Heat Transfer* [7].

Laminar flow

1. For practical purposes, hydrodynamic and thermal fully developed flow can be said to occur for

$$\frac{x}{D_H} \gtrsim 0.04 \text{ Re} \qquad \text{HFD} \tag{6-22}$$

$$\frac{x}{D_H} \gtrsim 0.04 \text{ Re Pr} \quad \text{TFD} \tag{6-23}$$

(The hydrodynamic entrance length is often approximated by $0.05 \text{ Re } D_H$ and values as large as $0.125 \text{ Re Pr } D_H$ are sometimes used for the thermal entrance length [8].)

2.   For HFD and TFD laminar flow,

$$f = \frac{C_1}{\text{Re}} \qquad \text{Nu} = \frac{hD_H}{k} = C_2 \qquad (6\text{-}24, 25)$$

where C_1 and C_2 are dependent upon geometry, and C_2 is dependent upon the thermal boundary conditions. Representative values of C_1 and C_2 are given in Table 6-1 and in Fig. 6-2. Note that $C_1 = 16$ for flow in a circular tube, with $C_2 = 3.66$ for uniform wall-temperature heating and $C_2 = 4.36$ for uniform wall-flux heating.

TABLE 6-1   HFD and TFD laminar flow: coefficients for Eqs. (6-24) and (6-25)

Geometry	C_1	C_2* Uniform T_s	Uniform q_0''
Square tube	14.2	2.95	3.65
Circular tube	16.0	3.66	4.36
Infinite parallel plates	24.0	7.50	8.25

*C_2 is given in the literature for other types of thermal boundary conditions.

3.   The friction factor f for developing flow lies above the HFD value. Theoretical predictions by Langhaar [11] for laminar developing flow are shown in Fig. 6-3. The fact that f is constant for $x/(D\,\text{Re})$ greater than a value of the order 0.04 reinforces the criterion for HFD conditions given by Eq. (6-22).

4.   Theoretical and empirical correlations are available in the literature for the coefficient of heat transfer h for thermal developing laminar flow in circular tubes and other geometries. Typical of the design equations that are available is the following popular correlation by Hausen [12] for HFD flow in a circular tube with uniform wall temperature heating:

$$\overline{\text{Nu}} = 3.66 + \frac{0.0668 \dfrac{\text{Re Pr}}{x/D}}{1 + 0.04\left(\dfrac{\text{Re Pr}}{x/D}\right)^{2/3}} \qquad (6\text{-}26)$$

For small values of x/D, this equation reduces to

$$\overline{\text{Nu}} = 1.67\left(\frac{\text{Re Pr}}{x/D}\right)^{1/3} \qquad \frac{x/D}{\text{Re Pr}} \lesssim 0.01 \qquad (6\text{-}27)$$

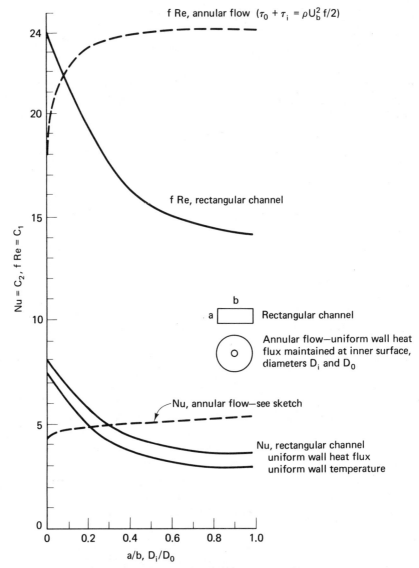

FIGURE 6-2 Nusselt number and Fanning friction factor for laminar fully developed flow. (From Kays and Clark [9] and Landberg et al. [10].)

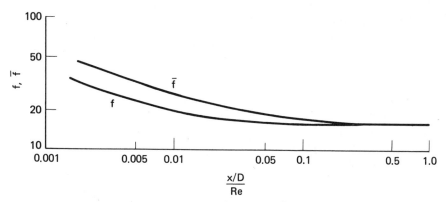

FIGURE 6-3 Predictions for coefficients of friction by Langhaar [11] for hydrodynamic developing laminar flow in a circular tube.

These equations are shown in Fig. 6-4. For thermal developing HFD laminar flow with uniform wall flux heating in very short tubes, Nu can be approximated by [13]

$$Nu = 1.30\left(\frac{Re\ Pr}{x/D}\right)^{1/3} \qquad \frac{x/D}{Re\ Pr} \lesssim 0.01 \qquad (6\text{-}28)$$

Correlations are also available for combined hydrodynamic and thermal developing laminar flows. For example, Kays [14] has developed a correlation for developing laminar flow of air in a circular pipe with

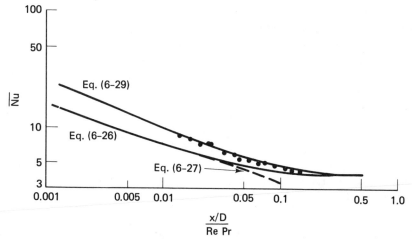

FIGURE 6-4 Correlations and experimental data for thermal developing flow with uniform wall-temperature heating. (Data for air (Pr=0.72) from Kays [14].)

uniform wall temperature of the form

$$\overline{Nu} = 3.66 + \frac{0.104 \,(\text{Re Pr } D/x)}{1 + 0.016(\text{Re Pr } D/x)^{0.8}} \tag{6-29}$$

This equation is compared with Eq. (6-26) and with experimental data for developing flow in Fig. 6-4. Note that the Nusselt number is larger for hydrodynamic developing flow than for HFD flow. Another correlation for developing flow with constant wall temperature which has been widely used for liquids takes the basic form [15]

$$\overline{Nu} = 1.86\left(\frac{\text{Re Pr}}{x/D}\right)^{1/3} \qquad \frac{x/D}{\text{Re Pr}} \gtrsim 0.01 \tag{6-30}$$

Equations (6-27), (6-28), and (6-30) are most accurate for $x/D \gtrsim 0.01$ Re Pr. However, for preliminary design calculations, these equations can be extended to $x/D \gtrsim 0.04$ Re Pr, and Nu can be approximated by the TFD value for $x/D \gtrsim 0.04$ Re Pr.

Turbulent flow

1. Fully turbulent flow generally occurs for $\text{Re} \gtrsim 10^4$ and transitional turbulent flow occurs for $2000 \gtrsim \text{Re} \gtrsim 10^4$.
2. Hydrodynamic and thermal developed turbulent flow occur over much shorter distances than for laminar flow. For example, HFD conditions occur for x/D greater than a value of the order of 10 to 20. TFD conditions occur over about the same distance for gases such as air, with the length of the thermal entrance region decreasing as the Prandtl number Pr increases, and vice versa.
3. The Fanning friction factor for fully developed and fully turbulent flow in smooth tubes is well represented by the Prandtl/Nikuradse equation [16]

$$\sqrt{\frac{2}{f}} = 2.46 \ln\left(\text{Re}\sqrt{\frac{f}{2}}\right) + 0.292 \tag{6-31}$$

This implicit equation has been shown to be very closely approximated by the explicit Filonenko formula [17],

$$f = (1.58 \ln \text{Re} - 3.28)^{-2} \tag{6-32}$$

Equation (6-32) is compared with experimental data in Fig. 6-5.

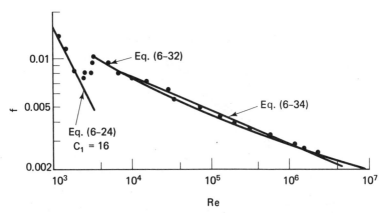

FIGURE 6-5 Correlations and experimental data for friction factor f for HFD flow in circular tubes.

For the Nusselt number, the following correlation by Petukhov and Kirillov [18] and White[1] [19] is recommended:

$$Nu = \frac{(f/2)\text{Re Pr}}{1.07 + 12.7\sqrt{f/2}\,(\text{Pr}^{2/3} - 1)} \tag{6-33}$$

This equation is reported to have an accuracy of 5 to 6% in the ranges $10^4 \gtrsim \text{Re} \gtrsim 5 \times 10^6$ and $0.5 \gtrsim \text{Pr} \gtrsim 200$, and 10% for the same Reynolds number range and for $200 \gtrsim \text{Pr} \gtrsim 2000$ [20]. Equation (6-33) is shown in Fig. 6-6, with f given by Eq. (6-32). Notice that the Nusselt number is much larger for turbulent flow than for laminar flow.

Simpler but less accurate correlations for the Fanning friction factor f and the Nusselt number Nu for fully developed and fully turbulent flow ($\text{Re} \gtrsim 10^4$) in smooth tubes are given by

$$f = 0.046\,\text{Re}^{-0.2} \tag{6-34}$$

$$Nu = \frac{f}{2}\,\text{Re Pr}^n \tag{6-35}$$

or

$$Nu = 0.023\,\text{Re}^{0.8}\,\text{Pr}^n \tag{6-36}$$

where $n = 0.5$ for $0.5 \gtrsim \text{Pr} \gtrsim 5.0$ and $n = \frac{1}{3}$ for $\text{Pr} \gtrsim 5.0$. The earliest correlations of the form of Eq. (6-36) were developed by Dittus and Boelter [21]

---

[1]The coefficients in the White correlation differ very slightly.

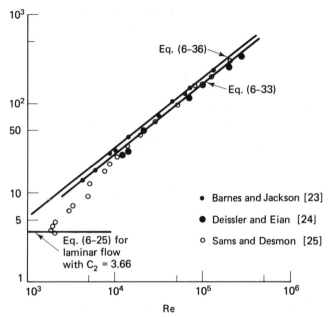

FIGURE 6-6 Correlations and experimental data for Nusselt number for TFD turbulent flow of air.

in 1930 ($n = 0.4$ for heating and $n = 0.3$ for cooling) and Colburn [22] in 1933 ($n = \frac{1}{3}$). With n set equal to $\frac{1}{3}$, Eq. (6-36) is known as the *Colburn analogy*. Equations (6-34) and (6-36) are compared with experimental data in Figs. 6-5 and 6-6.

With Reynolds number Re and Nusselt number Nu expressed in terms of the hydraulic diameter D_H, both f and Nu are essentially independent of cross-sectional geometry for HFD and TFD fully turbulent flow. Consequently, the correlations above apply to noncircular systems as well as to circular tubes. In addition, for TFD fully turbulent conditions, Nu is essentially independent of the form of the thermal boundary conditions, except for low-Prandtl-number fluids such as liquid metals. Hence, these equations can be used for uniform wall flux, uniform wall temperature, and other thermal boundary conditions for TFD flow.

    4. Experimental and theoretical studies have recently been conducted for HFD transitional turbulent flow by Patel and Head [26], Lawn [27], Thomas and Kakarala [28], and others. These studies indicate that the correlations for f and Nu begin to separate from the correlations for fully turbulent flow at $\text{Re} \gtrsim 10^4$ and approach the correlations for laminar conditions as Re falls toward a value of the order of 2000. This trend in the Nusselt number is clearly seen in Fig. 6-6. However, as a first approximation, the Fanning friction factor for HFD transitional turbulent flow is

sometimes given

$$f = 0.079 \, Re^{-0.25} \tag{6-37}$$

and Eq. (6-33) or Eq. (6-35) is utilized for Nu.

5. The friction factor is dependent on x/D_H and Re for developing turbulent flow. Correlations for f and $\bar{f}$ for developing flows are available in [29].

6. One of the most convenient and widely used correlations for TFD flow in short smooth tubes is given by [7][2]

$$\overline{Nu} = Nu\left(1 + \frac{C}{x/D}\right) \qquad \frac{x}{D} \gtrsim 10 \tag{6-38}$$

where Nu is evaluated for fully developed conditions and $C = 1.4$ for the case in which HFD conditions exist at the entrance and $C = 6$ when no hydrodynamic calming section is used. This equation has been extensively used in the thermal analysis of heat exchangers.

Correlations are also available for thermal developing flow. For example, Kreith [4] and McAdams [30] recommend

$$\overline{Nu} = Nu\left[1 + \left(\frac{D}{x}\right)^{0.7}\right] \qquad 2 \gtrsim \frac{x}{D} \gtrsim 20 \tag{6-39}$$

where Nu is evaluated for fully developed conditions.

7. For liquid metals that have quite low Prandtl numbers, large amounts of scatter exist in the published data for Nusselt number. A correlation developed by Subbotin et al. [31] which is frequently utilized for design purposes for uniform heat-flux applications takes the form

$$Nu = 5.0 + 0.025 \, (Re \, Pr)^{0.8} \tag{6-40}$$

where $(Re \, Pr) \gtrsim 100$. This equation represents the mean of most of the data in the literature.

EXAMPLE 6-2

Determine the friction factor and the coefficient of heat transfer for fully developed flow of air in a 10-cm-diameter circular pipe for flow rates of 0.1 m/s and 10 m/s, assuming approximate isothermal conditions at 27°C and atmospheric pressure.

---

[2]Although this equation was developed for turbulent flow of air, it is also often used as a first approximation for other fluids, except for liquid metals.

Solution

The pertinent properties of air at 27°C are $\rho = 1.18$ kg/m^3, $\nu = 15.7 \times 10^{-6}$ m^2/s, $k = 0.0262$ W/(m °C), and Pr $= 0.708$.

The Reynolds numbers for these two flow rates are

$$\text{Re}_1 = \frac{DU_b}{\nu} = \frac{(0.1 \text{ m})(0.1 \text{ m/s})}{15.7 \times 10^{-6} \text{ m}^2/\text{s}}$$

$$= 637 \qquad \text{for } U_b = 0.1 \frac{\text{m}}{\text{s}}$$

$$\text{Re}_2 = 6.37 \times 10^4 \qquad \text{for } U_b = 10 \frac{\text{m}}{\text{s}}$$

Hence, the flow is laminar for the first case and turbulent for the second.

For laminar HFD tube flow to occur, x/D must be greater than Re/20 = 31.9. On the other hand, HFD turbulent flow can occur for $x/D \gtrsim 10$. The Fanning friction factors are calculated for fully developed conditions as follows:

$$f_1 = \frac{16}{\text{Re}_1} = \frac{16}{637} = 0.0251 \qquad \text{from Eq. (6-24)}$$

$$f_2 = (1.58 \ln \text{Re} - 3.28)^{-2} = 4.96 \times 10^{-3} \qquad \text{from Eq. (6-32)}$$

These results can also be obtained from Fig. 6-5. For short tubes that do not satisfy these x/D requirements, relationships that apply to the developing region must be used.

To achieve TFD tube flow, x/D must be greater than Re Pr/10 = 46 for laminar conditions and 10 for turbulent flow. To obtain the coefficients of heat transfer, we first calculate the Nusselt numbers as follows:

$$\text{Nu}_1 = 3.66 \qquad\qquad (6\text{-}25)$$

$$\text{Nu}_2 = \frac{(f/2)\text{Re Pr}}{1.07 + 12.7\sqrt{f/2} \ (\text{Pr}^{2/3} - 1)} \qquad\qquad (6\text{-}33)$$

$$= \frac{(4.96 \times 10^{-3}/2)(6.37 \times 10^4)(0.708)}{1.07 + 12.7\sqrt{4.96 \times 10^{-3}/2} \ (0.708^{2/3} - 1)} = 119$$

The coefficients h_1 and h_2 are now obtained.

$$h_1 = \frac{k}{D} \text{Nu}_1 = \frac{[0.0262 \text{ W}/(\text{m } °\text{C})](3.66)}{0.1 \text{ m}}$$

$$= 0.960 \frac{\text{W}}{\text{m}^2 \, °\text{C}}$$

$$h_2 = \frac{k}{D} \text{Nu}_2 = 31 \frac{\text{W}}{\text{m}^2 \, °\text{C}}$$

EXAMPLE 6-3

Determine the mean coefficient of heat transfer for a 10-m/s flow of air in a short circular tube of 5-m length and 10-cm diameter. The mean fluid temperature is about 27°C.

Solution

Based on Example 6-2, we know that the flow is turbulent with $\text{Re} = 6.37 \times 10^4$ and $\text{Nu} = 119 [h = 31 \text{ W}/(\text{m}^2 \, °\text{C})]$ for $x/D \gtrsim 10$, where the flow is TFD.

For our problem, $L/D = 50$. Therefore, we utilize Eq. (6-38) with $C = 6$ to calculate $\bar{h}$.

$$\overline{\text{Nu}} = \text{Nu}\left(1 + \frac{6}{x/D}\right) = 119(1.12) = 133$$

or

$$\bar{h} = 31 \frac{\text{W}}{\text{m}^2 \, °\text{C}} (1.12) = 34.7 \frac{\text{W}}{\text{m}^2 \, °\text{C}}$$

Effects of property variation

The significance of property variation caused by temperature should always be kept in mind when analyzing convection-heat-transfer processes. The two primary effects which are brought about by temperature-induced property change are (1) direct effects which are not related to gravity and (2) indirect effects which involve the creation of buoyancy forces by variations in the density with temperature.

For many situations, the changes in fluid properties with temperature are small over moderate temperature bands. For such rather standard situations, uniform property analyses are generally utilized with the fluid properties evaluated at the arithmetic average inlet and outlet bulk stream temperatures. In addition, the direct effects of property variation on the momentum and heat transfer are approximately accounted for by utilizing correction factors for the friction factor and Nusselt number. For liquids, where variations in the viscosity are generally the main consideration, these correction factors take the form

$$\frac{f}{f_{cp}} = \left(\frac{\mu_s}{\mu_b}\right)^m \tag{6-41}$$

$$\frac{Nu}{Nu_{cp}} = \left(\frac{\mu_b}{\mu_s}\right)^n \tag{6-42}$$

where μ_s is evaluated at the wall temperature T_s, and the other properties are evaluated at the bulk stream temperature T_b. On the other hand, the variations in density and thermal conductivity as well as viscosity are usually the most important factors for gases. Because these three properties are related to the absolute temperature for gases, the following correction factors are commonly used:

$$\frac{f}{f_{cp}} = \left(\frac{T_s}{T_b}\right)^m \tag{6-43}$$

$$\frac{Nu}{Nu_{cp}} = \left(\frac{T_s}{T_b}\right)^n \tag{6-44}$$

Representative values of m and n are given in Table 6-2. Hence, the

TABLE 6-2   Variable property conditions—coefficients associated with Eqs. (6-41) through (6-44)

Fluid	Condition	m	n	References
Liquid	Laminar			
	Cooling	0.50	0.14	15, 24
	Heating	0.58	0.14	15, 24
	Turbulent[a]			
	Cooling	0.24	0.25	8, 20
	Heating	b	0.11	8, 20
Gas	Laminar	1.0	0	8
	Turbulent			
	Cooling	−0.38	−0.36	20
	Heating	−0.52	−0.5	20

[a]The Nu correction for turbulent flow of liquids is recommended by Petukhov [20] for the ranges $0.08 \gtrsim \mu_s/\mu_b \gtrsim 40$; $2 \gtrsim Pr \gtrsim 140$; and $10^4 \gtrsim Re \gtrsim 1.25 \times 10^5$.

[b]For heating, Petukhov [20] recommends $f/f_{cp} = (7 - \mu_b/\mu_s)/6$.

relationships for friction factor and Nusselt number that are presented throughout this chapter for ideal isothermal conditions may be adjusted to account for mild property variations by merely multiplying by the appropriate viscosity or temperature ratio correction factor.

The assumption of uniform properties and the use of correction factors for the friction factors and Nusselt number cannot be utilized when dealing with large temperature differences or fluids which experience large property variation. For example, the use of a uniform property analysis for engine oil, and particularly liquid carbon dioxide near its critical point (304 K), will produce significant errors for even moderate temperature differences because of the change in specific heat with temperature. For such cases, analyses must be developed which account for the property variation or detailed experimental measurements must be made.

The second indirect type of temperature-induced property-variation effect is natural convection. The topic of combined forced/natural convection heat transfer will be discussed in Chap. 7.

EXAMPLE 6-4

To illustrate the direct type of temperature-induced property-variation effect, compare the actual velocity distribution for laminar flow with the ideal isothermal velocity profile.

Solution

Referring to Tables A-C-3 through A-C-5, we observe that the viscosity is directly proportional to temperature for gases and inversely proportional for liquids. As a result, when a gas is heated or when a liquid is cooled, the fluid near the wall is more viscous than the fluid farther away from the wall. Consequently, we would expect the actual velocity distribution for these conditions to be lower relative to the isothermal velocity profile in the region near the wall and larger in the region away from the wall. This perspective is shown in Fig. E6-4 for laminar tube flow. Of course, the opposite result would be expected for the cooling of a gas or heating of a liquid.

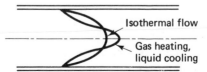

FIGURE E6-4 Influence of property variation on velocity distribution for laminar tube flow.

EXAMPLE 6-5

Air at 20°C and 1 atm pressure enters a 10-cm-diameter tube with a bulk velocity of 0.1 m/s. The tube wall is maintained at 75°C and the outlet temperature is 34°C. Determine the friction factor and the coefficient of heat transfer for fully developed conditions.

Solution

To approximately account for the effect of property variations on the coefficients of friction and heat transfer, we will evaluate the fluid properties at the arithmetic average bulk stream temperature and we will employ the correction factors given by Eqs. (6-43) and (6-44),

$$\frac{f}{f_{cp}} = \left(\frac{T_s}{T_b}\right)^m \tag{6-43}$$

$$\frac{Nu}{Nu_{cp}} = \left(\frac{T_s}{T_b}\right)^n \tag{6-44}$$

The arithmetic average bulk stream temperature is 27°C. It follows from Example 6-2 that $Re = 6.37 \times 10^4$, $f_{cp} = 4.96 \times 10^{-3}$, and $Nu_{cp} = 119$. Referring to Table 6-2, we find that $m = -0.52$ and $n = -0.5$ for turbulent flow and heating. Setting $T_s = 348$ K and $T_b = 300$ K, we have

$$f = 4.96 \times 10^{-3} \left(\frac{348 \text{ K}}{300 \text{ K}}\right)^{-0.52} = (4.96 \times 10^{-3})(0.926) = 4.59 \times 10^{-3}$$

and

$$Nu = (119)\left(\frac{348 \text{ K}}{300 \text{ K}}\right)^{-0.5} = (119)(0.928) = 111$$

It follows that the coefficient of heat transfer is

$$h = \left(31 \frac{W}{m^2 \, °C}\right)(0.928) = 28.8 \frac{W}{m^2 \, °C}$$

Enhancement of heat transfer

The correlations for heat transfer and friction presented earlier in this section only apply to tubes and channels with smooth surfaces. The heat transfer in industrial applications is often enhanced by the use of fins, rough surfaces, and other techniques. A review of the various methods of enhancement has been developed by Bergles [32]. For example, heat transfer for turbulent tube flow has been increased by as much as 400% by the use of rough surfaces. However, large increases in the friction factor have also been found to occur, such that care must be taken to utilize surfaces that are efficient as far as both heat transfer and pressure drops are concerned.

Thermal analyses of systems employing enhancement features such as rough surfaces or fins require the use of special design correlations. Correlations for heat transfer are generally provided by the commercial manufacturers of rough surface or finned tubes. When such information is not available for tubes with rough surfaces, the coefficient of heat transfer can be estimated by the use of Eq. (6-33), with f evaluated on the basis of the well-known Moody chart, which is shown in Fig. 6-7.

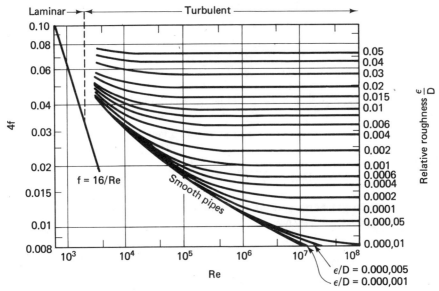

FIGURE 6-7 Moody friction factor chart for rough surfaces. (From Moody [33].)

Other factors

The coefficients of friction and heat transfer are also sometimes influenced by other factors such as unsteady operation, nonuniform distributions in wall temperature or heat flux, viscous dissipation, phase change (see Chap. 8), and mass transfer through the wall. As indicated earlier, when the convection process is complicated by factors such as these, their influence on the coefficients should be accounted for. Consequently, one must be prepared to search the literature for appropriate correlations, or even to undertake or commission an experimental or theoretical study.

6-2-5 Lumped Analysis—Momentum Transfer

In the practical lumped analysis of momentum transfer for internal flow systems such as the circular tube shown in Fig. 6-8, we apply Newton's second law of motion to the lumped volume $A\,dx$. For the case of HFD flow, Newton's second law reduces to the following form for steady flow (see Prob. 6-19):

$$\sum F_x = \frac{d}{dt}(mu) = 0 \tag{6-45}$$

Therefore, a force balance on the lumped volume $A\,dx$ for a horizontal circular-tube geometry gives

$$(PA)|_x - \tau_s p\,dx - (PA)|_{x+dx} = 0$$

or

$$\frac{dP}{dx} = -\tau_s \frac{p}{A} \tag{6-46}$$

where $A = \pi r_0^2$ and $p = 2\pi r_0$. This same equation also applies to noncircular-tube and parallel-plate geometries. With the wall shear stress τ_s expressed in terms of f, the pressure drop is given by

$$\frac{dP}{dx} = -\rho U_b^2 \frac{f}{2} \frac{p}{A} \tag{6-47}$$

Because the right-hand side of this equation is independent of x, dP/dx can be replaced by $\Delta P/\Delta x$.

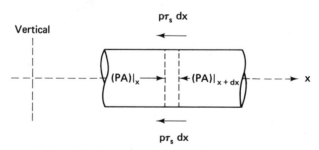

FIGURE E6-8 Predictions for bulk stream temperature T_b.

For unsteady or developing flows, the term $d(mu)/dt$ (or ma_x) in the Newton second law is not zero. The problem of analyzing these types of flows is further complicated by the fact that the friction factor is a function of t or x. Nevertheless, practical lumped analyses can sometimes be made for these more complex flows, as discussed in [8] and [12].

EXAMPLE 6-6

Water at 40°F is heated in a 0.787-in.-diameter tube with a relative roughness of 0.008. The wall temperature is 80°F and outlet bulk stream temperature is 60°F. Determine the pressure drop per foot and the Nusselt number in the fully developed region for a bulk stream velocity of 3.28 ft/s.

Solution

The arithmetic average bulk stream temperature is 50°F. Referring to Table A-C-3, the pertinent properties at this temperature are $Pr = 9.4$,

$$\rho = 999 \frac{kg}{m^3} \frac{0.0624 \, lb_m/ft^3}{1 \, kg/m^3} = 62.3 \frac{lb_m}{ft^3}$$

$$c_p = 4.20 \frac{kJ}{kg \, °C} \frac{0.239 \, Btu/(lb_m \, °F)}{1 \, kJ/(kg \, °C)} = 1.0 \frac{Btu}{lb_m \, °F}$$

$$\mu = 1.31 \times 10^{-3} \frac{kg}{m \, s} \frac{0.672 \, lb_m/(ft \, s)}{kg/(m \, s)} = 8.8 \times 10^{-4} \frac{lb_m}{ft \, s}$$

$$\nu = \frac{\mu}{\rho} = \frac{8.8 \times 10^{-4} \, lb_m/(ft \, s)}{62.3 \, lb_m/ft^3} = 1.41 \times 10^{-5} \frac{ft^2}{s}$$

The viscosity at 80°F is 8.6×10^{-4} kg/(m s)$= 5.78$ lb$_m$/(ft s). The Reynolds number for these conditions is

$$Re = \frac{U_b D}{\nu} = \frac{(3.28 \, ft/s)(0.787/12)ft}{1.41 \times 10^{-5} \, ft^2/s}$$

$$= 15,300$$

Utilizing the Moody chart of Fig. 6-7, we find

$$f_{cp} = \frac{0.04}{4} = 0.01$$

for $\varepsilon/D = 0.008$. To account for property variation, we utilize the correction recommended in Table 6-2.

$$f = f_{cp} \frac{7 - \mu_b/\mu_s}{6}$$

$$= 0.01 \frac{7 - 8.8 \times 10^{-4}/5.78 \times 10^{-4}}{6}$$

$$= (0.01)(0.913) = 9.13 \times 10^{-3}$$

Therefore, ΔP is given by Eq. (6-47).

$$\frac{\Delta P}{\Delta x} = -\rho U_b^2 \frac{f}{2} \frac{p}{A}$$

$$= -62.3 \frac{\text{lb}_m}{\text{ft}^3} \left(3.28 \frac{\text{ft}}{\text{s}}\right)^2 \frac{0.00913}{2} \frac{4(12)}{0.787 \text{ ft}}$$

$$= -187 \frac{\text{lb}_m}{\text{ft}^2 \text{ s}^2} \frac{\text{lb}_f \text{ s}^2}{32.2 \text{ ft lb}_m}$$

$$= -5.79 \frac{\text{lb}_f}{\text{ft}^3} = -4.02 \times 10^{-2} \frac{\text{psi}}{\text{ft}} = -909 \frac{\text{Pa}}{\text{m}}$$

where 1 pascal (Pa) = 1 N/m². The corresponding pressure drop for a smooth tube would be

$$\frac{\Delta P}{\Delta x} = -2.95 \times 10^{-2} \frac{\text{psi}}{\text{ft}} = -667 \frac{\text{Pa}}{\text{m}}$$

To obtain the Nusselt number for the rough surface, we substitute $f = 0.01$ into Eq. (6-33).

$$\text{Nu}_{cp} = \frac{(0.01/2)(15300)(9.4)}{1.07 + 12.7\sqrt{0.01/2} \, (9.4^{2/3} - 1)} = 172$$

Utilizing Eq. (6-42) to correct for the effect of variable properties on Nu, we have

$$\text{Nu} = \text{Nu}_{cp} \left(\frac{\mu_b}{\mu_s}\right)^{0.11}$$

$$= 172 \left(\frac{8.8 \times 10^{-4}}{5.78 \times 10^{-4}}\right)^{0.11} = (172)(1.05) = 180$$

On the other hand, for a smooth tube Nu is equal to about 128. Thus, the rough surface is about 30% more effective in transferring heat. But, this augmentation in the heat transfer is paid for by an increase in pressure drop of about 50%.

6-2-6   Lumped Analysis—Heat Transfer

Practical lumped analyses are developed in this section for the overall heat-transfer performance of single fluid internal convection flow systems such as the circular tube or channel shown in Fig. 6-9. The inlet bulk

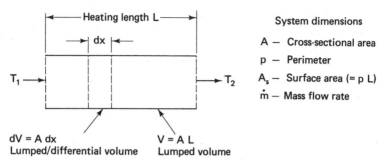

FIGURE 6-9 Tube or channel flow with uniform wall temperature or heat flux around the perimeter.

temperature, fluid properties, mass flow rate, and system dimensions (or
outlet bulk temperature T_2) are assumed to be known. The main objective
of these analyses is to predict the overall rate of heat transfer q_c and the
outlet temperature T_2 (or the system dimensions) in terms of the input
conditions and the boundary conditions of interest. The concepts developed in this section will provide the framework for the practical analysis of
heat exchangers in Chap. 9.

The two basic thermal boundary conditions to be considered include
specified wall heat flux and specified wall temperature, with emphasis
given to uniform property steady state conditions.

Specified wall heat flux

For the case in which the wall heat flux q_c'' is specified as a function
of x, the overall rate of heat transfer is simply

$$q_c = \int_{A_s} q_c'' \, dA_s = p \int_0^L q_c'' \, dx \qquad (6\text{-}48)$$

For uniform wall heat flux (i.e., $q_c'' = q_0''$), q_c is given by

$$q_c = q_0'' A_s = q_0'' pL \qquad (6\text{-}49)$$

Applying the first law of thermodynamics to the fluid volume LA
shown in Fig. 6-9, we obtain

$$\dot{H}_1 + q_c = \dot{H}_2 \qquad (6\text{-}50)$$

For a single-phase fluid with uniform specific heat, this equation takes the
form

$$q_c = \dot{m} c_p (T_2 - T_1) \qquad (6\text{-}51)$$

Although this completes the analysis for q_c and T_2 for this simple boundary condition, predictions are obtained for the local bulk stream temperature T_b and local wall temperature T_s as follows: Applying the first law of thermodynamics to the lumped/differential fluid volume $A \, dx$, we have

$$\dot{H}_b|_x + q_c'' p \, dx = \dot{H}_b|_{x+dx} \tag{6-52}$$

Eliminating $\dot{H}_b|_{x+dx}$ by use of the derivative, Eq. (6-52) reduces to

$$q_c'' p = \frac{d\dot{H}_b}{dx} \tag{6-53}$$

For no phase change and uniform specific heat, $d\dot{H}_b = \dot{m}c_p \, dT_b$ such that Eq. (6-53) reduces to

$$q_c'' p = \dot{m}c_p \frac{dT_b}{dx} \tag{6-54}$$

The boundary condition is

$$T_b = T_1 \qquad \text{at} \quad x=0 \tag{6-55}$$

To obtain an expression for T_b, we integrate Eq. (6-54) over the length o to x to obtain

$$T_b - T_1 = \frac{p}{\dot{m}c_p} \int_0^x q_c'' \, dx \tag{6-56}$$

For uniform wall-flux heating, this equation reduces to

$$T_b - T_1 = \frac{q_0'' p}{\dot{m}c_p} x = \frac{4q_0'' D_H}{k} \frac{x/D_H}{\mathrm{Re} \, \mathrm{Pr}} \tag{6-57}$$

This expression is shown in Fig. 6-10. Notice that the bulk stream temperature T_b is linear over the entire length of the system. In addition, because T_b is independent of the coefficient of heat transfer, the predictions for T_b are the same for laminar and turbulent flow.

An expression is obtained for the unknown surface temperature T_s by combining Eq. (6-56) with the defining equation for the coefficient of heat transfer (the general Newton law of cooling); that is,

$$q_c'' = h(T_s - T_b) \tag{6-58}$$

$$T_s - T_1 = \frac{q_c''}{h} + \frac{p}{\dot{m}c_p} \int_0^x q_c'' \, dx \tag{6-59}$$

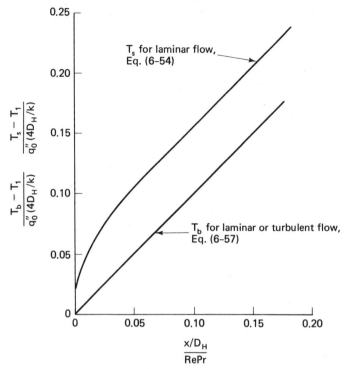

FIGURE 6-10 Bulk stream and wall temperatures for flow in a circular tube with uniform wall heat flux; $D_h = D$.

For a uniform wall heat flux, our prediction for T_s takes the form

$$T_s - T_1 = \frac{4q_0'' D_H}{k} \left(\frac{1}{4\,\mathrm{Nu}} + \frac{x/D_H}{\mathrm{Re}\,\mathrm{Pr}} \right) \tag{6-60}$$

Note that T_s is a function of the coefficient of heat transfer. Accordingly, to obtain calculations for T_s, h or Nu must be specified. For example, for HFD laminar thermal developing flow in a circular tube, Nu can be approximated by Eq. (6-28) in the entrance region and by Eq. (6-25) (Nu = 4.36) in the TFD region, such that Eq. (6-60) becomes

$$\frac{T_s - T_1}{4q_0'' D/k} = 0.192 \left(\frac{x/D}{\mathrm{Re}\,\mathrm{Pr}} \right)^{1/3} + \frac{x/D}{\mathrm{Re}\,\mathrm{Pr}} \qquad \frac{x/D}{\mathrm{Re}\,\mathrm{Pr}} \lesssim 0.04 \tag{6-61a}$$

$$= 0.0573 + \frac{x/D}{\mathrm{Re}\,\mathrm{Pr}} \qquad \frac{x/D}{\mathrm{Re}\,\mathrm{Pr}} \gtrsim 0.04 \tag{6-61b}$$

This equation is shown in Fig. 6-10. T_s is seen to be nonlinear in the thermal entrance region, but is linear in the TFD region. Similar calculations can be made for turbulent flow. Because the Nusselt number for turbulent flow is much larger than for laminar flow, we conclude on the basis of Eq. (6-60) that the wall will be much cooler for turbulent flow.

EXAMPLE 6-7

Air at 27°C and atmospheric pressure flowing at a rate of 10 m/s enters a 10-cm-diameter tube of 1 m length. The heat flux from the surface of the tube is specified by

$$q_c'' = q_0'' \left(1 - \cos\frac{4\pi x}{L}\right)$$

where $q_0'' = 2.71$ kW/m². Determine the outlet temperature T_2, the local bulk stream temperature T_b, and the surface temperature T_s at the exit.

Solution

Because the outlet temperature T_2 is unknown, we will evaluate the properties at the inlet temperature. The properties of air at 27°C are $\rho = 1.18$ kg/m³, $c_p = 1.01$ kJ/(kg °C), $\mu = 1.98 \times 10^{-5}$ kg/(m s), $k = 0.0262$ W/(m °C), and Pr = 0.708.

To obtain the outlet temperature for approximate uniform property conditions, we utilize Eq. (6-51) which was obtained by performing an energy balance on the lumped control volume AL; i.e.,

$$q_c = \dot{m}c_p(T_2 - T_1) \tag{6-51}$$

where

$$q_c = p\int_0^L q_c'' \, dx$$

$$= q_0'' p \int_0^L \left(1 - \cos\frac{4\pi x}{L}\right) dx$$

$$= q_0'' pL = \left(2.71\frac{\text{kW}}{\text{m}^2}\right)(\pi\,0.1\text{ m})(1\text{ m})$$

$$= 0.851 \text{ kW}$$

Calculating the mass flow rate, we have

$$\dot{m} = \rho A U_b = 1.18\frac{\text{kg}}{\text{m}^3}\,\frac{\pi(0.1\text{ m})^2}{4}\,\frac{10\text{ m}}{\text{s}}$$

$$= 0.0927\frac{\text{kg}}{\text{s}}$$

Following through with the calculation for T_2,

$$T_2 = \frac{q_c}{\dot{m}c_p} + T_1$$

$$= \frac{0.851 \text{ kW}}{(0.0927 \text{ kg/s})[1.01 \text{ kJ/(kg °C)}]} + 27°C$$

$$= 36.1°C$$

Referring to Table A-C-5, we see that the variations in properties of air over the temperature range from 27°C to 36.1°C are very small. Consequently, our use of a uniform property analysis with the properties evaluated at 27°C is quite reasonable.

The bulk stream temperature T_b is obtained from Eq. (6-56) which was developed by applying the first law of thermodynamics to the lumped/differential element $A\ dx$.

$$T_b - T_1 = \frac{p}{\dot{m}c_p} \int_0^x q_c'' \, dx \tag{6-56}$$

$$= \frac{q_0'' pL}{\dot{m}c_p} \left(\frac{x}{L} - \frac{1}{4\pi} \sin\frac{4\pi x}{L} \right)$$

$$= \frac{\pi(0.1 \text{ m})(1 \text{ m})(2.71 \text{ kW/m}^2)}{(9.27 \times 10^{-2} \text{ kg/s})[1.01 \text{ kJ/(kg °C)}]} \left(\frac{x}{L} - \frac{1}{4\pi} \sin\frac{4\pi x}{L} \right)$$

$$T_b = 27°C + 9.09°C \left(\frac{x}{L} - \frac{1}{4\pi} \sin\frac{4\pi x}{L} \right)$$

This equation is shown in Fig. E6-7. For the purpose of comparison, T_b is also shown for a uniform heat flux of 2.71 kW/m².

To estimate the local wall temperature T_s at $x = L$, we utilize the general Newton law of cooling,

$$q_c'' = h(T_s - T_b) \tag{6-13}$$

Because the coefficient depends on whether the flow is laminar or turbulent, we calculate the Reynolds number.

$$\text{Re} = \frac{DU_b}{\nu} = \frac{(0.1 \text{ m})(10 \text{ m/s})}{1.68 \times 10^{-5} \text{ m}^2/\text{s}} = 5.95 \times 10^4$$

The flow is fully turbulent. In the region for which $x/L \gtrsim 10$, the flow can be assumed to be thermal fully developed and h can be calculated from Eq. (6-33) for constant property conditions, with f given by Eq.

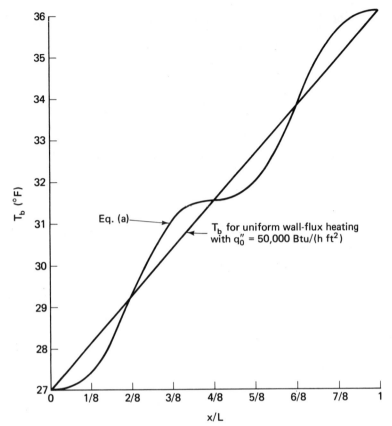

FIGURE E6-7   Predictions for bulk stream temperature T_b.

(6-32); i.e.,

$$f = (1.58 \ln \mathrm{Re} - 3.28)^{-2} = 5.04 \times 10^{-3}$$

$$\mathrm{Nu} = \frac{(f/2)\mathrm{Re}\,\mathrm{Pr}}{1.07 + 12.7\sqrt{f/2}\,(\mathrm{Pr}^{2/3} - 1)}$$

$$= \frac{(5.04 \times 10^{-3}/2)(5.95 \times 10^4)(0.708)}{1.07 + 12.7\sqrt{5.04 \times 10^{-3}/2}\,(0.708^{2/3} - 1)}$$

$$= 113$$

$$h = \frac{(113)\left[0.0262\ \mathrm{W/(m\ ^\circ C)}\right]}{0.1\,\mathrm{m}} = 29.6\frac{\mathrm{W}}{\mathrm{m}^2\,^\circ \mathrm{C}}$$

It follows that the surface temperature at the exit is approximately

$$T_s = \frac{q_c''}{h} + T_b$$

$$= \frac{2.71 \times 10^3 \text{ W/m}^2}{29.6 \text{ W/(m}^2 \text{ °C)}} + 36.1°\text{C}$$

$$= 128°\text{C}$$

Although the evaluation of the properties at the inlet temperature can be expected to provide reasonable accuracy, the fairly large difference between the wall temperature and the bulk temperature should be accounted for in the correlation for the coefficient of heat transfer. By utilizing Eq. (6-44) to correct for h, the wall temperature at the exit is found to be equal to 135°C (see Prob. 6-34).

Up to this point we have restricted ourselves to uniform property steady state conditions. This practical thermal analysis approach is extended to specified wall flux heating with variable specific heat in Example 6-8. To analyze unsteady processes, we must account for the rate of change of energy stored within our lumped/differential control volume $A\,dx$. This point is expanded upon in Prob. 6-35.

EXAMPLE 6-8

Engine oil flowing at a rate of 0.25 kg/s is to be heated from 20°C to 140°C in a 10-cm-diameter tube with uniform wall flux maintained over a 10 m length. Determine the level of heat flux q_0'' required to perform this task and obtain the local bulk stream temperature T_b.

Solution

To obtain q_c, we apply the first law of thermodynamics to the lumped control volume AL.

$$\dot{H}_1 + \dot{q}_c = \dot{H}_2 \tag{6-50}$$

Referring to Table A-C-3, we find that the specific heat varies by about 27% for the temperature range from 20 to 140°C. To account for this rather large variation in c_p on the outlet temperature T_2, we replace the enthalpy rates $\dot{H}_1$ and $\dot{H}_2$ by

$$\dot{H}_1 = \dot{m}c_{p1}T_1$$

$$\dot{H}_2 = \dot{m}c_{p2}T_2$$

Substituting into Eq. (6-50), we have

$$q_c = \dot{m}(c_{p2}T_2 - c_{p1}T_1) \tag{a}$$

where the temperatures must now be expressed in absolute units. It follows that

$$q_c = 0.25\frac{\text{kg}}{\text{s}}\left[\left(2.4\frac{\text{kJ}}{\text{kg K}}\right)(413\text{ K}) - \left(1.88\frac{\text{kJ}}{\text{kg K}}\right)(293\text{ K})\right]$$
$$= 110\text{ kW}$$

and

$$q_0'' = \frac{q_c}{\pi DL} = \frac{110\text{ kW}}{\pi(0.1\text{ m})(10\text{ m})}$$
$$= 35\frac{\text{kW}}{\text{m}^2}$$

For the purpose of comparison, we calculate the heat flux based on a uniform property analysis with c_p evaluated at the arithmetic mean of the inlet and outlet temperatures; i.e.,

$$q_c = \dot{m}c_p(T_2 - T_1) \tag{6-51}$$
$$= \left(0.25\frac{\text{kg}}{\text{s}}\right)\left(2.13\frac{\text{kJ}}{\text{kg °C}}\right)(140°\text{C} - 20°\text{C})$$
$$= 63.9\text{ kW}$$
$$q_0'' = \frac{63.9\text{ kW}}{\pi(0.1\text{ m})(10\text{ m})} = 20.3\frac{\text{kW}}{\text{m}^2}$$

This uniform property calculation is 42% below the correct value! This result underscores the fact that simple uniform property analyses based on arithmetic average fluid properties can lead to significant errors for situations such as this in which the fluid experiences large property variations.

The local bulk stream temperature T_b is obtained by applying the energy conservation principle to the lumped/differential control volume $A\,dx$, with the result

$$q_c''p = \frac{d\dot{H}_b}{dx} \tag{6-53}$$
$$= \frac{d}{dx}(\dot{m}c_pT_b) \tag{b}$$

where $\dot{m}$ is constant but c_p is temperature dependent. Integrating, we have

$$\int_{c_{p1}T_1}^{c_p T_b} d(c_p T_b) = \frac{q_0'' p}{\dot{m}} \int_0^x dx$$

$$c_p T_b = c_{p1} T_1 + \frac{q_0'' p}{\dot{m}} x$$

$$= \left(1.88 \frac{\text{kJ}}{\text{kg K}}\right)(293 \text{ K}) + 35 \frac{\text{kW}}{\text{m}^2} \frac{\pi(0.1 \text{ m})}{0.25 \text{ kg/s}} x$$

$$= 551 \frac{\text{kJ}}{\text{kg}} + 44 \frac{\text{kJ}}{\text{kg m}} x$$

The distance x can be calculated for various values of T_b by combining this equation with the information provided in Table A-C-3 pertaining to the dependence of c_p on temperature. The calculations for x are shown in Table E6-8. The resulting bulk stream temperature distribution is shown in Fig. E6-8. For comparison, Eq. (6-57) which was obtained from the uniform property analysis is also shown. For this particular problem, the actual bulk stream temperature is slightly nonlinear.

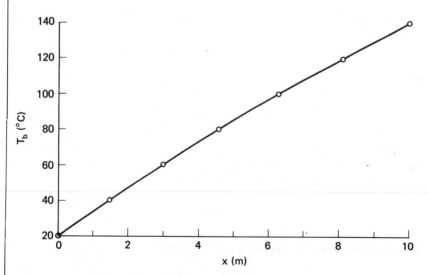

FIGURE E6-8 Predictions for bulk stream temperature T_b.

TABLE E6-8   Bulk stream temperature for fluid with variable specific heat.

T_b (°C)	T_b (K)	c_p [kJ/(kg K)]	$c_p T_b$ (kJ/kg)	x (m)
20	293	1.88	551	0
40	313	1.96	613	1.42
60	333	2.05	683	2.99
80	353	2.13	752	4.57
100	373	2.22	828	6.30
120	393	2.31	908	8.11
140	413	2.40	991	10.0

Specified wall temperature

To develop expressions for the two primary parameters q_c and T_2 for the case in which the wall temperature is specified, two independent equations must be written. Following the pattern established in the analysis of the uniform wall-heat-flux problem, these two independent equations are developed by applying the first law of thermodynamics to (1) the lumped fluid volume LA, and (2) the lumped/differential element $A\,dx$ (see Fig. 6-9). Focusing attention on single phase fluids with uniform properties, the development of an energy balance on the lumped fluid volume LA gives rise to an expression for q_c of the form of Eq. (6-50).

$$q_c = \dot{H}_2 - \dot{H}_1 = \dot{m}c_p(T_2 - T_1) \tag{6-62}$$

Second, an energy balance on the fluid volume $A\,dx$ is given by Eq. (6-52),

$$\dot{H}_b\big|_x + q_c'' \, p \, dx = \dot{H}_b\big|_{x+dx} \tag{6-52}$$

which reduces to

$$q_c'' p = \frac{d\dot{H}_b}{dx} = \dot{m}c_p \frac{dT_b}{dx} \tag{6-63}$$

Because the wall temperature is specified in this problem, we express q_c'' in terms of T_s and h by means of the general Newton law of cooling. With this input for q_c'', Eq. (6-63) takes the form

$$hp(T_s - T_b) = \dot{m}c_p \frac{dT_b}{dx} \tag{6-64}$$

The boundary condition is given by Eq. (6-55).

$$T_b = T_1 \quad \text{at } x = 0 \tag{6-55}$$

Solution for Uniform Wall Temperature   For a uniform wall temperature boundary condition ($T_s = T_0$), Eq. (6-64) can be written as

$$\frac{d(T_b - T_0)}{T_b - T_0} = -\frac{ph}{\dot{m}c_p}dx \qquad (6\text{-}65)$$

Integrating this equation, we obtain

$$\ln\frac{T_b - T_0}{T_1 - T_0} = -\frac{p}{\dot{m}c_p}\int_0^x h\,dx = -\frac{\bar{h}px}{\dot{m}c_p} \qquad (6\text{-}66)$$

where the mean coefficient of heat transfer $\bar{h}$ is defined by Eq. (6-14). The rearrangement of this equation gives

$$\frac{T_b - T_0}{T_1 - T_0} = \exp\left(-\frac{\bar{h}px}{\dot{m}c_p}\right) \qquad (6\text{-}67)$$

or

$$\frac{T_b - T_1}{T_0 - T_1} = 1 - \exp\left(-\frac{\bar{h}px}{\dot{m}c_p}\right) \qquad (6\text{-}68)$$

As shown in Fig. 6-11, this equation indicates that T_b asymptotically approaches T_0 as x increases.

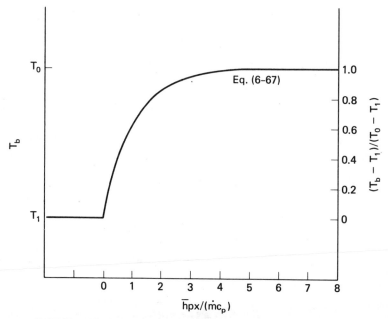

FIGURE 6-11   Bulk stream temperature for flow with uniform wall temperature.

Setting x equal to L in Eq. (6-68), we obtain an expression for the bulk outlet temperature T_2 of the form

$$\frac{T_2 - T_1}{T_0 - T_1} = 1 - \exp\left(-\frac{\bar{h}A_s}{\dot{m}c_p}\right) \tag{6-69}$$

where $A_s = pL$. Finally, the coupling of Eqs. (6-62) and (6-69) gives rise to an expression for q_c of the form

$$q_c = \dot{m}c_p(T_0 - T_1)\left[1 - \exp\left(-\frac{\bar{h}A_s}{\dot{m}c_p}\right)\right] \tag{6-70}$$

In the limit, as the dimensionless parameter $\bar{h}A_s/(\dot{m}c_p)$ increases, the outlet temperature T_2 approaches T_0 and the total rate of heat transfer q_c approaches a maximum value of $\dot{m}c_p(T_0 - T_1)$. This thermodynamic limit occurs at a value of $\bar{h}A_s/(\dot{m}c_p)$ of the order of 4 to 5.

To obtain calculations for q_c, T_2, and T_b for a given flow condition, we must utilize appropriate correlations for the mean coefficient of heat transfer.

Although the theoretical aspects of the analysis for uniform wall temperature heating are complete, the predictions for q_c are now put into the effectiveness and log mean temperature difference (LMTD) formats. As we shall see in Chap. 9, these concepts are commonly used in the analysis and design of heat exchangers.

Effectiveness.   The *effectiveness* ε of a heat transfer system is defined by

$$\varepsilon = \frac{q_c}{q_{c,\,\text{max}}} \tag{6-71}$$

where the maximum possible rate of heat transfer $q_{c,\,\text{max}}$ for uniform wall temperature heating is given by

$$q_{c,\,\text{max}} = \dot{m}c_p(T_0 - T_1) \tag{6-72}$$

Introducing Eq. (6-62),

$$q_c = \dot{m}c_p(T_2 - T_1) \tag{6-62}$$

we see that the effectiveness also represents the temperature ratio $(T_2 - T_1)/(T_0 - T_1)$. Referring to Eq. (6-69), ε is given by

$$\varepsilon = \frac{T_2 - T_1}{T_0 - T_1} = 1 - \exp\left(-\frac{\bar{h}A_s}{\dot{m}c_p}\right) \tag{6-73}$$

This equation is shown in Fig. 6-12 in terms of $\bar{h}A_s/(\dot{m}c_p)$.

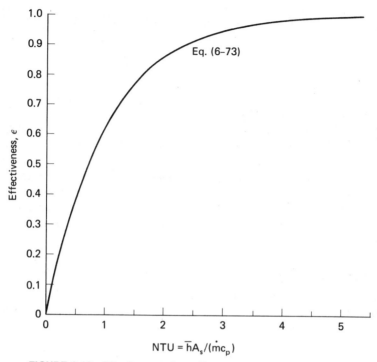

FIGURE 6-12 Effectiveness for uniform wall temperature heating.

The dimensionless parameter $\bar{h}A_s/(\dot{m}c_p)$ is sometimes referred to as the *number of transfer units* NTU. The NTU provides a comparison between the thermal capacity $\bar{h}A_s$ of the system and the capacity rate $\dot{m}c_p$. The larger the NTU, the closer the heat-transfer system comes to approaching its thermodynamic limit.

Log mean temperature difference. In order to put the q_0 predictions for uniform wall temperature heating into the LMTD form, we replace the capacity rate $\dot{m}c_p$ in Eq. (6-62) by utilizing Eq. (6-69); that is,

$$\dot{m}c_p = -\frac{\bar{h}A_s}{\ln\dfrac{T_2 - T_0}{T_1 - T_0}} = \frac{\bar{h}A_s}{\ln\dfrac{T_0 - T_1}{T_0 - T_2}} \tag{6-74}$$

This substitution gives rise to the famous LMTD equation

$$q_c = \bar{h}A_s \text{ LMTD} \tag{6-75}$$

where

$$\text{LMTD} = \frac{T_2 - T_1}{\ln \dfrac{T_0 - T_1}{T_0 - T_2}} = \frac{\Delta T_1 - \Delta T_2}{\ln \dfrac{\Delta T_1}{\Delta T_2}} \qquad (6\text{-}76)$$

and $\Delta T_1 = T_0 - T_1$ and $\Delta T_2 = T_0 - T_2$.

The LMTD format is most useful for making design calculations in order to size a heat-transfer system (i.e., to determine the necessary thermal capacity $\bar{h}A_s$) when the inlet, outlet, and surface temperatures are specified. However, for situations in which the outlet temperature is not known, use of the LMTD approach to evaluate T_2 involves time-consuming iterations. For such cases, the more efficient approach generally is to simply utilize Eqs. (6-69) and (6-70) directly, or to use the effectiveness concept.

EXAMPLE 6-9

Water at 20°C enters a 10-cm-diameter tube of 10-m length. The mass flow rate of the water is 10 kg/s and the wall temperature is 80°C. Determine the outlet temperature of the water.

Solution

Because the outlet temperature is not known, we will utilize the effectiveness approach, with the properties evaluated at 20°C. The required properties of water at 15°C are $\rho = 1000 \text{ kg/m}^3$, $c_p = 4.18 \text{ kJ/(kg °C)}$, $\nu = 1.01 \times 10^{-6} \text{ m}^2/\text{s}$, $k = 0.597 \text{ W/(m °C)}$, and $\text{Pr} = 7.02$. Calculating $q_{c,\,\text{max}}$, we obtain

$$q_{c,\,\text{max}} = \dot{m}c_p(T_0 - T_1)$$

$$= \left(10\,\frac{\text{kg}}{\text{s}}\right)\left(4.18\,\frac{\text{kJ}}{\text{kg °C}}\right)(80°\text{C} - 20°\text{C})$$

$$= 2510 \text{ kW}$$

Thus, the actual rate of heat transfer is given by

$$q_c = \varepsilon q_{c,\,\text{max}} = \varepsilon\, 2510 \text{ kW} \qquad (a)$$

where the effectiveness ε is given by Eq. (6-73) or Fig. 6-12. To obtain ε, we must evaluate the parameter $\bar{h}A_s/(\dot{m}c_p)$, which necessitates the calculation of $\bar{h}$.

The Reynolds number is

$$Re = \frac{U_b D}{\nu} = \frac{\dot{m}D}{\mu A} = \frac{\dot{m}}{\mu} \frac{4}{\pi D}$$

$$= \frac{4(10 \text{ kg/s})}{\pi(0.1 \text{ m})\left[1.01 \times 10^{-3} \text{ kg/(m s)}\right]}$$

$$= 1.26 \times 10^5$$

such that the flow is turbulent. Utilizing Eqs. (6-32) and (6-33) to approximate Nu for isothermal conditions, we have

$$f_{cp} = (1.58 \ln 1.26 \times 10^5 - 3.28)^{-2} = 4.29 \times 10^{-3}$$

$$Nu_{cp} = \frac{(4.29 \times 10^{-3}/2)(1.26 \times 10^5)(7.02)}{1.07 + 12.7\sqrt{4.29 \times 10^{-3}/2} \ (7.02^{2/3} - 1)} = 719$$

To correct for nonuniform viscosity, we use Eq. (6-42).

$$Nu = Nu_{cp}\left(\frac{\mu_b}{\mu_s}\right)^{0.11}$$

$$= 719\left(\frac{1.01 \times 10^{-3}}{3.54 \times 10^{-4}}\right)^{0.11}$$

$$= 807$$

Approximating $\bar{h}$ by h, we obtain

$$\bar{h} = \frac{k}{D} Nu = \frac{\left[0.597 \text{ W/(m °C)}\right](807)}{0.1 \text{ m}} = 4820 \frac{W}{m^2 C}$$

The number of transfer units $\bar{h}A_s/(\dot{m}c_p)$ is then found to be

$$\frac{\bar{h}A_s}{\dot{m}c_p} = \frac{\left[4.82 \text{ kW/(m}^2 \text{ °C)}\right](\pi)(0.1 \text{ m})(10 \text{ m})}{(10 \text{ kg/s})\left[4.18 \text{ kJ/(kg °C)}\right]} = 0.362$$

Utilizing Eq. (6-73), we determine that $\varepsilon \simeq 0.304$.
Returning to Eq. (a), we obtain

$$q_c = 0.304 \ (2510 \text{ kW}) = 762 \text{ kW}$$

To obtain the outlet temperature T_2, we write

$$q_c = \dot{m}c_p(T_2 - T_1)$$

$$T_2 = \frac{762 \text{ kW}}{(10 \text{ kg/s})[4.18 \text{ kJ}/(\text{kg }°\text{C})]} + 20°\text{C}$$

$$= 18.2°\text{C} + 20°\text{C} = 38.2°\text{C}$$

Referring to Table A-C-3, we find that the specific heat of water at 38.2°C is equal to the value at 20°C to within three significant figures. Therefore, our assumption of uniform specific heat is quite adequate. However, we note that the viscosity μ changes considerably over this temperature range. Therefore, our analysis can be refined by approximating the properties at the arithmetic average of the inlet and outlet temperatures. This refinement is considered in Prob. 6-39.

EXAMPLE 6-10

Air at 7 atm flowing at a rate of 0.1 kg/s is to be cooled from 400 to 300 K in a 10-cm-diameter tube with uniform wall temperature of 250 K. Determine the length of the tube.

Solution

Referring to Table A-C-5, the pertinent fluid properties at the arithmetic average inlet and outlet temperature of 350 K are $\rho = 0.998$ kg/m^3, $c_p = 1.01$ kJ/(kg K), $\mu = 2.08 \times 10^{-5}$ kg/(m s), $\nu = 2.08 \times 10^{-5}$ m^2/s, $k = 0.03$ W/(m K) and Pr = 0.697.

Noting that the specific heat is essentially uniform over the temperature range from 300 to 400 K, we utilize Eq. (6-62) to calculate q_c.

$$q_c = \dot{m}c_p(T_2 - T_1) = \left(0.1 \frac{\text{kg}}{\text{s}}\right)\left(1.01 \frac{\text{kJ}}{\text{kg K}}\right)(300 \text{ K} - 400 \text{ k}) = -10.1 \text{ kW}$$

Because the inlet and outlet temperatures are known, we can easily utilize the LMTD relationship

$$q_c = \bar{h}A_s \text{ LMTD} \qquad\qquad (6\text{-}75)$$

where

$$\text{LMTD} = \frac{\Delta T_1 - \Delta T_2}{\ln\dfrac{\Delta T_1}{\Delta T_2}} = \frac{(250 \text{ K} - 400 \text{ K}) - (250 \text{ K} - 300 \text{ K})}{\ln\dfrac{250 \text{ K} - 400 \text{ K}}{250 \text{ K} - 300 \text{ K}}} = -91 \text{ K}$$

To obtain $\bar{h}$, we calculate the Reynolds number.

$$\mathrm{Re} = \frac{D U_b}{\nu} = \frac{\dot{m} D}{A\mu} = 0.1 \, \frac{\mathrm{kg}}{\mathrm{s}} \, \frac{4}{\pi(0.1 \, \mathrm{m})\left[2.08 \times 10^{-5} \, \mathrm{kg/(m \, s)}\right]} = 6.12 \times 10^4$$

Thus, the flow is turbulent. Utilizing Eqs. (6-32) and (6-33) to calculate Nu for uniform property conditions, we have

$$f_{cp} = (1.58 \ln 6.12 \times 10^4 - 3.28)^{-2} = 5.01 \times 10^{-3}$$

$$\mathrm{Nu}_{cp} = \frac{(5.01 \times 10^{-3}/2)(6.12 \times 10^4)(0.697)}{1.07 + 12.7\sqrt{5.01 \times 10^{-3}/2} \,(0.697^{2/3} - 1)} = 114$$

The property effect associated with the wall to bulk stream temperature difference is corrected for by using Eq. (6-44).

$$\mathrm{Nu} = 114\left(\frac{250}{350}\right)^{-0.36} = 129$$

Approximating $\bar{h}$ by h, we obtain

$$\bar{h} = \mathrm{Nu}\,\frac{k}{D} = \frac{129\left[0.03 \, \mathrm{W/(m \, °C)}\right]}{0.1 \, \mathrm{m}} = 38.7\,\frac{\mathrm{W}}{\mathrm{m}^2 \, °C}$$

The length can now be calculated from Eq. (6-75).

$$A_s = \frac{q_c}{\bar{h}\,\mathrm{LMTD}} = \frac{-10.1 \, \mathrm{kW}}{\left[38.7 \, \mathrm{W/(m^2 \, °C)}\right](-91 \, \mathrm{K})} = 2.87 \, \mathrm{m}^2$$

$$L = \frac{2.87 \, \mathrm{m}^2}{\pi\,0.1 \, \mathrm{m}} = 9.14 \, \mathrm{m}$$

Solution for Nonuniform Conditions   Whereas an analytical solution was easily developed for the situation in which the wall temperature and fluid properties are uniform, analytical solutions are not so readily obtained for nonuniform wall temperature heating or variable property conditions. For these more general conditions, the energy balances on the lumped elements AL and $A\,dx$ take the forms

$$q_c = \dot{H}_2 - \dot{H}_1 = \dot{m}(c_{p2}T_2 - c_{p1}T_1) \qquad (6\text{-}77)$$

and

$$hp(T_s - T_b) = \dot{m}\frac{d}{dx}(c_p T_b) \qquad (6\text{-}78)$$

Equation (6-78) can be put into a practical finite difference form which accounts for variations in h, c_p and T_s. The numerical solution of this equation is illustrated in Example 6-11 for the case of nonuniform wall temperature heating. The use of this approach in solving problems with variable c_p is considered in Prob. 6-57. [The use of Eq. (6-77) has been illustrated in Example 6-8.] This approach can be extended to unsteady conditions by accounting for the change of energy stored within the fluid per unit time, as discussed in the context of uniform wall flux heating in Prob. 6-35.

EXAMPLE 6-11

Develop a numerical finite difference formulation for convection heat transfer in a tube with nonuniform coefficient of heat transfer h, specific heat c_p, and wall temperature T_s. Then solve Example 6-9 for the case in which h and c_p are essentially uniform and T_s is given by

$$T_s = 20°C + 80°C\left(1 - \cos\frac{4\pi x}{L}\right) \qquad (a)$$

Solution

To develop an analysis for the bulk stream temperature T_b for nonuniform conditions, we first write the energy balance for the lumped/differential volume $A\,dx$.

$$\dot{H}_b|_x + q_c'' p\,dx = \dot{H}_b|_{x+dx} \qquad (6\text{-}52)$$

$$q_c'' p = \frac{d\dot{H}_b}{dx} \qquad (b)$$

For situations in which the specific heat is a function of temperature, this equation becomes

$$q_c'' p = \frac{d}{dx}(\dot{m}c_p T_b) \qquad (c)$$

which is identical to Eq. (b) in Example 6-8.

Introducing the general Newton law of cooling and taking $\dot{m}$ as constant, we obtain

$$hp(T_s - T_b) = \dot{m}\frac{d}{dx}(c_p T_b) \qquad (d)$$

The solution to this equation for the case in which c_p and T_s are constant is given by Eq. (6-67).

To develop a solution to Eq. (d) for the case in which h, c_p, and T_s are nonuniform, we write this equation in finite-difference format as follows:

$$h_m p(T_{sm} - T_{bm}) = \frac{\dot{m}}{\Delta x}(c_{pm+1} T_{bm+1} - c_{pm} T_{bm}) \qquad (e)$$

or

$$T_{bm+1} = \frac{1}{c_{pm+1}}\left[\frac{h_m p \, \Delta x}{\dot{m}} T_{sm} + \left(c_{pm} - \frac{h_m p \, \Delta x}{\dot{m}}\right)T_{bm}\right] \qquad (f)$$

where m ranges from 1 to M as shown in Fig. E6-11a. With increment Δx taken sufficiently small, c_{pm+1} can be approximated by c_{pm}, such that Eq. (f) reduces to

$$T_{bm+1} = \frac{h_m p \, \Delta x}{\dot{m} c_{pm}} T_{sm} + \left(1 - \frac{h_m p \, \Delta x}{\dot{m} c_{pm}}\right)T_{bm} \qquad (g)$$

Convection with T_s and
$\overline{h}$ specified

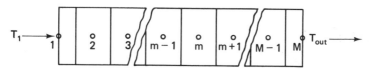

FIGURE E6-11a Finite-difference grid for flow system.

Finally, applying the first law of thermodynamics to the lumped element AL, we have

$$q_c = \dot{H}_{bM} - \dot{H}_{b1} = \dot{m}(c_{pM} T_{bM} - c_{p1} T_{b1}) \qquad (h)$$

For situations in which c_p is uniform, this equation reduces to Eq. (6-62).

This completes our formulation for nonuniform conditions. To develop numerical calculations for the case in which h and c_p are uniform and T_s is given by Eq. (a), Eqs. (g) and (h) are written in FORTRAN as follows:

```
T(J+1)
=s/((H*P/MDOT*CP))*DX*TS+T(J)*(1.-(H*P/(MDOT*CP)*DX))
```
                                                                        (i)
```
QO=MDOT*CP*(T(M)-T(1))
```
                                                                        (j)

where $D = 0.1$ m, $L = 10$ m, $h = H = 4820$ W/(m² °C), $p = P = \pi D$, $\dot{m} =$ MDOT $= 10$ kg/s, and $c_p = CP = 4.18$ kJ/(kg °C). A simple FORTRAN program is given in Appendix A-E-4 which solves these equations for any specified wall temperature. Calculations are shown for T_b in Fig. E6-11b for T_s specified by Eq. (a). The solution is found to converge for values of M of the order of 100. The outlet temperature is 44.3°C and the total rate of heat transfer q_0 is 1020 kW.

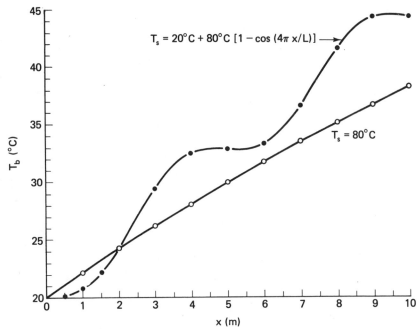

FIGURE E6-11b   Predictions for bulk stream temperature.

To serve as a check and for purpose of comparison, the program is also run with T_s set equal to 80°C. The calculations for this uniform wall temperature condition are also shown in Fig. E6-11b. The outlet temperature is 38.3°C and the rate of heat transfer is 763 kW. These results are consistent with the solution obtained in Example 6-9.

6-3 EXTERNAL FLOW

6-3-1 Introduction

We now consider external forced flow over flat plates, cylinders, and spheres, as illustrated in Figs. 6-13 and 6-14. The velocity and temperature of the fluid approaching the bodies in question are U_∞ and T_∞.

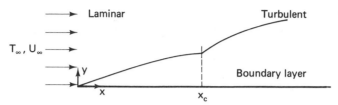

FIGURE 6-13 Longitudinal flow over a flat plate or cylinder.

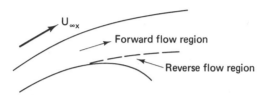

FIGURE 6-14 Crossflow over a cylindrical body.

Similar to the situation found in internal flow systems, a hydrodynamic boundary layer develops along the wall. Outside the HBL ($y > \delta$), the fluid velocity is within 1% of the free stream velocity U_∞. The HBL associated with external flow continues to grow as x increases, as illustrated in Fig. 6-13. The HBL is generally laminar for small values of x, but develops into turbulence at some critical value of x ($= x_c$) for large-enough flow rates.

Thermal boundary layers (TBL) also develop along surfaces that are heated or cooled. The temperature of the fluid varies from T_s at the wall ($y = 0$) to within 1% of the free stream temperature T_∞ at y equal to Δ.

For uniform property analyses, the properties for external flows are generally evaluated at the free stream temperature T_∞.

Because of significant differences between external longitudinal flow over flat plates and cylinders and flow normal to such bodies, these two situations will be discussed separately.

6-3-2 Longitudinal Flow over Flat Plates and Cylinders

Longitudinal flow over flat plates and cylinders is characterized by a free stream velocity that is independent of x ($U_\infty = $ constant), and a zero pressure drop along the body. The hydrodynamic properties are often presented in the form of a Reynolds number Re_x, which is defined as

$$\mathrm{Re}_x = \frac{U_\infty x}{\nu} \tag{6-79}$$

The flow is generally laminar for $0 < \mathrm{Re}_x \gtrsim 10^5$, transitional turbulent for $10^5 \gtrsim \mathrm{Re}_x \gtrsim 5 \times 10^5$, and fully turbulent for $5 \times 10^5 \gtrsim \mathrm{Re}_x$. It should be mentioned that other characteristic lengths are sometimes utilized in the definition of the Reynolds number for this type of flow. This point will be touched upon in Chap. 10 in the context of the theoretical analysis of boundary layer flow over a flat plate.

Coefficients of friction and heat transfer

The definitions for the Fanning friction factor f_x and coefficient of heat transfer h_x for flow over flat plates and cylinders are given by Eqs. (5-4) and (5-9) with $U_F = U_\infty$ and $T_F = T_\infty$; that is,

$$\tau_s = \rho U_\infty^2 \frac{f_x}{2} \tag{6-80}$$

$$q_c'' = h_x(T_s - T_\infty) \tag{6-81}$$

The standard definitions for $\bar{f}$ and $\bar{h}$ given by Eqs. (5-6) and (5-11) also apply.

Useful correlations for f_x for laminar and turbulent boundary layer flow over a flat plate are given by

$$f_x = 0.664 \, \mathrm{Re}_x^{-0.5} \qquad \begin{array}{l} \text{laminar flow} \\ 0 < \mathrm{Re}_x \gtrsim \mathrm{Re}_{x_c} \end{array} \tag{6-82}$$

after Blasius [34], and

$$f_x = 0.059 \, \mathrm{Re}_x^{-0.2} \qquad \begin{array}{l} \text{turbulent flow} \\ \mathrm{Re}_{x_c} \gtrsim \mathrm{Re}_x \gtrsim 10^7 \end{array} \tag{6-83}$$

which is developed in [8], [35], and others.

A somewhat less convenient but more accurate expression for f_x for turbulent flow recommended by White [19] takes the form

$$f_x = \frac{0.455}{\left[\ln(0.06 \, \mathrm{Re}_x)\right]^2} \qquad \begin{array}{l} \text{turbulent flow} \\ \mathrm{Re}_{x_c} \gtrsim \mathrm{Re}_x \gtrsim 10^9 \end{array} \tag{6-84}$$

These equations are compared with experimental data in Fig. 6-15. The local Fanning friction factor has not been universally characterized for transitional turbulent boundary layer flow. Therefore, Eqs. (6-83) and (6-84) are generally utilized in both the transitional and fully turbulent regions.

To obtain an expression for the mean friction factor for laminar conditions, we integrate Eq. (6-82) in accordance with the defining equation for $\bar{f}$, Eq. (5-6), to obtain

$$\bar{f} = 1.33 \, \mathrm{Re}_L^{-0.5} \qquad \text{laminar flow} \tag{6-85}$$

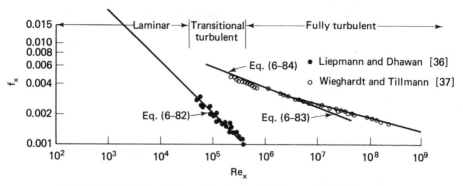

FIGURE 6-15 Correlations for f_x for laminar and turbulent boundary layer flow.

Similarly, for the case in which turbulent flow is assumed to occur over the entire plate, the coupling of Eqs. (5-6) and (6-83) gives

$$\bar{f} = 0.0738 \, Re_L^{-0.2} \qquad \text{turbulent flow} \qquad (6-86)$$

White [19] has recommended an approximate correlation for $\bar{f}$ for fully turbulent flow of the form of Eq. (6-84).

$$\bar{f} = \frac{0.523}{\left[\ln \left(0.06 \, Re_L \right) \right]^2} \qquad (6-87)$$

The coefficients h_x and $\bar{h}$ for external flows are usually expressed in terms of the Nusselt number ($Nu_x = h_x x / k$, $\overline{Nu} = \bar{h} L / k$) or Stanton number [$St_x = h_x / (\rho c_p U_\infty)$, $\overline{St} = \bar{h} / (\rho c_p U_\infty)$].

A convenient theoretical correlation has been developed for the local Nusselt number for laminar boundary layer flow with uniform wall temperature heating which takes the form of the Colburn analogy [22]; that is,

$$Nu_x = \frac{f_x}{2} Re_x \, Pr^{1/3} \qquad (6-88)$$

or

$$Nu_x = 0.332 \sqrt{Re_x} \, Pr^{1/3} \qquad (6-89)$$

Based on our defining equation for the mean coefficient of heat transfer, Eq. (5-11), an expression can be developed for the mean Nusselt number

$\overline{\mathrm{Nu}}$ for uniform wall-temperature heating as follows:

$$\bar{h} = \frac{1}{L} \int_0^L h_x \, dx = \frac{1}{L} \int_0^L \mathrm{Nu}_x \frac{k}{x} \, dx \tag{6-90}$$

$$\overline{\mathrm{Nu}} = \frac{\bar{h} L}{k} = \int_0^L 0.332 \sqrt{\frac{U_\infty}{\nu}} \ \mathrm{Pr}^{1/3} \, x^{-1/2} \, dx$$

$$= 0.664 \sqrt{\mathrm{Re}_L} \ \mathrm{Pr}^{1/3} \tag{6-91}$$

or

$$\overline{\mathrm{Nu}} = \frac{\bar{f}}{2} \mathrm{Re}_L \ \mathrm{Pr}^{1/3} \tag{6-92}$$

An approximate theoretical correlation has also been developed for laminar boundary layer flow with uniform wall-flux heating of the form (Chap. 10)

$$\mathrm{Nu}_x = 0.417 \sqrt{\mathrm{Re}_x} \ \mathrm{Pr}^{1/3} \tag{6-93}$$

Numerous correlations have been developed for the Nusselt number for turbulent boundary layer flow. One of the more simple and useful correlations for fully turbulent flow over a flat plate with uniform wall temperature or uniform wall-flux heating is given by [8]

$$\mathrm{Nu}_x = \frac{f_x}{2} \mathrm{Re}_x \ \mathrm{Pr}^n \tag{6-94}$$

or

$$\mathrm{Nu}_x = 0.0295 \ \mathrm{Re}_x^{0.8} \ \mathrm{Pr}^n \tag{6-95}$$

where $n = 0.5$ for $0.5 \gtrsim \mathrm{Pr} \gtrsim 5.0$ and $n = \frac{1}{3}$ for $\mathrm{Pr} \gtrsim 5.0$. [Note the similarity between this correlation and the correlation for TFD fully turbulent tube flow given by Eqs. (6-35) and (6-36).] For the case in which turbulence essentially occurs over the entire length of the plate, a mean Nusselt number can be obtained from Eq. (6-95), which takes the form

$$\overline{\mathrm{Nu}} = 0.0369 \ \mathrm{Re}_L^{0.8} \ \mathrm{Pr}^n \tag{6-96}$$

or

$$\overline{\mathrm{Nu}} = \frac{\bar{f}}{2} \mathrm{Re}_L \ \mathrm{Pr}^n \tag{6-97}$$

For improved accuracy, the following relationship developed by White [19] is recommended for $0.5 \gtrsim \mathrm{Pr}$:

$$\mathrm{Nu}_x = \frac{(f_x/2)\mathrm{Re}_x \ \mathrm{Pr}}{1 + 12.8 \sqrt{f_x/2} \ (\mathrm{Pr}^{0.68} - 1)} \tag{6-98}$$

For all practical purposes, this equation is equivalent to the Petukhov–Kirillov correlation, Eq. (6-33), which was recommended for TFD fully turbulent tube flow. Notice that Eq. (6-98) reduces to Eq. (6-94) as the Prandtl number approaches unity.

The Nu_x correlations for laminar and turbulent boundary layer flow over a flat plate with uniform wall-temperature heating are compared in Fig. 6-16. Once again, we see that the heat transfer is much greater for turbulent flow than for laminar flow.

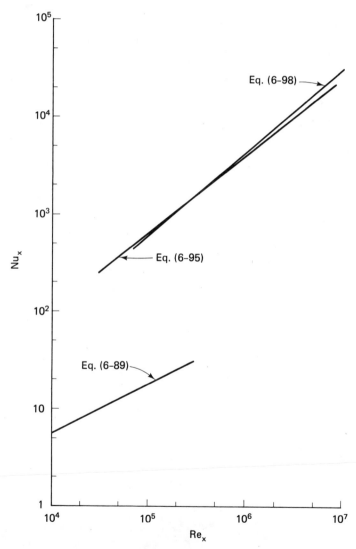

FIGURE 6-16 Comparison of heat-transfer coefficients for laminar and turbulent flow over a flat plate, Pr = 5.0.

Correlations have also been developed for the coefficients of heat transfer for laminar and turbulent boundary layer flow over flat plates with nonuniform wall heating. This subject is introduced in [8] and [38] and in Chap. 10. Correlations are available in the literature for low-Prandtl-number fluids such as liquid metals [8].

Correction factors of the type introduced for tube flow are often utilized to account for the effect of property variations on the friction factor and coefficient of heat transfer. As a matter of fact, Eqs. (6-41) and (6-42) and the values of m and n given in Table 6-2 are recommended by Kays [8] for boundary flow of liquids. Similarly, Eqs. (6-43) and (6-44) are sometimes used for gases with m and n being dependent upon the situation. Alternatively, Eckert [39] has suggested that the effect of property variation for boundary layer flow of gases be accounted for by evaluating the properties at the film temperature T_f $[\equiv(T_s + T_\infty)/2]$.

As in the case of internal flow, the heat transfer is often enhanced for flow over flat plates and cylinders by the use of rough surfaces or fins. But for such surfaces, the manufacturers' design correlations for $\bar{h}$ and $\bar{f}$ must be employed.

Once again, we should be reminded that the coefficients of friction and heat transfer are also sometimes influenced by other factors, such as nonuniform heating, viscous dissipation, phase change and unsteady operation. When the convection process is complicated by factors such as these, their influence on the coefficients should be accounted for.

Lumped analysis—momentum transfer

Based on the definition for wall shear stress τ_s, the differential shear force dF_x which is caused by fluid flowing over the surface of a flat plate or cylinder at any axial location x is given by

$$dF_x = \tau_s \, dA_s = \tau_s p \, dx \tag{6-99}$$

It follows that the total shear force acting on the body is

$$F_L = \int_0^L \tau_s p \, dx \tag{6-100}$$

Introducing the Fanning friction factor and noting that U_∞ is constant for longitudinal flow over surfaces, we obtain

$$F_L = \int_0^L \rho U_\infty^2 \frac{f_x}{2} p \, dx$$

$$= \rho \frac{U_\infty^2}{2} p \int_0^L f_x \, dx$$

$$= \rho U_\infty^2 \frac{\bar{f}}{2} A_s \tag{6-101}$$

Lumped analysis—heat transfer

To determine the total convection-heat-transfer rate from the surface of a flat plate or cylinder of length L, we first express the rate of heat transfer from a differential surface area dA_s $(=p\,dx)$ in terms of the local heat flux q_c''.

$$dq_c = q_c''\,dA_s = q_c''p\,dx \qquad (6\text{-}102)$$

Thus, the total rate of convection heat transfer over the entire surface is

$$q_c = p\int_0^L q_c''\,dx \qquad (6\text{-}103)$$

where the evaluation of the integral depends on the form of the thermal boundary condition.

Specified Wall Heat Flux For the case in which the heat flux is specified, Eq. (6-103) can be integrated directly to obtain q_c. For example, for a uniform wall heat flux $(q_c'' = q_0'')$, q_c is simply given by

$$q_c = q_0''pL = A_s q_0'' \qquad (6\text{-}104)$$

To obtain the local wall temperature, we introduce the coefficient of heat transfer h_x by utilizing the general Newton law of cooling, Eq. (6-81),

$$q_c'' = h_x(T_s - T_\infty) \qquad (6\text{-}81)$$

Solving for T_s, we have

$$T_s - T_\infty = \frac{q_c''}{h_x} = q_c''\frac{x/k}{\mathrm{Nu}_x} \qquad (6\text{-}105)$$

To calculate T_s, Eq. (6-105) must be coupled with the appropriate correlations for Nu_x, as illustrated in Example 6-12. Notice that the wall operates at much lower temperature for turbulent flow than for laminar flow (assuming negligible viscous dissipation effects).

Specified Wall Temperature With the wall temperature specified as a function of x, h_x is immediately introduced into Eq. (6-103) by utilizing the general Newton law of cooling; that is,

$$q_c = p\int_0^L h_x(T_s - T_\infty)\,dx \qquad (6\text{-}106)$$

For a uniform wall-temperature boundary condition $(T_s = T_0)$, Eq. (6-106) gives rise to an expression for q_c in terms of the mean coefficient of heat transfer $\bar{h}$ of the form

$$q_c = (T_0 - T_\infty)p \int_0^L h_x \, dx$$

$$= \bar{h} A_s (T_0 - T_\infty) \tag{6-107}$$

This equation is recognized as the simplified form of Newton's law of cooling that was introduced in Chap. 1.

EXAMPLE 6-12

Air at 20 °C and 1 atm flows at a rate of 3 m/s over a plate of 1-m length. The plate is heated electrically, with a uniform wall flux of 1000 W/m². Determine the approximate temperature along the plate.

Solution

The local surface temperature T_s of the plate is given by Eq. (6-105),

$$T_s = T_\infty + \frac{q_c''}{k} \frac{x}{\mathrm{Nu}_x}$$

where Nu_x is dependent on location x and on whether the flow is laminar or turbulent. To see whether the flow becomes turbulent, we calculate Re_L.

$$\mathrm{Re}_L = \frac{L U_\infty}{\nu} = \frac{(1 \text{ m})(3 \text{ m/s})}{15.7 \times 10^{-6} \text{ m}^2/\text{s}} = 1.91 \times 10^5$$

Because Re_L is only slightly greater than 10^5, the flow may or may not be turbulent.

Assuming that viscous dissipation effects are secondary for this problem, Nu_x is approximated by Eq. (6-93) for laminar flow, such that T_s becomes

$$T_s = T_\infty + \frac{q_c'' \nu}{0.417 k U_\infty} \frac{\mathrm{Re}_x^{0.5}}{\mathrm{Pr}^{1/3}}$$

$$= 20 \,°\text{C} + \frac{(1000 \text{ W/m}^2)(15.7 \times 10^{-6} \text{ m}^2/\text{s})}{(0.417)[0.0262 \text{ W/(m °C)}](3 \text{ m/s})(0.71)^{1/3}} \mathrm{Re}_x^{0.5}$$

$$= 20 \,°\text{C} + 0.537 \, \mathrm{Re}_x^{0.5} \tag{a}$$

This equation is shown in Fig. E6-12. Notice that T_s increases with x by the factor $x^{0.5}$.

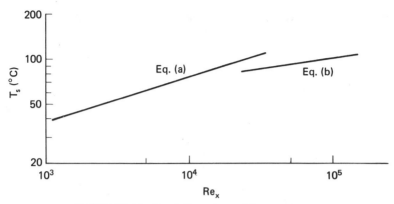

FIGURE E6-12 Predictions for wall temperature.

If the flow is tripped such that turbulence occurs over most of the plate, Nu_x can be approximated by Eq. (6-95). For this situation T_s is given by

$$T_s = T_\infty + \frac{q_c''}{0.0295} \frac{\nu}{kU_\infty} \frac{Re_x^{0.2}}{Pr^{0.5}}$$

$$= 20\,°C + 7.98\ Re_x^{0.2}\ °C \qquad (b)$$

This equation is also shown in Fig. E6-12. We observe that the plate is much cooler if turbulent flow can be induced.

6-3-3 Flow Normal to Cylinders and Spheres

Flow across cylinders and spheres is complicated by the fact that both the velocity at the outer edge of the boundary layer and the pressure drop are functions of the location on the body. Under certain conditions, a reversal of the flow occurs very near the wall on the back side of such bodies as is shown in Fig. 6-17. This separated flow phenomenon further complicates the analysis of this type of problem.

Although a local Reynolds number is sometimes written for flow normal to a cylinder or sphere in terms of the distance along the curved

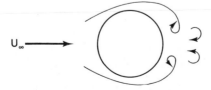

FIGURE 6-17 Sketch of crossflow over a circular cylinder.

surface, a more convenient overall Reynolds number is given by

$$Re = \frac{U_\infty D_N}{\nu} \tag{6-108}$$

where the characteristic length D_N is the height of the body as seen by the onrushing fluid. For example, for circular cylinders and spheres, D_N is simply equal to the diameter.

Coefficients of drag and heat transfer

The net force acting on a body in crossflow is the result of both the total wall shear stress and overall pressure drop. Because it is this drag force F_D that is generally needed in design, a drag coefficient is defined as

$$F_D = \frac{\rho U_\infty^2}{2} C_D A_N \tag{6-109}$$

where A_N is the area of the body normal to the direction of flow. Figure 6-18 shows the coefficients of drag for flow across cylinders and spheres.

Although empirical correlations are available for the local coefficient of heat transfer h_θ for flow normal to cylinders and spheres, practical

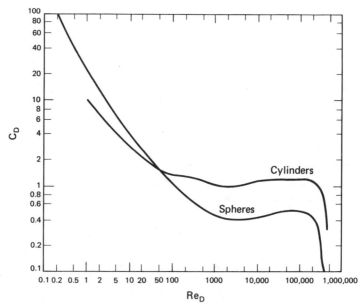

FIGURE 6-18 Representative drag coefficients for circular cylinders and spheres in crossflow.

considerations have resulted in the use of a mean coefficient $\bar{h}$ which is defined by

$$\bar{h} = \frac{1}{2\pi} \int_0^{2\pi} h_\theta \, d\theta \tag{6-110}$$

Accordingly, for standard uniform wall-temperature heating, the rate of heat transfer q_c is given by the simple Newton law of cooling,

$$q_c = \bar{h} A_s (T_s - T_\infty) \tag{6-111}$$

Numerous empirical correlations are available in the literature for the mean Nusselt number $\overline{\text{Nu}}$ for flow across cylinders. One of the more recently developed correlations for gas or liquid flow over a circular cylinder with uniform wall temperature is given by [40]

$$\overline{\text{Nu}} = (0.4 \, \text{Re}^{0.5} + 0.06 \, \text{Re}^{2/3}) \text{Pr}^{0.4} \left(\frac{\mu_\infty}{\mu_0} \right)^{0.25} \tag{6-112}$$

where $\overline{\text{Nu}} = \bar{h} D / k$, $10 \gtrsim \text{Re} \gtrsim 10^5$, $0.67 \gtrsim \text{Pr} \gtrsim 300$, and $0.25 \gtrsim \mu_\infty / \mu_0 \gtrsim 5.2$. This correlation has been reported by Whitaker [41] to lie within $\pm 25\%$ of the experimental data.

A similar correlation has been recommended by Whitaker [40] for flow over a sphere, which takes the form

$$\overline{\text{Nu}} = 2 + (0.4 \, \text{Re}^{0.5} + 0.06 \, \text{Re}^{2/3}) \text{Pr}^{0.4} \left(\frac{\mu_\infty}{\mu_0} \right)^{0.25} \tag{6-113}$$

where $3.5 \gtrsim \text{Re} \gtrsim 7.6 \times 10^4$, $0.71 \gtrsim \text{Pr} \gtrsim 380$, and $1 \gtrsim \mu_\infty / \mu_0 \gtrsim 3.2$. For higher Reynolds numbers, the following correlation by Achenbach [42] is recommended for gases:

$$\text{Nu} = 430 + a \, \text{Re} + b \, \text{Re}^2 + c \, \text{Re}^3 \tag{6-114}$$

where $a = 5 \times 10^{-3}$, $b = 0.025 \times 10^{-9}$, $c = -3.1 \times 10^{-17}$, $4 \times 10^5 \gtrsim \text{Re} \gtrsim 5 \times 10^6$, and $\text{Pr} \simeq 0.71$.

Correlations are available for crossflow over noncircular and nonspherical bodies in [6], [7], and [43]. Furthermore, design correlations are available for finned tubes which are widely employed in industry.

Lumped analyses

The use of Eqs. (6-109) and (6-111) directly give the total drag and rate of convection heat transfer from bodies with uniform surface temperature in crossflow.

EXAMPLE 6-13

Air at atmospheric pressure and 35 °C flows across a 1-cm-diameter wire at a velocity of 100 m/s. 2000 W/m is generated in the wire per unit length of the wire by electric current. Estimate the mean temperature difference between the wire and the air.

Solution

Although the heat-transfer correlations found in the literature were developed for uniform wall-temperature heating, these correlations can be used to obtain an approximate solution for the surface temperature for this uniform heat-flux condition. As a first estimate, we evaluate the fluid properties at a film temperature T_f of 350 K; that is, $\rho = 0.998$ kg/m³, $\mu = 2.08 \times 10^{-5}$ kg/(m s), $k = 0.03$ W/(m °C), and Pr $= 0.697$. Calculating the Reynolds number, we have

$$\text{Re} = \frac{U_\infty D}{\nu} = \frac{(100 \text{ m/s})(0.01 \text{ m})(0.998 \text{ kg/m}^3)}{2.08 \times 10^{-5} \text{ kg/(m s)}}$$

$$= 4.8 \times 10^4$$

To obtain the mean Nusselt number, we use Eq. (6-112),

$$\overline{\text{Nu}} = (0.4 \text{ Re}^{0.5} + 0.06 \text{ Re}^{2/3})\text{Pr}^{0.4}\left(\frac{\mu_\infty}{\mu_0}\right)^{0.25} \qquad (6\text{-}112)$$

Approximating μ_∞ / μ_0 by unity, we obtain

$$\overline{\text{Nu}} = 144$$

or

$$\bar{h} = \frac{k}{D}(144) = \frac{[0.03 \text{ W/(m °C)}](144)}{0.01 \text{ m}}$$

$$= 432 \frac{\text{W}}{\text{m}^2 \text{ °C}}$$

An estimate can now be made for the surface temperature by using the Newton law of cooling.

$$T_0 - T_\infty = \frac{q_c}{\bar{h} A_s} = \frac{q_c}{L} \frac{1}{\bar{h} p}$$

$$= \frac{2000 \text{ W/m}}{[432 \text{ W/(m}^2 \text{ °C})] \pi (0.01 \text{ m})}$$

$$T_0 = 147 \text{ °C} + 35 \text{ °C} = 182 \text{ °C}$$

To refine our prediction for wall temperature, μ_0 can be evaluated with T_0 approximated by 182°C and the other fluid properties can be reevaluated at a film temperature of 381 K (see Prob. 6-68). (Higher-order refinements in the property values are not justified because of the approximate nature of the correlation used for the Nusselt number.

6-4 FLOW ACROSS TUBE BANKS

6-4-1 Introduction

As we shall see in Chap. 9, flow across tube bundles is extremely important in heat-exchanger applications. The flow and heat-transfer characteristics are dependent upon the number of tube rows crossed, arrangement of the tubes in the bank, the relative spacing, the flow rates, directions of flow, and the fluid. Two standard arrangements for direct crossflow are shown in Fig. 6-19(a) and (b).

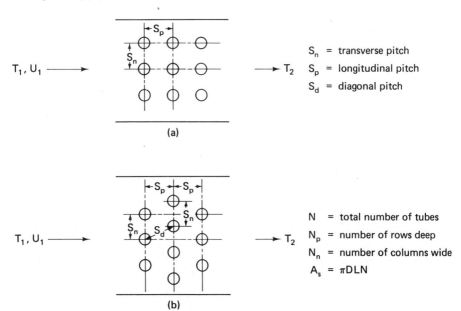

S_n = transverse pitch
S_p = longitudinal pitch
S_d = diagonal pitch

N = total number of tubes
N_p = number of rows deep
N_n = number of columns wide
A_s = πDLN

FIGURE 6-19 Tube bank arrangements. (a) In line. (b) Staggered.

Figure 6-20 shows representative flow patterns for water flowing across tubes arranged in a staggered array. The flow is seen to be confined by this geometry and should be classified as internal flow. As in the case of internal flow in tubes, annuli, and channels and between parallel plates, the bulk stream velocity U_b and the bulk stream temperature T_b are the

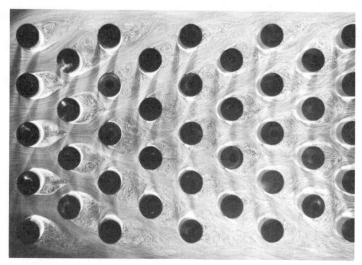

FIGURE 6-20 Flow patterns for flow over tube bundle. (Courtesy of H. Werle, ONERA, Paris.)

key velocity and temperature for characterizing flow across tube banks; recall that U_b is defined in terms of the mass flow rate $\dot{m}$ by Eq. (6-2),

$$\dot{m} = \rho A U_b \tag{6-2}$$

and T_b is defined by Eq. (6-11) in terms of the rate of energy carried by the fluid (enthalpy),

$$\dot{H}_b = \dot{m} c_p T_b \tag{6-11}$$

For this particular application, the cross-sectional area A varies with x, such that U_b is also a function of x. To overcome this complexity, the characteristic reference velocity U_F which is utilized for flow over tube banks is often set equal to the bulk stream velocity $U_{b,\,max}$ that occurs at the minimum free area A_{min} available for fluid flow. Referring to Fig. 6-19, $A_{min} = (S_n - D)L$ for in-line tube systems. For staggered arrangements, the minimum free-flow area may occur between adjacent tubes in a row or between diagonally opposed tubes. Hence, A_{min} is the smaller of the two values $(S_n - D)L$ and $(S_d - D)L$. The Reynolds number based on $U_{b,\,max}$ for flow over tube banks is defined as

$$\mathrm{Re}_{max} = \frac{G_{max}D}{\mu} = \frac{U_{b,\,max}D}{\nu} \tag{6-115}$$

where D is the diameter of an individual tube, and the maximum mass velocity G_{max} is given by

$$G_{\text{max}} = \frac{\dot{m}}{A_{\text{min}}} = \rho U_{b,\text{ max}} \qquad (6\text{-}116)$$

Based on this definition for the Reynolds number, the flow has been reported to be laminar for Re_{max} less than a value of the order of 200, and the flow is transitional turbulent for $200 \gtrsim \text{Re}_{\text{max}} \gtrsim 6000$.

A Reynolds number Re_1 based on the bulk stream velocity U_1 of the entering fluid is also sometimes used; that is,

$$\text{Re}_1 = \frac{U_1 D}{\nu} = \frac{G_1 D}{\mu} \qquad (6\text{-}117)$$

6-4-2 Correlations for Pressure Drop and Heat Transfer

With the characteristic reference velocity taken as U_1, the pressure drop in tube banks is given by

$$\Delta P = \rho U_1^2 \frac{\bar{f}}{2} N_p = \frac{G_1^2}{\rho} \frac{\bar{f}}{2} N_p \qquad (6\text{-}118)$$

where $\bar{f}$ is a mean *coefficient of friction / drag* and N_p is the number of tube rows deep; with the mass velocity G_1 given in $\text{kg}/(\text{m}^2 \text{ s})$ and ρ in kg/m^3, ΔP is in N/m^2. Correlations have been developed for $\bar{f}$ by Zukauskas [44], the Babcock and Wilcox Co. [45], and others. For example, the Babcock and Wilcox (B & W) $\bar{f}$ correlation for crossflow of gas over in-line tube bank arrangements is given in terms of Re_1 in Fig. 6-21(a). For tube banks with less than 10 rows deep, B & W recommended the use of the *depth-factor correction* F_d given in Fig. 6-21(b), where

$$\bar{f} = F_d \bar{f}|_{10 \text{ rows}} \qquad (6\text{-}119)$$

Useful correlations have also been developed for $\bar{f}$ in terms of the alternative Reynolds number Re_{max} by Jacob [46], Bergelin et al. [47], and others.

Although some data are available for local coefficients of heat transfer for tube bank flow, because of the complexity of these types of systems, heat-transfer correlations have been developed directly for uniform wall temperature heating in terms of a mean coefficient $\bar{h}$ and the log mean temperature difference LMTD by

$$q_c = \bar{h} A_s \text{ LMTD} = \bar{h} A_s \frac{T_2 - T_1}{\ln \dfrac{T_0 - T_1}{T_0 - T_2}} \qquad (6\text{-}120)$$

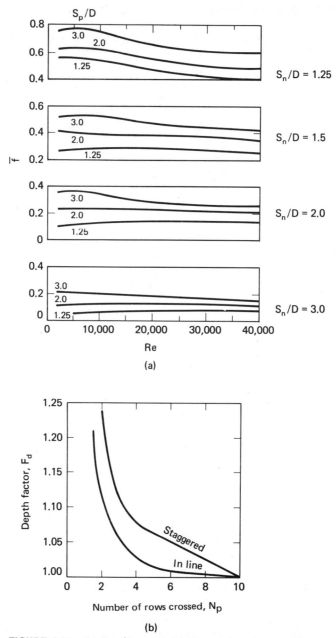

FIGURE 6-21 Friction/drag correlations for crossflow of gas over in-line tube bank. (a) Coefficient of friction/drag for 10 or more rows. (b) Depth factor for less than 10 rows. (From *Steam—Its Generation and Use*. Copyright 1975 by Babcock and Wilcox Company. Used with permission.)

Empirical correlations have been developed for $\bar{h}$ by Bergelin et al. [47], Kays and Lo [48], Grimson [49], Whitaker [40], B & W [45], and others. For example, the B & W heat-transfer correlation for crossflow of gas over tube banks is given by

$$\overline{Nu} = 0.287 \, Re_{max}^{0.61} \, Pr^{1/3} \, F_a \qquad (6\text{-}121)$$

where F_a is the *arrangement factor*. A B & W correlation for F_a is given in Fig. 6-22(a) for several in-line tube bank systems. For tube banks with less than 10 rows deep, the *heat-transfer depth factor F_h* shown in Fig. 6-22(b) is used. F_h is defined by

$$\overline{Nu} = F_h \, \overline{Nu}\big|_{10 \text{ rows}} \qquad (6\text{-}122)$$

In addition, data have been taken for the heat transfer associated with tube banks for cases in which the direction of flow ranged between zero to 90 degrees from the tube axis. The maximum rates of heat transfer were found to occur for pure crossflow. As a result, baffles are generally utilized in heat-exchanger applications involving tube banks in order to achieve better crossflow patterns. For situations in which direct crossflow is not actually achieved, the coefficient of heat transfer can be as much as 50 to 60% below the value for ideal crossflow. Design correlations are available from the various manufacturers for finned tube banks.

It should be mentioned that most of the heat-transfer correlations which are available for tube-bank flow have been obtained for essentially uniform wall-temperature conditions. Nevertheless, these correlations are generally employed for other types of boundary conditions, with the accuracy expected to be best for fully turbulent flow.

For large temperature differences, correction factors of the type introduced in Sec. 6-2-4 are often utilized to account for the effects of variable properties. For example, Whitaker [40] recommends a correction factor for Nusselt number of the form

$$\frac{Nu}{Nu_{cp}} = \left(\frac{\mu_b}{\mu_s}\right)^{0.14} \qquad (6\text{-}123)$$

6-4-3 Lumped Analyses

The development of predictions for the pressure drop for crossflow over tube banks merely requires the use of Eq. (6-118).

Referring to Fig. 6-19, lumped analyses are now developed for the heat transfer for situations in which uniform wall-heat-flux and uniform wall-temperature boundary conditions are maintained at the surfaces of the tubes.

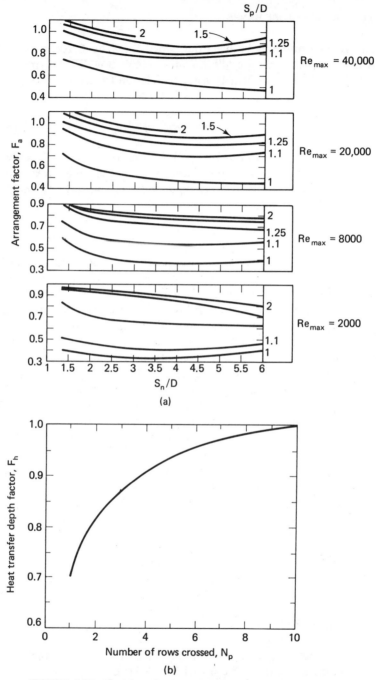

FIGURE 6-22 Heat-transfer correlations for crossflow of gas over in-line tube bank. (a) Arrangement factor for 10 or more rows. (b) Heat-transfer depth factor for less than 10 rows. (From *Steam—Its Generation and Use*. Copyright 1975 by Babcock and Wilcox Company. Used with permission.)

Uniform wall heat flux

The total rate of heat transfer for a uniform wall-heat-flux condition is simply given by

$$q_c = \int_{A_s} q_0'' \, dA_s = q_0'' A_s \tag{6-124}$$

where A_s is the total surface area of all the tubes in the bank.

The outlet temperature T_2 is obtained by performing an energy balance on the entire system (see Fig. 6-19), with the result (for a single-phase fluid)

$$T_2 - T_1 = \frac{q_c}{\dot{m}c_p} = \frac{q_0'' A_s}{\dot{m}c_p} \tag{6-125}$$

where $A_s = \pi DLN$ and $\dot{m} = \rho U_1 [S_n(N_n - 1) + 2\delta]$.

As illustrated in Example 6-15, even though the standard correlations for $\bar{h}$ were developed for uniform wall-temperature heating, these correlations can be used to estimate the mean surface temperature of the rods.

Uniform wall temperature

By definition, q_c is expressed directly in terms of $\bar{h}$ by Eq. (6-120) for uniform wall-temperature heating. However, the outlet temperature T_2 is also unknown. To obtain a second equation involving q_c and T_2, we write an energy balance for the entire system. This step gives

$$q_c = \dot{m}c_p(T_2 - T_1) \tag{6-126}$$

Eliminating q_c between Eqs. (6-120) and (6-126), it follows that

$$\frac{\bar{h}A_s}{\ln \dfrac{T_0 - T_1}{T_0 - T_2}} = \dot{m}c_p \tag{6-127}$$

Solving for T_2, we have

$$\frac{T_2 - T_0}{T_1 - T_0} = \exp\left(-\frac{\bar{h}A_s}{\dot{m}c_p}\right) \tag{6-128}$$

Finally, an expression is written for q_c of the form

$$q_c = \dot{m}c_p(T_2 - T_1)$$

$$= \dot{m}c_p(T_0 - T_1)\left[1 - \exp\left(-\frac{\bar{h}A_s}{\dot{m}c_p}\right)\right] \tag{6-129}$$

Parenthetically, Eqs. (6-128) and (6-129) are identical to the results given by Eqs. (6-69) and (6-70), respectively, for internal flow with uniform wall-temperature heating. This is because of the use of the LMTD in our defining equation for $\bar{h}$ for tube bank flow.

EXAMPLE 6-14

Air with a mass velocity of 8 kg/(m² s) is to be heated in a 2.5-m-long tube bank with 50 tubes arranged in-line. The 3-cm-diameter tubes are placed 5 tubes deep with longitudinal and transverse pitches of 4.5 cm. The air enters at 1 atm and 27 °C and the tubes are maintained at 98 °C. Determine the temperature of the exiting air and the total rate of heat transfer and pressure drop per unit tube bank length.

Solution

The properties of air at the entering fluid temperature are $\rho = 1.18$ kg/m³, $c_p = 1.01$ kJ/(kg °C), $\mu = 1.85 \times 10^{-5}$ kg/(m s), $k = 0.0262$ W/(m °C), and $Pr = 0.708$.

To determine the coefficient $\bar{f}$, we first calculate the Reynolds number Re_1.

$$Re_1 = \frac{G_1 D}{\mu} = \frac{8 \text{ kg/(m}^2 \text{ s})(0.03 \text{ m})}{1.85 \times 10^{-5} \text{ kg/(m s)}}$$

$$= 1.30 \times 10^4$$

The pitch ratios are

$$\frac{S_p}{D} = \frac{4.5 \text{ cm}}{3 \text{ cm}} = 1.5 \qquad \frac{S_n}{D} = 1.5$$

Referring to Fig. 6-21(a) and (b), we have

$$\bar{f}|_{10 \text{ rows}} = 0.53 \qquad F_d = 1.02$$

such that

$$\bar{f} = (0.53)(1.02) = 0.54$$

The pressure drop is now calculated on the basis of Eq. (6-118).

$$\Delta P = \frac{G_1^2}{\rho} \frac{\bar{f}}{2} N_p = \frac{\left[8 \text{ kg/(m}^2 \text{ s})\right]^2 (0.54/2)(5)}{1.18 \text{ kg/m}^3}$$

$$= 73.2 \frac{\text{N}}{\text{m}^2} = 0.0106 \frac{\text{lb}_f}{\text{in.}^2}$$

To determine the Nusselt number, we must calculate Re_{max}. The maximum flow rate is simply equal to the entering fluid velocity U_1 times the ratio of the minimum flow area to the total frontal area, $S_n/(S_n - D)$; that is,

$$U_{b,\text{max}} = U_1 \frac{S_n}{S_n - D}$$

or

$$G_{\text{max}} = G_1 \frac{S_n}{S_n - D}$$

It follows that

$$\text{Re}_{\text{max}} = \frac{U_{\text{max}}D}{\nu} = \frac{G_{\text{max}}D}{\mu}$$

$$= \text{Re}_1 \frac{S_n}{S_n - D} = (1.30 \times 10^4)\frac{4.5}{4.5-3}$$

$$= 3.90 \times 10^4$$

Turning to Fig. 6-22(a) and (b), we find that the arrangement factor F_a is about 1.05 and the heat-transfer depth factor F_h is about 0.93. The mean Nusselt number is obtained from Eqs. (6-121) and (6-122).

$$\overline{\text{Nu}} = 0.287 \, \text{Re}_{\text{max}}^{0.61} \, \text{Pr}^{1/3} \, F_a F_h$$

$$= 0.287(3.9 \times 10^4)^{0.61}(0.708)^{1/3}(1.05)(0.93)$$

$$= 158$$

Thus, the mean coefficient of heat transfer can be calculated.

$$\bar{h} = \overline{\text{Nu}} \, \frac{k}{D} = \frac{(158)[0.0262 \, \text{W}/(\text{m} \, °\text{C})]}{0.03 \, \text{m}}$$

$$= 138 \frac{\text{W}}{\text{m}^2 \, °\text{C}}$$

Equation (6-128) is utilized to calculate the outlet temperature T_2.

$$\frac{T_2 - T_0}{T_1 - T_0} = \exp\left(-\frac{\bar{h}A_s}{\dot{m}c_p}\right) \qquad (6\text{-}128)$$

where the mass flow rate is given by

$$\dot{m} = \rho A_1 U_1 = G_1 A_1$$

A_1 is equal to $(N_n + 1)S_n L$ for a total clearance on both sides equal to S_n, and the surface area A_s is equal to πDNL. Calculating the thermal capacity $\bar{h}A_s$ and the capacity rate $\dot{m}c_p$, we have

$$\bar{h}A_s = \bar{h}\pi DNL$$

$$= \pi\left(138\frac{W}{m^2\,°C}\right)(0.03\text{ m})(50)(2.5\text{ m})$$

$$= 1.63\frac{kW}{°C}$$

$$\dot{m}c_p = G_1(N_n + 1)S_n Lc_p$$

$$= \left(8\frac{kg}{m^2\,s}\right)(11)(0.045\text{ m})(2.5\text{ m})\left(1.01\frac{kJ}{kg\,°C}\right)$$

$$= 10\frac{kW}{°C}$$

The number of thermal units is

$$\text{NTU} = \frac{\bar{h}A_s}{\dot{m}c_p} = \frac{1.63\text{ kW/°C}}{10\text{ kW/°C}} = 0.163$$

Substituting this result into Eq. (6-128), we have

$$T_2 = 98\,°C + (27°C - 98°C)\exp(-0.163)$$

$$= 37.7\,°C$$

Finally, Eq. (6-129) is used to calculate the total rate of heat transfer per unit length L.

$$q_c = \dot{m}c_p(T_0 - T_1)\left[1 - \exp\left(-\frac{\bar{h}A_s}{\dot{m}c_p}\right)\right]$$

$$= \left(10\frac{kW}{°C}\right)(98\,°C - 27\,°C)[1 - \exp(-0.163)]$$

$$= 107\text{ kW}$$

Now that an estimate is available for T_2, the analysis can be refined by evaluating the properties at the arithmetic average of the inlet and outlet temperatures. In addition, the correction factor given by Eq. (6-123) can be utilized to improve the estimate for $\bar{h}$. These refinements are considered in Prob. 6-68.

EXAMPLE 6-15

The tubes in the bank described in Example 6-14 are to be heated electrically with a resultant uniform flux of 7.5 kW/m². Utilize a uniform property analysis to estimate the outlet air temperature and the approximate surface temperature of the tubes at the outlet. The inlet temperature is 27°C and the pressure is 1 atm.

Solution

The total rate of heat transfer to the air is

$$q_c = q_c'' A_s = 7.5 \frac{kW}{m^2} \pi DNL$$

$$= 7.5 \frac{kW}{m^2} (11.8 \text{ m}^2) = 88.5 \text{ kW}$$

Performing an energy balance on the system and assuming uniform properties, we obtain the outlet air temperature.

$$q_c = \dot{m} c_p (T_2 - T_1)$$

$$T_2 = \frac{88.5 \text{ kW}}{10 \text{ kW/°C}} + 27°C = 35.8°C$$

To estimate the average tube surface temperature, we utilize the general Newton law of cooling,

$$q_c'' = h(T_s - T_b)$$

with h assumed to be approximately equal to the value of $\bar{h}$ obtained in Example 6-14; that is,

$$T_s = \frac{q_c''}{h} + T_b$$

$$T_{s2} = \frac{7.5 \text{ kW/m}^2}{0.138 \text{ kW/(m}^2 \text{ °C)}} + 35.8°C$$

$$= 90.1°C$$

PROBLEMS

6-1. Utilize the dimensional analysis approach to determine the pertinent dimensionless groups for TFD heat transfer in a long tube.

6-2. The Filonenko equation for f, Eq. (6-32), is sometimes written in the form $4f = (1.82 \log \text{Re} - 1.64)^{-2}$. Demonstrate that these two equations are equivalent.

6-3.   The Prandtl/Nikuradse equation for f, Eq. (6-31), is implicit. Use this equation to develop an explicit relationship for Re in terms of f.

6-4.   Show that Eq. (6-31) can be written as

$$\sqrt{\frac{1}{4f}} = 2 \log\left(\text{Re}\sqrt{4f}\right) - 0.8$$

6-5.   Compare Eqs. (6-31) and (6-32) by plotting these equations on log–log paper. Use the explicit form of Eq. (6-31) developed in Prob. 6-3.

6-6.   Air enters a square channel with cross-sectional area of 100 cm² at a velocity of 15 m/s and a temperature of 27 °C. Determine the friction factor and coefficient of heat transfer for fully developed conditions.

6-7.   Solve Prob. 6-6 for the case in which the entering velocity is only 0.05 m/s.

6-8.   Determine the friction factor and the coefficient of heat transfer for fully developed annular flow of water for radii of 1 cm and 5 cm. The liquid enters at a velocity of 10 cm/s and a temperature of 60 °C. The outer surface is insulated and a uniform heat flux is maintained at the inner surface.

6-9.   Solve Prob. 6-8 for the case in which U_b is equal to 0.1 cm/s.

6-10.   Develop an expression for Nu for combined hydrodynamic and thermal developing flow in a circular tube with uniform wall-temperature heating which is consistent with the correlation for $\overline{\text{Nu}}$ given by Eq. (6-30).

6-11.   The effect of property variation on the velocity distribution for laminar flow of gas in a hot tube is shown in Example 6-4. Sketch in a representative velocity distribution for the case in which a gas is cooled.

6-12.   Estimate the friction factor and the coefficient of heat transfer for Prob. 6-6 for the case in which the outlet air temperature is 77 °C and the wall temperature is 127 °C.

6-13.   Estimate the friction factor and the coefficient of heat transfer for Prob. 6-8 if the outlet water temperature is 15 °C and the wall temperature is 5 °C.

6-14.   Solve Example 6-3 for the case in which the wall temperature is 227 °C.

6-15.   Determine the mean coefficient of heat transfer for HFD flow of water in a 10-cm-diameter tube with a short heating section of only 1-m length. The water enters at a velocity of 0.10 m/s and a temperature of 300 K.

6-16.   Compare the correlations for Nusselt number given by Eqs. (6-33) and (6-35); let f be given by (a) Eq. (6-32) and (b) Eq. (6-34). Make your comparison by plotting these two equations on a log–log graph in terms of Nu vs. Pr for Re$= 10^4$ and 10^5 over the range of $0.5 \lesssim \text{Pr} \lesssim 10^3$.

6-17.   Oil at 50 °C flowing at a rate of 0.25 kg/s enters a 1-cm-diameter tube of 1-m length. The mean wall temperature is 100 °C and the oil exits at 70 °C. Determine the mean coefficient of heat transfer over the entire length.

6-18.   Solve Example 6-2 for a tube with relative roughness of 0.01.

6-19. Show that Newton's second law of motion reduces to the following form for HFD steady flow:

$$\Sigma F_x = 0$$

6-20. For HFD steady flow in a tube, the shear stress is given by Eq. (6-46). Show that τ_s is given by

$$\tau_s = -\frac{A}{p}\left(\frac{\partial P}{\partial x} + \rho \frac{dU_b}{dt}\right)$$

for HFD unsteady flow.

6-21. Water at 15 °C is pumped by a pulsatile flow pump in a long 0.5-mm-diameter tube at 0.1 cycle/s with a bulk flow rate given by $U_b(t) = 20[1 + 0.25 \sin(\omega t)]$ m/s. At such low frequencies $[D_H^2 \omega/(4\nu) \gtrsim 0.1]$, the flow is classified as quasi-steady and the instantaneous friction factor can be approximated by $f(t) = 16/\text{Re}(t)$ for laminar conditions and $f(t) = 0.046[\text{Re}(t)]^{-0.2}$ for turbulent conditions. Show that this quasi-steady criterion is satisfied and develop predictions for the instantaneous pressure drop. Plot your results for $U_b(t)$, $f(t)$ and $dP(t)/dx$ over a full cycle.

6-22. Determine the effect of gravity on the pressure drop for both upward and downward HFD flow in a vertical tube.

6-23. Determine the effect of gravity on the pressure drop in fully developed upward flow of (a) air, (b) water, and (c) liquid mercury. The fluid temperature is 27°C and $\text{Re} = 10^5$.

6-24. Water at 10 °C enters an electrically heated 2-cm-diameter tube of 1 m length. The bulk flow rate is 100 cm/s. Determine the bulk stream temperature T_b, outlet temperature T_2, and wall temperature for uniform wall flux heating with $q_0'' = 100$ kW/m^2.

6-25. Solve Prob. 6-24 for nonuniform wall flux heating with $q_s'' = 100 \sin(\pi x/L)$ kW/m^2.

6-26. Referring to Example 6-2, determine the outlet temperature T_2 for a length of 5 m for both flow rates, assuming uniform wall flux heating with $q_0'' = 1$ kW/m^2.

6-27. Water at 50 °F is heated in a 4-ft-long annulus with 1-in. and 2-in. radii. A uniform wall flux of 10,000 Btu/(h ft^2) is maintained along the inner surface and the outer surface is insulated. Determine the outlet temperatures T_2 for bulk flow rates of 0.1 ft/s and 1 ft/s. Also determine the temperatures of the heating surface at the outlet.

6-28. Air at 15 °C is heated in a 10-m-long 10-cm-square duct with uniform wall flux equal to 100 W/m^2. Determine the outlet temperature T_2 for a flow rate of 20 m/s.

6-29. Solve Prob. 6-28 for a flow rate of 0.1 m/s and for the case in which only one side is uniformly heated and the other sides are insulated.

6-30. Air at 81 °F enters a 4-in.-diameter tube with a mass flow rate of 1 kg/s. Determine the length of tube required to bring the air to a temperature of 90 °F for nonuniform wall flux heating of $20x/D$ Btu/(h ft^2).

6-31. Referring to the tables in Appendix C, we observe that the variation in specific heat over moderate temperature ranges is small for most common fluids. However, c_p for liquid CO_2 varies from 1.84 kJ/(kg °C) at -50 °C to 36.4 kJ/(kg °C) at 30 °C. Obviously, a uniform-property analysis for heat transfer to liquid carbon dioxide would be highly unreliable for even moderate temperature differences. To illustrate this point, consider the flow of liquid CO_2 in a 2-cm-diameter tube of 1-m length. The mass flow rate is 0.01 kg/s and the inlet temperature is -20 °C. (a) Determine the uniform wall flux q_0'' required to produce an outlet temperature of 20 °C. (b) Compare your solution with the results of a uniform property analysis.

6-32. Referring to Prob. 6-31, develop an analysis for the bulk stream temperature that accounts for property variation. Compare this result with predictions for T_b obtained from a uniform property analysis.

6-33. Engine oil at 0 °C is heated in a 3-cm-diameter tube of 2-m length. Assuming uniform wall flux heating of 2.4 kW/m^2, determine the outlet temperature T_2 if the mass flow rate is 0.1 kg/min. Account for the variation of properties in your analysis and compare your result with a constant property analysis.

6-34. Refine the calculation for T_s in Example 6-7 by correcting for the effect of property variation on h.

6-35. Air at 27 °C and 2 atm flows in a 1-m-long 2-m-diameter thin-wall tube. The Reynolds number is 1.27×10^4 and the surface temperature is initially 27 °C. An electrical switch is then thrown which initiates a uniform wall flux of 1600 W/m^2. First determine the steady state outlet temperature. Then develop an approximate lumped formulation for the unsteady part of the process.

6-36. Develop a solution to the unsteady problem of Prob. 6-35 by the Laplace transform method which is introduced in [50].

6-37. Develop an approximate solution to the unsteady problem of Prob. 6-35 by utilizing a lumped/differential volume that moves at a velocity of U_b with the flow stream.

6-38. Air at 27 °C enters a 10-cm-diameter circular tube of 10-m length with surface at 100 °C. Determine the outlet temperature T_2 and the total rate of heat transfer for a bulk flow rate of 0.2 m/s.

6-39. Refine the solution to Example 6-9 by evaluating the properties at the arithmetic average of the inlet and outlet temperatures.

6-40. Water at 10 °C enters a 2-cm-diameter 10-m-long tube. The bulk flow rate is 100 cm/s. Determine the bulk stream temperature T_b, the outlet temperature T_2, and the total rate of heat transfer q_c for a uniform wall temperature of 75 °C.

6-41. Referring to Example 6-6, estimate the length of tube required to bring the fluid to 60 °F.

6-42. Water at 40 °F is heated in a 20-ft-long annulus with 1-in. and 2-in. radii. The temperature of the inner surface is 150 °F and the outer surface is insulated. Estimate the outlet temperature and total rate of heat transfer for a bulk flow rate of 4 ft/s.

6-43. Determine the length of tube for the system described in Prob. 6-40 which is required to bring the outlet temperature to 50 °C.

6-44. For situations such as found in Example 6-9 in which the temperature differences are relatively small, approximate solutions can be developed by setting the fluid temperature T_F in the one-dimensional Newton law of cooling equal to the arithmetic average of the inlet and outlet temperatures. Thus, for a uniform wall temperature condition, $T_0 - T_F$ is approximated by $T_0 - (T_1 + T_2)/2$. Develop an approximate solution to Example 6-9 by utilizing this approach.

6-45. Show that the approximate solution approach introduced in Prob. 6-44 for convection heat transfer is within 1% of the LMTD solution for $0.75 \gtrsim \Delta T_1/\Delta T_2 \gtrsim 1.5$.

6-46. Solve Example 6-10 by the approximate method introduced in Prob. 6-44.

6-47. Water at 20 °C with a mass flow rate of 10 kg/s enters a 10-cm-diameter tube with surface at 80 °C. Determine the length of tube required to bring the water to 38.2 °C. Compare this result with Example 6-9.

6-48. Air at 27 °C and 1 atm enters a 10-cm-diameter tube with a mass flow rate of 0.5 kg/s. Determine the outlet temperature of the fluid for a tube length of 2 m and a surface temperature of 100 °C.

6-49. Air at 27 °C and 1 atm enters a 10-cm-diameter tube with a mass flow rate of 0.0025 kg/s. Determine the length of tube required to bring the air to a temperature of 35 °C if the surface temperature is 75 °C.

6-50. Air at 77 °C and 1 atm enters a 10-cm-diameter tube with a surface temperature of 27 °C. Determine the length of tube required to bring the air temperature to 50 °C if the mass flow rate is 5 kg/s.

6-51. A convenient expression has been developed for the total rate of heat transfer q_c over the length of a tube with uniform wall temperature by combining Eqs. (6-62) and (6-69) [i.e., Eq. (6-70)]. On the other hand, an expression can also be developed for q_c by integrating the Newton law of cooling. Compare these two approaches. Explain why the indirect approach which leads to Eq. (6-70) is preferred.

6-52. Water at 20 °C flowing at a rate of 10 kg/s enters a 10-m-long 10-cm-diameter tube with surface temperature given by $T_s = (15 + 60x/L)$°C. Determine the outlet temperature and the overall rate of heat transfer.

6-53. Solve Prob. 6-52 for a sinusodial wall temperature given by $T_s = 60 \sin (\pi x/L)$ °C.

6-54. Engine oil at 60 °C flowing at a rate of 1 kg/s is to be heated in a 10-cm-diameter tube of 5-m length with uniform wall temperature of 140 °C. Develop a lumped variable-property analysis to obtain the local bulk stream temperature T_b and outlet temperature T_2.

6-55. Demonstrate that the LMTD equation [Eq. (6-75)] does not apply to uniform wall flux heating.

6-56. Demonstrate that the LMTD equation is in error for fluids in which the specific heat changes significantly.

6-57. Air at 1000 K flowing at a rate of 0.1 kg/s is to be cooled in a 10-cm-diameter tube with uniform wall temperature of 250 K. Develop an analysis for this problem which accounts for the variation in specific heat with temperature.

6-58. Air at 27 °C and 1 atm flows at a rate of 2 m/s over a flat plate. Determine the mean friction factor $\bar{f}$ and the mean coefficient of heat transfer $\bar{h}$ for a plate length of 25 cm. Also determine the total rate of heat transfer per unit area from the plate if the wall temperature is 100 °C.

6-59. Solve Prob. 6-58 for the case in which the free stream velocity is 20 m/s and the plate length is 2.5 m.

6-60. Air at 10 °C and 1 atm flows at a rate of 35 m/s over a flat plate. The plate is 1 m long and is maintained at 60 °C. Calculate the mean friction factor $\bar{f}$ and the mean coefficient of heat transfer $\bar{h}$. Also determine the total drag force on the plate and the total rate of heat transfer.

6-61. Air at 20 °C and 2 atm flows at a rate of 3.5 m/s over a 10-cm-long flat plate. A uniform wall heat flux of 1 kW/m² is maintained along the plate. Determine the temperature of the plate at any axial location.

6-62. Water at 10 °C flows with a free stream velocity of 1 m/s over a flat plate. Determine the total rate of heat transfer from the plate if the surface is at 80 °C.

6-63. Air at 35 °C and 1 atm flows at a velocity of 50 m/s across a 5-mm-diameter wire. The wire temperature is maintained at 100 °C. Determine the total rate of heat transfer per unit length of wire.

6-64. A 1-cm-diameter wire generates 390 W/m. Determine the flow rate of air required to maintain the wire at 100 °C if the air temperature is 27 °C.

6-65. Water at 35 °C flows at a rate of 4 m/s over a 1-cm-diameter sphere. Calculate the drag and rate of heat transfer if the surface temperature is 75 °C.

6-66. Air at 10 °C and 1 atm flows at a rate of 5 kg/s over a bank of 75 tubes. The 2.54-cm-diameter tubes are arranged in line five rows deep, with a pitch of 4 cm and a surface temperature of 65 °C. Determine the total pressure drop and rate of heat transfer per unit length and the outlet temperature.

6-67. Air at 75 °F and 1 atm flows at a rate of 1000 ft³/min over a tube bank, with 200 electrical heating rods arranged in line ten deep. The rods are 5 ft long and 1 in. in diameter, and the pitch is 1.5 in. Electrical heating produces a uniform flux from each rod of 1 kW/m². Determine the outlet temperature and estimate the surface temperature of the rods at the exit.

6-68. Refine the analysis of Example 6-13 by correcting for effects of property variation.

6-69. Refine the analysis of Example 6-14 by correcting for effects of property variation.

6-70. Refine the analysis of Example 6-15 by correcting for effects of property variation.

REFERENCES

[1] BUCKINGHAM, E., "On Physically Similar Systems; Illustrations of the Use of Dimensional Analysis," *Phys. Rev.,* **4**, 1914, 345.

[2] BRIDGEMAN, P. W., *Dimensional Analysis.* New Haven, Conn.: Yale University Press, 1931.

[3] LANGHAAR, H. L., *Dimensional Analysis and Theory of Models.* New York: John Wiley & Sons, Inc., 1951.

[4] KREITH, F., *Principles of Heat Transfer,* 3rd ed. New York: Intext Press, Inc., 1973.

[5] KARLEKAR, B. V., and R. M. DESMOND, *Engineering Heat Transfer.* St. Paul, Minn.: West Publishing Co., 1977.

[6] *Engineering Sciences Data.* London: Heat Transfer Subsciences, Technical Editing and Production Ltd., 1970.

[7] ROHSENOW, W. M., and J. P. HARTNET, *Handbook of Heat Transfer.* New York: McGraw-Hill Book Co., 1973.

[8] KAYS, W. M., *Convective Heat and Mass Transfer.* New York: McGraw-Hill Book Company, 1966.

[9] KAYS, W. M., and S. H. CLARK, TR No. 17, Department of Mechanical Engineering, Stanford University, Stanford, Calif., 1953.

[10] LANDBERG, R. E., W. C. REYNOLDS, and W. M. KAYS, *NASA TN—1972.* Washington, D.C., 1963.

[11] LANGHAAR, H. L., "Steady Flow in the Transition Length of a Straight Tube," *J. Appl. Mech.,* **9**, 1942, A55–A58.

[12] HAUSEN, H., "Darstellung des Wärmeüberganges in Rohren durch verallgemeinerte Potenzbeziehungen," *VDIZ,* **4**, 1943, 91.

[13] SELLARS, J. R., M. TRIBUS, and T. S. KLEIN, *Trans. ASME,* **78**, 1956, 441.

[14] KAYS, W. M., "Numerical Solutions for Laminar-Flow Heat Transfer in Circular Tubes," *Trans. ASME,* **77**, 1955, 1265.

[15] SEIDER, E. M., and C. E. TATE, "Heat Transfer and Pressure Drop of Liquids in Tubes," *Ind. Eng. Chem.,* **28**, 1936, 1429.

[16] NIKURADSE, J., "Wärmeübergang in Rohrleitungen" *Forsch. Arb. Ing. Wes.,* 1932, 356.

[17] WEBB, R. L., "A Critical Evaluation of Analytical Solutions and Reynolds Analogy Equations for Turbulent Heat and Mass Transfer in Smooth Tubes," *Wärme- und Stoffübertragung,* **4**, 1971. 197–204.

[18] PETUKHOV, B. S., and V. V. KIRILLOV, "Heat Exchange for Turbulent Flow of Liquid in Tubes," *Teploenerg.*, **4**, 1958.

[19] WHITE, F. M., *Viscous Fluid Flow*. New York: McGraw-Hill Book Company, 1974.

[20] PETUKHOV, B. S., "Heat Transfer and Friction in Turbulent Pipe Flow with Variable Physical Properties," in *Advances in Heat Transfer*. New York: Academic Press, Inc., 1970, 504–576.

[21] DITTUS, F. W., and L. M. K. BOELTER, "Heat Transfer in Automobile Radiators of the Tubular Type," Univ. of California-Berkeley, Pub. Eng. **2**, 1930, 443.

[22] COLBURN, A. P., "A Method of Correlating Forced Convection Heat Transfer Data and a Comparison with Fluid Friction," *Trans. AIChE*, **29**, 1933, 1974.

[23] BARNES, J. F., and J. D. JACKSON, "Heat Transfer to Air, Carbon Dioxide and Helium Flowing Through Smooth Circular Tubes under Conditions of Large Surface/Gas Temperature Ratio," *J. Mech. Eng. Sci.*, **3**, 1961, 303.

[24] DEISSLER, R. G., and C. S. EIAN, "Analytical and Experimental Investigation of Heat Transfer with Variable Fluid Properties," *NACA TN 2629*, 1952.

[25] SAMS, E. W., and L. G. DESMON, "Heat Transfer from High Temperature Surface to Fluids," *NACA Memo E9D12*, 1949.

[26] PATEL, V. C., and M. R. HEAD, "Some Observations on Skin Friction and Velocity Profiles in Fully Developed Pipe and Channel Flow," *J. Fluid Mech.*, **38**, 1969, 181.

[27] LAWN, C. J., "Turbulent Heat Transfer at Low Reynolds Numbers," *J. Heat Transfer*, **91**, 1969, 532.

[28] THOMAS, L. C., and C. R. KAKARALA, "A Unified Model for Turbulent and Laminar Momentum Transfer: Channel Flow," *J. Appl. Mech. Trans.*, **43**, 1976, 8; CANCAM, Fredericton, N. B., Canada, 1975.

[29] DEISSLER, R. G., "Analysis of Turbulent Heat Transfer and Flow in the Entrance Regions of Smooth Passages," *NACA TN 3016*, 1953, 88.

[30] MCADAMS, W. M., *Heat Transmission*, 3rd ed. New York: McGraw-Hill Book Company, 1954.

[31] SUBBOTIN, V. I., A. K. PAPOVYANTS, P. L. KIRILLOV, and N. N. IVANOVSKII, "A Study of Heat Transfer to Molten Sodium in Tubes," *Soviet J. Atomic Energy*, **13**, 1962, 380.

[32] BERGLES, A. E., "Enhancement of Heat Transfer," *Sixth Int. Heat Transfer Conf., Toronto*, **6**, 1978, 89–108.

[33] MOODY, F. F., Friction Factors for Pipe Flow, *Trans. ASME*, **6**, 1944, 671.

[34] BLASIUS, H., "Das Ähnlichkeitsgesetz bei Reibungsvorgangen in Flüssigkeiten," *Forsch. Gebiete Ingenieurw.*, **131**, 1913.

[35] SCHLICTING, H., *Boundary Layer Theory*, 6th ed. New York: McGraw-Hill Book Company, 1968.

[36] LIEPMANN, H. W., and S. DHAWAN, "Direct Measurements of Local Skin Friction in Low-Speed and High-Speed Flow," *Proc. First U.S. Nat. Cong. Appl. Mech.*, 1951.

[37] WIEGHARDT, K., and W. TILLMANN, "On the Turbulent Friction Layer for Rising Pressure," *NACA TM 1314*, 1951.

[38] ECKERT, E. R. G., and R. M. DRAKE, JR., *Analysis of Heat and Mass Transfer.* New York: McGraw-Hill Book Company, 1972.

[39] ECKERT, E. R. G., *J. Aero. Sci.,* 1955, 585–587.

[40] WHITAKER, S., *Elementary Heat Transfer Analysis.* New York: Pergamon Press, Inc., 1976.

[41] WHITAKER, S., "Forced Convection Heat Transfer Correlations for Flow in Pipes, Past Flat Plates, Single Cylinders, Single Spheres and for Flow in Packed Beds and Tube Bundles," *AIChE J.,* **18**, 1972, 361.

[42] ACHENBACH, E., "Heat Transfer from Spheres Up to $Re = 6 \times 10^6$," *Sixth Int. Heat Transfer Conf., Toronto,* **5**, 1978, 341–346.

[43] JACOB, M., *Heat Transfer*, Vol. 1. New York: John Wiley & Sons, Inc., 1949.

[44] ZUKAUSKAS, A., "Heat Transfer in Banks of Tubes in Crossflow of Fluid," *'Mintis' Vilnius,* **1968**, 124–125.

[45] *Steam—Its Generation and Use,* 38th ed. New York: The Babcock and Wilcox Company, 1975.

[46] JACOB, M., "Heat Transfer and Flow Resistance in Crossflow of Gases over Tube Banks," *Trans. ASME,* **60**, 1938, 384.

[47] BERGELIN, O. P., G. A. BROWN, and S. C. DOBERSTEIN, "Heat Transfer and Fluid Friction During Flow across Banks of Tubes," *Trans. ASME,* **74**, 1952.

[48] KAYS, W. M., and R. K. LO, "Basic Heat Transfer and Flow Friction Data for Gas Flow Normal to Banks of Staggered Tubes: Use of a Transient Technique," *Stanford Univ. Tech. Rept. 15,* 1952.

[49] GRIMISON, E. D., "Correlation and Utilization of New Data on Flow of Gases over Tube Banks," *Trans. ASME,* **59**, 1937, 538.

[50] KREYSZIG, E., *Advanced Engineering Mathematics,* 3rd ed. New York: John Wiley & Sons, Inc., 1972.

CHAPTER 7

CONVECTION HEAT TRANSFER:
Practical Thermal Analysis—
NATURAL CONVECTION

7-1 INTRODUCTION

As indicated in Chap. 5, flow caused by temperature-induced density gradients within the fluid is known as *natural convection*. The most familiar natural-convection flow fields occur as a result of the influence of gravity on fluids in which density gradients have been thermally established. For example, when a vertical cold plate is placed in warm stationary fluid, the temperature of the fluid near the wall will be decreased by conduction heat transfer. As the temperature of the fluid falls, its density will, of course, increase. This difference in density will eventually lead to the downward flow of the heavier cold fluid near the plate and upward flow of the lighter warm fluid. Similarly, the placement of a hot plate in a cool motionless fluid will result in the upward flow of the light warm fluid near the plate. Examples of natural convection in gravitational force fields include the cooling of electrical devices such as power transistors and transformers, the heating or cooling of building walls on windless days, and the heating of a pan of water.

Another very important type of natural convection flow field occurs in the presence of centrifugal forces which are also proportional to fluid density. This type of natural convection is commonly used to cool rotating components such as turbine blades.

In this chapter we deal with the practical thermal analysis of natural convection. Consequently, we will employ local and mean coefficients of heat transfer. (The theoretical analysis of a basic natural convection system will be developed in Chap. 10.) For systems in which the local coefficient of heat transfer h can be conveniently obtained, h is defined by the general Newton law of cooling,

$$q_c'' = \frac{dq_c}{dA_s} = h(T_s - T_F) \tag{7-1}$$

and $\bar{h}$ is defined in terms of h by

$$\bar{h} = \frac{1}{A_s} \int_{A_s} h \, dA_s \tag{7-2}$$

For more complex systems in which the use of a local coefficient is not practical, $\bar{h}$ is defined by

$$q_c = \bar{h} A_s \overline{(T_s - T_F)} \tag{7-3}$$

where $\overline{T_s - T_F}$ depends upon the system geometry and thermal boundary conditions. The local and mean coefficients of heat transfer for natural convection systems are generally expressed in terms of the Nusselt number. For example, for flow over a flat plate, $\mathrm{Nu}_x = h_x x / k$ and $\overline{\mathrm{Nu}} = \bar{h} L / k$.

The *Grashof number* Gr_δ and Prandtl number Pr are the key parameters for characterizing natural convection processes. The Grashof number is defined by

$$\mathrm{Gr}_\delta = \frac{g \beta \delta^3}{\nu^2} (T_s - T_F) \tag{7-4}$$

The reference fluid temperature T_F and the characteristic length δ are dependent upon the system geometry. The *coefficient of thermal expansion* β is given in Table A-C-3 for several liquids. For ideal gases β can be shown to be equal to $1/T$, where T is the absolute temperature of the gas. Because the Grashof number represents the ratio of buoyant to viscous forces, this dimensionless parameter is the primary variable in natural convection flows for determining whether the flow is laminar or turbulent. The product Gr_δ Pr is also frequently encountered in natural convection systems. This parameter is known as the *Rayleigh number* Ra_δ.

An instrument known as the *Mach–Zehnder interferometer* is often utilized to study natural convection flows. This optical instrument produces interference fringes which are the result of changes in the index of refraction that are caused by small density differences within the fluid. Consequently, lines of constant density and constant temperature can be determined by the use of this instrument. For example, Fig. 7-1 shows photographs of the fringe pattern for natural convection flow over a vertical heated flat plate. This figure clearly indicates the instantaneous flow pattern for this geometry. Because of the streamline pattern observed in Fig. 7-1(a), the flow can be assumed to be laminar over this part of the plate. On the other hand, the pattern observed in Fig. 7-1(b) is clearly nonstreamline, which is characteristic of turbulent flow.

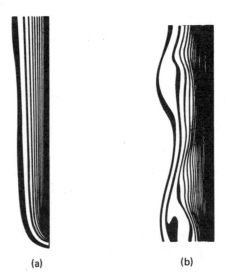

(a)   (b)

FIGURE 7-1 Interferograms for natural convection flow of air over a vertical flat plate. (Courtesy of E. R. G. Eckert and E. E. Soehngen.) (a) Laminar flow. (b) Turbulent flow.

As suggested in Chap. 5, the same classifications found in forced convection pertaining to the geometry of the fluid–solid interface, the nature of the path followed by individual elements of fluid, the type of boundary conditions, and so on, also apply to natural convection systems. Consideration is now given to the four following types of natural convection systems: external flow, internal flow, flow in enclosed spaces, and combined natural and forced convection. Lumped analyses of these types of systems will be developed in examples in each section.

7-2 EXTERNAL FLOW

Our attention is first focused on flow over a vertical flat plate, which is the classic example of natural convection heat transfer. Consideration will then be given to a general class of external flow processes, which includes cylindrical, spherical, and rectangular solid geometries.

The reference temperature T_F for external natural convection flow is equal to the free stream temperature T_∞ which is uniform. The fluid properties for these systems are generally evaluated at the film temperature T_f or free stream temperature T_∞.

7-2-1 Vertical Flow over a Flat Plate

Similar to the situation encountered in forced-convection boundary layer flow over a flat plate, hydrodynamic and thermal boundary layers develop along the wall of a vertical flat plate in a natural convection flow field. As illustrated in Fig. 7-2 for natural convection flow over a heated plate, the thicknesses of the boundary layers increase with x. The flow is always laminar toward the front of the plate, but develops into turbulence at a point downstream at which the Grashof number $\mathrm{Gr}_x[\equiv g\beta x^3(T_s - T_\infty)/\nu^2]$ is equal to about 10^9.

For natural convection flow over vertical flat plates, Eq. (7-1) applies. Thus, the rate of heat transfer q_c is given by

$$q_c = \int_{A_s} h(T_s - T_\infty)\,dA_s \qquad (7-5)$$

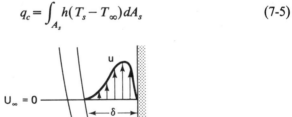

FIGURE 7-2 Representative temperature and velocity profiles for laminar natural convection flow over a heated vertical flat plate.

For uniform wall-temperature heating, we have

$$q_c = \bar{h} A_s (T_0 - T_\infty) \tag{7-6}$$

where $\bar{h}$ is defined by Eq. (7-2).

Correlations have been developed for h_x and $\bar{h}$ for natural convection flow over a vertical plate at uniform temperature of the form

$$\mathrm{Nu}_x = \frac{h_x x}{k} = 0.443 (\mathrm{Gr}_x \, \mathrm{Pr})^{1/4} = 0.443 \, \mathrm{Ra}_x^{1/4} \tag{7-7}$$

and

$$\overline{\mathrm{Nu}} = \frac{\bar{h} L}{k} = 0.59 \, \mathrm{Ra}_L^{1/4} \tag{7-8}$$

in the laminar region for which $10^4 \gtrsim \mathrm{Ra}_x \gtrsim 10^9$ [1], and

$$\overline{\mathrm{Nu}} = 0.1 \, \mathrm{Ra}_L^{1/3} \tag{7-9}$$

in the turbulent zone [2,3]. These equations are shown to be in good agreement with experimental data in Fig. 7-3.

Correlations have also been developed for natural convection from a vertical plate with uniform wall flux heating by Vliet [5], and Vliet and Liu [6]. These correlations take the form

$$\mathrm{Nu}_x = 0.528 \, \mathrm{Ra}_x^{1/4} \tag{7-10}$$

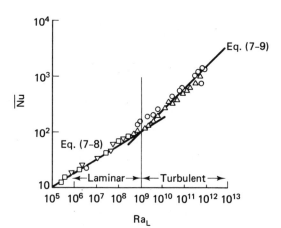

FIGURE 7-3 Experimental data and correlations for natural convection on a vertical flat plate. (From Eckert and Jackson [4].)

for laminar flow $(1.67 \times 10^4 < Ra_x < 1.05 \times 10^9)$, and

$$Nu_x = 0.484 \, Ra_x^{0.282} \tag{7-11}$$

for turbulent flow $(4.17 \times 10^{10} < Ra_x < 5.32 \times 10^{12})$.

Once the coefficients of heat transfer h_x and $\bar{h}$ are known for convection over vertical flat plates, the lumped analysis becomes identical to the analysis of forced convection flow over flat plates.

EXAMPLE 7-1

With Nu_x for laminar natural convection on a vertical flat plate given by Eq. (7-7), show that $\overline{Nu}$ is indeed given by Eq. (7-8).

Solution

Our problem is to obtain an expression for $\bar{h}$. Therefore, we first write an expression for h_x; that is,

$$h_x = 0.443 \frac{k}{x} \, Pr^{1/4} \left[\frac{g\beta x^3}{\nu^2} (T_0 - T_\infty) \right]^{1/4}$$

or

$$h_x = 0.443 k \, Pr^{1/4} \left[\frac{g\beta}{\nu^2} (T_0 - T_\infty) \right]^{1/4} x^{-1/4}$$

Utilizing the defining equation for $\bar{h}$ [Eq.(7-2)], we obtain

$$\bar{h} = 0.443 k \, Pr^{1/4} \left[\frac{g\beta}{\nu^2} (T_0 - T_\infty) \right]^{1/4} \frac{1}{L} \int_0^L x^{-1/4} \, dx$$

Because

$$\frac{1}{L} \int_0^L x^{-1/4} \, dx = \frac{4}{3} L^{-1/4}$$

we obtain

$$\frac{\bar{h} L}{k} = 0.591 \, Ra_L^{1/4}$$

which is consistent with Eq. (7-8).

EXAMPLE 7-2

The power amplifier shown in Fig. E7-2 is mounted vertically in air at 25°C. The case is made of anodized aluminum with a surface area of about 3800 mm² and a height of 40 mm. Determine the coefficient of

FIGURE E7-2   Power amplifier.

convection for natural convection cooling with a case temperature of 125°C. Also estimate the power dissipation from the unit.

Solution

Because of the small thickness of the power amplifier, we will utilize a flat-plate approximation in our analysis.

The properties of the air are evaluated at the film temperature T_f of 75°C by referring to Table A-C-5 [i.e., $\nu = 2.06 \times 10^{-5}$ m^2/s, Pr $= 0.697$, and $k = 0.0299$ W/(m °C)]. The coefficient of thermal expansion β is approximated by

$$\beta = \frac{1}{T_f} = \frac{1}{348 \text{ K}} = \frac{2.87 \times 10^{-3}}{\text{K}}$$

Next, the Grashof and Raleigh numbers are calculated for a plate height of 40 mm.

$$\text{Gr}_L = \frac{(9.81 \text{ m/s}^2)(2.87 \times 10^{-3}/\text{K})(40 \times 10^{-3} \text{ m})^3(125°C - 25°C)}{(2.06 \times 10^{-5} \text{ m}^2/\text{s})^2}$$

$$= 4.24 \times 10^5$$

$$\text{Ra}_L = \text{Gr}_L \text{ Pr} = (4.24 \times 10^5)(0.697)$$

$$= 2.95 \times 10^5$$

Because the Grashof number is well below 10^9, the flow is judged to be laminar. Based on Eq. (7-8) the mean Nusselt number is given by

$$\overline{Nu} = 0.591 \, Ra_L^{1/4} = 0.591(2.95 \times 10^5)^{1/4}$$
$$= 13.8$$

The mean coefficient of heat transfer is then

$$\bar{h} = \overline{Nu} \, \frac{k}{L} = \frac{(13.8)[0.0299 \, W/(m \, ^\circ C)]}{0.06 \, m}$$
$$= 6.88 \, \frac{W}{m^2 \, ^\circ C}$$

The total rate of heat transfer q from the surface is

$$q = q_c + q_R$$

where q_c is given by Eq. (7-6). Because the plate is made of black anodized aluminum, we will utilize the blackbody approximation for q_R given by Eq. (1-24), with T_R set equal to the ambient temperature. Substituting for q_c and q_R, we have

$$q = \bar{h} A_s (T_0 - T_\infty) + \sigma A_s F_{s-R} (T_0^4 - T_R^4)$$
$$= \left(6.88 \, \frac{W}{m^2 \, ^\circ C}\right)(3.8 \times 10^{-3} \, m^2)(125^\circ C - 25^\circ C)$$
$$+ \left(5.67 \times 10^{-8} \, \frac{W}{m^2 \, K^4}\right)(3.8 \times 10^{-3} \, m^2)(1)[(398 \, K)^4 - (298 \, K)^4]$$
$$= 2.61 \, W + 3.71 \, W = 6.32 \, W$$

The natural-convection heat transfer accounts for about 41% of the total power dissipated by the power amplifier.

EXAMPLE 7-3

A vertical plate 10 cm high and 5 cm wide is cooled by natural convection. The rate of heat transfer is 5.55 W and the air temperature is 38°C. Estimate the maximum temperature of the plate.

Solution

Because the surface temperature T_s is unknown, we will evaluate the fluid properties at 38°C; that is, $\nu = 1.67 \times 10^{-5} \, m^2/s$, Pr = 0.72, $k =$

0.0266 W/(m °C), and

$$\beta \approx \frac{1}{T_\infty} = \frac{1}{311\ \text{K}} = \frac{3.22 \times 10^{-3}}{\text{K}}$$

Notice that the Grashof number cannot be determined directly because T_s is unknown. Taking the heat flux to be essentially uniform, and assuming negligible thermal radiation, we write

$$q_c'' = h_x(T_s - T_\infty) = \frac{5.55\ \text{W}}{0.005\ \text{m}^2} = 1110 \frac{\text{W}}{\text{m}^2}$$

or

$$T_s - T_\infty = \frac{q_c''}{h_x} = \frac{q_c''}{\text{Nu}_x} \frac{x}{k}$$

Assuming that the flow is laminar, Eq. (7-10) enables us to obtain an expression for T_s of the form

$$T_s - T_\infty = q_c'' \frac{x}{k} \frac{1}{0.528} \frac{1}{\text{Ra}_x^{1/4}}$$

where

$$\text{Gr}_x = \frac{g\beta x^3}{\nu^2}(T_s - T_\infty)$$

Solving for $T_s - T_\infty$, we obtain

$$(T_s - T_\infty)^{5/4} = q_c'' \frac{x}{k} \frac{1}{0.528} \left(\frac{1}{\text{Pr}} \frac{\nu^2}{g\beta x^3} \right)^{1/4}$$

Calculating the parameter $g\beta/\nu^2$, we have

$$\frac{g\beta}{\nu^2} = \frac{(9.81\ \text{m/s}^2)(3.22 \times 10^{-3}/\text{K})}{(1.67 \times 10^{-5}\ \text{m}^2/\text{s})^2} = \frac{1.13 \times 10^8}{\text{m}^3\ \text{K}}$$

Thus, $T_s - T_\infty$ is given by

$$(T_s - T_\infty)^{5/4} = \frac{(1110\ \text{W/m}^2)x^{1/4}}{[0.0266\ \text{W/(m °C)}](0.528)\{(0.72)[1.13 \times 10^8/(\text{m}^3\ \text{K})]\}^{1/4}}$$

$$= 831 \frac{°\text{C}^{5/4}}{\text{m}^{1/4}} x^{1/4}$$

$$T_s - T_\infty = 217°\text{C} \left(\frac{x}{\text{m}} \right)^{1/5}$$

The maximum temperature is at the top of the plate ($x = 0.1$ m).

$$T_{\max} - T_\infty = 137°C$$
$$T_{\max} = 175°C$$

The local maximum Grashof number is therefore given by

$$Gr_x = \frac{g\beta}{\nu^2} x^3 (T_s - T_\infty)$$

with $x = 0.1$ m; that is,

$$Gr_{\max} = \frac{1.13 \times 10^8}{m^3\,K} (0.1\ m)^3 (137°C)$$
$$= 1.55 \times 10^7$$

Because $Gr_{\max} < 10^9$, the flow is indeed laminar.

The solution can be refined by evaluating the properties at a film temperature, with T_s equal to the temperature at $x = L/2$.

7-2-2 General External Natural Convection Flows

For the more complex external natural-convection flows that are encountered in practice, the rate of heat transfer is generally given directly by Eq. (7-3). An empirical correlation has been developed for the mean coefficient of heat transfer for external natural convection flows over various common surfaces, which takes the form

$$\overline{Nu} = A + B\,Ra_\delta^m \qquad (7\text{-}12)$$

For this general situation, $T_s - T_F$ is equal to $T_0 - T_\infty$ such that Eqs. (7-3) and (7-6) are equivalent. The coefficients A, B, and m and the characteristic length δ are given in Table 7-1 for some of the more important geometries. Applications involving these geometries include steam and water pipes, air ducts, room walls, floors and ceilings, and cryogenic containers.

A correlation has been developed for natural convection flow over an inclined flat plate with the lower surface heated of the form [12]

$$\overline{Nu} = 0.56(Ra_L \sin\theta)^{1/4} \qquad 10^5 \gtrsim Ra_L \sin\theta \gtrsim 10^{11} \qquad \pi/90 \gtrsim \theta \gtrsim \pi/2$$
$$(7\text{-}13)$$

where θ is the angle of inclination with respect to the horizontal.

TABLE 7-1   Natural convection: external flow-coefficients for Eq. (7-12)

Geometry	Ra_δ	B	m	A	References/comments
Vertical cylinder (or plate)— uniform wall temperature					[2], [3], [8]
					$\delta = L, (D/L \gtrsim 35\mathrm{Gr}_L^{-\frac{1}{4}})$
					See [8] for lower Ra_L
	10^4–10^9	0.59	$\frac{1}{4}$	0	Laminar
	10^9–10^{12}	0.10	$\frac{1}{3}$	0	Turbulent
Horizontal cylinder—uniform wall temperature					[8]
					$\delta = D$
					See [8] for lower Ra_D
	10^4–10^9	0.53	$\frac{1}{4}$	0	Laminar
	10^9–10^{12}	0.13	$\frac{1}{3}$	0	Turbulent
Horizontal plate with area A and perimeter p—uniform wall flux					[8]-[11]
Hot surface up (or cold surface down)					$\delta = L$ for square
					$\delta = (L + w)/2$ for rectangle
					$\delta = 0.9D$ for circular disk
					$\delta = A/p$ for nonsymmetrical surface.
	10^5–10^7	0.54	$\frac{1}{4}$	0	Laminar
	10^7–10^{11}	0.15	$\frac{1}{3}$	0	Turbulent
Hot surface down (or cold surface up)					
	10^5–10^{11}	0.27	$\frac{1}{4}$	0	Laminar
Rectangular solid with average horizontal dimension L_h and average vertical dimension L_v— uniform wall temperature					[12]
					$1/\delta = 1/L_h + 1/L_v$
	10^4–10^9	0.55	$\frac{1}{4}$	0	Laminar
Sphere—uniform wall temperature					[13]
					$\delta = D$
	1–10^5	0.43	$\frac{1}{4}$	2	Laminar

Further information is available in the literature on topics such as natural convection associated with rotating bodies [14]. Also, design correlations are available from the manufacturers of natural-convection finned surfaces systems. For comprehensive reviews of external natural convection flow, papers by Ostrach [15] and Gebhart [16] and [17] are recommended.

EXAMPLE 7-4

Estimate the coefficient of heat transfer for the power amplifier of Example 7-2 if it is mounted horizontally.

Solution

For a flat surface mounted horizontally, $\overline{Nu}$ is given by Eq. (7-12) and Table 7-1. Thus, for the device mounted with the hot surface up, we have

$$\overline{Nu} = 0.54 \, Ra_L^{1/4} = 17$$

and

$$\bar{h} = \frac{(17)[0.0299 \, W/(m \, °C)]}{0.06 \, m} = 8.47 \frac{W}{m^2 \, °C}$$

On the other hand, with the hot surface facing down, we obtain

$$\overline{Nu} = 0.27 \, Ra_L^{1/4} = 8.5$$

and

$$\bar{h} = \frac{(8.5)[0.0299 \, W/(m \, °C)]}{0.06 \, m} = 4.24 \frac{W}{m^2 \, °C}$$

Clearly, one should avoid mounting the power amplifier with the hot surface facing down, if at all possible.

EXAMPLE 7-5

Determine the mean coefficient of heat transfer for natural convection from the surface of the cabinet shown in Fig. E7-5. The cabinet is mounted on a vertical wall. Its surface temperature is 125°C and the ambient temperature is 25°C.

Solution

The properies of air at T_f are $\nu = 2.06 \times 10^{-5} \, m^2/s$, $Pr = 0.697$, $k = 0.0299 \, W/(m \, °C)$, and $\beta = 2.87 \times 10^{-3}/K$. Referring to Table 7-1, the

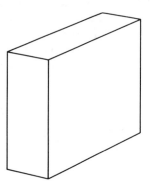

Surrounding walls
at 25°C

Cabinet dimensions
0.418 m x 0.318 m x 0.16 m

Total surface area for
heat transfer—A_s = 0.268 m^2

FIGURE E7-5   Cabinet mounted on a vertical wall.

characteristic length δ for this rectangular solid is given by

$$\frac{1}{\delta} = \frac{1}{L_h} + \frac{1}{L_v}$$

$$= \frac{1}{(0.16 \text{ m} + 0.318 \text{ m})/2} + \frac{1}{0.418 \text{ m}}$$

$$\delta = 0.152 \text{ m}$$

Calculating the Grashof and Raleigh numbers, we have

$$\text{Gr}_\delta = \frac{g\beta\delta^3}{\nu^2}(T_0 - T_F)$$

$$= \frac{(9.81 \text{ m/s}^2)(2.87 \times 10^{-3}/\text{K})(0.152 \text{ m})^3(125°\text{C} - 25°\text{C})}{(2.06 \times 10^{-5} \text{ m}^2/\text{s})^2}$$

$$= 2.33 \times 10^7$$

$$\text{Ra}_\delta = (2.33 \times 10^7)(0.697) = 1.62 \times 10^7$$

Utilizing Eq. (7-12) and the coefficients given in Table 7-1, we calculate the mean Nusselt number and the mean coefficient of heat transfer.

$$\overline{\text{Nu}} = 0.55 \text{ Ra}_\delta^{1/4} = 0.55(1.62 \times 10^7)^{1/4}$$

$$= 34.9$$

$$\bar{h} = \overline{\text{Nu}} \frac{k}{\delta} = (34.9)\frac{0.0299 \text{ W}/(\text{m °C})}{0.152 \text{ m}}$$

$$= 6.87 \frac{\text{W}}{\text{m}^2 \text{ °C}}$$

7-3 INTERNAL FLOW

Internal natural convection flows are often encountered in heat-transfer applications involving fin units, fireplaces, and solar heating units, as well as other systems. To illustrate, the schematic of a natural circulation solar water heater is shown in Fig. 7-4. With the storage tank located above the solar collector, water circulates by natural convection when solar energy is captured by the collector.

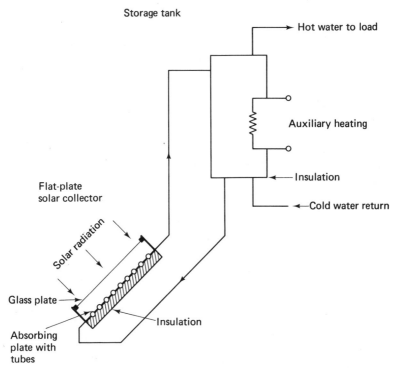

FIGURE 7-4 Schematic of a natural-convection solar heating system.

    Internal natural convection flow systems are complicated by the fact that the bulk stream temperature T_b generally varies with axial location x. The flow pattern depends upon the relative values of the wall temperatures and the temperature of entering fluid T_1. For example, for flow between parallel plates with uniform surface temperatures T_0 and T_w, fluid rises in the vicinity of the warmer wall and falls in the region of the cooler surface for $T_0 > T_1 > T_w$. On the other hand, with T_0 and T_w equal and greater than T_1, fluid rises throughout the entire system. These two situations are illustrated in Fig. 7-5(a) and (b) for laminar flow conditions. Such

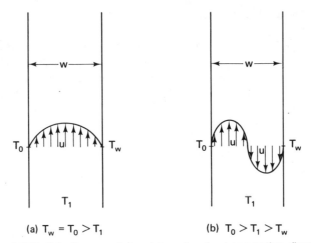

(a) $T_w = T_0 > T_1$ (b) $T_0 > T_1 > T_w$

FIGURE 7-5 Representative internal natural-convection flow patterns for parallel-plate geometry.

parallel-plate geometries are often used to approximate cooling fins in transformers, radiators, and other industrial devices, and in natural-circulation solar flat-plate collectors.

Correlations have been developed on the basis of Eq. (7-3) for internal natural convection flow with uniform wall temperatures T_0 and T_w, where[1] $\overline{T_s - T_F} = \overline{T}_s - T_1 = (T_0 + T_w)/2 - T_1$; that is,

$$q_c = \bar{h} A_s \left(\overline{T}_s - T_1 \right) \tag{7-14}$$

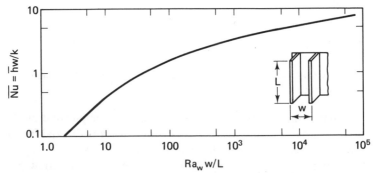

FIGURE 7-6 Natural-convection heat transfer from vertical parallel plates with uniform surface temperatures. (From Elenbaas [19].)

---

[1] Correlations are also found in the literature with $\overline{T_s - T_F}$ set equal to $T_0 - T_w$.

A $\overline{\text{Nu}}$ correlation developed by Elenbaas [19] for laminar natural convection flow between vertical parallel plates with uniform wall temperature heating ($T_0 = T_w > T_1$) is shown in Fig. 7-6. The characteristic length δ for this situation is w. Heat-transfer correlations for other cases involving internal natural convection flow are available in the literature [20]–[23]. For example, correlations have been developed for natural convection flow between inclined parallel plates by Tabor [22] and by Dropkin and Somerscales [23].

EXAMPLE 7-6

Determine the total rate of heat transfer via natural convection between vertical parallel plates which are 2 in. apart, 3 ft high, and 3 ft wide. The walls are maintained at 273°F and the air temperature is 68°F.

Solution

The total rate of heat transfer to the fluid is given by

$$q_c = \bar{h} A_s (\overline{T}_s - T_1) \tag{7-14}$$

To obtain $\bar{h}$, we must calculate Gr_w. The properties are evaluated at the average temperature $(\overline{T}_s + T_1)/2$, which is 171°F. Utilizing Table A-C-5, we have $\text{Pr} = 0.697$, $k = 0.03$ W/(m °C) = 0.0173 Btu/(h ft °F), $\nu = 20.8 \times 10^{-6}$ m²/s = 2.24×10^{-4} ft²/s; in addition, we have

$$\beta \simeq \frac{1}{631°R} = \frac{1.58 \times 10^{-3}}{°R}$$

Calculating the Grashof number, we have

$$\text{Gr}_w = \frac{g\beta w^3}{\nu^2} (\overline{T}_s - T_1)$$

$$= \frac{(32.2 \text{ ft/s}^2)(1.58 \times 10^{-3}/°R)(2/12 \text{ ft})^3}{(2.24 \times 10^{-4} \text{ ft}^2/\text{s})^2} (273°F - 68°F)$$

$$= 9.62 \times 10^5$$

It follows that

$$\text{Ra}_w \frac{w}{L} = 9.62 \times 10^5 (0.697) \left(\frac{2/12}{3}\right) = 3.72 \times 10^4$$

Referring to Fig. 7-6, we have

$$\overline{\text{Nu}} = 7$$

To calculate $\bar{h}$, we write

$$\bar{h} = \overline{\text{Nu}}\,\frac{k}{w} = \frac{7[0.0173\ \text{Btu}/(\text{h ft °F})]}{(2/12)\ \text{ft}}$$

$$= 0.727\,\frac{\text{Btu}}{\text{h ft}^2\ \text{°F}}$$

The total rate of heat transfer is

$$q_c = \left(0.727\,\frac{\text{Btu}}{\text{h ft}^2\ \text{°F}}\right)(9\ \text{ft}^2)(273\text{°F} - 68\text{°F})$$

$$= 1340\,\frac{\text{Btu}}{\text{h}} = 393\ \text{W}$$

7-4 FLOW IN ENCLOSED SPACES

A typical enclosed natural-convection flow system and flow pattern is shown in Fig. 7-7. This type of natural circulation is particularly important in the cooling of electronic devices. In addition, the development of natural-convection flow in enclosures is a factor in the use of air gaps to insulate building walls and cryogenic chambers, and in the design of solar flat-plate collectors.

Correlations have been developed for natural-convection heat transfer in enclosures with uniform wall temperatures T_0 and T_w. These correlations for $\bar{h}$ are based on Eq. (7-3) with $\overline{T_s - T_F} = T_0 - T_w$; that is,

$$q_c = \bar{h} A_s (T_0 - T_w) \tag{7-15}$$

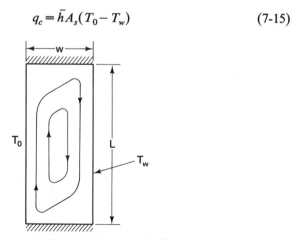

FIGURE 7-7 Representative natural-convection flow pattern for an enclosed vertical space; $\delta + w, T_0 > T_w$.

The mean coefficient $\bar{h}$ is generally given by an empirical expression of the form

$$\frac{\bar{h}\delta}{k} = \overline{\mathrm{Nu}} = C\left(\frac{L}{\delta}\right)^{m}\mathrm{Ra}_{\delta}^{n} \qquad (7\text{-}16)$$

where $\mathrm{Gr}_{\delta} = g\,\beta\delta^{3}(T_{0} - T_{w})/\nu^{2}$; δ and A_{s} are dependent upon the geometry (see Table 7-2). The properties are generally evaluated at the average temperature $(T_{0} + T_{w})/2$ for internal flows. The coefficients C, m, and n are given in Table 7-3 for parallel vertical plates and annuli (with the enclosing end surfaces insulated), and concentric spheres.

In regard to natural convection between horizontal parallel-plate enclosures, the flow patterns depend upon whether the hotter plate is on the top or the bottom. In the first case, the low-density fluid lies above the heavier fluid, such that no buoyancy effects occur. For this situation, we have pure conduction heat transfer in the fluid, such that $\overline{\mathrm{Nu}} = 1$.

TABLE 7-2 δ, A_{s}, and R_{c} for natural convection in enclosed spaces with uniform wall temperatures T_{0} and T_{w}

Geometry	δ	A_{s}	R_{c}
Parallel plane walls	w	$A_{0} = A_{w}$	$\dfrac{w}{A_{0}k\ \overline{\mathrm{Nu}}}$
Vertical concentric annuli	$r_{w} - r_{0}$	$\dfrac{A_{w} - A_{0}}{\ln(A_{w}/A_{0})}$	$\dfrac{\ln(r_{w}/r_{0})}{2\pi Lk\ \overline{\mathrm{Nu}}}$
Concentric spheres	$r_{w} - r_{o}$	$\sqrt{A_{w}A_{0}}$	$\dfrac{r_{w} - r_{0}}{4\pi r_{w}r_{0}k\ \overline{\mathrm{Nu}}}$

TABLE 7-3 Natural convection: vertical enclosed spaces—coefficients for Eq. (7-16)

Geometry	Fluid	Ra_{δ}	Pr	C	n	m	L/δ	References and comments
Vertical plates, $\delta = w$;	Gas	$<2\times10^{3}$	0.5–2	1	0	0	—	[24], [25]
		2×10^{3}–2×10^{5}	0.5–2	0.197	$\frac{1}{4}$	$-\frac{1}{9}$	11–42	[26]
Annuli,	Liquid	2×10^{5}–10^{7}	0.5–2	0.073	$\frac{1}{3}$	$-\frac{1}{9}$	11–42	
$\delta = r_{0} - r$		10^{3}–10^{7}	1–2×10^{4}	$0.42\,\mathrm{Pr}^{0.012}$	$\frac{1}{4}$	-0.3	10–40	[27], [28] Uniform
		10^{6}–10^{9}	1–20	0.046	$\frac{1}{3}$	0	1–40	q_{s}'' or T_{s}
Concentric spheres, $\delta = r_{o} - r_{i}$	Gas or liquid	10^{2}–10^{9}	0.7–4×10^{3}	0.228	0	0.226	—	[29], [30]

For the second case, the heavier fluid is on the top. For values of $Gr_\delta (= Gr_w)$ less than a critical value, the buoyancy forces are not large enough to cause the fluid to turn over. The critical value of Ra_w for gases is about 1700. For this case, we have stability with no natural convection currents. The flow pattern for values of Ra_w between 1700 and 3.2×10^5 is laminar and takes the form of cells of circulating fluid. For larger values of Ra_w, the flow becomes turbulent and the cellular pattern no longer exists. An empirical equation of the form of Eq. (7-16) with $n = 0$ is recommended for natural convection between horizontal plates heated from below; that is,

$$\overline{Nu} = C\, Ra_w^m \tag{7-17}$$

The values of C and m and the ranges in Ra_w are shown in Table 7-4.

TABLE 7-4   Natural convection: horizontal enclosed surfaces —coefficients for Eq. (7-17)

Fluid	Ra_w	Pr	C	m	References
Plates heated from below:					
$\delta = w$					[24]–[26]
Gas	$< 1.7 \times 10^3$		1	0	[31]–[33]
	$1.7 \times 10^3 – 7.0 \times 10^3$	0.5–2	0.059	0.4	
	$7.0 \times 10^3 – 3.2 \times 10^5$	0.5–2	0.212	$\frac{1}{4}$	
	$3.2 \times 10^5 <$	0.5–2	0.061	$\frac{1}{3}$	
Liquid	$< 1.7 \times 10^3$		1	0	[25],[32]–[36]
	$1.7 \times 10^3 – 6.0 \times 10^3$	1–5000	0.012	0.6	
	$6.0 \times 10^3 – 3.7 \times 10^4$	1–5000	0.375	0.2	
	$3.7 \times 10^4 – 10^8$	1–20	0.13	0.3	
	$10^8 <$	1–20	0.057	$\frac{1}{3}$	
Concentric annuli:					
Liquid	$6.0 \times 10^3 – 10^6$	1–5000	0.11	0.29	[37]–[39]
Gas	$10^6 – 10^8$	1–5000	0.40	0.20	

For natural convection in inclined enclosures such as the one shown in Fig. 7-8, studies by Hollands et al. [34], [40], and [41], Catton et al. [42], and Ayyaswamy and Catton [43] indicate that the Nusselt number correlations for vertical enclosures can be utilized, with the gravitational acceleration g replaced by its directional component along the surface of the plate, $g \sin \theta$. Utilizing this substitution, for $2 \times 10^3 < Ra_\delta < 2 \times 10^5$ we have

$$\overline{Nu} = \overline{Nu}_{\theta = \pi/2} (\sin \theta)^{1/4} \tag{7-18}$$

This equation has been reported to correlate experimental data in the region $0 \gtrsim \theta \gtrsim \pi/2$ (i.e., heating from below). For $\pi/2 < \theta < \pi$, in which the

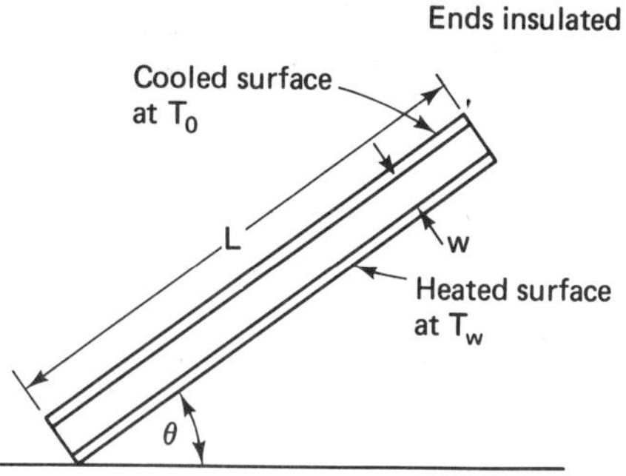

FIGURE 7-8 Natural convection in an enclosed inclined rectangular space.

upper surface is heated, Catton [44] recommends the correlation by Arnold et al. [45],

$$\mathrm{Nu} = 1 + (\mathrm{Nu}_{\theta = \pi/2} - 1)\sin\theta \qquad (7\text{-}19)$$

Incidentally, q_c is also sometimes expressed in terms of a thermal resistance for natural convection in enclosures as

$$q_c = \frac{T_0 - T_w}{R_c} \qquad (7\text{-}20)$$

where $R_c = \delta/(\overline{\mathrm{Nu}}\,kA_s)$; specific expressions for R_c are given in Table 7-2. Because of the similarity between these expressions for R_c and the expressions developed in Chap. 2 for R_k for one-dimensional conduction heat transfer, the product $\overline{\mathrm{Nu}}\,k$ is sometimes referred to as the *apparent thermal conductivity* k_e; that is,

$$k_e = \overline{\mathrm{Nu}}\,k \qquad (7\text{-}21)$$

For $\overline{\mathrm{Nu}} = 1$ or $k_e = k$, we have pure conduction heat transfer through the fluid with no natural convection effect.

For more information on the topic of natural convection in enclosures, review articles by Catton [44] and Ostrach [46] are suggested.

EXAMPLE 7-7

Two concentric blackbody spheres with 2-cm and 5-cm diameters are separated by air at 0.5 atm. The surface temperatures are equal to 154°C and 0°C. Determine the total rate of heat transfer in this system.

Solution

The total rate of heat transfer is given by

$$q = q_c + q_R$$

The rate of heat transfer by natural convection is calculated as follows:

$$q_c = \bar{h} A_s (T_0 - T_w) \tag{a}$$

where $A_s = \sqrt{A_w A_s}$ according to Table 7-2. Utilizing Eq. (7-16) and Table 7-3, $\bar{h}$ is given by

$$\bar{h} = \overline{Nu} \frac{k}{\delta} = 0.228 \, Ra_\delta^{0.226} \frac{k}{\delta} \tag{b}$$

for the range $10^2 \gtrsim Ra_\delta \gtrsim 10^9$ and $0.7 \gtrsim Pr \gtrsim 4 \times 10^3$, where

$$Gr_\delta = \frac{g\beta}{\nu^2} (r_0 - r_i)^3 (T_0 - T_w)$$

The properties evaluated at $(T_0 + T_w)/2$ ($= 77°C$) are $\nu = 20.8 \times 10^{-6}$ m²/s, $Pr = 0.697$, $k = 0.03$ W/(m °C). β is approximately

$$\beta = \frac{1}{350 \, K} = \frac{2.86 \times 10^{-3}}{K}$$

Substituting into Eq. (c), we have

$$Gr_\delta = \frac{(9.81 \text{ m/s}^2)(2.86 \times 10^{-3}/K)(0.05 \text{ m} - 0.02 \text{ m})^3(154°C - 0°C)}{(20.8 \times 10^{-6} \text{ m}^2/s)}$$

$$= 3.0 \times 10^8$$

and

$$Ra_\delta = (3.0 \times 10^8)(0.697) = 2.09 \times 10^8$$

Because Ra_δ and Pr are within the region of applicability of Eq. (b), we proceed with the calculation of $\bar{h}$.

$$\bar{h} = 0.228(2.09 \times 10^8)^{0.226} \frac{0.03 \text{ W/(m °C)}}{0.03 \text{ m}}$$

$$= 17.3 \frac{W}{m^2 \, °C}$$

The rate of convection heat transfer is then

$$q_c = 17.3 \frac{W}{m^2 \, °C} \left[4\pi(0.05)(0.02) \, m^2 \right] (154°C - 0°C)$$
$$= 33.4 \, W$$

The radiation heat transfer is given by Eq. (1-24),

$$q_R = \sigma A_s F_{s-R} \left(T_s^4 - T_R^4 \right)$$
$$= \left(5.67 \times 10^{-8} \frac{W}{m^2 \, K^4} \right) (4\pi)(0.02 \, m)^2 (1) \left[(427 \, K)^4 - (273 \, K)^4 \right]$$
$$= 7.89 \, W$$

Thus, the total heat exchange rate is

$$q = q_c + q_R = 33.4 \, W + 7.89 \, W$$
$$= 41.3 \, W$$

The natural convection accounts for about 81% of this total!

7-5   COMBINED NATURAL AND FORCED CONVECTION

Strictly speaking, natural convection occurs in any nonisothermal forced convection system in which gravitational or centrifugal force fields are present. For example, natural convection superimposed upon forced convection in vertical tubes brings about an increase in heat-transfer rate for laminar upward flow and a decrease for laminar downward flow. That is, the heat transfer is enhanced for laminar flow in vertical tubes when the buoyancy forces are in the direction of flow. On the other hand, the opposite has recently been found to be true for turbulent flow in vertical tubes. In fact, severe deteriorations in the rate of heat transfer have been reported for turbulent upward vertical flow of high-pressure supercritical fluids in nuclear reactors and fossil-fired steam generators [47] and [48].

As a guide in determining the significance of natural convection in forced flow fields, the buoyancy forces are generally small and can be neglected for situations in which $Gr_\delta \ll Re_\delta^2$. However, for cases in which Gr_δ and Re_δ^2 are of the same order of magnitude, both natural convection and forced convection are usually significant. This point is reinforced by the theoretical and experimental results shown in Fig. 7-9 for upward laminar forced convection flow over a vertical heated flat plate with uniform wall temperature. Notice that the theoretical predictions and data

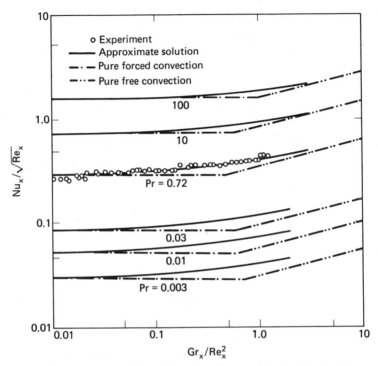

FIGURE 7-9 Local Nusselt number for combined natural and forced convection from an isothermal vertical plate. (From Lloyd and Sparrow [49].)

heat-transfer and temperature distribution for various standard systems for air approach the limiting solution for pure forced convection as the parameter Gr_x/Re_x^2 falls toward a value of the order of 0.02. The natural convection effect is the order of 10% for Gr_x/Re_x^2 equal to about 0.225. The threshold value of Gr_x/Re_x^2 for which natural convection is important is seen to be proportional to the Prandtl number Pr, such that natural convection effects would be expected to be more significant for liquid metals than for high-Prandtl-number fluids. The theoretical predictions and experimental data are also seen to approach limiting curves for pure natural convection as the parameter Gr_x/Re_x^2 approaches a value of the order of 10.

Several dimensionless number criteria have been developed to establish the limits between forced and combined convection-heat-transfer regimes. For example, Metais and Eckert [51] published preliminary limit criterion charts in 1964 for vertical and horizontal tube flow of moderate-

Prandtl-number fluids. Natural convection limit criterion have also been developed for liquid metals [52]–[54] and for supercritical fluids [55]–[57].

For information concerning Nusselt number correlations for combined and natural convection, one can refer to [50],[58]–[61].

PROBLEMS

7-1. Determine the Grashof number for natural convection flow of air at 50°C and 1 atm over a vertical flat plate 35 cm in length for the following surface temperatures: (a) −46°C; (b) 54°C. Classify the nature of the flow for each of these conditions.

7-2. Determine the Grashof number for natural convection flow of water at 50°C and 1 atm over a vertical flat plate 1.5 cm in length for the following surface temperatures: (a) −10°C; (b) 150°C. Classify the nature of the flow for each of these conditions.

7-3. Sketch representative hydrodynamic and thermal boundary layers and velocity and temperature distributions for natural convection flow over a cold vertical flat plate.

7-4. A 1-m-long flat plate is inclined at an angle of $\pi/4$ from the horizontal. The air temperature is 40°C and the lower surface of the plate is at 100°C. Determine the mean Nusselt number for natural convection flow.

7-5. A vertical surface 4 m high and 2 m wide exchanges 10 W/m² of thermal radiation with a nearby furnace. The metal is insulated on the back side and painted black so that all the net incoming radiation is transferred away by natural convection. Determine the average surface temperature of the plate if the surrounding air is at 27°C.

7-6. A 25-cm-diameter horizontal pipe is surrounded by air at 30°C. Determine the rate of natural convection heat transfer for a surface temperature of 200°C and a length of 1 m.

7-7. A 25-cm-diameter vertical pipe is surrounded by air at 30°C. Determine the rate of natural convection heat transfer for a surface temperature of 200°C and a length of 1 m.

7-8. A thin 10-cm-diameter disk is maintained at 100°C in a large container of water at 20°C. Determine the rate of heat input into the plate necessary to balance the natural convection heat transfer from the top and bottom surfaces.

7-9. Determine the rate of natural convection heat transfer from the plate of Example 7-3 if its surface is at a uniform temperature of 175°C.

7-10. Determine the rate of natural convection heat transfer from the plate of Prob. 7-9 if the fluid is water at 20 atm.

7-11. A 5-cm-diameter 2-m-long electrical cable generates 5 W. The surrounding air temperature is 50°C. Estimate the surface temperature of the cable if it is horizontal.

7-12. Solve Prob. 7-11 for a vertical orientation.

7-13. Determine the rate of natural convection heat transfer from a 3-in.-O.D. thin-walled horizontal pipe carrying steam at 212°F. The pipe is surrounded by air at 70°F.

7-14. Solve Prob. 7-13 if the surrounding fluid is water at 70°F.

7-15. Explain why the surface of a bridge always freezes faster than the surface of a road.

7-16. A cylindrical heating element 2.54 cm in diameter and 0.5 m long is placed vertically in a container of water at 20°C. The electrically generated heat transfer is 2 kW. Estimate the average surface temperature of the element.

7-17. Solve Prob. 7-16 for a horizontal orientation.

7-18. One surface of a 1-m by 1-m thin plate is maintained at 90°C. The other face is insulated. Calculate the rate of natural convection heat transfer to air at 0°C for the following arrangements: (a) vertical plate, and (b) inclined plate with angle of inclination with the horizontal equal to $\pi/4$ and with the heated surface facing down.

7-19. Solve Prob. 7-18 for (a) horizontal plate with heated surface up, and (b) horizontal plate with heated surface down.

7-20. The surfaces of a 1-m by 1-m thin plate are maintained at 0°C. Calculate the rate of natural convection heat transfer to air at 90°C if the plate is vertical.

7-21. Solve Prob. 7-20 if the plate is horizontal with both sides cooled by convection.

7-22. Determine the rate of natural convection heat transfer from two surfaces of a vertical plate with 10-cm height and 4-cm width. The surrounding air temperature is 27°C and the plate surface temperature is 150°C.

7-23. Two vertical plates with 10-cm height and 4-cm width are separated by a distance of 2 cm. The two facing surfaces are at 150°C and the air temperature is 27°C. Determine the rate of natural convection heat transfer from these two surfaces to the air. Compare this answer to the result for Prob. 7-22.

7-24. Air at atmospheric pressure is enclosed in a rectangular space which consists of two vertical plates with 10-cm height and 4-cm width, which are 2 cm apart. These two plates are at 0°C and 50°C, respectively. The four other plates that complete the enclosure are insulated. Determine the rate of heat transfer via natural convection between the two plates.

7-25. Determine the total rate of heat transfer between the two plates in Prob. 7-24 by both thermal radiation and natural convection. Assume that all the surfaces are black.

7-26. Determine the rate of heat transfer by natural convection in the enclosure of Prob. 7-24 if the 10-cm-high and 2-cm-wide end plates are maintained at 0°C and 50°C, with the other four surfaces insulated.

7-27. Determine the rate of heat transfer by natural convection in the enclosure of Prob. 7-24 if the 4-cm by 2-cm top and bottom plates are maintained at 0°C and 50°C, respectively, with the other four surfaces insulated.

7-28. Air at 1 atm is contained between two horizontal parallel plates which lie 2 cm apart. The lower plate temperature is 100°C and the upper plate is at 25°C. Determine the heat transfer flux via natural convection between these two plates. What would be the maximum thermal radiation heat flux between these plates?

7-29. A double-pane glass window consists of two 2-mm-thick plates separated by a 5-mm air gap. Determine the rate of natural convection heat transfer across the air gap if the inside surface temperatures of the glass plates are 35°C and 55°C, and the 3-m by 3-m plates are mounted vertically.

7-30. Resolve Prob. 7-29 if the temperature of the outer surfaces of the glass plates are specified as 35°C and 55°C.

7-31. A cryogenic chamber consists of a 2-m-I.D. spherical shell with a 1-cm-thick wall. An outer 2.12-m-I.D. shell surrounds the inner shell, with air at 0.1 atm in the enclosed space. Determine the rate of heat transfer by natural convection if the two surfaces in contact with the enclosed gas are at 40°C and 14°C. (Refer back to Example 1-6 for a related problem.)

7-32. A square flat-plate solar collector with a 1-m^2 surface area is inclined at an angle of 35° with the horizontal. The bottom collector plate is at a temperature of 180°C and the upper glass plate surface is at 35°C. The glass plate and blackbody collector plates are separated by a distance of 4 cm. The pressure within the enclosure is 0.2 atm. Determine the rate of heat loss across the air gap by natural convection.

7-33. Air at 35°C is forced upward along a vertical heated flat plate 1 m high by 10 m wide with a surface temperature of 150°C. Determine the free stream velocity above which the effects of natural convection can be neglected.

7-34. The coefficient of thermal expansion β is defined by $\beta = 1/V \; \partial V/\partial T|_p$. Demonstrate that β is equal to $1/T$ for ideal gases.

REFERENCES

[1] BAYLEY, F. J., J. M. OWEN, and A. B. TURNER, *Heat Transfer.*, London: Thomas Nelson and Sons Ltd., 1972.

[2] BAYLEY, F. J., "An Analysis of Turbulent Free Convection Heat Transfer," *Proc. Inst. Mech. Eng., London,* **160**, 1955, 361.

[3] WARNER, C. Y., and V. S. ARPACI, "An Investigation of Turbulent Natural Convection in Air at Low Pressure along a Vertical Heated Flat Plate," *Int. J. Heat Mass Transfer,* **11**, 1968, 397–406.

[4] ECKERT, E. R. G., and T. W. JACKSON, "Analysis of Turbulent Free Convection Boundary Layer on a Flat Plate," *NACA Rept. 1015,* 1951.

[5]   VLIET, G. C., "Natural Convection Local Heat Transfer on Constant Heat Flux Inclined Surfaces," *J. Heat Transfer*, **91**, 1969, 511.

[6]   VLIET, G. C., and C. K. LIU, "An Experimental Study of Turbulent Natural Convection Boundary Layers," *J. Heat Transfer*, **91**, 1969, 517.

[7]   ECKERT, E. R. G., and R. M. DRAKE, JR., *Analysis of Heat and Mass Transfer.* New York: McGraw Hill Book Company, 1972.

[8]   McADAMS, W. H., *Heat Transmission*, 3rd ed. New York: McGraw-Hill Book Company, 1954.

[9]   LLOYD, J. R., and W. R. MORAN, "Natural Convection Adjacent to Horizontal Surface of Various Planforms," *ASME Paper No. 74-WA/HT-66*, 1974.

[10]   FUJI, I., and H. IMURA, "Natural Convection Heat Transfer from a Plate with Arbitrary Inclination," *Int. J. Heat Mass Transfer*, **15**, 1972, 755.

[11]   SINGH, S. N., R. C. BIRKEBAK, and R. M. DRAKE, "Laminar Free Convection Heat Transfer from Downward-facing Horizontal Surfaces of Finite Dimensions," *Progr. Heat Mass Transfer*, **2**, 1969, 87.

[12]   KING, W. J., "The Basic Laws and Data of Heat Transmission," *Mech. Eng.*, **54**, 1932, 347.

[13]   YUGE, T., "Experiments on Heat Transfer from Spheres Including Combined Natural and Forced Convection," *J. Heat Transfer*, **82**, 1960, 214–220.

[14]   ANDERSON, J. T., and O. A. SAUNDERS, "Convection from an Isolated Heated Horizontal Cylinder Rotating about Its Axis," *Proc. Roy. Soc.*, **A, 217**, 1953, 555.

[15]   OSTRACH, S., in *High Speed Aerodynamics and Jet Propulsion,* Princeton, N. J.: Princeton University Press, 1964.

[16]   GEBHART, B., *Heat Transfer.* New York: McGraw-Hill Book Company, 1971.

[17]   GEBHART, B., "Natural Convection Flows and Stability," in *Advances in Heat Transfer.* New York: Academic Press, Inc., 1973.

[18]   DUFFIE, J. A., and W. A. BECKMAN, *Solar Energy Thermal Processes.* New York: John Wiley & Sons, Inc., 1974.

[19]   ELENBAAS, W., "Dissipation of Heat by Free Convection—Part II," *Philips Res. Rept.*, **3**, 1948, 450–465.

[20]   STARNER, K. E., and H. N. McMANUS, JR., "An Experimental Investigation of Free-Convection Heat Transfer from Rectangular-Fin Arrays," *J. Heat Transfer*, **84**, 1963, 273.

[21]   SOBEL, N., F. LANDIS, and W. K. MUELLER, *Proc. Third Int. Heat Transfer Conf.*, **2**, 1966, 121.

[22]   TABOR, H., "Radiation, Convection and Conduction Coefficients in Solar Collection," *Bull. Res. Council Israel*, **6C**, 1958, 155.

[23]   DROPKIN, D., and E. SOMERSCALES, "Heat Transfer by Natural Convection in Liquids Confined by Two Parallel Plates Which Are Inclined at Various Angles with Respect to the Horizontal," *J. Heat Transfer*, **87**, 1965, 77.

[24]   JAKOB, M., "Free Convection through Enclosed Plane Gas Layers," *Trans. ASME*, **68**, 1946, 189.

[25] JAKOB, M., *Heat Transfer*, vol. 1. New York: John Wiley & Sons, Inc., 1949.

[26] GRAFF, J. G. A., and E. F. M. VAN DER HELD, "The Relation between the Heat Transfer and Convection Phenomena in Enclosed Plain Air Layers," *Appl. Sci. Res.*, **3**, 1952, 393.

[27] MACGREGOR, R. K., and A. P. EMERY, "Free Convection through Vertical Plane Layers: Moderate and High Prandtl Number Fluids," *J. Heat Transfer*, **91**, 1969, 391.

[28] EMERY, A., and N. C. CHU, "Heat Transfer across Vertical Layers," *J. Heat Transfer*, **87**, 1965, 110.

[29] WEBER, N., R. E. ROWE, E. H. BISHOP, and J. A. SCANLAN, "Heat Transfer by Natural Convection between Vertically Eccentric Spheres," *ASME Paper 72-WA/HT-2*, 1972.

[30] SCANLAN, J. A., E. H. BISHOP, and R. E. POWE, "Natural Convection Heat Transfer between Concentric Spheres," *Int. J. Heat Mass Transfer*, **13**, 1970, 1857.

[31] O'TOOLE, J., and P. L. SILVESTON, "Correlation of Convective Heat Transfer in Confined Horizontal Layers," *Chem, Eng. Progr. Symp.*, **57**, 1961, 81.

[32] GOLDSTEIN, R. J., and T. Y. CHU, "Thermal Convection in a Horizontal Layer of Air," *Progr. Heat Mass Transfer*, **2**, 1969, 55.

[33] GLOBE, S., and D. DROPKIN, "Natural-Convection Heat Transfer in Liquids Confined by Two Horizontal Plates and Heated from Below," *J. Heat Transfer*, **81**, 1959, 24–28.

[34] HOLLANDS, K. G. T., G. D. RAITHBY, and L. KONICEK, "Correlation Equations for Free Convection Heat Transfer in Horizontal Layers of Air and Water," *Int. J Heat Mass Transfer*, **18**, 1975, 879.

[35] SCHMIDT, E., "Free Convection in Horizontal Fluid Spaces Heated from Below," *Proc. Int. Heat Transfer Conf., Boulder, Colo., ASME*, 1961.

[36] CLIFTON, J. V., and A. J. CHAPMAN, "Natural Convection on a Finite Size Horizontal Plate," *Int. J. Heat Mass Transfer*, **12**, 1969, 1573

[37] KRASSHOLD H., "Wärmeabgabe von zylindrischen Flüssigkeitsschichten bei natürlicren Konvecktion," *Forsch. Gebiete Ingenieurw.*, **2**, 1931, 165.

[38] BECKMANN, W., "Die Wärmeübertragung in zylindrischen Gasschichten bei natürlicher Konvektion," *Forsch. Gebiete Ingenieurw.*, **2**, 1931, 186.

[39] LIU, C. Y., W. K. MUELLER, and F. LANDIS, "Natural Convection Heat Transfer in Long Horizontal Cylindrical Annuli, *Int. Dev. Heat Transfer*, **5**, 1961, 976.

[40] HOLLANDS, K. G. T., T. E. UNNY, and G. D. RAITHBY, "Free Convective Heat Transfer across Inclined Air Layers, *ASME Paper 75-HT-55*, 1975.

[41] RAITHBY, G. D., and K. G. T. HOLLANDS, "A General Method of Obtaining Approximate Solutions to Laminar and Turbulent Free Convection Problems," in *Advances in Heat Transfer*. New York: Academic Press, Inc., 1974.

[42] CATTON, I., P. S. AYYASWAMY, and R. M. CLEVER, "Natural Convection in a Finite, Rectangular Slot Arbitrarily Oriented with Respect to the Gravity Vector," *Int. J. Heat Mass Transfer*, **17**, 1974, 173.

[43] AYYASWAMY, P. S., and I. CATTON, "The Boundary-Layer Regime for Natural Convection in a Differentially Heated, Tilted Rectangular Cavity," *J. Heat Transfer*, **95**, 1973, 543.

[44] CATTON, I., "Natural Convection in Enclosures," *Sixth Int. Heat Transfer Conf., Toronto*, **6**, 1978, 13–32.

[45] ARNOLD, J. N., P. N. BONAPARTE, O. CATTON, and D. K. EDWARDS, "Experimental Investigation of Natural Convection in a Finite Rectangular Region Inclined at Various Angles from 0° to 180°," *Proc, HTFMI*, Stanford, Calif.: Stanford University Press, 1974.

[46] OSTRACH, J., "Natural Convection in Enclosures," in *Advances in Heat Transfer*. New York: Academic Press, Inc., 1972.

[47] SHITSMAN, M. E., "Natural Convection Effect on Heat Transfer to Turbulent Water Flow in Intensively Heated Tubes at Supercritical Pressure," Symp. Heat Transfer and Fluid Dynamics of Near Critical Fluids, *Proc. Inst. Mech. Eng.*, **182**, Part 3I, 1968.

[48] JACKSON, J. D., and K. EVANS-LUTTERODT, "Impairment of Turbulent Forced Convection Heat Transfer to Supercritical Pressure CO_2 Caused by Buoyancy Forces," University of Manchester, England, *Res. Rept. N-E-2*, 1968.

[49] LLOYD, J. R., and E. M. SPARROW, "Combined Forced and Free Convection Flow on Vertical Surfaces," *Int. J. Heat Mass Transfer*, **13**, 1970, 434–438.

[50] KLIEGEL, J. R., "Laminar Free and Forced Convection Heat Transfer from a Vertical Flat Plate," Ph.D. Thesis, University of California, Berkeley, 1959.

[51] METAIS, B., and E. R. G. ECKERT, "Forced, Mixed and Free Convection Regimes," *J. Heat Transfer*, **86**, 1964, 295–296.

[52] BUHR, H. D., A. D. CARR, and R. R. BALZHISER, "Temperature Profiles in Liquid Metals and the Effects of Superimposed Free Convection in Turbulent Flow," *Int. J. Heat Mass Transfer*, **11**, 1968, 641.

[53] BUHR, H. A., E. A. HORSTEN, and A. D. CARR, "The Distribution of Turbulent Velocity and Temperature Profiles on Heating, for Mercury in a Vertical Pipe," Natl. Heat Transfer Conf., Denver, Colo., *ASME 72-HT-21*, 1972.

[54] SESONSKI, A., L. L. EYLER, and G. A. KLEIN, "Effects of Free Convection on Turbulent Flow of Mercury Heat Transfer," *ANS Trans.*, **21**, 1975, 410.

[55] SHIRALKAR, B., and P. GRIFFITH, "The Effect of Swirl, Inlet Conditions, Flow Direction, and Tube Diameter on the Heat Transfer to Fluids at Supercritical Pressure," *J. Heat Transfer*, **92**, 1970, 465.

[56] HALL, W. B., "Heat Transfer Near the Critical Point," in *Advances in Heat Transfer*. New York: Academic Press, Inc., 1971.

[57] KAKARALA, C. R., "Development of Turbulent Wall Layer Models for Momentum and Heat Transfer in Tube Flow," Ph.D. Dissertation, University of Akron, Akron, Ohio, 1976.

[58] COLLIS, D. C., and M. J. WILLIAMS, "Two-Dimensional Convection from Heated Wires at Low Reynolds Numbers," *J. Fluid Mech.*, **6**, 1959, 357–384.

[59] OOSTHUIZEN, P. H., and S. MADAN, "Combined Convective Heat Transfer from Horizontal Cylinders in Air," *J. Heat Transfer*, **92**, 1970, 194–196.

[60] GEBHART, B., T. AUDUNSON, and L. PERA, "Forced, Mixed and Natural Convection from Long Horizontal Wires, Experiments at Various Prandtl Numbers," *Fourth Int. Heat Transfer Conf., Paris*, IV, Sec. 3.2, 1970.

[61] BROWN, C. K., and W. H. GAUVIN, "Combined Free and Forced Convection," Pts. I and II, *Can. J. Chem. Eng.*, **43**, 1965, 306.

CHAPTER 8

CONVECTION HEAT TRANSFER:
Practical Thermal Analysis—
BOILING AND CONDENSATION

8-1 INTRODUCTION

The study of two- and three-phase substances is one of the main topics covered in introductory courses in thermodynamics. Based on our background in thermodynamics, we know that a phase change occurs when a single-phase substance is brought to the saturation state. For example, when the temperature of a subcooled liquid is raised to the saturation temperature T_{sat}, vaporization or *boiling* occurs. On the other hand, when the temperature of a superheated vapor is lowered to T_{sat}, *condensation* occurs. The thermodynamic study of boiling and condensation as well as melting and freezing is developed in the context of idealistic equilibrium conditions. In practice, however, heat-transfer processes involving phase change occur under nonequilibrium conditions in which the difference between the wall surface temperature T_s and T_{sat} is not zero. The study of boiling and condensation heat transfer deals with such nonequilibrium liquid to vapor and vapor to liquid phase-change processes.

Familiar engineering applications of boiling and condensation heat transfer occur in power or refrigeration cycles. In addition, because of the large heat fluxes that can be accomplished in processes that involve phase change, boiling and condensation processes are also used in compact heat exchangers.

It should be mentioned that two-phase heat-transfer processes such as boiling and condensation are considerably more involved than are single-phase convection-heat-transfer processes because of the complicating effects of factors such as property variation, surface tension, latent heat of vaporization, and surface conditions. However, because we will be dealing with the practical thermal analysis of the simplest types of boiling and condensation-heat-transfer processes, the complexities that will be encountered in our introductory treatment of this topic will be minimized.

The change of phase between a liquid and its vapor produces quite distinctive and opposite effects, depending upon whether we are dealing with boiling or condensation. Therefore, these two phenomena will be considered separately.

8-2 BOILING HEAT TRANSFER

8-2-1 Introduction

Boiling heat transfer occurs when a fluid is exposed to a surface with temperature sufficiently greater than the saturation temperature T_{sat}.

The two types of boiling that are found in practice are *pool boiling* and *flow boiling*. Pool boiling occurs in the absence of bulk fluid flow, such as in the boiling of a pan of water. This type of boiling can also be easily produced by submerging an electrically heated wire or a steam-heated tube in a liquid, as shown in Figs. 8-1 and 8-2. Common examples of flow boiling processes include fossil fuel or nuclear steam generators and refrigerant evaporators.

Pool and flow boiling processes are further classified according to whether the bulk liquid temperature is equal to or less than the saturation temperature. If the bulk temperature of the liquid is below T_{sat}, the process is referred to as *subcooled* boiling. Otherwise, the process is known as *saturated* boiling.

8-2-2 Pool Boiling

Boiling curve for pool boiling

For a given fluid, operating pressure and heating surface, the heat flux that occurs during pool boiling is dependent upon the excess temperature $T_s - T_{sat}$. To illustrate this point, the convection heat flux q_c'' is plotted in terms of $T_s - T_{sat}$ in Fig. 8-3 for pool boiling of water. This general

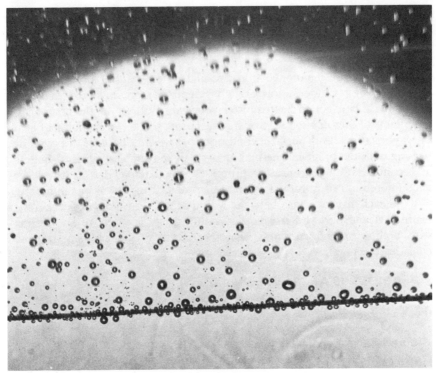

FIGURE 8-1 Nucleate pool boiling on an electrically heated wire. (Courtesy of Professor E. Hahne.)

boiling curve applies to pool boiling of water at any subcritical pressure. Scales are also shown on this figure for pool boiling of water at atmospheric pressure [1],[2]. Similarly shaped pool boiling curves have been developed for many other fluids and operating conditions. Although the general shape of the boiling curve is widely agreed upon, considerable controversy exists concerning the quantitative aspects of these curves [3], especially for fluids other than water.

The temperature of the fluid that is in contact with the wall is T_s. Thus, although the bulk liquid temperature is always less than or equal to T_{sat}, the fluid very near the heating surface is superheated, with the degree of superheat being equal to the *excess temperature* $T_s - T_{sat}$. As we move along the boiling curve from left to right, the degree of superheat of fluid in contact with the heating surface increases.

Boiling curves such as this exhibit five rather distinct regimes. In regime I, in which $T_s - T_{sat}$ is small, fluid motion occurs by means of natural convection. The natural convection circulates the slightly superheated liquid from the vicinity of the heating surface to the surface of the

$$D = 19 \text{ mm}$$
$$q''_o = 38.8 \text{ kW/m}^2$$
$$T_s - T_F = 138°C$$

FIGURE 8-2   Film pool boiling of methanol on a vertical steam-heated copper tube. (Courtesy of Professor J. W. Westwater.)

pool, where evaporation occurs. No boiling per se occurs in this *natural convection regime*.

At the somewhat higher values of excess temperature that are associated with regime II, liquid is transformed into vapor nuclei at various preferential sites on the heating surface. These nucleation sites then produce vapor bubbles which eventually break away from the heating surface and condense before reaching the surface of the pool. In this *subcooled nucleate boiling* region, the bulk water temperature is below T_{sat}.

In regime III, the nucleate boiling process produces larger more plentiful bubbles that break off and rise all the way to the surface of the pool. In this region, the bulk temperature of the fluid is equal to T_{sat}. The

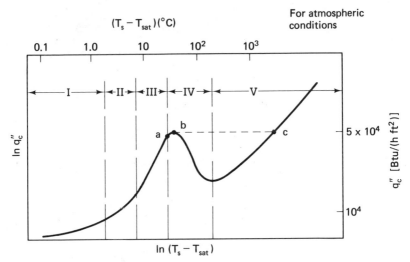

FIGURE 8-3 Representative boiling curve for pool boiling of water.

heat flux increases rapidly with increase in excess temperature in this saturated *nucleate boiling regime* until a *departure from nucleate boiling* (DNB) occurs at point *a*. This point is also known as the *critical heat flux*.

Both the subcooled and saturated nucleate boiling regimes are characterized by very high heat transfer coefficients. Consequently, the temperature difference $T_s - T_{sat}$ is generally fairly small in these regions, and no danger of tube wall destruction exists.

In regime IV, which is immediately beyond the nucleate boiling regions, the heat flux q_c'' reaches a maximum value at point *b*, after which a dramatic decrease in heat flux occurs. This degradation in heat flux is brought about by the fact that the inflow of cooler liquid toward the heating surface is inhibited by vapor rising from the surface. In effect, a vapor film is intermittently built up and partially destroyed in this *transition* (or *nucleate/film*) *boiling regime*, such that partial nucleate boiling continues to occur. In practice, operation within regime IV does not occur unless it is possible to control the excess temperature, such as in a tube heated by steam.

In regime V, a continuous stable vapor film covers the heating surface. Because the vapor has a much lower thermal conductivity than the liquid, very large temperature differences occur across the vapor film. In this high-temperature *film boiling regime*, the heat flux is seen to once again increase with increasing excess temperature. This increase in q_c'' can be credited to the large driving potential $T_s - T_{sat}$ and to the effects of

FIGURE 8-4 Several regimes of pool boiling of isopropanol on a copper fin. (Courtesy of Professor J. W. Westwater.)

thermal radiation. The surface temperature T_s continues to increase with q_c'' until the melting point of the material is reached and *burnout* occurs.

Several regimes of pool boiling have been captured by camera for boiling on a copper rod immersed in a fluid as shown in Fig. 8-4. The fin is heated at the wall to the right, such that the excess temperature is a minimum at the left end of the rod and increases as we move toward the wall. Both nucleate boiling regimes II and III are seen to be active over the left one-third of the rod, with transition and film boiling occurring over the remainder of the rod.

The boiling curve shown in Fig. 8-3 can be produced by gradually increasing the surface temperature T_s. By slowly increasing T_s we can reach and maintain steady-state boiling at any point along this curve. However, for the case in which the electrical power input $\dot{W}$ to the heater is slowly raised, stable operation between the points b and c cannot be maintained. For this situation, increasing the heat flux just beyond the critical heat flux, point b, results in a rapid often destructive transition to film boiling at point c, where the excess temperature may be sufficiently high to cause meltdown of the boiling surface. This point is expanded upon in the following example.

EXAMPLE 8-1

Utilize the lumped-analysis approach to explain the rapid unstable transition from point b to point c on the boiling curve for the case in which the electrical power input $\dot{W}$ to the heater is gradually increased.

Solution

To aid in our understanding of this point, we write the lumped energy balance for the heating surface as follows:

$$\sum \dot{E}_i = \sum \dot{E}_o + \frac{\Delta E_s}{\Delta t}$$

$$\dot{W} = q_c'' A_s + \rho V c_v \frac{dT_s}{dt}$$

Based on this equation, we conclude that the surface temperature T_s will increase with increase in power $\dot{W}$ to the heater. The boiling heat flux q_c'' responds to any change in T_s in accordance with the boiling curve. Putting this information together, in regimes I, II, and III, where q_c'' increases with increasing T_s, a small increase in $\dot{W}$ will produce an increase in T_s, which in turn will cause q_c'' to increase until a new steady-state surface temperature is reached and $q_c'' A_s$ is equal to $\dot{W}$. However, if after reaching the peak heat flux at point b, we again increase $\dot{W}$ by a very small amount, an unstable sequence of events is set in motion in which the initial increase in T_s produces a *decrease* in q_c''! This degradation in the heat flux q_c'' then stimulates the rate of increase in surface temperature dT_s/dt. As T_s quickly increase with time at this fixed value of $\dot{W}$, q_c'' will change roughly in accordance with the boiling curve until burnout occurs or equilibrium is reached near point c. At this point, the rate of heat generation within the heater $\dot{W}$ is again in balance with the rate of heat transfer from the surface $q_c'' A_s$. Because point b on the boiling curve is associated with this sometimes tragic event, it is often referred to as the *burnout point*.

Correlations for pool boiling

Following the pattern established in the analysis of single-phase heat-transfer processes, the rate of heat transfer associated with regimes II through V of pool boiling is given in terms of the mean coefficient of boiling heat transfer $\bar{h}_B$ as[1]

$$q_c'' = \bar{h}_B(T_s - T_{\text{sat}}) \tag{8-1}$$

Note that it is the excess temperature $T_s - T_{\text{sat}}$ that is utilized in this defining equation. Unlike the situation encountered in forced convection in which $\bar{h}$ is often independent of $T_s - T_F$, $\bar{h}_B$ is always a function of

---

[1] The standard coefficient of heat transfer $\bar{h}$ is generally utilized in the correlation of data in the natural convection regime.

$T_s - T_{sat}$. Consequently, some experimentalists report their pool boiling heat-transfer data directly in terms of q_c'' instead of $\bar{h}_B$.

Despite the fact that considerable controversy exists concerning the accuracy of available correlations for pool boiling, various design curves have been used to adequately, although perhaps not optimally, design boiling-heat-transfer equipment. Because of the burnout threat in many boiling-heat-transfer applications, the safety of the design comes first and the question of efficiency comes second.

Empirical design correlations for q_c'' or $\bar{h}_B$ have been developed for regimes II, III, and V of pool boiling. Some data even exist for the transition boiling regime [4], but general correlations are not presently available for this zone. It should be noted that the geometrical orientation of the surface is relatively unimportant for pool boiling. Therefore, the same correlations are generally used for horizontal, vertical, or inclined surfaces.

Experimental curves for nucleate pool boiling of water on a 0.024-in.-diameter platinum wire are shown in Fig. 8-5 for various saturation pressures. The points at which the maximum heat flux occurs are also

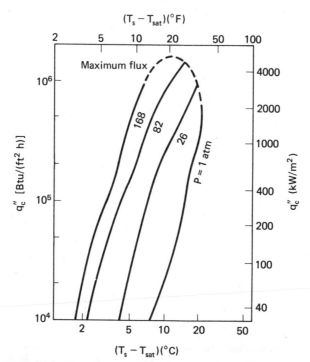

FIGURE 8-5 Nucleate pool boiling of water at various pressures on electrically heated platinum wire. (From Addoms [7].)

shown. Experimental data for $T_s - T_{sat}$ vs. q_c'' for nucleate pool boiling have been correlated by Rohsenow [5], Fritz [6], and many others. For example, the popular correlation by Rohsenow takes the form

$$\frac{c_{p\ell}(T_s - T_{sat})}{h_{\ell v}\,\mathrm{Pr}_\ell^n} = C_{sf}\left[\frac{q_c''}{\mu_\ell h_{\ell v}}\sqrt{\frac{\sigma}{g(\rho_\ell - \rho_v)}}\,\right]^{1/3} \tag{8-2}$$

where the subscripts ℓ and v designate saturated liquid and saturated vapor, $h_{\ell v}$ is the enthalpy of vaporization, σ is the surface tension of the liquid–vapor interface, and C_{sf} and n are empirical constants. Table 8-1 gives values of C_{sf} for various surface–fluid combinations. For example, C_{sf} is equal to 0.013 for a copper–water combination. The Prandtl number exponent n varies between 0.8 to 2.0, depending upon the surface contamination. For a clean surface, $n \simeq 1.7$. The surface tension σ is a function of both the fluid and temperature, as shown in Table A-C-6 of the Appendix.

It should be emphasized that these and other correlations for nucleate pool boiling only apply to that part of the boiling curve for which $T_s - T_{sat}$ is less than the critical excess temperature. The maximum heat flux attainable with nucleate boiling is generally quite crucial to the

TABLE 8-1 Values of the coefficient C_{sf} in Eq. (8-2) for various liquid–surface combinations

Liquid–surface combination	C_{sf}
Water–copper	0.0130
Water–scored copper	0.0068
Water–emery-polished copper	0.0128
Water–emery-polished, paraffin-treated copper	0.0147
Water–chemically etched stainless steel	0.0133
Water–mechanically polished stainless steel	0.0132
Water–ground and polished stainless steel	0.0080
Water–Teflon-pitted stainless steel	0.0058
Water–platinum	0.0130
Water–brass	0.0060
Benzene–chromium	0.0100
Ethyl alcohol–chromium	0.0027
Carbon tetrachloride–copper	0.0130
Carbon tetrachloride–emery-polished copper	0.0070
n-Pentane–emery-polished copper	0.0154
n-Pentane–emery-polished nickel	0.0127
n-Pentane–emery-rubbed copper	0.0074
n-Pentane–lapped copper	0.0049

Source: From [7]–[11].

designer because of the importance of avoiding the burnout dangers of the film boiling regimes, particularly in high-performance constant-heat flux systems. The critical heat flux for saturated nucleate boiling can be approximated by empirical curves developed by Zuber [12], Rohsenow and Griffith [13], Kutateladze [14], and others. The Zuber correlation is given by

$$q''_{\text{crit}} = \frac{\pi}{24} \rho_v h_{\ell v} \left(\sigma g \frac{\rho_\ell - \rho_v}{\rho_v^2} \right)^{1/4} \left(\frac{\rho_\ell + \rho_v}{\rho_\ell} \right)^{1/2}$$

(8-3)

For subcooled nucleate boiling, q''_{crit} can be approximated by [15]

$$q''_{\text{crit}} = q''_{\text{crit, sat}} \left\{ 1 + \frac{2k_\ell(T_{\text{sat}} - T_\ell)}{\sqrt{\pi \alpha_\ell C}} \frac{24}{\pi h_{\ell v} \rho_v} \left[\frac{\rho_v^2}{\sigma g(\rho_\ell - \rho_v)} \right]^{1/4} \right\}$$

(8-4)

where

$$C = \frac{\pi}{3} \sqrt{2\pi} \left[\frac{\sigma}{g(\rho_\ell - \rho_v)} \right]^{1/2} \left[\frac{\rho_v^2}{\sigma g(\rho_\ell - \rho_v)} \right]^{1/4}$$

(8-5)

and $q''_{\text{crit, sat}}$ is determined from Eq. (8-3). Although these correlations can be utilized as a first approximation for most fluid–surface combinations, practical experience indicates that q''_{crit} is dependent upon the surface material.

According to Eq. (8-3), q''_{crit} can be increased by increasing the gravitational field or by selecting a fluid with a large value of $h_{\ell v}$. For example, in the selection of a fluid, $h_{\ell v}$ is higher for water than other common liquids, such that water will sustain the highest critical heat flux. Further, because the fluid properties are functions of pressure, q''_{crit} can be increased to a maximum for any given fluid by optimizing the pressure. The variation of q''_{crit} with pressure is indicated in Fig. 8-5 for water. The optimum pressure for water is about 100 atm and the maximum heat flux is approximately 3.8×10^6 W/m^2 [16].

A popular correlation for film boiling on the surface of a horizontal tube has been developed by Bromley [17] which takes the form

$$\bar{h}_B = 0.62 \left[\frac{g(\rho_\ell - \rho_v)h_{\ell v}k_v^3}{D \nu_v (T_s - T_{\text{sat}})} \right]^{1/4}$$

(8-6)

where D is the tube diameter and $h_{\ell v}$ is evaluated at T_{sat}. To account for the effects of thermal radiation that sometimes occurs in high-temperature film boiling, a correlation has been proposed of the form

$$\bar{h}_{B,R} = \bar{h}_B + \bar{h}_R \tag{8-7}$$

The coefficient without thermal radiation $\bar{h}_B$ is given by Eq. (8-6) and $\bar{h}_R$ is approximated by

$$\bar{h}_R = \frac{\sigma(T_s^4 - T_{sat}^4)}{T_s - T_{sat}} \tag{8-8}$$

for an opaque graybody surface, assuming that the saturated liquid which lies over the vapor film can be approximated as a blackbody.

EXAMPLE 8-2

Water at atmospheric conditions is boiled by an electrically heated 5-mm-diameter platinum wire. Determine the maximum wall temperature and heat flux that can be attained in the safe nucleate boiling regime.

Solution

The saturation temperature is, of course, 100°C. Referring to Fig. 8-5, we see that the critical heat flux for atmospheric conditions for boiling from a platinum wire is approximately 1400 kW/m² and T_s is 120°C. To check this value, we use the Zuber correlation, Eq. (8-3). The pertinent properties at 100°C are $\rho_\ell = 961$ kg/m³, $\rho_v = 0.586$ kg/m³, $h_{\ell v} = 2260$ kJ/kg, $\sigma = 5.88 \times 10^{-2}$ N/m, $Pr_\ell = 1.74$, $c_p = 4.22$ kJ/(kg °C), and $\mu_\ell = 2.83 \times 10^{-4}$ kg/(m s). Calculating q''_{crit}, we have

$$q''_{crit} = \frac{\pi}{24} \rho_v h_{\ell v} \left(\frac{\rho_\ell + \rho_v}{\rho_\ell} \right)^{1/2} \left(\sigma g \frac{\rho_\ell - \rho_v}{\rho_v^2} \right)^{1/4}$$

$$= \frac{\pi}{24} \left(0.586 \, \frac{\text{kg}}{\text{m}^3} \right) \left(2260 \, \frac{\text{kJ}}{\text{kg}} \right) \left(\frac{961 + 0.586}{961} \right)^{1/2}$$

$$\times \left[\left(5.88 \times 10^{-2} \frac{\text{N}}{\text{m}} \right) \left(9.81 \frac{\text{m}}{\text{s}^2} \right) \frac{(961 - 0.586) \, \text{kg/m}^3}{(0.586 \, \text{kg/m}^3)^2} \right]^{1/4}$$

$$= 1.1 \times 10^6 \, \frac{\text{W}}{\text{m}^2}$$

Note that this value is approximately 30% below the value obtained from Fig. 8-5. However, it is almost identical to the result obtained from the Kutateladze correlation (see Prob. 8-1).

T_s can be calculated from Eq. (8-2),

$$T_s - T_{sat} = h_{\ell v} \, Pr_\ell^n \frac{C_{sf}}{c_{p\ell}} \left[\frac{q''_c}{\mu_\ell h_{\ell v}} \sqrt{\frac{\sigma}{g(\rho_\ell - \rho_v)}} \right]^{1/3}$$

where $n = 1.7$ and $C_{sf} = 0.013$. Following through with the calculation, we obtain

$$T_s - T_{sat} = 28.4°C$$

or

$$T_s = 128°C$$

Thus, we have a discrepancy between the correlation of Fig. 8-5 and the result obtained from Rohsenow's correlation of the order of about 30% for $T_s - T_{sat}$. This is typical of the inconsistencies found among various boiling-heat-transfer correlations.

EXAMPLE 8-3

Saturated water at 130°F is boiled by a 3.28 ft long 0.816-in.-diameter copper tube with surface temperature equal to 150°F. Determine the rate of heat transfer and the rate of evaporation.

Solution

The excess temperature is

$$T_s - T_{sat} = 150°F - 130°F = 20°F$$

The properties of saturated water at 130°F are $h_{fv} = 873$ Btu/lb_m, $c_{pl} = 0.998$ Btu/(lb_m °F), $\sigma = 4.59 \times 10^{-3}$ lb_f/ft, $\rho_l = 61.6$ lb_m/ft^3, $\mu_l = 3.45 \times 10^{-4}$ lb_m/(ft s), and $Pr_l = 3.3$. Utilizing the Rohsenow correlation, we calculate the heat flux as follows:

$$q_c'' = \left[\frac{c_{pl}(T_s - T_{sat})}{h_{fv} Pr_l^n C_{sf}} \right]^3 \mu_l h_{fv} \sqrt{\frac{g(\rho_l - \rho_v)}{\sigma}}$$

$$= \left\{ \frac{[0.998 \text{ Btu}/(lb_m \text{ °F})](20°F)}{(873 \text{ Btu}/lb_m)(3.3)^{1.7}(0.013)} \right\}^3 \left(3.45 \times 10^{-4} \frac{lb_m}{\text{ft s}} \right) \left(873 \frac{\text{Btu}}{lb_m} \right)$$

$$\times \left\{ \frac{(32.2 \text{ ft/s}^2)(61.6 \text{ } lb_m/\text{ft}^3)}{(4.59 \times 10^{-3} \text{ } lb_f/\text{ft})[32.2 \text{ ft } lb_m/(lb_f \text{ s}^2)]} \right\}^{1/2}$$

$$= 0.430 \frac{\text{Btu}}{\text{s ft}^2} = 1550 \frac{\text{Btu}}{\text{h ft}^2}$$

$$= 4.89 \frac{\text{kW}}{\text{m}^2}$$

Thus, the total rate of heat transfer is

$$q_c = q_c'' \pi D L$$

$$= \left(1550 \frac{\text{Btu}}{\text{h ft}^2} \right)(\pi)\left(\frac{0.816 \text{ ft}}{12} \right)(3.28 \text{ ft})$$

$$= 1090 \frac{\text{Btu}}{\text{h}} = 0.318 \text{ kW}$$

To calculate the rate of evaporation, we write

$$\dot{m} = \frac{q_c}{h_{\ell v}}$$

$$= \frac{1090 \text{ Btu/h}}{873 \text{ Btu/lb}_\text{m}} = 1.25 \frac{\text{lb}_\text{m}}{\text{h}}$$

Enhancement of pool boiling

The heat-transfer correlations given in the previous section apply to smooth surfaces. Many methods have been utilized for producing improvements in pool boiling heat transfer. Perhaps the most common approaches involve the use of rough surfaces and extended surfaces. Several other methods include surface treatment, liquid additives, and mechanical agitation or vibration. A survey of various methods of heat-transfer enhancement is provided by Bergles [18]. The main idea in using rough or extended surfaces is to increase the number of nucleation sites. The Thermoexcel-E surface shown in Fig. 8-6 is a recent entry in the commercial development of machined rough surfaces. The effect of these enhancement techniques on the boiling heat flux is shown in Fig. 8-7. The use of properly designed fins or rough surfaces such as Thermoexcel-E has also been reported to substantially increase the critical heat flux and delay the onset of DNB [18], [19].

8-2-3   Flow Boiling

Physical description of flow boiling

Flow boiling occurs in internal and external flow forced and natural convection systems. External flow boiling over a cylinder or plate is much like pool boiling because the free stream temperature and quality are

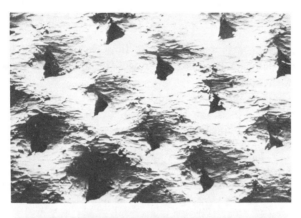

(b)

FIGURE 8-6 Thermoexcel-E surface for enhanced boiling. (a) Plan view. (b) Cross section. (Courtesy of Hitachi, Ltd.)

Refrigerant, Freon R-12, boiling temperature, 0°C

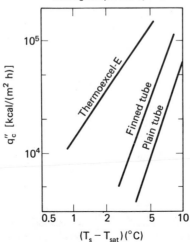

FIGURE 8-7 Nucleate pool boiling heat-transfer performance of several surfaces. (From Arai et al. [19].)

essentially constant with respect to the flow direction. But internal flow boiling is more complex because of changes in the bulk stream temperature and quality with respect to x.

In flow boiling processes, liquid (or a liquid–vapor mixture) enters a heating chamber such as the tube shown in Fig. 8-8. Beyond the axial location at which the boiling process is initiated (point a), the quality of the

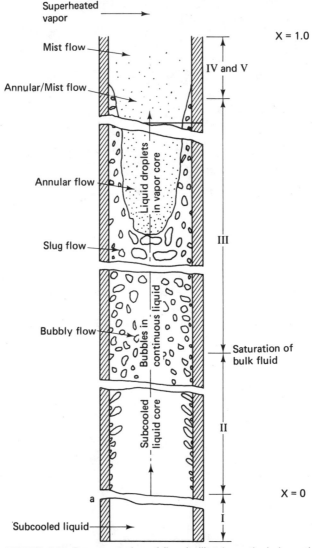

FIGURE 8-8 Representation of flow boiling in vertical channel. (From *Boiling Heat Transfer and Two-Phase Flow* by L. S. Tong. Copyright 1965 by John Wiley & Sons, Inc. Reprinted by permission of John Wiley & Sons, Inc.)

liquid–vapor mixture increases with distance until total vaporization occurs. Thus, for the case in which a subcooled liquid enters the heating chamber, the bulk fluid temperature increases to a value somewhat less than T_{sat}, at which point subcooled nucleate boiling is initiated. The bulk temperature then quickly reaches T_{sat} and the quality goes from zero to unity. After all the liquid has been vaporized, the entire bulk fluid becomes superheated. Strictly speaking, T_{sat} varies slightly with x because of pressure losses within the chamber. However, the assumption that T_{sat} is essentially constant is generally quite adequate.

Boiling curve for flow boiling

A representative boiling curve for internal flow boiling in heated tubes or channels is shown in Fig. 8-9. Notice that the boiling curves for flow boiling and pool boiling are of the same general form. But, for flow boiling, the bulk stream temperature T_b, is sometimes used instead of T_{sat}. Points on the flow boiling curve represent the difference in $T_s - T_b$ or $T_s - T_{sat}$ at various axial locations x on the surface vs. the local heat flux q_c''. The actual numerical values of q_c'' and the temperature difference are a function of pressure, geometry, mean velocity, thermal boundary conditions, and degree of subcooling.

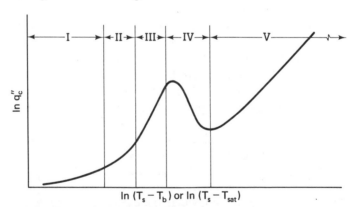

FIGURE 8-9 Representative boiling curve for internal flow boiling.

It should be noted that the coefficient of heat transfer for flow boiling is defined in the traditional sense; that is,

$$dq_c = h \, dA_s (T_s - T_b) \tag{8-9}$$

The same five basic regimes found in pool boiling also occur in flow boiling: I, natural convection; II, subcooled nucleate boiling; III, saturated nucleate boiling; IV, transition boiling; and V, film boiling. However, for internal flow boiling processes, because the fluid is totally confined within the tube or channel with no free interface from which vapor can escape, rather distinctive two-phase flow patterns are associated with each of these regimes. Take, for example, the situation in which a subcooled liquid enters a heated vertical tube in which T_s is greater than T_{sat}. Natural convection (or combined forced/natural convection) occurs in region I where $T_s - T_b$ is small. But, as in the case of pool boiling, no actual boiling takes place in this zone. When the temperature difference $T_s - T_b$ is sufficiently large to produce regime II subcooled nucleate boiling, the bubbles form at nucleation sites but do not carry far into the main stream because the local bulk temperature T_b is still below T_{sat}. However, as T_b approaches T_{sat}, more and more bubbles appear in the bulk stream. Because of the presence of discrete bubbles in the flow stream, regime II is sometimes called the *bubbly flow regime*.

The saturated nucleate boiling regime for flow boiling is often subdivided into two subregions. In the first part of regime III, the bubbles that occur throughout the bulk stream begin to coalesce into slugs of vapor. Although the quality is quite low in this region, the percent of the volume that is occupied by vapor is as large as 50%. Consequently, the bulk flow rate increases. This portion of the saturated nucleate boiling zone is sometimes called the *slug flow regime*. In the second part of regime III, the tube wall is covered by a thin liquid film with vapor and liquid droplets flowing in the center of the tube. Although nucleate boiling occurs in this region, vapor is also believed to be generated by vaporization at the liquid–vapor interface. Large increases in the heat flux continue to occur in this zone, until the point of departure from nucleate boiling (DNB) occurs. The coefficients of heat transfer in the vicinity of the DNB can be 20 to 50 times as large as the coefficient for single-phase flow. This part of the saturated nucleate boiling regime is also known as the *annular flow regime*.

As in the case of pool boiling, the heat flux in flow boiling processes experiences a peak, followed by a drastic decrease in regime IV. In this transition boiling region, a wall-drying process occurs in which a vapor film with large thermal resistance gradually develops on the surface. Region IV is also sometimes known as the *annular/mist regime*.

In the film boiling regime, the wall is completely dried out with liquid droplets being dispersed throughout the bulk of the fluid. Heat is transferred from the wall to the vapor and then from the vapor to the liquid until complete vaporization occurs. Regime V is also known as the *mist flow regime*.

As in the case of pool boiling, the coefficients of heat transfer are quite high in the subcooled and saturated nucleate flow boiling regimes. Hence, the wall temperature is usually not too much larger than the bulk fluid temperature or saturation temperature and the walls are not in danger of meltdown or burnout. However, because of the greatly reduced coefficients of heat transfer in the regimes IV and V, these regions are avoided in certain applications. For example, fossil-fuel-fired boilers or nuclear reactors are usually operated in the nucleate boiling regimes to avoid the onset of DNB. The coefficient of heat transfer attained in these systems is sufficiently high to permit satisfactory operation as long as fouling is prevented by proper water treatment. On the other hand, steam generators for pressurized water reactor systems (i.e., water to boiling water heat exchangers) and certain types of process heat exchangers are sometimes operated in regimes IV and V. In these cases, the temperature of the heat source is within the safe operating range of the equipment. In once-through boilers in which superheated steam is produced, the wall heat flux is generally reduced in the section of the heating chamber where regimes IV and V occur.

The effect of fluid enthalpy (or quality) and heat flux on the wall temperature T_s and bulk fluid temperature T_b and on the location of the DNB point are indicated in Fig. 8-10 for an upward flow of water in a vertical uniformly heated tube. Notice that T_b increases until the fluid reaches the saturation temperature. T_b then drops slightly in the region

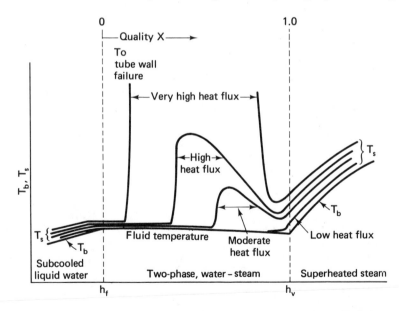

Enthalpy

FIGURE 8-10 Bulk fluid and tube wall temperature for flow boiling of water. (From *Steam—Its Generation and Use.* Copyright 1975 by Babcock and Wilcox Company. Used with permission.)

where steam is being generated because of a small pressure drop. In the single-phase superheat region, T_b once again increases. Except for the low-heat-flux case, the behavior in surface temperature T_s is quite another matter. For moderate heat flux, T_s remains somewhat above T_b until the DNB point associated with regime IV is reached in the region of high steam quality, and a degradation in heat transfer occurs. In this nucleate/film boiling regime, T_s experiences a rather pronounced increase to a mild peak, and then decreases as the steam quality approaches 100%. T_s once again gradually increases in the superheat region. For higher heat fluxes, the DNB point and regime IV boiling is reached at lower steam quality and the peak in T_s is higher. As the heat flux increases, the peak in T_s increases and moves in the direction of lower and lower bulk stream enthalpy, until the metal melts or the tube or channel ruptures.

EXAMPLE 8-4

Sketch a boiling curve for external forced convection flow boiling on an electrically heated wire.

Solution

For the case is which the free stream velocity is zero, free convection and pool boiling control. Therefore, we start with the pool boiling curve shown in Fig. E8-4a. As U_∞ increases, the coefficient of heat transfer in the single-phase region increases. This effect would be expected to carry over to regimes II and III. Therefore, boiling curves are sketched in for low, moderate, and high values of U_∞, which should account for the qualitative effects of forced convection.

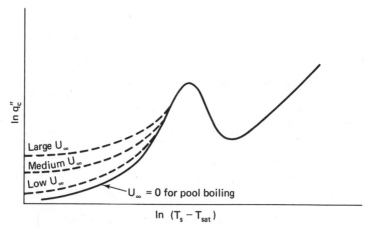

FIGURE E8-4a Boiling curve for flow boiling.

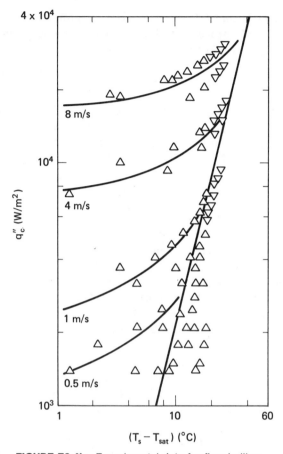

FIGURE E8-4b Experimental data for flow boiling.

To reinforce this result, experimental data reported by Lung et al. [21] are shown in Fig. E8-4b. Our external flow boiling curves are in general agreement with these data. Note that the heat-transfer process is essentially controlled by boiling for larger values of $T_s - T_{sat}$ and by forced convection for smaller values.

Correlations for flow boiling

Some success has been achieved in developing correlations for the coefficient of heat transfer and the peak heat flux for flow boiling processes. In particular, fairly simple correlations are available for the local subcooled nucleate boiling that occurs in regime II and for the full nucleate boiling of regime III.

In regimes II and III, where nucleate boiling occurs, the total heat flux q_c'' is artificially broken into boiling- and nonboiling-convection components. This idea is reflected in the following equation:

$$q_c'' = q_c''|_{\text{boiling}} + q_c''|_{\text{nonboiling}} \qquad (8\text{-}10)$$

where $q_c''|_{\text{boiling}}$ is approximated by the correlations for two-phase nucleate pool boiling and $q_c''|_{\text{nonboiling}}$ is obtained from the standard correlations for single-phase convection. In regime III, where the heat-transfer process is essentially governed by the boiling mechanism, Eq. (8-10) can be approximated by

$$q_c'' = q_c''|_{\text{boiling}} \qquad (8\text{-}11)$$

In this region, the heat-transfer flux is essentially independent of the velocity and forced or natural convection effects. These assumptions are consistent with the data shown in Example 8-4.

Empirical burnout or peak heat-flux correlations have been developed for various flow boiling systems. A comprehensive survey of the work done in this area has been presented by Hewitt [22].

Empirical correlations have also been developed for flow boiling heat transfer associated with regimes IV and V. However, the quantitative treatment of this two-phase flow subject is complex and lies beyond the scope of our introductory study. Several references on the subject of two-phase flow include the works by Tong [20], Hsu and Graham [23], Collier [24], and Butterworth and Hewitt [26].

Enhancement of flow boiling

Techniques similar to those used in pool boiling are used for improving flow boiling characteristics [18]. One of the more successful methods involves the use of internally ribbed tubes such as the one shown in Fig. 8-11. This type tube has been reported to suppress the onset of DNB and to permit operation at higher heat fluxes than is possible with smooth tubes

FIGURE 8-11   Single-lead ribbed tube. (From *Steam—Its Generation and Use*. Copyright 1975 by Babcock and Wilcox Company. Used with permission.)

[25], [26]. However, because of the expense of manufacturing internally ribbed tubes, the Babcock and Wilcox Company recommends their use only in high-pressure systems (above 150 atm).

8-3 CONDENSATION HEAT TRANSFER

8-3-1 Introduction

Condensation heat transfer occurs when a vapor is exposed to a surface with temperature less than the saturation temperature. The liquid condensate that is formed flows down the surface under the influence of gravity. If the liquid does not wet the surface, the process is called *dropwise condensation*. Under these conditions, droplets form on the surface as shown in Fig. 8-12.

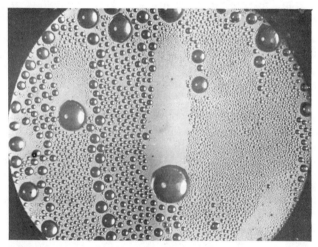

FIGURE 8-12 Dropwise condensation of steam under ideal conditions. (Courtesy of Professor M. N. Ozisik.)

These drops generally flow over the surface along random paths. For those liquids that wet the surface, the condensate forms a film that builds up as it flows down the surface. This *film condensation* process is illustrated in Fig. 8-13. Because the existence of such a film increases the resistance to heat transfer, dropwise condensation is as much as 10 times as effective as film condensation. However, most surfaces become wetted when exposed to a condensing vapor over an extended length of time. Consequently, film condensation is generally encountered in industrial applications and is usually planned for in design work.

It should be mentioned that the rate of heat transfer is generally lower when a vapor exists in the presence of a noncondensable gas.

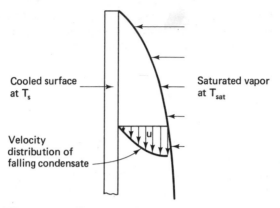

Cooled surface at T_s

Saturated vapor at T_{sat}

Velocity distribution of falling condensate

FIGURE 8-13   Film of condensate on a vertical surface.

Therefore, noncondensable gases are sometimes vented to improve the efficiency of the condensation-heat-transfer process.

8-3-2   Film Condensation

Reynolds number for film condensation

As in single-phase forced-convection processes, the heat transfer in film condensation is dependent upon whether the flow within the film is laminar or turbulent. To provide a criterion to establish whether the flow is laminar or turbulent, a Reynolds number for condensate liquid film flow has been defined which takes the form

$$\text{Re}_f = \frac{D_H U_b}{\nu_\ell} = \frac{4A}{p} \frac{U_b}{\nu_\ell} \qquad (8\text{-}12)$$

where U_b is the bulk velocity flow rate of condensate and $A = \delta p$; δ is the local thickness of the film and p is the perimeter. Based on the principle of continuity, U_b can be expressed in terms of the condensate mass flow rate $\dot{m}$ as

$$\dot{m} = \rho_\ell A U_b \qquad (8\text{-}13)$$

Thus, Re_f can be written in the form

$$\text{Re}_f = \frac{4\dot{m}}{p\mu_\ell} \qquad (8\text{-}14)$$

Notice the Re_f increases as the film grows. For vertical tubes and plates, the critical Reynolds number is approximately 2000.

The properties of the liquid film are generally evaluated at the film temperature T_f,

$$T_f = \frac{T_s + T_v}{2} \tag{8-15}$$

Coefficient of condensation heat transfer

Consistent with the traditional approach to correlating heat transfer data for single-phase convection processes, the local rate of heat transfer for film condensation is generally expressed in terms of a *coefficient of condensation heat transfer* h_C.

$$dq_c = h_C \, dA_s (T_{sat} - T_s) \tag{8-16}$$

For a uniform wall-temperature boundary condition, the total rate of heat transfer over the entire surface is given by

$$q_c = \bar{h}_C A_s (T_{sat} - T_s) \tag{8-17}$$

where the mean coefficient of condensation heat transfer is defined by

$$\bar{h}_C = \frac{1}{L} \int_o^L h_C \, dx \tag{8-18}$$

A useful relationship between $\bar{h}_C$ and Re_f can be developed by equating the rate of convection heat transfer from the film to the wall q_c, to the rate of energy transfer to the film by the condensation process, $\dot{m} h_{\ell v}$; that is,

$$\bar{h}_C A_s (T_{sat} - T_s) = \dot{m} h_{\ell v} \tag{8-19}$$

or

$$\mathrm{Re}_f = \frac{4\dot{m}}{p\mu_\ell} = \frac{4\bar{h}_C A_s (T_{sat} - T_s)}{p\mu_\ell h_{\ell v}} \tag{8-20}$$

Thus, with $\bar{h}_C$ known on the basis of empirical or theoretical correlations, Re_f can be calculated. Several design correlations are now presented for the coefficient of condensation heat transfer associated with external film condensation. For information on the more complex topic of film condensation inside tubes and channels, the review by Collier [24] is recommended.

EXAMPLE 8-5

Develop approximate expressions for the thickness of film condensate.

Solution

Referring to the defining equation for Re_f

$$\mathrm{Re}_f = \frac{4A}{p}\frac{U_b}{\nu_\ell} = 4\delta\frac{U_b}{\nu_\ell} \tag{8-12}$$

we have

$$U_b\delta = \frac{\nu_\ell \,\mathrm{Re}_f}{4}$$

Although we know the product $U_b\delta$ we do not know δ or U_b individually.

However, δ can be estimated by performing a simple thermal analysis. Assuming as a first approximation that the temperature distribution across the film is linear, we write

$$q_c = q_y|_0$$

$$h_C\,dA_s(T_s - T_{\mathrm{sat}}) = -k_\ell\,dA_s\frac{\partial T}{\partial y}\bigg|_0 = -k_\ell\,dA_s\frac{T_{\mathrm{sat}} - T_s}{\delta}$$

or

$$\delta = \frac{k_\ell}{h_C} \tag{a}$$

Thus, we can approximate δ once correlations are available for h_C.

Vertical Surface   In his pioneering work in 1916, Nusselt [27] developed the following theoretical expression for laminar film condensation on a plane vertical surface:

$$h_C = \left[\frac{g\rho_\ell(\rho_\ell - \rho_v)k_\ell^3 h_{\ell v}}{4x\mu_\ell(T_{\mathrm{sat}} - T_s)}\right]^{1/4} \tag{8-21}$$

Based on this famous result for the local coefficient of condensation heat transfer, an expression can be written for the mean $\bar{h}_C$:

$$\bar{h}_C = 0.943\left[\frac{g\rho_\ell(\rho_\ell - \rho_v)k_\ell^3 h_{\ell v}}{\mu_\ell L(T_{\mathrm{sat}} - T_s)}\right]^{1/4} \tag{8-22}$$

As pointed out by Bell and Panchal [28], Nusselt's equation has been tested under conditions closely satisfying his assumptions and has been generally confirmed. Nevertheless, McAdams [29] and others recommended coefficients that are about 15% above the Nusselt theory.

By coupling Eqs. (8-20) and (8-22), we obtain an expression for the Reynolds number Re_f of the form

$$Re_f = 3.77 \left[\frac{L(T_{sat} - T_s)}{\mu_\ell h_{\ell v}} \right]^{3/4} \left[\frac{g \rho_\ell (\rho_\ell - \rho_v) k_\ell^3}{\mu_\ell^2} \right]^{1/4} \qquad (8\text{-}23)$$

This equation permits us to calculate Re_f. If Re_f is less than 2000, h_C and $\bar{h}_C$ can then be calculated by the use of Eqs. (8-21) and (8-22). Taking one step further, we couple Eqs. (8-22) and (8-23) to obtain

$$\bar{h}_C = 1.47 \, Re_f^{-1/3} \left[\frac{g \rho_\ell (\rho_\ell - \rho_v) k_\ell^3}{\mu_\ell^2} \right]^{1/3} \qquad (8\text{-}24)$$

This expression is shown in Fig. 8-14.

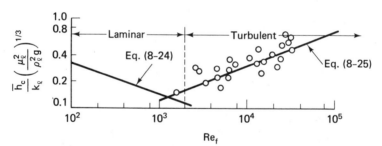

FIGURE 8-14 Coefficient of condensation heat transfer for vertical surface. (Data by Kirkbride [30].)

For turbulent flow conditions that occur for Re_f greater than 2000, the following correlation for $\bar{h}_C$ by Kirkbride [30] has been recommended for vertical surfaces:

$$\bar{h}_C = 0.0076 \, Re_f^{0.4} \left[\frac{g \rho_\ell (\rho_\ell - \rho_v) k_\ell^3}{\mu_\ell^2} \right]^{1/3} \qquad (8\text{-}25)$$

Coupling this expression with Eq. (8-20), we have

$$Re_f = 0.00296 \left[\frac{L(T_{sat} - T_s)}{\mu_\ell h_{\ell v}} \right]^{1.67} \left[\frac{g \rho_\ell (\rho_\ell - \rho_v) k_\ell^3}{\mu_\ell^2} \right]^{0.556} \qquad (8\text{-}26)$$

Once Re_f is calculated by the use of Eq. (8-26), $\bar{h}_C$ can be obtained from Eq. (8-25). Equation (8-25) and experimental data are shown in Fig. 8-14.

Equations (8-21) through (8-26) can also be used for film condensation on the outside of vertical tubes if the tube diameter is large compared to the film thickness δ. It should also be noted that because ρ_ℓ is generally so much larger than ρ_v, the term $\rho_\ell - \rho_v$ appearing in condensation-heat-transfer correlations can be approximated by ρ_ℓ for many situations.

Inclined Surface   For plates that are inclined by an angle θ with the horizontal, the equations above can be utilized as a first approximation for film condensation on the upper surface by replacing the gravitational acceleration by its component parallel to the surface, $g \sin \theta$. However, these equations cannot be extended to inclined tubes.

Horizontal Tube   For laminar film condensation on the outside of horizontal tubes, $\bar{h}_C$ can be approximated by (16)

$$\bar{h}_C = 0.725 \left[\frac{g\rho_\ell(\rho_\ell - \rho_v)k_\ell^3 h_{\ell v}}{D\mu_\ell(T_{\text{sat}} - T_s)} \right]^{1/4} \tag{8-27}$$

Combining this expression with Eq. (8-20), we have

$$\bar{h}_C = 1.2 \, \text{Re}_f^{-1/3} \left[\frac{g\rho_\ell(\rho_\ell - \rho_v)k_\ell^3}{\mu_\ell^2} \right]^{1/3} \tag{8-28}$$

where

$$\text{Re}_f = 4.56 \left[\frac{L(T_{\text{sat}} - T_s)}{\mu_\ell h_{\ell v}} \right]^{3/4} \left[\frac{g\rho_\ell(\rho_\ell - \rho_v)k_\ell^3}{(L/D)^3 \mu_\ell^2} \right]^{1/4} \tag{8-29}$$

A conservative estimate can be made for laminar film condensation on the outside of vertical banks of N tubes in line by simply replacing D with ND.

EXAMPLE 8-6

Steam at a saturation temperature of 54.4°C condenses on the surface of a 1-m-long 2-cm-diameter vertical tube with a surface temperature of 43.3°C. Assuming film condensation, determine the total rate of heat transfer and the total mass flow rate of condensate over the entire tube length.

Solution

The film temperature is

$$T_f = \frac{43.3°C + 54.4°C}{2} = 48.9°C$$

The properties evaluated at this temperature are $h_{\ell v}=2370$ kJ/kg, $k_\ell=0.642$ W/(m °C), $\rho_\ell=988$ kg/m³, and $\mu_\ell=0.558\times10^{-3}$ kg/(m s). Assuming for the moment that the flow is laminar, and that the flat-plate corrections can be used, the Reynolds number Re_f is given by Eq. (8-23),

$$\mathrm{Re}_f=3.77\left[\frac{L(T_{\mathrm{sat}}-T_s)}{\mu_\ell h_{\ell v}}\right]^{3/4}\left[\frac{g\rho_\ell(\rho_\ell-\rho_v)k_\ell^3}{\mu_\ell^2}\right]^{1/4} \qquad (8\text{-}23)$$

Because $\rho_\ell\gg\rho_v$, we have

$$\mathrm{Re}_f=3.77\left\{\frac{1\ \mathrm{m}(54.4°\mathrm{C}-43.3°\mathrm{C})}{[0.588\times10^{-3}\ \mathrm{kg}/(\mathrm{m\ s})](2370\ \mathrm{kJ/kg})}\right\}^{3/4}$$

$$\times\left\{\frac{(9.81\ \mathrm{m/s^2})\ (988\ \mathrm{kg/m^3})^2[0.642\ \mathrm{W}/(\mathrm{m\ °C})]^3}{[0.588\times10^{-3}\ \mathrm{kg}/(\mathrm{m\ s})]^2}\right\}^{1/4}$$

$$=165$$

Because Re_f is considerably below 2000, we conclude that the flow is indeed laminar.

The condensate mass flow rate is immediately calculated from Eq. (8-14).

$$\dot{m}=\frac{p\mu_\ell\,\mathrm{Re}_f}{4}$$

$$=\frac{\pi(0.02\ \mathrm{m})[0.558\times10^{-3}\ \mathrm{kg}/(\mathrm{m\ s})](165)}{4}$$

$$=1.45\times10^{-3}\frac{\mathrm{kg}}{\mathrm{s}}$$

To calculate $\bar{h}_C$, we utilize Eq. (8-24).

$$\bar{h}_C=1.47\,\mathrm{Re}_f^{-1/3}\left[\frac{g\rho_\ell(\rho_\ell-\rho_v)k_\ell^3}{\mu_\ell^2}\right]^{1/3}$$

$$=1.47(165)^{-1/3}\left\{\frac{(9.81\ \mathrm{m/s^2})(988\ \mathrm{kg/m^3})^2[0.642\ \mathrm{W}/(\mathrm{m\ °C})]^3}{[0.558\times10^{-3}\ \mathrm{kg}/(\mathrm{m\ s})]^2}\right\}^{1/3}$$

$$=5390\frac{\mathrm{W}}{\mathrm{m^2\ °C}}$$

The total rate of heat transfer over the tube is

$$q_c = \bar{h}_C A_s (T_{sat} - T_s)$$

$$= 5390 \frac{W}{m^2 \, ^\circ C} (\pi)(1 \text{ m})(0.02 \text{ m})(54.4 ^\circ C - 43.3 ^\circ C)$$

$$= 3.76 \text{ kW} \quad \cdot$$

As seen in Prob. 8-28 the local film thickness δ for this situation is given by

$$\delta = 1.63 \times 10^{-4} (x/L)^{1/4} \text{ m}$$

Because δ is much less than the tube diameter D, we are justified in utilizing the flat-plate correlation.

Referring back to Example 8-3 on boiling heat transfer, we see that for the same saturation temperature and temperature difference of $11.1 ^\circ C$ ($= 52 ^\circ F$), the heat-transfer rate from the vertical surface is eleven times larger for this laminar filmwise condensation than for the boiling. However, at high pressures, boiling heat transfer is much more effective than laminar film condensation (see Probs. 8-26 and 8-27).

EXAMPLE 8-7

Referring back to Example 8-6 in which 1.45×10^{-3} kg/s of condensate was produced on a 1-m-long vertical 2-cm-diameter tube, determine the length of a horizontal 2-cm-diameter tube which is required to produce the same condensate mass flow rate.

Solution

First, by comparing Eqs. (8-22) and (8-27), we have

$$\frac{h_{horiz}}{h_{vert}} = \frac{0.725}{0.943} \left(\frac{L}{D} \right)^{1/4} = 0.769 \left(\frac{L}{D} \right)^{1/4}$$

Thus, we see immediately that laminar film condensation is more effective for horizontal tubes if the length L is greater than $2.87D$.

Because $\dot{m}$ is known, we will utilize Eq. (8-19) to solve for the unknown length.

$$L = \frac{A_s}{\pi D} = \frac{\dot{m} h_{\ell v}}{\pi D \bar{h}_C (T_{sat} - T_s)} \tag{a}$$

Assuming laminar flow, Eq. (8-27) is used to calculate $\bar{h}_C$.

$$\bar{h}_C = 0.725 \left\{ \frac{(9.81 \text{ m s}^2)(988 \text{ kg/m}^3)^2 [0.642 \text{ W/(m °C)}]^3 (2370 \text{ kJ/kg})}{(0.02 \text{ m})[0.558 \times 10^{-3} \text{ kg/(m s)}](54.4 \text{ °C} - 43.3 \text{°C})} \right\}^{1/4}$$

$$= 1.08 \times 10^4 \frac{\text{W}}{\text{m}^2 \text{ °C}}$$

Notice that this coefficient is over twice as large as the coefficient obtained for the vertical tube! Substituting into Eq. (a), we have

$$L = \frac{(1.45 \times 10^{-3} \text{ kg/s})(2370 \text{ kJ/kg})}{\pi(0.02 \text{ m})[1.08 \times 10^4 \text{ W/(m}^2 \text{ °C)}](54.4\text{°C} - 43.3\text{°C})}$$

$$= 0.456 \text{ m}$$

To determine whether the flow is actually laminar, we calculate the film Reynolds number. Utilizing Eq. (8-14),

$$\text{Re}_f = \frac{4 \dot{m}}{p\mu_f} \qquad (8\text{-}14)$$

With $p = 2L$ for this horizontal tube geometry, we obtain

$$\text{Re}_f = \frac{4(1.45 \times 10^{-3} \text{ kg/s})}{2(0.456 \text{ m})[0.558 \times 10^{-3} \text{ kg/(m s)}]}$$

$$= 11.4$$

Thus, the flow is laminar and Eq. (8-27) does indeed apply.

It follows that horizontal tube arrangements are often used in condenser design. However, it should be noted that $\bar{h}_C$ can be greater for turbulent film condensation on vertical tubes than for laminar film condensation on horizontal tubes.

Enhancement of film condensation

The key to improving film condensation heat transfer is to decrease the liquid film thickness, which acts as a thermal resistance, and to stimulate turbulence. To accomplish these objectives, rough surfaces and finned tubes have been designed which improve the condensate mass flow $\dot{m}$. Figure 8-15 shows the heat-transfer coefficient for film condensation on horizontal smooth, finned, and rough surfaces. The high heat fluxes

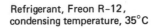

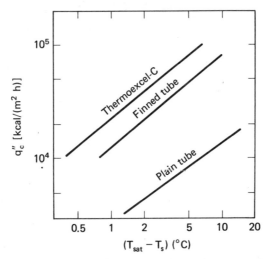

FIGURE 8-15   Condensation-heat-transfer performance of several surfaces. (From Arai et al. [19].)

obtained by the Thermoexcel-C saw teeth surface have been attributed to a superior condensate dropping ability [31].

For more information on the enhancement of film condensation for both internal and external flow systems, the review article by Bergles [18] is suggested.

8-3-3   Dropwise Condensation

Experimental data for dropwise condensation of steam are shown in Fig. 8-16. Note the very high coefficients of heat transfer that are achieved. As a matter of fact, dropwise condensation is one of the most effective of all known heat-transfer mechanisms. Consequently, efforts are continuously being made to harness this dynamic mechanism for industrial use. Various methods have been utilized to attempt to achieve consistent dropwise condensation, such as the use of polymer surface coatings. However, aside from several surface/fluid combinations that are naturally nonwetting (e.g., stainless steel/mercury), dropwise condensation has only been achieved with consistency in steam under carefully controlled conditions. But, as indicated by Tanasawa [32] in a review of this subject, the rapid progress in materials technology will most assuredly clear the way for effective industrial use of dropwise condensation.

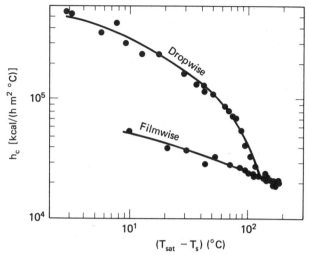

FIGURE 8-16 Heat-transfer coefficients for condensation of steam on a short vertical copper surface. 1 kcal/(h m² °C)= 1.16 W/(m² °C). (From Takeyama and Shimizu [33].)

EXAMPLE 8-8

The *heat pipe* is a recently developed device that utilizes evaporation and condensation to transfer heat extremely effectively. A basic heat pipe system is shown in Fig. E8-8a. A heat pipe consists of a closed pipe lined with a wicking material. A condensible fluid is contained within the pipe, with gas filling the hollow core and liquid permeating the wicking material. When one end of the pipe is exposed to a temperature above the saturation temperature T_{sat} of the fluid, and the other end is exposed to a temperature below T_{sat}, the fluid within the pipe circulates as shown in Fig. E8-8a. This circulation is caused by (1) the evaporation of liquid in the heated portion of the wick, (2) the condensation of fluid

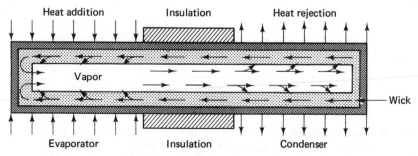

FIGURE E8-8a Basic heat pipe.

back into the cooled section of the wick, and (3) capillary liquid flow within the wick.

Because of the effectiveness of the vaporization and condensation mechanisms and the capillary pumping action, properly designed heat pipes are capable of transferring tens and even hundreds of times as much heat as can be transferred in solid metal bars of equal size. To illustrate, the temperature distribution and heat flux in a sodium−stainless steel heat pipe is shown in Fig. E8-8b. Let us compare the heat transfer in this system with conduction heat transfer in a solid stainless steel 1-in.-diameter 18.5-in.-long rod.

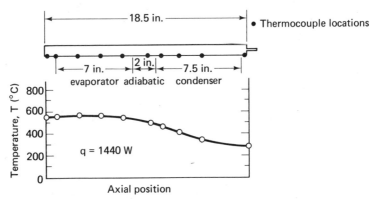

FIGURE E8-8b Axial temperature distributions for sodium heat pipe. (After Dzakowic et al. [34].)

Solution

Referring to Fig. E8-8b, the heat pipe transfers 1440 W across a temperature difference of approximately 270°C. The rate of heat transfer in a stainless steel bar with this temperature drop is

$$q = \frac{kA}{L}(T_1 - T_2)$$

$$= \left(54\,\frac{\text{W}}{\text{m °C}}\right)\frac{\pi(1\text{ in.})^2}{4(18.5\text{ in.})}(270°C)$$

$$= 619\,\frac{\text{W in.}}{100\text{ cm}}\,\frac{2.54\text{ cm}}{\text{in.}}$$

$$= 15.7\text{ W}$$

Comparing this rate of heat transfer with the 1440 W transferred in the heat pipe, we have a factor of about 90.

PROBLEMS

8-1. The Kutateladze [14] correlation for peak boiling heat flux is given by $q''_{crit} = Kh_{fv}\sqrt{\rho_v}\,[\sigma g(\rho_\ell - \rho_v)]^{\frac{1}{4}}$; the average value of K is 0.14, but this parameter ranges from 0.13 to 0.19 for various surface conditions. Determine q''_{crit} for Example 8-2 by utilizing this correlation and compare the result obtained by the use of the Zuber correlation.

8-2. Water at atmospheric pressure is boiled on a mechanically polished stainless steel surface which is heated electrically. Determine the boiling heat flux if the surface temperature is 110°C. Also calculate the critical excess temperature.

8-3. Solve Prob. 8-2 if a scored copper surface is used.

8-4. Water at 100 atm is boiled on a Teflon-pitted stainless steel surface. Determine the boiling heat flux if the excess temperature is 10°C.

8-5. Water at atmospheric pressure is boiled on an electrically heated copper plate. Determine the surface temperature if the power dissipated per unit surface area is 440 kW/m².

8-6. Water at atmospheric pressure is evaporated from a boiling pan of water at a rate of 5×10^{-4} kg/s. The bottom of the copper pan is flat with a 10 cm diameter. Determine the heat flux and the approximate temperature of the bottom surface of the pan.

8-7. A 1-cm-diameter 2-m-long vertical tube is to be used to boil water at 1 atm. Determine the rate of heat transfer and the rate of evaporation if the surface temperature of the tube is 108°C.

8-8. Water at a saturation temperature of 160°C is boiled by a 1-m-long 2-cm-diameter copper tube with surface temperature at 170°C. Determine the rate of heat transfer and the rate of evaporation. Compare these results with predictions for 1 atm and the same excess temperature.

8-9. As seen in Fig. 8-5, the boiling heat flux is dependent on pressure. The optimal pressure for water is approximately 100 atm. Show that the critical heat flux at this pressure is about 4×10^6 W/m².

8-10. Saturated water at 100 atm is boiled by a 1-m-long 2-cm-diameter copper tube with excess surface temperature of 11.1°C. Determine the rate of heat transfer and the rate of evaporation. Compare these results with predictions for 1 atm and the same excess temperature.

8-11. Develop Eqs. (8-23) and (8-24).

8-12. Develop Eqs. (8-28) and (8-29).

8-13. Develop Eq. (8-22).

8-14. Develop Eq. (8-26).

8-15. Steam at a saturation pressure of 0.1 MPa condenses on the outside surface of a 2-cm-O.D. 4-m-long vertical tube with surface temperature equal to 90°F. Assuming that filmwise condensation occurs, determine the total rate of heat transfer and the total rate of condensation.

8-16. Determine the length of tube in Prob. 8-15 required to produce a condensate mass flow rate of 3.15×10^{-3} kg/s.

8-17. Solve Prob. 8-15 for the case of a horizontal tube.

8-18. Solve Prob. 8-16 for the case of a horizontal tube.

8-19. Ammonia at a saturation temperature of $-20°C$ condenses on the surface of a vertical plate 1 m high by 2 m wide. The surface temperature of the plate is $-30°C$. Determine the total rate of heat transfer and the total mass flow rate of condensate over the entire surface.

8-20. Referring to Prob. 8-15, determine the rate of heat transfer from the upper surface if the plate is inclined at an angle of $\pi/4$ rad with the horizontal.

8-21. Freon-12 at a saturation temperature of $-10°C$ condenses on the surface of a 1-m-long 2-cm-diameter vertical tube with a surface temperature of $-15°C$. Determine the total mass flow rate of a condensate.

8-22. Determine the condensation rate for Prob. 8-21 if the tube is horizontal.

8-23. Fifty tubes of 2 cm diameter are arranged in an in-line array with five tubes deep. The tubes, which are at 90°C, are exposed to steam at atmospheric conditions. Determine the rate of condensation heat transfer and the mass flow rate of condensate.

8-24. A 1-cm-diameter 2-m-long vertical tube is to be used to condense steam at 1 atm. Determine the rate of heat transfer and the condensate mass flow rate if the surface temperature of the tubes is 90°C. Compare these results to the predictions for boiling heat transfer in Prob. 8-2. Also determine whether the flow is laminar or turbulent at the bottom of the tube.

8-25. Determine the mass flow rate of condensate for Prob. 8-24 if the tube is horizontal.

8-26. The optimal boiling pressure for water is approximately 100 atm. The boiling-heat-transfer rate at this pressure for an excess temperature of 11.1°C is determined in Prob. 8-10 for a 1-m-long 2-cm-diameter copper tube. For purpose of comparison, determine the rate of condensation heat transfer for a horizontal tube, at a temperature of 11.1°C below the saturation temperature for 100 atm.

8-27. Demonstrate that $\bar{h}_C$ is only slightly dependent on pressure for laminar film condensation, but that it increases with pressure for turbulent conditions.

8-28. Develop an expression for the thickness of laminar film condensation on a vertical flat plate. Then show that δ for Example 8-6 is given by $1.63 \times 10^{-4}(x/L)^{1/4}$ m.

8-29. Steam at atmospheric pressure is to be condensed on a horizontal 2-cm-diameter tube with a surface temperature of 98°C. Determine the tube length required to produce 10^{-5} kg/s condensate.

8-30. Referring back to Chap. 4, show that Eq. (8-8) takes the following more general form for a diffuse opaque nongraybody surface (for $F_{s-R} = 1$):

$$\bar{h}_R = \frac{\sigma \alpha_s \left[(\varepsilon_s/\alpha_s) T_s^4 - T_{sat}^4 \right]}{T_s - T_{sat}}$$

REFERENCES

[1] FARBER, E.A., and R.L. SCORAH, "Heat Transfer to Boiling Water under Pressure," *Trans. ASME*, **70**, 1948, 369–384.

[2] NUKYIYAMA, S., "Maximum and Minimum Values of Heat Transmitted from a Metal to Boiling Water under Atmospheric Pressure," *Japan Soc. Mech. Eng.*, **37**, 1934, 367–394.

[3] COOPER, M. G., "Nucleate Boiling," *Sixth Int. Heat Transfer Conf., Montreal*, **6**, 1978, 463–471.

[4] HAHNE, E., and U. GRIGULL, *Heat Transfer in Boiling*. Washington, D.C.: Hemisphere Publishing Corporation, 1977.

[5] ROHSENOW, W. M., "A Method of Correlating Heat Transfer Data for Surface Boiling Liquids," *Trans. ASME*, **74**, 1952, 969–975.

[6] FRITZ, W. "Grundlagen der Wärmeübertragung beim Verdampfen von Flüssigkeiten," *Chem. Eng. Tech.*, **11**, 1963, 753.

[7] ADDOMS, J. N., *Heat Transfer at High Rates to Water Boiling Outside Cylinders,"* D.Sc. Thesis, Department of Chemical Engineering, Massachusetts Institute of Technology, 1948.

[8] PIRET, E. L., and H. S. ISBIN, "Natural Circulation Evaporation Two-Phase Heat Transfer," *Chem. Eng. Prog.*, **50**, 1954, 305.

[9] CICHELLI, M. T., and C. F. BONILLA, "Heat Transfer to Liquids Boiling under Pressure," *Trans. AIChE*, **41**, 1945, 755–787.

[10] CRYDER, D. S., and A. C. FINALBARGO, "Heat Transmission from Metal Surfaces to Boiling Liquids: Effect of Temperature of the Liquid on Film Coefficient," *Trans. AIChE*, **33**, 1937, 346–362.

[11] VACHON, R. I., G. H. NIX, and G. E. TANGER, "Evaluation of Constants for the Rohsenow Pool-Boiling Correlation," *J. Heat Transfer*, **90**, 1968, 239–247.

[12] ZUBER, N., "On the Stability of Boiling Heat Transfer," *Trans. ASME*, **80**, 1958, 711.

[13] ROHSENOW, W., and P. GRIFFITH, "Correlations of Maximum Heat Transfer Data for Boiling of Saturated Liquids," *Chem. Eng. Progr. Symp. Ser.*, **52**, 1956, 47.

[14] KUTATELADZE, S. S., "Heat Transfer in Condensation and Boiling," *USAEC Rept. AEC–tr–3770*, 1952.

[15] ZUBER, N., M. TRIBUS, and J. W. WESTWATER, "The Hydrodynamic Crisis in Pool Boiling of Saturated and Subcooled Liquids," *Proc. Int. Conf. Developments in Heat Transfer*, ASME, New York, 1962, pp. 230–236.

[16] KREITH, F., *Principles of Heat Transfer*, 3rd ed. New York: Intext Press, Inc., 1973.

[17] BROMLEY, L. A., "Heat Transfer in Stable Film Boiling," *Chem. Eng. Progr.*, **46**, 1950, 221.

[18] BERGLES, A. E., "Enhancement of Heat Transfer," *Sixth Int. Heat Transfer Conf., Toronto*, **6**, 1978, 89–108.

[19] ARAI, N., T. FUKUSHIMA, A. ARAI, T. NAKAJIMA, K. FUJIE, and Y. NAKAYAMA, "Heat Transfer Tubes Enhancing Boiling and Condensation in Heat Exchangers of a Refrigeration Machine, *ASHRAE J.*, **83**, 1977, 58–70.

[20] TONG, L. S., *Boiling Heat Transfer and Two-Phase Flow*. New York: John Wiley & Sons, Inc., 1965.

[21] LUNG, H., K. LATSCH, and H. RAMPF, "Boiling Heat Transfer to Subcooled Water in Turbulent Annular Flow," in *Heat Transfer in Boiling*, by E. Hahne and U. Grigull. Washington, D.C.: Hemisphere Publishing Company, 1977.

[22] HEWITT, G. F., "Critical Heat Flux in Flow Boiling," *Sixth Int. Heat Transfer Conference, Toronto*, **6**, 1978, 143–172.

[23] HSU, Y. Y., and R. W. GRAHAM, *Transport Processes in Boiling and Two-Phase Systems*. New York: McGraw-Hill Book Company, 1976.

[24] COLLIER, J. G., *Convective Boiling and Condensation*. New York: McGraw-Hill Book Company, 1972.

[25] ACKERMAN, J. W., "Pseudoboiling Heat Transfer to Supercritical Pressure Water in Smooth and Ribbed Tubes," *J. Heat Transfer*, **92**, 1970, 490–498.

[26] BUTTERWORTH, D., and G. F. HEWITT, *Two-Phase Flow and Heat Transfer*. Oxford: Oxford University Press, 1977.

[27] NUSSELT, W., "Die Oberflächenkondensation des Wasserdampfes," *Z. Ver. Deut. Ing.*, **60**, 1916, 541–569.

[28] BELL, K. J., and C. B. PANCHAL, "Condensation", *Sixth Int. Heat Transfer Conference, Toronto*, **6**, 1978, 361–375.

[29] MCADAMS, W. H., *Heat Transmission*, 3rd ed. New York: McGraw-Hill Book Company, 1954.

[30] KIRKBRIDE, C. G., "Heat Transfer by Condensing Vapors," *Trans. AIChE*, **30**, 1934, 170–186.

[31] NAKAYAMA, W., T. DAIKOKU, H. KUWAHARA, and K. KAKIZAKI, "High -Performance Heat Transfer Surface Thermoexcel," *Hitachi Rev.*, **24**, 1975, 329–334.

[32] TANASAWA, I. T., "Dropwise Condensation—The Way to Practical Applications," *Sixth Int. Heat Transfer Conf., Toronto*, **6**, 1978, 393–405.

[33] TAKEYAMA, T., and S. SHIMIZU, "On the Transition of Dropwise—Film Condensation," *Fifth Int. Heat Transfer Conf., Tokyo*, **III**, 1974, 274.

[34] DZAKOWIC, G. S., et al., "Experimental study of Vapor Velocity Limit in a Sodium Heat Pipe," *ASME Paper 69-HT-21, National Heat Transfer Conference, Minneapolis. Minn.*, 1969.

CHAPTER 9

CONVECTION HEAT TRANSFER:
Practical Thermal Analysis—
HEAT EXCHANGERS

9-1 INTRODUCTION

Convective heat exchangers are devices that transfer heat via convection and conduction between two fluids which are separated by a wall. Examples of convective heat exchangers include automobile radiators, aircraft oil coolers, and refrigeration and power generation condensers and evaporators. In this chapter we utilize the principles developed in Chaps. 5 through 8 in the practical thermal analysis of various standard types of convective heat exchangers.

9-2 TYPES OF CONVECTIVE HEAT EXCHANGERS

The *double-pipe* system shown in Fig. 9-1 is the simplest type of convective heat exchanger. As seen in this figure, double-pipe heat exchangers involve combinations of tube and annular flow, with the tube and annular flows being in the same direction for parallel-flow arrangements and in opposing

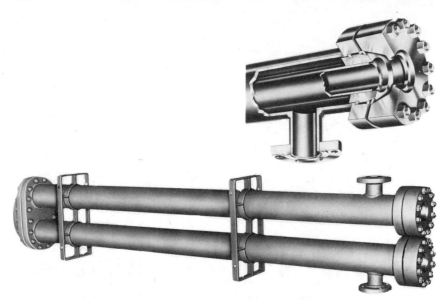

FIGURE 9-1   Double-pipe hairpin heat exchanger with fintube.
(Courtesy of Brown Fintube Company.)

directions for counterflow. However, because of the rather small surface
areas which are available for the transfer of heat, double-pipe heat exchangers are generally reserved for applications that require low to moderate heat-transfer rates.

    In applications that require the transfer of high rates of heat, other
kinds of heat exchangers that provide large surface areas are utilized. Two
types of convective heat exchangers that fall into this category are the
shell-and-tube and crossflow configurations. Typical *shell-and-tube heat
exchangers* are shown in Figs. 5-3 and 9-2. These types of heat exchangers
are generally classified according to the number of tube and shell passes.
The shell-and-tube heat exchanger shown in Fig. 5-3 has one shell pass and

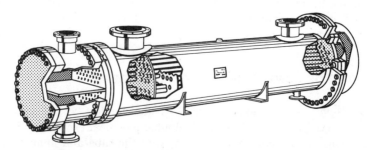

FIGURE 9-2   Typical shell-and-tube heat exchanger—single
pass on shell side, two passes on tube side. (Courtesy of
Conseco.)

one tube pass, whereas the heat exchanger pictured in Fig. 9-2 is constructed with a single shell pass and double tube pass. Heat exchangers with multiple shell and tube passes are often employed in industry.

Crossflow heat exchangers consist of a number of interconnected tubes or passageways which are separated by one or more channels, as illustrated in Fig. 9-3. Heat exchangers of this type are often used when one of the fluids is a gas, and are categorized according to whether the gas flowing across the tube or passageways is *mixed* òr *unmixed*.

FIGURE 9-3 Crossflow air-cooled heat exchanger. (Courtesy of Yuba Heat Transfer Corporation.)

Shell-and-tube and crossflow heat exchangers that provide a large surface area/volume ratio are known as *compact heat exchangers*. These types of exchangers are particularly important in aircraft and spaceflight applications in which size and weight are critical. Representative compact crossflow heat exchangers which were designed for use in aircraft are shown in Fig. 9-4.

The tube diameter and tube bank pitch in shell-and-tube and crossflow heat exchangers should be held to a minimum for most effective heat transfer [1], [2]. However, one must also consider the cost of fabrication, installation, maintenance (cleaning, repairing, etc.) and pressure drops. With these design considerations in mind, one strives to minimize operating cost as well as the initial cost of the unit. Recommendations for minimum tube diameter and pitch specifications are provided by the Tubular Exchanger Manufacturers Association.

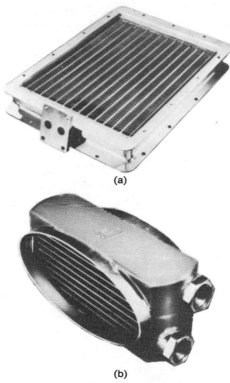

(a)

(b)

FIGURE 9-4 Compact heat exchangers for use in aircraft. (a) Plate-fin air-oil cooler. (b) Air-oil temperature regulator. (Courtesy of Garrett AiResearch.)

The rate of heat transfer in heat exchangers is also often enhanced by the use of longitudinal, circumferential, or spiraled fins, spines, ribs, grooves, and so on. Several of these types of surfaces are shown in Fig. 9-5. Improvements in heat-transfer coefficients of the order 500 to 700% have been reported for such extended surface systems [3]. Tubes with rough surfaces are also sometimes used to enhance the heat transfer.

(a)   (b)

(c)   (d)

FIGURE 9-5 Examples of fintubes. (a) Continuous longitudinal fintube. (b) Perforated longitudinal fintube. (Designed to promote circulation.) (c) Internal–external fintube. (d) Circumferential fintube. (Courtesy of Brown Fintube Company.)

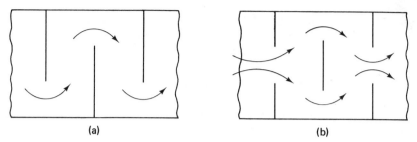

(a)   (b)

FIGURE 9-6 Typical baffle arrangements. (a) Segmental. (b) Disk and doughnut.

Baffles are generally used to increase the heat-transfer performance in commercial shell-and-tube heat exchangers. The use of baffles also provides a means of strengthening the mechanical structure of the exchanger. Two common baffle arrangements are illustrated in Fig. 9-6. Whereas these arrangements do improve the overall heat-transfer efficiency because of the mixing and approximate crossflow pattern that is established, low heat-transfer rates occur in the regions of relative stagnant flow. Another type of baffle which consists of vertical and horizontal rods which provide support and containment for the tubes has recently been introduced. This rodbaffle system has been reported to produce uniform, free-flowing turbulence, thereby eliminating dead or low flow areas in the bundle and increasing the effective heat-transfer area [4].

9-3 EVALUATION, SELECTION, AND DESIGN

As suggested in Chap. 1, the objective of the evaluation function is to determine the total rate of heat transfer and the temperature distribution that can be produced by an existing or proposed system under given operating conditions (i.e., fluids, mass flow rates, and inlet temperatures). The evaluation of heat-exchanger performance provides the basis of (1) prescribing changes in the operating conditions of existing equipment that will bring about improved operation; (2) determining when an existing unit must be cleaned, overhauled, modified, or replaced; and (3) selection of new equipment to perform a specified task. In regard to the selection of new equipment, it is generally most economical to use the manufacturers' standard units for situations in which service conditions allow. Such units have been predesigned with power requirements, size, cleaning, maintenance, fabrication procedures, and cost in mind. For situations in which standard units do not adequately satisfy the requirements, designs can be developed for the modification of a standard unit or for the construction of a nonstandard made-to-order heat exchanger. The objective of the thermal design function is to determine the heat-transfer surface area required to transfer a specified rate of heat for given fluids, mass flow rates, and temperatures. Of course, the full design of heat exchangers also includes mechanical and economical design considerations.

The practical lumped analysis approach provides the basis for evaluation, selection, and thermal design of heat exchangers. Lumped analyses will be developed momentarily for double pipe, shell-and-tube, and crossflow heat exchangers. But first, we want to consider the overall coefficient of heat transfer.

9-4 OVERALL COEFFICIENT OF HEAT TRANSFER

Unlike the single fluid processes studied in the previous sections, the surface temperatures or heat fluxes are not specified for heat exchangers. However, this complication can be overcome by returning to the electrical analogy developed earlier for heat transfer between two fluids separated by a wall. Referring to Fig. 9-7, the differential rate of heat transfer dq_c across the cylindrical wall of radii r_i and r_o and length dx can be written as

$$dq_c = \frac{T_{ib} - T_{ob}}{\dfrac{1}{h_i p_i \, dx} + \dfrac{\ln(r_o/r_i)}{2\pi k \, dx} + \dfrac{1}{h_o p_o \, dx}} \tag{9-1}$$

where T_{ib} and T_{ob} are the local bulk stream temperatures of the tubular and annular fluids, respectively. Observe that T_{ib} and T_{ob} are functions of

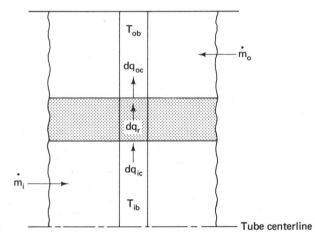

FIGURE 9-7 Sketch for overall heat-transfer coefficient U ($dq_c = dq_{ci} = dq_r = dq_{co}$).

x, except for the case of two-phase flow in condensers and evaporators. Rearranging Eq. (9-1), we obtain

$$dq_c = \frac{p_i\,dx(T_{ib} - T_{ob})}{\dfrac{1}{p_ih_i} + p_i\dfrac{\ln(r_o/r_i)}{2\pi k} + \dfrac{p_i}{p_oh_o}} = U\,dA_i(T_{ib} - T_{ob}) \qquad (9\text{-}2)$$

where $dA_i(=p_i\,dx)$ has been taken as the reference surface area; the local *overall coefficient of heat transfer* U is defined by

$$
\begin{aligned}
U &= \frac{1}{\dfrac{1}{h_i} + p_i\dfrac{\ln(r_o/r_i)}{2\pi k} + \dfrac{p_i}{p_oh_o}} \\[2mm]
&= \frac{1}{\dfrac{1}{h_i} + \dfrac{r_i}{k}\ln\!\left(\dfrac{r_o}{r_i}\right) + \dfrac{1}{h_o}\dfrac{r_i}{r_o}}
\end{aligned} \qquad (9\text{-}3)
$$

(U is sometimes defined with respect to the outside surface area dA_o instead of dA_i.) U is dependent on x for situations in which h_i and/or h_o are functions of x.

Similar to the defining equation for $\bar{h}$, the mean overall coefficient of heat transfer is given by

$$\bar{U} = \frac{1}{A_s}\int_{A_s} U\,dA_s \qquad (9\text{-}4)$$

For a two-dimensional system, $\bar{U}$ is given by

$$\bar{U} = \frac{1}{L}\int_0^L U\,dx \qquad (9\text{-}5)$$

As we shall see, it is the mean overall coefficient $\bar{U}$ that is generally called for in our analysis of heat exchangers. For the two limiting cases in which either the inside or outside thermal resistance controls, $\bar{U}$ can be approximated by $\bar{h}_i$ or $\bar{h}_o$, respectively. Aside from these two important simple cases, $\bar{U}$ usually cannot be obtained analytically for heat-exchanger applications. For cases in which both h_i and h_o are of the same order of magnitude and one or both of these coefficients depend strongly on x, numerical techniques are often required. However, as a first approximation in design work, U and $\bar{U}$ are sometimes calculated by evaluating the coefficients h_i and h_o at the midpoint of the exchanger, for which case $\bar{U} \simeq U$. Alternatively, $\bar{U}$ can be approximated by Eq. (9-3) by setting h_i and

h_o equal to $\bar{h}_i$ and $\bar{h}_o$ at the outlets. Representative values of $\bar{U}$ are given in Table 9-1 for various situations found in practice. In evaluating U or $\bar{U}$, the local convection coefficients h_i and h_o are determined by the means described in Chaps. 6 through 8.

TABLE 9-1   Overall heat-transfer coefficients—representative values

| | $\bar{U}$ | |
Application	[$Btu/(h\,ft^2\,{}^\circ F)$]	[$W/(m^2\,{}^\circ C)$]
Steam condenser	200–10^3	10^3–6×10^3
Freon 12 condenser with water coolant	50–200	300–10^3
Water-to-water	100–300	600–2×10^3
Water-to-oil	20–60	100–400
Water-to-gasoline	60–90	300–500
Steam-to-light fuel oil	30–60	200–400
Steam-to-heavy fuel oil	10–30	60–200
Steam-to-gasoline	50–200	300–10^3
Finned-tube heat exchanger; water in tubes, air over tubes	5–10	30–60
Finned-tube heat exchanger; steam in tubes, air over tubes	5–50	30–300

Source: Based on [7].
1 W/(m² °C)=0.1761 Btu/(h ft² °F)

It should be noted that the overall coefficient of heat transfer is often reduced by fouling deposits such as dirt that accumulate on the heat-exchanger walls. For example, Fig. 9-8 shows severe fouling on the superheater tubes of a boiler. The effect of such deposits is usually accounted for by the *fouling factor* F_f, which is defined by

$$F_f = \frac{1}{U} - \frac{1}{U_{\text{clean}}} \tag{9-6}$$

or

$$U = \cfrac{1}{\cfrac{1}{h_i} + \cfrac{r_i}{k}\ln\left(\cfrac{r_o}{r_i}\right) + \cfrac{1}{h_o}\cfrac{r_i}{r_i} + F_f} \tag{9-7}$$

Representative fouling factors are given in Table 9-2. A review of the recent literature on fouling in heat exchangers is provided by Epstein [5].

Corrosion in heat exchangers is another problem that is encountered in practice, especially in the chemical industry. To overcome this problem,

FIGURE 9-8 Ash deposit fouling on secondary superheater tubes. (From *Steam—Its Generation and Use*. Copyright 1975 by Babcock and Wilcox Company. Used with permission.)

TABLE 9-2 Fouling factors—representative values

	F_f	
Type fluid	*(h ft^2 °F/Btu)*	*(m^2 °C/W)*
Seawater:		
Below 50°C	5×10^{-4}	9×10^{-5}
Above 50°C	1×10^{-3}	2×10^{-4}
Treated boiler feedwater above 50°C	1×10^{-3}	2×10^{-4}
Fuel oil	5×10^{-3}	9×10^{-4}
Quenching oil	4×10^{-3}	7×10^{-4}
Alcohol vapors	5×10^{-4}	9×10^{-5}
Steam, non-oil-bearing	5×10^{-4}	9×10^{-5}
Industrial air	2×10^{-3}	4×10^{-4}
Refrigerating liquid	1×10^{-3}	2×10^{-4}

Source: Based on [6].
1 m^2 °C/W = 5.679 h ft^2 °F/Btu

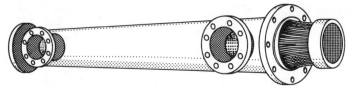

FIGURE 9-9 Shell-and-tube heat exchanger constructed of corrosive-resistant materials. (Courtesy of DuPont De Nemours International S.A.)

glass and plastic tubes and shells are often used. For example, the shell-and-tube heat exchanger shown in Fig. 9-9 is constructed of an outer shell of carbon steel or epoxy resin with Teflon tubes. The excellent thermal conduction properties of borosilicate materials makes the use of glass quite attractive in applications involving corrosive chemicals.

EXAMPLE 9-1

Freon F-12 at $-20°C$ flows in the annulus of a small double-pipe heat exchanger at a rate of 0.265 kg/s. Hot water at 98°C passes through the tube with a mass flow rate of 0.035 kg/s. The heat exchanger is constructed of thin-walled copper tubing with 2-cm inside diameter, 3-cm outside diameter, and 3-m length. Estimate the approximate overall coefficient of heat transfer.

Solution

The properties of the hot water at 98°C are $\rho = 961$ kg/m^3, $\mu = 2.83 \times 10^{-4}$ kg/(m s), $k = 0.68$ W/(m °C), Pr = 1.74, and $c_p = 4.21$ kJ/(kg °C), and the properties of the Freon F-12 at $-20°C$ are $\rho = 1.46 \times 10^3$ kg/m^3, $\mu = 3.43 \times 10^{-4}$ kg/(m s), $k = 0.071$ W/(m °C), Pr = 4.4, and $c_p = 0.907$ kJ/(kg °C).

The Reynolds number of the tubular fluid is calculated first.

$$\text{Re}_i = \frac{D_i U_b}{\nu} = \frac{\dot{m} D_i}{A \mu} = \frac{\dot{m} D_i}{\mu \pi D_i^2 / 4}$$

$$= \frac{4(0.035 \text{ kg/s})}{\left[2.83 \times 10^{-4} \text{ kg/(m s)}\right](\pi)(0.02 \text{ m})}$$

$$= 7870$$

Thus, the flow is transitional turbulent. Because $L/D \ (=150)$ is much greater than 10, we assume that the flow is fully developed with negligible entrance effects. To approximate the coefficient of heat transfer h_i, we utilize Eqs. (6-33) and (6-37).

$$f = 0.079\,(7870)^{-0.25} = 8.39 \times 10^{-3}$$

$$Nu_i = \frac{(8.39 \times 10^{-3}/2)(7870)(1.74)}{1.07 + 12.7\sqrt{8.39 \times 10^{-3}/2}\ (1.74^{2/3} - 1)} = 40$$

$$h_i = Nu_i \frac{k}{D_i} = 40 \frac{0.68\ \text{W/(m }^\circ\text{C)}}{0.02\ \text{m}}$$

$$= 1360 \frac{\text{W}}{\text{m}^2\ ^\circ\text{C}}$$

For the Freon flowing in the annulus, we have

$$Re_o = \frac{\dot{m}_o D_H}{A_o\,\mu} = \frac{\dot{m}(D_o - D_i)}{\mu\pi\left(D_o^2 - D_i^2\right)/4}$$

$$= \frac{4\dot{m}}{\mu\pi(D_o + D_i)}$$

$$= \frac{4(0.265\ \text{kg/s})}{\left[3.43 \times 10^{-4}\ \text{kg/(m s)}(\pi)(0.05\ \text{m})\right]} = 1.97 \times 10^4$$

Utilizing Eqs. (6-32) and (6-33), the Nusselt number and coefficient of heat transfer are calculated for this fully turbulent condition.

$$f_o = (1.58 \ln 1.97 \times 10^4 - 3.28)^{-2} = 6.56 \times 10^{-3}$$

$$Nu_o = \frac{(6.56 \times 10^{-3}/2)(1.97 \times 10^4)(4.4)}{1.07 + 12.7\sqrt{6.56 \times 10^{-3}/2}\ (4.4^{2/3} - 1)} = 124$$

$$h_o = Nu_o \frac{k}{D_H} = 124 \frac{0.071\ \text{W/(m }^\circ\text{C)}}{0.03\ \text{m} - 0.02\ \text{m}} = 880 \frac{\text{W}}{\text{m}^2\ ^\circ\text{C}}$$

The overall coefficient of heat transfer is now obtained by using Eq. (9-3) with $p = p_o \simeq p_i$.

$$\frac{1}{U} = \frac{1}{h_i} + \frac{1}{h_o} = \left(\frac{1}{1360} + \frac{1}{880}\right) \frac{\text{m}^2\ ^\circ\text{C}}{\text{W}}$$

$$U = 534 \frac{\text{W}}{\text{m}^2\ ^\circ\text{C}}$$

If the outlet temperatures and wall temperatures are known, our calculations for h_i and h_o can be refined by evaluating the properties of each fluid at the arithmetic average of its inlet and outlet temperatures and by utilizing the correction factors introduced in Chap. 6. (See Probs. 9-2 and 9-6.)

EXAMPLE 9-2

A double-pipe heat exchanger is constructed of 0.113-in.-thick steel tubing with 0.824-in.-I.D. inner tube and 1.05-in. outer tube. The inside and outside coefficients of heat transfer are 200 Btu/(h ft² °F) and 1000 Btu/(h ft² °F), respectively, and the fouling factor is 5.67×10^{-4} (h ft² °F)/Btu. Calculate the overall coefficient of heat transfer.

Solution

For this case in which the coefficients of heat transfer are independent of x, we have

$$U = \cfrac{1}{\cfrac{1}{h_i} + \cfrac{r_i}{k} \ln\left(\cfrac{r_o}{r_i}\right) + \cfrac{1}{h_o} \cfrac{r_i}{r_o} + F_f}$$

$$= \cfrac{\text{Btu/(h ft² °F)}}{\cfrac{1}{200} + \cfrac{0.412 \ln(0.525/0.412)}{12(26)} + \cfrac{1}{1000} \cfrac{0.412}{0.525} + 5.67 \times 10^{-4}}$$

$$= \cfrac{\text{Btu/(h ft² °F)}}{0.005 + 0.00032 + 0.000785 + 0.000567}$$

$$= 150 \frac{\text{Btu}}{\text{h ft² °F}}$$

EXAMPLE 9-3

Air at 27°C and 1 atm with a mass velocity of 8 kg/(m² s) is to be heated in a crossflow heat exchanger with fifty 3-cm-diameter 2.5-m-long tubes. The tubes are placed five tubes deep in an in-line array with longitudinal and transverse pitches of 1.5 cm. Hot water at 98°C enters the tubes with a mass flow rate of 0.5 kg/s. Estimate the mean overall coefficient of heat transfer for this application.

Solution

Referring back to Example 6-14, the mean coefficient of heat transfer $\bar{h}_o$ for flow of air over the tube bank is approximately 138 W/(m² °C).

To determine the coefficient of heat transfer for the tubular water flow, we must calculate the Reynolds number. The bulk flow rate U_{ib} is

$$U_{ib} = \frac{\dot{m}_i}{\rho_i A_i} = \frac{0.5 \text{ kg/s}}{(961 \text{ kg/m}^3)(\pi/4)(0.03 \text{ m})^2(50)}$$

$$= 0.0147 \frac{\text{m}}{\text{s}}$$

and Re_i becomes

$$\text{Re}_i = \frac{D_i U_{ib}}{\nu} = \frac{(0.03 \text{ m})(0.0147 \text{ m/s})}{0.294 \times 10^{-6} \text{ m}^2/\text{s}}$$

$$= 1500$$

Thus, the flow is laminar and the mean Nusselt number can be approximated by Eq. (6-30),

$$\overline{\text{Nu}} = 1.86 \left(\frac{\text{Re Pr}}{x/D} \right)^{1/3} \tag{6-30}$$

Setting x equal to L, the mean Nusselt number for the water at the outlet of the heat exchanger is

$$\overline{\text{Nu}}_i = 1.86 \left[\frac{(1500)(1.74)}{(2.5 \text{ m})/(0.03 \text{ m})} \right]^{1/3} = 5.87$$

Thus, $\bar{h}_i$ is

$$\bar{h}_i = \overline{\text{Nu}}_i \frac{k}{D_i} = 5.87 \frac{0.68 \text{ W/(m }^\circ\text{C)}}{0.03 \text{ m}}$$

$$= 133 \frac{\text{W}}{\text{m}^2 \,^\circ\text{C}}$$

Assuming that the thermal resistance of the tube wall and fouling is negligible, the mean overall coefficient of heat transfer is approximated by

$$\frac{1}{\overline{U}} = \left(\frac{1}{133} + \frac{1}{138} \right) \frac{\text{m}^2 \,^\circ\text{C}}{\text{W}}$$

$$\overline{U} = 67.7 \text{ W/(m}^2 \,^\circ\text{C)}$$

As in the previous example, this approximation for $\overline{U}$ can be refined to account for property variation with temperature once the outlet temperatures and the wall temperatures are known.

9-5   DOUBLE-PIPE HEAT EXCHANGERS

9-5-1   Introduction

A practical lumped analysis involving the use of the overall coefficient of heat transfer U is developed in this section for double-pipe heat exchangers. Our main objective is to develop relationships for the overall rate of heat transfer q_c and the outlet temperatures of the tubular and annular fluids in terms of the system parameters (tubular and annular fluid inlet temperatures, mass flow rates $\dot{m}_i$ and $\dot{m}_o$, flow directions, and system dimensions.) This approach is patterned after the analysis developed in Sec. 6-2-6 for uniform wall-temperature heating of single-fluid systems. The predictions for q_c resulting from this lumped approach will be put into effectiveness and LMTD formats which are commonly utilized in the heat-exchanger industry.

9-5-2   Lumped Analysis

Referring to the parallel-flow double-pipe heat exchanger shown in Fig. 9-10, two independent equations are developed by applying the first law of thermodynamics to the lumped fluid volumes $A_i L$ and $A_o L$ in the tube and annulus. A third independent equation is developed by applying the first law of thermodynamics to the lumped/differential fluid volumes $A_i\, dx$ and $A_o\, dx$.

First, lumped energy balances are made on the fluids in the tube and annulus. To develop these energy balances, we arbitrarily locate the origin of the x axis at the tube inlet, with q_c taken as positive if heat is transferred

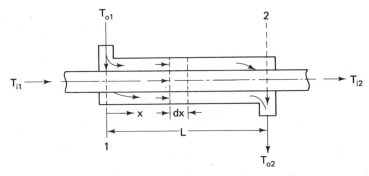

FIGURE 9-10   Parallel-flow double-pipe heat exchanger.

from the tubular fluid to the annular fluid. (Other orientations could just as easily be used. As a matter of fact, heat-exchanger analyses are often developed with reference to the hot and cold fluids rather than with respect to the tubular and annular fluids.) Assuming that the specific heat is uniform and that no phase change occurs, our energy balance is developed as follows:

$$\dot{H}_{i1} = q_c + \dot{H}_{i2} \tag{9-8}$$

or

$$q_c = \dot{m}_i c_{pi}(T_{i1} - T_{i2}) \tag{9-9}$$

in the tube, and

$$\dot{H}_{o1} + q_c = \dot{H}_{o2} \tag{9-10}$$

or

$$q_c = \dot{m}_o c_{po}(T_{o2} - T_{o1}) \tag{9-11}$$

in the annulus. (See Example 9-7 for the case in which one fluid is condensing.)

Next, applying the first law of thermodynamics to the fluid volumes $A_i \, dx$ and $A_o \, dx$, we obtain (for $p_i \simeq p_o \simeq p$)

$$\dot{H}_{ib}|_x = q_c'' p \, dx + \dot{H}_{ib}|_{x+dx} \quad \text{tube} \tag{9-12}$$
$$\dot{H}_{ob}|_x + q_c'' p \, dx = \dot{H}_{ob}|_{x+dx} \quad \text{annulus} \tag{9-13}$$

such that

$$\frac{d\dot{H}_{ib}}{dx} = -q_c'' p \quad \text{tube} \tag{9-14}$$

$$\frac{d\dot{H}_{ob}}{dx} = q_c'' p \quad \text{annulus} \tag{9-15}$$

By expressing q_c'', $d\dot{H}_{ib}$, and $d\dot{H}_{ob}$ in terms of T_{ib}, T_{ob}, and the overall coefficient of heat transfer U, we obtain the following expressions for single-phase fluids with uniform specific heats:

$$\frac{dT_{ib}}{T_{ib} - T_{ob}} = -\frac{Up}{\dot{m}_i c_{pi}} dx \quad \text{tube} \tag{9-16}$$

and

$$\frac{dT_{ob}}{T_{ib} - T_{ob}} = \frac{Up}{\dot{m}_o c_{po}} dx \quad \text{annulus} \tag{9-17}$$

Combining these two equations, we obtain

$$\frac{d(T_{ib} - T_{ob})}{T_{ib} - T_{ob}} = -\eta U p \, dx \qquad \text{tube–annulus} \qquad (9\text{-}18)$$

where

$$\eta = \frac{1}{\dot{m}_i c_{pi}} + \frac{1}{\dot{m}_o c_{po}} \qquad (9\text{-}19)$$

Introducing the capacity rates $C_i (\equiv \dot{m}_i c_{pi})$ and $C_o (\equiv \dot{m}_o c_{po})$, which are both positive for parallel flow, η becomes

$$\eta = \frac{1}{C_i} + \frac{1}{C_o} \qquad (9\text{-}20)$$

The boundary condition for the dependent variable $T_{ib} - T_{ob}$ is

$$T_{ib} - T_{ob} = T_{i1} - T_{o1} \qquad \text{at } x = 0 \qquad (9\text{-}21)$$

The solution of Eqs. (9-18) and (9-21) takes the form

$$\ln \frac{T_{ib} - T_{ob}}{T_{i1} - T_{o1}} = -\eta \overline{U} p x \qquad (9\text{-}22)$$

or

$$\frac{T_{ib} - T_{ob}}{T_{i1} - T_{o1}} = \exp\left(-\eta \overline{U} p x\right) \qquad (9\text{-}23)$$

where $\overline{U}$ is the mean overall coefficient of heat transfer over the length x,

$$\overline{U} = \frac{1}{x} \int_o^x U \, dx \qquad (9\text{-}24)$$

Setting $T_{ib} - T_{ob} = T_{i2} - T_{o2}$ at $x = L$, Eq. (9-23) gives

$$\frac{T_{i2} - T_{o2}}{T_{i1} - T_{o1}} = \exp\left(-\eta \overline{U} A_s\right) \qquad (9\text{-}25)$$

where the product $\overline{U} A_s$ represents the thermal capacity of the heat exchanger.

We now have three equations, Eqs. (9-9), (9-11), and (9-25), such that predictions can be obtained for any three dependent variables for parallel-flow double-pipe heat exchangers.

Because the inlet temperatures are generally specified, the independent variables are taken as T_{i1} and T_{o1} for parallel-flow. Therefore, in order to express q_c in terms of these two variables, Eqs. (9-9) and (9-11) are substituted into Eq. (9-25) to eliminate the outlet temperatures T_{i2} and T_{o2}; that is,

$$\frac{q_c}{C_o} + T_{o1} + \frac{q_c}{C_i} - T_{i1} = (T_{o1} - T_{i1})\exp\left(-\eta \overline{U} A_s\right) \qquad (9\text{-}26)$$

or

$$q_c = \frac{T_{i1} - T_{o1}}{\eta}\left[1 - \exp\left(-\eta \overline{U} A_s\right)\right] \qquad (9\text{-}27)$$

To this point, our attention has been focused on parallel-flow arrangements. This analysis can be easily adapted to the counterflow double-pipe heat exchanger shown in Fig. 9-11 by merely recognizing that the capacity rate $\dot{m}_o c_{po}$ for this system is negative. Thus, Eqs. (9-8) through (9-27) can be applied to counterflow systems by setting

$$\dot{m}_o c_{po} = -C_o \qquad (9\text{-}28)$$

and

$$\eta = \frac{1}{C_i} - \frac{1}{C_o} \qquad (9\text{-}29)$$

However, for counterflow systems, the inlet annular temperature T_{o2} is generally specified and the outlet temperature T_{o1} is unknown. Therefore, we eliminate T_{o1} in Eq. (9-27) by utilizing Eq. (9-11) to obtain

$$q_c = \frac{(T_{i1} - T_{o2})\left[1 - \exp\left(-\eta \overline{U} A_s\right)\right]}{1/C_i - (1/C_o)\exp\left(-\eta \overline{U} A_s\right)} \qquad (9\text{-}30)$$

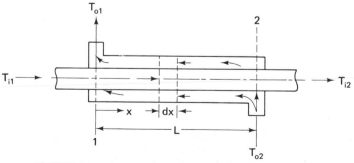

FIGURE 9-11 Counterflow double-pipe heat exchanger.

As indicated in Sec. 6-2-6 heat-exchanger performance is generally presented in terms of the effectiveness ε or the log mean temperature difference LMTD. Therefore, these representations of the performance of double-pipe heat exchangers are presented in the next two sections.

Effectiveness method

The *effectiveness* ε is defined by Eq. (6-71),

$$\varepsilon = \frac{q_c}{q_{c,\,max}} \tag{9-31}$$

For parallel-flow or counterflow heat exchangers, the maximum rate of heat transfer $q_{c,max}$ would occur if the outlet temperature of the fluid with the smaller value of $\dot{m}c_p$ were to be equal to the inlet temperature of the other fluid. By referring to Eqs. (9-9) and (9-11), we see that if the fluid with larger absolute value of $\dot{m}c_p$ were to experience this maximum temperature change, that the temperature change of the other fluid would be even greater, which is impossible. Consequently, we can express the maximum possible rate of heat transfer for either a parallel-flow or a counterflow double-pipe heat exchanger by

$$q_{c,\,max} = C_{min}(T_{i,\,in} - T_{o,\,in}) \tag{9-32}$$

where C_{min} is the minimum absolute capacity rate $|(\dot{m}c_p)_{min}|$. The coupling of this expression with Eqs. (9-27) and (9-31) gives rise to the following relationship for the effectiveness of a parallel-flow double-pipe heat exchanger:

$$\varepsilon = \frac{(T_{i1} - T_{o1})\left[1 - \exp\left(-\eta\overline{U}A_s\right)\right]}{\eta C_{min}(T_{i1} - T_{o1})}$$

$$= \frac{1 - \exp\left(-\eta\overline{U}A_s\right)}{1 + C_{min}/C_{max}} \tag{9-33}$$

Similarly, the use of Eq. (9-30) for counterflow gives

$$\varepsilon = \frac{1 - \exp\left(-\eta\overline{U}A_s\right)}{C_{min}/C_i - (C_{min}/C_o)\exp\left(-\eta\overline{U}A_s\right)} \tag{9-34}$$

This equation reduces to

$$\varepsilon = \frac{1 - \exp\left(-\eta\overline{U}A_s\right)}{1 - (C_i/C_o)\exp\left(-\eta\overline{U}A_s\right)} \tag{9-35}$$

for $C_{min} = C_i$, and

$$\varepsilon = \frac{1 - \exp\left(-\eta \overline{U} A_s\right)}{C_o / C_i - \exp\left(-\eta \overline{U} A_s\right)} \tag{9-36}$$

for $C_{min} = C_o$. A careful comparison of these two equations reveals that they are equivalent.

The parameter $\eta \overline{U} A_s$ appearing in Eqs. (9-33) through (9-36) can be written in terms of the number of transfer units NTU as

$$\eta \overline{U} A_s = (\eta \, C_{min}) \, \text{NTU} \tag{9-37}$$

where NTU is defined as the ratio of the thermal capacity of the heat exchanger $\overline{U} A_s$ to the minimum capacity rate C_{min}; that is,

$$\text{NTU} = \frac{\overline{U} A_s}{C_{min}} \tag{9-38}$$

As in the case of the single-fluid/uniform-wall-temperature system discussed in Chap. 6, the NTU for a heat exchanger provides an index of the size of the exchanger. For the design of heat exchangers, care must be taken to maintain moderate values of NTU so as not to undersize or oversize the system.

It should be noted that the effectiveness can also be written as

$$\varepsilon = \frac{q_c}{q_{c,max}} = \frac{C_{min} |\Delta T_{b,min}|}{C_{min}(T_{i,in} - T_{o,in})} = \frac{|\Delta T_{b,min}|}{T_{i,in} - T_{o,in}} \tag{9-39}$$

where $|\Delta T_{b,min}|$ is the absolute temperature drop of the fluid associated with the minimum capacity rate C_{min}.

The effectiveness provides a particularly convenient means by which the performances of parallel-flow and counterflow heat exchangers can be compared. Such a comparison is shown in Fig. 9-12. Note that the counterflow arrangement is more effective than the parallel-flow system for $C_{min}/C_{max} > 0$. As a result, counterflow exchangers are generally utilized in practice. It is also observed that the maximum effectiveness is approached as $C_{min}/C_{max} \rightarrow 0$. This limiting case actually corresponds to a constant temperature for the fluid with the larger capacity rate. For such conditions, the coefficient of heat transfer associated with the large flow rate is generally quite high, such that $\overline{U}$ essentially reduces to $\overline{h}_{min}$ for a thin-walled tube. Hence, we have in effect heat transfer in a tube or annulus with uniform wall temperature. In accordance with this perspective, we find that Eqs. (9-33) and (9-34) reduce to Eq. (6-73) as C_{min}/C_{max} approaches zero and $\overline{U} = \overline{h}$.

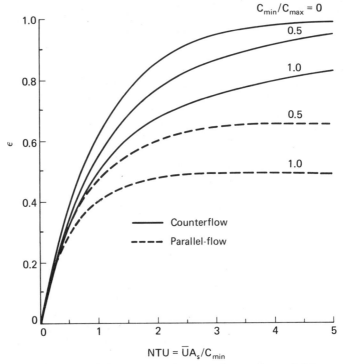

FIGURE 9-12 Effectiveness of counterflow and parallel-flow double-pipe heat exchangers. (Counterflow and parallel-flow curves are same for $C_{min}=0$.)

As mentioned in Chap. 6, the effectiveness format is particularly useful in evaluating the performance of existing heat exchangers with unknown outlet temperatures. Similarly, the effectiveness can be conveniently utilized when a heat exchanger has been tested for certain conditions but is to be used under a different set of service conditions. In addition, the effectiveness approach can be utilized for design-type calculations in which the thermal capacity $\overline{U}A_s$ is unknown and the outlet temperatures are specified. However, this design function is more conveniently handled by the LMTD approach.

EXAMPLE 9-4

Freon F-12 at $-20°C$ flowing at a rate of 0.265 kg/s is heated in a double-pipe heat exchanger. Hot water with a mass flow rate of 0.035 kg/s enters the tubes at a temperature of 98°C. The heat exchanger is constructed of thin-walled copper tubing with 2-cm inside diameter, 3-cm outside diameter and 3-m length. Estimate the total rate of heat transfer and the approximate bulk temperature distributions for a parallel-flow arrangement.

Solution

Our problem is represented in Fig. E9-4. The surface area of this exchanger is only 0.188 m² ($A_s = \pi D_i L$). Referring to Example 9-1, the overall coefficient of heat transfer is approximately 534 W/(m² °C). Assuming that $\overline{U}$ is approximately equal to U, the thermal capacity $\overline{U}A_s$ is taken as 100 W/°C.

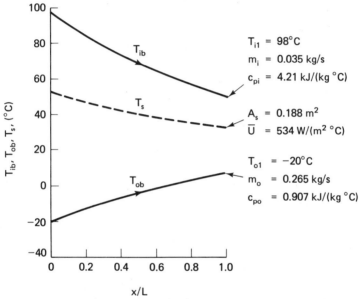

FIGURE E9-4 Predictions for bulk stream temperatures T_{ib} and T_{ob} for parallel-flow arrangement.

Because the inlet temperatures and the flow rates are given, the effectiveness approach will be utilized to determine the rate of heat transfer and the outlet temperatures. Therefore, calculations are first obtained for the parameters C_i, C_o, and NTU.

$$C_i = \dot{m}_i c_{pi} = \left(0.035\,\frac{\text{kg}}{\text{s}}\right)\left(4.21\,\frac{\text{kJ}}{\text{kg °C}}\right) = 0.147\,\frac{\text{kW}}{\text{°C}}$$

$$C_o = \dot{m}_o c_{po} = \left(0.265\,\frac{\text{kg}}{\text{s}}\right)\left(0.907\,\frac{\text{kJ}}{\text{kg °C}}\right) = 0.24\,\frac{\text{kW}}{\text{°C}}$$

$$C_{min} = C_i = 0.147\,\frac{\text{kW}}{\text{°C}} \qquad C_{max} = C_o = 0.24\,\frac{\text{kW}}{\text{°C}}$$

$$\frac{C_{min}}{C_{max}} = \frac{0.147}{0.24} = 0.613$$

$$NTU = \frac{\overline{U}A_s}{C_{min}} = \frac{100 \text{ W/°C}}{0.147 \text{ kW/°C}} = 0.68$$

$$q_{c,max} = C_{min}(T_{i,in} - T_{o,in})$$

$$= 0.147 \frac{kW}{°C}[98°C - (-20°C)]$$

$$= 17.3 \text{ kW}$$

For this parallel-flow arrangement, the effectiveness ε is given by Eq. (9-33) or by Fig. 9-12. For better accuracy, we utilize Eq. (9-33).

$$\varepsilon = \frac{1 - \exp[-NTU(1 + C_{min}/C_{max})]}{1 + C_{min}/C_{max}}$$

$$= \frac{1 - \exp[-0.68(1.61)]}{1.61} = 0.413$$

Utilizing Eq. (9-31), we obtain

$$q_c = \varepsilon q_{c,max} = (0.413)(17.3 \text{ kW}) \qquad \text{(a)}$$

$$= 7.14 \text{ kW}$$

The outlet temperatures are calculated as follows:

$$T_{o2} = \frac{q_c}{C_o} + T_{o1} = \frac{7.14 \text{ kW}}{0.24 \text{ kW/°C}} - 20°C \qquad \text{(b)}$$

$$= 9.75°C$$

$$T_{i2} = T_{i1} - \frac{q_c}{C_i} = 98°C - \frac{7.14 \text{ kW}}{0.147 \text{ kW/°C}} \qquad \text{(c)}$$

$$= 49.4°C$$

To get a better feel for the problem, expressions are developed for the bulk stream temperature distributions. To obtain predictions for T_{ib} and T_{ob}, we first substitute Eq. (9-23) into Eq. (9-16).

$$dT_{ib} = -(T_{i1} - T_{o1})[\exp(-\eta\overline{U}px)]\frac{Up}{C_i}dx \qquad \text{(d)}$$

Integrating from o to x with $\overline{U}$ and U set equal to 534 W/(m² °C) as a first approximation, we obtain

$$\frac{T_{ib} - T_{i1}}{T_{i1} - T_{o1}} = \frac{1}{\eta C_i}[\exp(-\eta\overline{U}px) - 1]$$

$$T_{ib} = T_{i1} + \frac{T_{i1} - T_{o1}}{\eta C_i}\left[\exp\left(-\eta C_{min}NTU\frac{x}{L}\right) - 1\right] \qquad \text{(e)}$$

where NTU$=0.68$ and

$$\eta C_i = \eta C_{min} = 1 + \frac{C_{min}}{C_{max}} = 1.61$$

Similarly, the coupling of Eqs. (9-23) and (9-17) gives rise to an expression for T_{ob} of the form

$$T_{ob} = T_{o1} + \frac{T_{i1} - T_{o1}}{\eta(\dot{m}c_p)_o}\left[1 - \exp\left(-\eta C_{min}\,\text{NTU}\frac{x}{L}\right)\right] \qquad (f)$$

where $(\dot{m}c_p)_o = C_o$ for parallel flow and

$$\eta C_o = \frac{C_o}{C_i} + 1 = 2.63$$

Thus, we have

$$T_{ib} = 98°\text{C} + \frac{(98°\text{C} + 20°\text{C})}{1.61}\left\{\exp\left[-1.61(0.68)\frac{x}{L}\right] - 1\right\} \qquad (g)$$

and

$$T_{ob} = -20°\text{C} + \frac{(98°\text{C} + 20°\text{C})}{2.63}\left\{1 - \exp\left[-1.61(0.68)\frac{x}{L}\right]\right\} \qquad (h)$$

These equations are plotted in Fig. E9-4.

To approximate the wall temperature T_s, we write

$$q_c'' = q_{ic}''$$
$$U(T_{ib} - T_{ob}) = h_i(T_{ib} - T_s)$$

Referring once again to Example 9-1, we approximate h_i by 1360 W/(m² °C) and U by 534 W/(m² °C), with the result

$$\frac{T_{ib} - T_s}{T_{ib} - T_{ob}} = \frac{U}{h_i} \simeq \frac{534}{1360} = 0.393$$

Predictions for T_s obtained by the use of this equation are shown in Fig. E9-4. The bulk stream temperatures and wall temperature at the mid-point are estimated to be $T_{ib} = 79.1°\text{C}$, $T_{ob} = -1.09°\text{C}$, and $T_s = 47.6°\text{C}$.

Now that the outlet fluid temperature, the bulk stream temperatures, and the wall temperature have been estimated, our analysis can be refined by approximating the properties at the arithmetic average of the inlet and outlet temperatures and by utilizing the property correction factors for h_i and h_o given by Eq. (6-42). This refinement is considered in Prob. 9-11.

EXAMPLE 9-5

Repeat Example 9-4 for a counterflow arrangement.

Solution

For counterflow, ε is given by Eq. (9-35).

$$
\begin{aligned}
\varepsilon &= \frac{1 - \exp\left[-\text{NTU}(1 - C_{\min}/C_{\max})\right]}{1 - (C_{\min}/C_{\max})\exp\left[-\text{NTU}(1 - C_{\min}/C_{\max})\right]} \\
&= \frac{1 - \exp\left[-0.68(0.387)\right]}{1 - 0.613\exp\left[-0.68(0.387)\right]} \\
&= 0.438
\end{aligned}
$$

Calculating q_c, we have

$$
q_c = 0.438(17.3 \text{ kW}) = 7.58 \text{ kW}
$$

Continuing with the calculation of the outlet temperatures, we obtain

$$
q_c = C_o(T_{o1} - T_{o2}) = C_i(T_{i1} - T_{i2})
$$

$$
T_{o1} = \frac{q_c}{C_o} + T_{o2} = \frac{7.58 \text{ kW}}{0.24 \text{ kW}/°C} - 20°C \tag{a}
$$

$$
= 11.6°C
$$

$$
T_{i2} = T_{i1} - \frac{q_c}{C_i} = 98°C - \frac{7.58 \text{ kW}}{0.147 \text{ kW}/°C} \tag{b}
$$

$$
= 46.4°C
$$

Notice that q_c is about 6% greater for counterflow than for parallel flow. Referring to Fig. 9-12, we observe that even greater benefit occurs from counterflow operation for larger values of NTU. (As a double check, our calculations can be confirmed by utilizing the LMTD approach —see Prob. 9-12.)

The bulk stream temperature distributions are given by Eqs. (e) and (f) in Example 9-4, with $(\dot{m}c_p)_o$ set equal to $-C_o$ and

$$
\eta = \frac{1}{C_i} - \frac{1}{C_o}
$$

such that

$$\eta C_i = \eta C_{min} = 1 - \frac{C_i}{C_o} = 1 - 0.613 = 0.387$$

and

$$-\eta C_o = -\frac{C_o}{C_i} + 1 = -0.633$$

It should also be noted that the outlet temperature T_{o1} is different for this counterflow arrangement than for the parallel-flow system.

Substituting into Eqs. (e) and (f) of Example 9-4, we have

$$T_{ib} = 98°C + \frac{(98°C - 11.6°C)}{0.387} \left\{ \exp\left[-0.387(0.68)\frac{x}{L} \right] - 1 \right\} \qquad (c)$$

and

$$T_{ob} = 11.6°C + \frac{(98°C - 11.6°C)}{-0.633} \left\{ 1 - \exp\left[-0.387(0.68)\frac{x}{L} \right] \right\} \qquad (d)$$

These distributions are shown in Fig. E9-5.

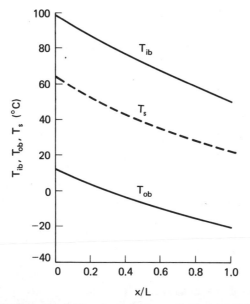

FIGURE E9-5 Predictions for bulk stream temperatures T_{ib} and T_{ob} for counterflow arrangement.

As shown in Example 9-4, the wall temperature T_s can be approximated by

$$\frac{T_{ib} - T_s}{T_{ib} - T_{ob}} = \frac{U}{h_i} \simeq 0.393 \tag{e}$$

This expression is shown in Fig. E9-5.

Log mean temperature difference method

In order to put the results of our analysis into the LMTD format, we utilize Eqs. (9-9), (9-11), and (9-25). Based on Eqs. (9-9) and (9-11), $\dot{m}_i c_{pi}$ and $\dot{m}_o c_{po}$ are expressed as

$$\dot{m}_i c_{pi} = \frac{q_c}{T_{i1} - T_{i2}} \tag{9-40}$$

$$\dot{m}_o c_{po} = \frac{q_c}{T_{o2} - T_{o1}} \tag{9-41}$$

where $\dot{m}_o c_{po}$ is negative for counterflow. Substituting these results into Eq. (9-25) gives

$$q_c = \bar{U} A_s \, \text{LMTD} \tag{9-42}$$

where

$$\begin{aligned}
\text{LMTD} &= \frac{(T_{i1} - T_{o1}) - (T_{i2} - T_{o2})}{\ln \dfrac{T_{i1} - T_{o1}}{T_{i2} - T_{o2}}} \\[2mm]
&= \frac{\Delta T_1 - \Delta T_2}{\ln \dfrac{\Delta T_1}{\Delta T_2}}
\end{aligned} \tag{9-43}$$

with $\Delta T_1 = T_{i1} - T_{o1}$, and $\Delta T_2 = T_{i2} - T_{o2}$.

This result applies to both parallel-flow and counterflow systems. Note the similarity between this equation and our LMTD expression for flow in a tube with uniform wall temperature given by Eq. (6-75).

EXAMPLE 9-6

A double-pipe heat exchanger is used to cool 55 lb_m/min of oil with a specific heat of 0.525 Btu/(lb_m °F) from 122°F to 104°F. A cooling fluid enters the exchanger at 68°F and exits at 77°F. The mean overall coefficient of heat transfer $\bar{U}$ is 88 Btu/(h ft^2 °F). Determine the heat-exchanger surface area A_s for both parallel flow and counterflow.

Solution

Because the exit temperatures are given, we utilize the LMTD approach. The total rate of heat transfer is

$$q_c = (\dot{m}c_p)_o \Delta T_o = (\dot{m}c_p)_i \Delta T_i$$
$$= \left(55\frac{\text{lb}_\text{m}}{\text{min}}\right)\left(0.525\frac{\text{Btu}}{\text{lb}_\text{m}\,{}^\circ\text{F}}\right)(122^\circ\text{F} - 104^\circ\text{F}) = 520\frac{\text{Btu}}{\text{min}} = 31{,}200\frac{\text{Btu}}{\text{h}}$$

Parallel flow

$$\text{LMTD} = \frac{(122^\circ\text{F} - 68^\circ\text{F}) - (104^\circ\text{F} - 77^\circ\text{F})}{\ln\dfrac{122^\circ\text{F} - 68^\circ\text{F}}{104^\circ\text{F} - 77^\circ\text{F}}} = 39^\circ\text{F}$$

$$q_c = \overline{U}A_s\,\text{LMTD}$$

$$A_s = \frac{q_c}{\overline{U}\,\text{LMTD}} = \frac{31{,}200\ \text{Btu/h}}{\left[88\ \text{Btu/(h ft}^2\ {}^\circ\text{F})\right](39^\circ\text{F})} = 9.09\ \text{ft}^2$$

Counterflow

$$\text{LMTD} = \frac{(122^\circ\text{F} - 77^\circ\text{F}) - (104^\circ\text{F} - 68^\circ\text{F})}{\ln\dfrac{122^\circ\text{F} - 77^\circ\text{F}}{104^\circ\text{F} - 68^\circ\text{F}}} = 40.3^\circ\text{F}$$

$$A_s = \frac{q_c}{\overline{U}\,\text{LMTD}} = 8.8\ \text{ft}^2$$

(This type of problem can also be solved by the effectiveness approach, as seen in Prob. 9-18.)

EXAMPLE 9-7

Saturated water with a quality of 0.1 and mass flow rate of 12 kg/min is to be cooled to 80°C at atmospheric pressure. Cooling water at 15°C and 20 kg/min mass flow rate is available. Determine the surface area required for a counterflow double-pipe heat exchanger if the mean overall coefficient of heat transfer is 400 W/(m² °C).

Solution

The enthalpy h_{i1} of the saturated water is calculated first. Referring to the steam tables, we find $h_{f_v} = 2260$ kJ/kg, $h_f = 419$ kJ/kg and the enthalpy at 80°C is 335 kJ/kg. It follows that

$$X_i = 0.1 = \frac{h_{i1} - h_\ell}{h_{\ell v}}$$

$$h_{i1} = 0.1\left(2260\frac{\text{kJ}}{\text{kg}}\right) + 419\frac{\text{kJ}}{\text{kg}} = 645\frac{\text{kJ}}{\text{kg}}$$

The rate of heat transfer q_{c1} required to completely condense 12 kg/min of this two-phase water is

$$q_{c1} = \dot{m}_i(h_{i1} - h_\ell) = 12\frac{\text{kg}}{\text{min}}(645 - 419)\frac{\text{kJ}}{\text{kg}}$$

$$= 2710\frac{\text{kJ}}{\text{min}} = 45.2 \text{ kW}$$

The energy required to subcool the saturated liquid water to 80°C is calculated as follows:

$$q_{c2} = \dot{m}_i(h_\ell - h_{i2}) = 12\frac{\text{kg}}{\text{min}}(419 - 335)\frac{\text{kJ}}{\text{kg}}$$

$$= 1010\frac{\text{kJ}}{\text{min}} = 16.8 \text{ kW}$$

Hence, the total rate of heat transfer q_c is

$$q_c = q_{c1} + q_{c2} = (2710 + 1010)\frac{\text{kJ}}{\text{min}}$$

$$= 3720\frac{\text{kJ}}{\text{min}} = 62 \text{ kW}$$

With this information and noting that the specific heat of water is about 4.18 kJ/(kg °C), we can obtain the cooling fluid outlet temperature.

$$q_c = (\dot{m}c_p)_o(T_{o2} - T_{o1})$$

$$T_{o2} = \frac{3720 \text{ kJ/min}}{(20 \text{ kg/min})[4.18 \text{ kJ/(kg °C)}]} + 15°C$$

$$= 59.5°C$$

Similarly, we can write the following expression for the temperature of the cooling water at the point at which the quality of the saturated fluid is zero:

$$q_{c2} = (\dot{m}c_p)_o (T_{o\ell} - T_{o2})$$

$$T_{o\ell} = \frac{q_{c2}}{(\dot{m}c_p)_o} + T_{o1}$$

$$= \frac{1010 \text{ kJ/min}}{(20 \text{ kg/min})[4.18 \text{ kJ/(kg °C)}]} + 15°C$$

$$= 27.1°C$$

The bulk temperatures of the two fluids are sketched in Fig. E9-7.

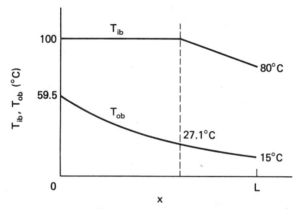

FIGURE E9-7   Predictions for bulk stream temperatures T_{ib} and T_{ob} for counterflow arrangement.

The surface area A_{s2} required to subcool the hot water is determined as follows:

$$\text{LMTD}_2 = \frac{(100°C - 27.1°C) - (80°C - 15°C)}{\ln \dfrac{100°C - 27.1°C}{80°C - 15°C}} = 68.9°C$$

$$q_{c2} = \overline{U} A_{s2} \text{ LMTD}_2$$

$$A_{s2} = \frac{(1010 \text{ kJ/min})[1 \text{ min}/(60 \text{ s})]}{[400 \text{ W}/(m^2 \text{ °C})](68.9°C)} = 0.611 \text{ m}^2$$

Similarly, to determine the area A_{s1} required to condense the vapor, we write

$$LMTD_1 = \frac{(100°C - 59.5°C) - (100°C - 27.1°C)}{\ln \dfrac{100°C - 59.5°C}{100°C - 27.1°C}} = 55.1°C$$

$$q_{c1} = \overline{U} A_{s1} \, LMTD_1$$

$$A_{s1} = 2.04 \text{ m}^2$$

The total surface area of the heat exchanger is

$$A_s = A_{s1} + A_{s2} = 2.04 \text{ m}^2 + 0.611 \text{ m}^2 = 2.65 \text{ m}^2$$

It should be noted that for two-phase problems such as this, one cannot calculate the surface area A_s directly from the overall LMTD. This is because of the change in the shape of the temperature difference curve in the two-phase and single-phase regions (see Fig. E9-7).

EXAMPLE 9-8

A double-pipe heat exchanger is utilized to heat water with a mass flow rate of 10 kg/s from 15°C to 33°C. The heating fluid enters at 75°C with a capacity rate of 25 kW/°C and the mean overall coefficient of heat transfer is 1570 W/(m² °C). Determine the necessary surface area for counterflow and parallel-flow operation.

Solution

Utilizing the LMTD equation, we have

$$A_s = \frac{q_c}{\overline{U} \, LMTD}$$

for both counterflow and parallel-flow arrangements. Approximating c_p by 4.19 kJ/(kg °C), the rate of heat transfer q_c is simply

$$\begin{aligned}
q_c &= (\dot{m}c_p)_i (T_{i2} - T_{i1}) \\
&= \left(10 \frac{\text{kg}}{\text{s}}\right)\left(4.18 \frac{\text{kJ}}{\text{kg °C}}\right)(33°C - 15°C) \\
&= 752 \text{ kW}
\end{aligned}$$

Similarly, the outlet temperature $T_{o,\text{out}}$ of the heating fluid can be calculated as

$$q_c = (\dot{m}c_p)_o (75°C - T_{o,\text{out}})$$

$$T_{o,\text{out}} = 75°C - \frac{752 \text{ kW}}{25 \text{ kW/°C}} = 44.9°C$$

Thus, we can calculate the LMTD.

Counterflow

$$\text{LMTD} = \frac{(44.9°C - 15°C) - (75°C - 33°C)}{\ln\dfrac{44.9°C - 15°C}{75°C - 33°C}} = 35.6°C$$

Parallel flow

$$\text{LMTD} = \frac{(75°C - 15°C) - (44.9°C - 33°C)}{\ln\dfrac{75°C - 15°C}{44.9°C - 33°C}} = 29.7°C$$

Substituting these inputs into Eq. (a), we obtain

$$A_s = \frac{752 \text{ kW}}{\left[1.57 \text{ kW}/(m^2 \text{ °C})\right]\text{LMTD}}$$

$$= 13.5 \text{ m}^2 \qquad \text{for counterflow}$$
$$= 16.1 \text{ m}^2 \qquad \text{for parallel flow}$$

Thus, the counterflow arrangement offers an approximate 20% savings on surface area. However, because of the rather large surface area required for this application, more-compact-type heat exchangers should be considered.

9-6 SHELL-AND-TUBE AND CROSSFLOW HEAT EXCHANGERS

9-6-1 Introduction

Shell-and-tube and crossflow heat exchangers are more complex than double-pipe exchangers and are more difficult to analyze. However, performance criteria have been developed for certain configurations which have both theoretical and experimental bases. These criteria have been developed in the effectiveness and LMTD formats. These practical methods of heat-exchanger analyses will now be considered.

9-6-2 Lumped Analysis

Effectiveness method

The effectiveness of a heat exchanger has already been defined in the context of simple tubular and double-pipe arrangements by Eq. (9-31),

$$\varepsilon = \frac{q_c}{q_{c,\text{max}}} \tag{9-31}$$

where

$$q_{c,\max} = C_{\min}(T_{i,\text{in}} - T_{o,\text{in}}) \tag{9-32}$$

For shell-and-tube and crossflow heat exchangers, the subscripts i and o now represent the tube side and shell or outer side, respectively. The effectiveness of representative shell-and-tube and crossflow heat-exchanger configurations is presented in Fig. 9-13(a) and (b). Similar correlations for ε are given in Appendix H for other shell-and-tube and crossflow heat exchangers. Fairly extensive listings of ε correlation curves are given by Kays and London [8] and Rohsenow and Hartnett [9].

Because q_c can be expressed in terms of NTU ($\equiv \overline{U}A_s / C_{\min}$) and the inlet temperatures, the effectiveness (or NTU) method is best suited for evaluating the performance of existing heat exchangers, but is of less use in developing the design of an exchanger.

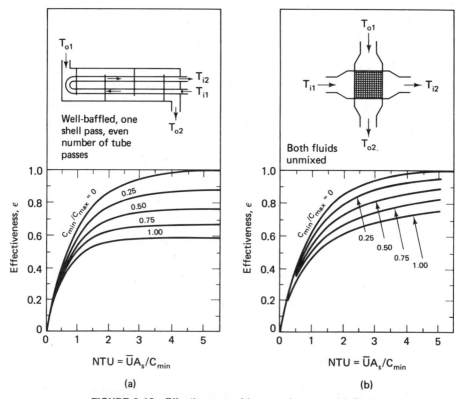

FIGURE 9-13   Effectiveness of heat exchangers. (a) Shell-and-tube system. (b) Crossflow system. (From *Compact Heat Exchangers* by W. M. Kays and A. L. London. Copyright 1964 by McGraw-Hill Book Company. Used with permission of McGraw-Hill Book Company.)

EXAMPLE 9-9

Air at 27°C and 1 atm with a mass velocity of 8 kg/(m² s) is to be heated in a tube bank with fifty 3-cm-diameter 2.5-m-long tubes. The tubes are placed 5 tubes deep in an in-line array with longitudinal and transverse pitches of 4.5 cm. Hot water at 98°C enters the tubes with a mass flow rate of 0.5 kg/s. Determine the rate of heat transfer and the outlet temperature.

Solution

Our problem is represented in Fig. E9-9. The properties of air at 27°C are $\rho = 1.18$ kg/m³, $c_p = 1.01$ kJ/(kg °C), $\mu = 1.85 \times 10^{-5}$ kg/(m s), $k = 0.0262$ W/(m °C), and Pr = 0.708, and the properties of water at 98°C are $\rho = 961$ kg/m³, $c_p = 4.21$ kJ/(kg °C), $\mu = 2.83 \times 10^{-4}$ kg/(m s), $k = 0.68$ W/(m °C), and Pr = 1.74. Referring back to Example 9-3, the overall coefficient of heat transfer $\bar{U}$ for this system is approximately 67.7 W/(m² °C).

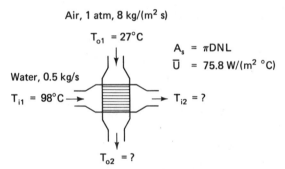

FIGURE E9-9 Crossflow arrangement with one fluid mixed and one fluid unmixed.

The effectiveness of a crossflow system such as this with one fluid mixed and one fluid unmixed is given in Fig. A-H-1(a) in the Appendix. To utilize the effectiveness approach, we calculate the capacity rates and NTU.

$$C_i = C_{\text{unmixed}} = (\dot{m}c_p)_i = \left(0.5 \frac{\text{kg}}{\text{s}}\right)\left(4.21 \frac{\text{kJ}}{\text{kg °C}}\right)$$

$$= 2.11 \frac{\text{kW}}{\text{°C}}$$

$$C_o = C_{\text{mixed}} = (\dot{m}c_p)_o = G_{\text{air}} A_1 c_{po}$$

$$= \left(8 \frac{\text{kg}}{\text{m}^2 \text{ s}}\right)(11)(0.045 \text{ m})(2.5 \text{ m})\left(1.01 \frac{\text{kJ}}{\text{kg °C}}\right)$$

$$= 10 \frac{\text{kW}}{\text{°C}}$$

$$C_{min} = C_i = 2.11 \frac{kW}{°C} \qquad \frac{C_{mixed}}{C_{unmixed}} = \frac{10}{2.11} = 4.74$$

$$\overline{U}A_s = \overline{U}\pi DNL$$

$$= \left(67.7 \frac{W}{m^2 \, °C}\right)(\pi)(0.03 \text{ m})(50)(2.5 \text{ m})$$

$$= 0.798 \frac{kW}{°C}$$

$$NTU = \frac{\overline{U}A_s}{C_{min}} = \frac{0.798 \text{ kW}/°C}{2.11 \text{ kW}/°C} = 0.378$$

Referring to Fig. A-H-1(a), we obtain an effectiveness ε of approximately 0.3. The total rate of heat transfer is calculated as follows:

$$q_{c,max} = C_{min}(T_{i,in} - T_{o,in})$$

$$= 2.11 \frac{kW}{°C}(98°C - 27°C)$$

$$= 150 \text{ kW}$$

$$q_c = \varepsilon q_{c,max} = (0.3)(150 \text{ kW})$$

$$= 45 \text{ kW}$$

The outlet fluid temperatures are now calculated.

$$q_c = C_i(T_{i1} - T_{i2})$$

$$T_{i2} = T_{i1} - \frac{q_c}{C_i} = 98°C - \frac{45 \text{ kW}/°C}{2.11 \text{ kW}/°C}$$

$$= 76.7°C$$

$$q_c = C_o(T_{o2} - T_{o1})$$

$$T_{o2} = \frac{q_c}{C_o} + T_{o1} = \left(\frac{45 \text{ kW}}{10 \text{ kW}/°C} + 27°C\right) = 31.5°C$$

The effect of fluid property variation with temperature can be approximately accounted for by resolving the problem with the thermal properties of the air and water evaluated at the arithmetic average of the inlet and outlet bulk stream temperatures. In addition, the wall temperature at the midpoint of the heat exchange can be estimated, such that $\overline{h}_i$ and $\overline{h}_o$ can be corrected for effects of property variation.

Note that our solution for q_c of 45 kW lies well below the value of 107 kW obtained in Example 6-14 for air flow over a tube bank with tubes maintained at a constant temperature of 98°C. This difference is attributed to the lower tube-wall temperature. As shown in Prob. 9-35, by increasing the mass flow rate of the heating fluid to 5 kg/s, the flow becomes turbulent and q_c is raised to approximately 100 kW.

EXAMPLE 9-10

An unmixed crossflow heat exchanger is used to heat 2.5 kg/s of air from 15°C to 30°C. Hot water enters at 52.5°C. The mean overall coefficient of heat transfer is 300 W/(m² °C) and the total surface area of the heat exchanger is 10 m². Determine the exit-water temperature and capacity rate of the water, and the heat-transfer rate.

Solution

An overview of our problem is shown in Fig. E9-10. Calculating C_o and q_c, we have

$$C_o = (\dot{m}c_p)_o = \left(2.5\,\frac{kg}{s}\right)\left(1.01\,\frac{kJ}{kg\,°C}\right)$$

$$= 2.53\,\frac{kW}{°C}$$

and

$$q_c = (\dot{m}c_p)_o(T_{o2} - T_{o1})$$

$$= 2.53\,\frac{kW}{°C}(30°C - 15°C) = 37.9\ kW$$

If C_o is less than C_i, ε and NTU are given by

$$\varepsilon = \frac{\Delta T_o}{\Delta T_{max}} = \frac{30°C - 15°C}{52.5°C - 15°C} = 0.4$$

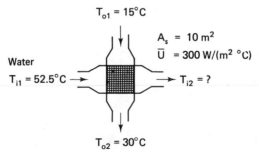

Air, 1 atm, 2.5 kg/s

$T_{o1} = 15°C$

$A_s = 10\ m^2$

$\bar{U} = 300\ W/(m^2\ °C)$

Water

$T_{i1} = 52.5°C$ → → $T_{i2} = ?$

$T_{o2} = 30°C$

FIGURE E9-10 Crossflow arrangement with both fluids unmixed.

and

$$NTU = \frac{\overline{U}A_s}{C_{min}} = \frac{\left[300W/(m^2\ {}^\circ C)\right](10\ m^2)}{2.53\ kW/{}^\circ C} = 1.19$$

With NTU equal to 1.19, we find that Fig. 9-13(b) calls for a value of C_o/C_i which is greater than unity. It follows that $\varepsilon \neq 0.4$, $NTU \neq 1.19$, and $C_{min} = C_i$.

We therefore write NTU and ε in terms of the unknown C_o as follows:

$$NTU = \frac{\overline{U}A_s}{C_{min}} = \frac{\overline{U}A_s}{C_i} = \frac{3\ kW/{}^\circ C}{C_i} \tag{a}$$

$$\varepsilon = \frac{\Delta T_i}{\Delta T_{max}} = \frac{\Delta T_i}{52.5{}^\circ C - 15{}^\circ C}$$

ΔT_i can be expressed in terms of C_i as

$$q_c = C_i\ \Delta T_i$$

or

$$\Delta T_i = \frac{q_c}{C_i}$$

where $q_c = 37.9$ kW. Therefore, we have

$$\varepsilon = \frac{37.9\ kW}{C_i(52.5{}^\circ C - 15{}^\circ C)}$$

$$= 1.01\frac{kW/{}^\circ C}{C_i} \tag{b}$$

We now have three unknowns, C_i, NTU, and ε, and three equations, Eqs. (a) and (b) and the equation represented by Fig. 9-13(b). Therefore, we utilize the following simple iteration scheme:

Iteration 1

Starting with $C_i = 2.25$ kW/°C, it follows that

$$\varepsilon = \frac{1.01\ kW/{}^\circ C}{2.25\ kW/{}^\circ C} = 0.449$$

$$\frac{C_i}{C_o} = \frac{2.25\ kW/{}^\circ C}{2.53\ kW/{}^\circ C} = 0.889$$

$$NTU = \frac{3\ kW/{}^\circ C}{2.25\ kW/{}^\circ C} = 1.33$$

However, from Fig. 9-13(b) we find that $\varepsilon \simeq 0.52$. Utilizing this value of ε, we compute a new value of C_i for our second iteration.

Iteration 2

Utilizing Eq. (b) with $\varepsilon = 0.52$, we have

$$C_i = \frac{1.01 \text{ kW/°C}}{0.52} = 1.94 \frac{\text{kW}}{\text{°C}}$$

$$\frac{C_i}{C_o} = \frac{1.94 \text{ kW/°C}}{2.53 \text{ kW/°C}} = 0.767$$

$$\text{NTU} = \frac{3 \text{ kW/°C}}{1.94 \text{ kW/°C}} = 1.55$$

From Fig. 4-13(b), we obtain $\varepsilon \simeq 0.60$.

Iteration 3

$$C_i = \frac{1.01 \text{ kW/°C}}{0.60} = 1.68$$

$$\frac{C_i}{C_o} = \frac{1.68 \text{ kW/°C}}{2.53 \text{ kW/°C}} = 0.664$$

$$\text{NTU} = \frac{3 \text{ kW/°C}}{1.68 \text{ kW/°C}} = 1.80$$

From Fig. 9-13(b), ε is approximately equal to 0.68.

Continuing this iterative process, ε eventually converges to a value of about 0.75 after a total of eight iterations. It follows that

$$C_i \simeq 1.33 \frac{\text{kW}}{\text{°C}}$$

The outlet temperature of the water is now obtained by writing

$$q_c = C_i(T_{i1} - T_{i2})$$

$$T_{i2} = T_{i1} - \frac{q_c}{C_i} = 52.5\text{°C} - \frac{37.9 \text{ kW}}{1.33 \text{ kW/°C}}$$

$$= 24\text{°C}$$

To evaluate the accuracy of our solution, this problem can be solved by the LMTD approach with T_{o1} set equal to 24°C and A_s unknown (see Example 9-11).

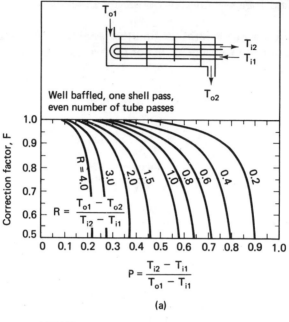

(a)

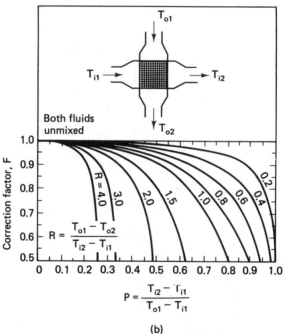

(b)

FIGURE 9-14 LMTD correction factors for heat exchangers. (a) Shell-and-tube system. (Courtesy of Tubular Exchanger Manufacturers Association.) (b) Crossflow system. (From Bowman et al. [10].)

Log mean temperature difference method

The performance of heat exchangers is often expressed in terms of the *log mean temperature difference* LMTD by

$$q_c = \overline{U} A_s [\, F(\text{LMTD}) \,] \qquad (9\text{-}44)$$

where F is a correction factor for the LMTD associated with a counterflow double-pipe heat exchanger. Correction factors are shown in Fig. 9-14(a) and (b) and in Fig. A-H-2 for basic shell-and-tube and crossflow arrangements. Tabulations of correction factors are available in [8]–[10] for various other heat exchangers.

In addition to being useful in the design of heat-exchanger units, this representation of the heat transfer provides a convenient comparison between the performance of the actual process and an ideal counterflow process.

EXAMPLE 9-11

An unmixed crossflow heat exchanger is to be constructed which will heat 2.5 kg/s of air from 15°C to 30°C. Hot water enters at 52.5°C. The mean overall coefficient of heat transfer is 300 W/(m² °C). Determine the required surface area to produce an outlet water temperature of 24°C.

Solution

Our problem is represented by the sketch shown in Fig. E9-11. To determine the unknown surface area A_s, we utilize Eq. (9-44),

$$A_s = \frac{q_c}{\overline{U}[\, F(\text{LMTD}) \,]}$$

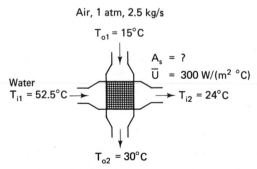

FIGURE E9-11 Crossflow arrangement with both fluids un-mixed.

where

$$q_c = (\dot{m}c_p)_o (T_{o2} - T_{o1})$$

$$= \left(2.5\frac{\text{kg}}{\text{s}}\right)\left(1.01\frac{\text{kJ}}{\text{kg °C}}\right)(30°\text{C} - 15°\text{C}) = 37.9 \text{ kW}$$

Calculating the log mean temperature difference, we have

$$\text{LMTD} = \frac{(52.5°\text{C} - 30°\text{C}) - (24°\text{C} - 15°\text{C})}{\ln\dfrac{52.5°\text{C} - 30°\text{C}}{24°\text{C} - 15°\text{C}}} = 14.7°\text{C}$$

To utilize Fig. 9-14(b), we calculate P and R.

$$P = \frac{T_{i2} - T_{i1}}{T_{o1} - T_{i1}} = \frac{24°\text{C} - 52.5°\text{C}}{15°\text{C} - 52.5°\text{C}} = 0.76$$

$$R = \frac{T_{o1} - T_{o2}}{T_{i2} - T_{i1}} = \frac{15°\text{C} - 30°\text{C}}{24.3°\text{C} - 52.5°\text{C}} = 0.526$$

Referring to Fig. 9-14(b), the correction factor F is approximately 0.85. Substituting into Eq. (a), we obtain

$$A_s = \frac{37.9 \text{ kW}}{\left[300 \text{ W}/(\text{m}^2 \text{ °C})\right]\left[0.85(14.7°\text{C})\right]}$$

$$= 10.1 \text{ m}^2$$

Compared to the 10-m^2 area given in Example 9-10, we have a difference of only 1%. This small discrepancy is a result of the limited accuracy with which the ε and F curves can be read.

EXAMPLE 9-12

A shell-and-tube heat exchanger with one shell pass and two tube passes heats water at 15°C with a mass flow rate of 0.796 kg/s. Heating oil $[c_p = 2.5 \text{ kJ}/(\text{kg °C})]$ enters the tubes at 80°C and exits at 35°C with a mass flow rate of 0.4 kg/s. Determine the surface area of the heat exchanger if the mean overall coefficient of heat transfer is 300 W/$(\text{m}^2 \text{ °C})$.

Solution

Referring to the sketch shown in Fig. E9-12, we have $T_{i1} = 80°\text{C}$, $T_{i2} = 35°\text{C}$, and $T_{o1} = 15°\text{C}$. The heat transfer q_c and outlet temperature

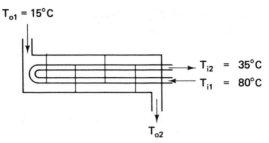

FIGURE E9-12 Shell-and-tube heat exchanger; one shell pass and two tube passes.

T_{o2} of the water are obtained from an energy balance as follows:

$$q_c = (\dot{m}c_p)_o (T_{o2} - T_{o1}) = (\dot{m}c_p)_i (T_{i1} - T_{i2})$$

$$= \left(0.4\frac{\text{kg}}{\text{s}}\right)\left(2.5\frac{\text{kJ}}{\text{kg °C}}\right)(80\text{°C} - 35\text{°C}) = 45 \text{ kW}$$

$$T_{o2} = \frac{(\dot{m}c_p)_i}{(\dot{m}c_p)_o}(T_{i1} - T_{i2}) + T_{o1}$$

$$= \frac{(0.4 \text{ kg/s})[2.5 \text{ kJ}/(\text{kg °C})]}{(0.796 \text{ kg/s})[4.19 \text{ kJ}/(\text{kg °C})]}(80\text{°C} - 35\text{°C}) + 15\text{°C}$$

$$= 28.5\text{°C}$$

Now that both outlet temperatures are known, we are able to calculate the LMTD for ideal counterflow conditions.

$$\text{LMTD} = \frac{(T_{i1} - T_{o2}) - (T_{i2} - T_{o1})}{\ln\dfrac{T_{i1} - T_{o2}}{T_{i2} - T_{o1}}}$$

$$= \frac{(80\text{°C} - 28.5\text{°C}) - (35\text{°C} - 15\text{°C})}{\ln\dfrac{80\text{°C} - 28.5\text{°C}}{35\text{°C} - 15\text{°C}}} = 33.3\text{°C}$$

To utilize Fig. 9-14(a), P and R are also calculated.

$$P = \frac{T_{i2} - T_{i1}}{T_{o1} - T_{i1}} = \frac{35\text{°C} - 80\text{°C}}{15\text{°C} - 80\text{°C}} = 0.692$$

$$R = \frac{T_{o1} - T_{o2}}{T_{i2} - T_{i1}} = \frac{15\text{°C} - 28.5\text{°C}}{35\text{°C} - 80\text{°C}} = 0.3$$

Based on Fig. 9-14(a), we obtain a correction factor F of about 0.9. To calculate A_s, we employ Eq. (9-44).

$$A_s = \frac{q_c}{\overline{U}[F(\text{LMTD})]}$$

$$= \frac{45 \text{ kW}}{[300 \text{ W}/(\text{m}^2 \text{ °C})][(0.9)(33.3°\text{C})]}$$

$$= 5 \text{ m}^2$$

9-7 NONUNIFORM CONDITIONS

Because of the large heat fluxes and temperature differences that are encountered in many modern heat exchanger applications, the effects of variable properties have recently acquired considerable practical importance. As we have already seen, to account for the direct effects of mild temperature induced variations in fluid properties, the properties are often evaluated at the arithmetic average of the inlet and outlet temperatures, and the coefficients of heat transfer are adjusted by the use of correction factors of the type introduced in Chap. 6. Alternatively, the Tubular Exchanger Manufacturers Association (TEMA) has developed charts for determining the average fluid temperature at which the properties are to be evaluated. These average temperature charts [13] account for the fractional change in the overall coefficient of heat transfer over the length of the exchanger and the temperature drop of each fluid.

For situations involving large temperature differences, numerical analyses are generally utilized that account for variations in the properties and overall coefficient of heat transfer with axial location. To illustrate, for a parallel-flow double-pipe system, Eq. (9-18),

$$\frac{d(T_{ib} - T_{ob})}{T_{ib} - T_{ob}} = -\eta U p \, dx \tag{9-18}$$

is put into the finite difference form

$$\frac{(T_{ib} - T_{ob})_{m+1} - (T_{ib} - T_{ob})_m}{(T_{ib} - T_{ob})_m} = -\eta_m U_m p \, \Delta x \tag{9-45}$$

or

$$(T_{ib} - T_{ob})_{m+1} = (T_{ib} - T_{ob})_m (1 - \eta_m U_m p \, \Delta x) \tag{9-46}$$

This equation is coupled with the finite-difference form of the defining equation for the overall coefficient of heat transfer; that is,

$$\Delta q_m = U_m \, p \, \Delta x (T_{ib} - T_{ob})_m \tag{9-47}$$

It follows that the total rate of heat transfer over the entire length of the heat exchanger is given by

$$q_c = \sum_{m=1}^{M} \Delta q_m \tag{9-48}$$

To account for the variable property effects on η_m and U_m, we must also calculate the nodal temperatures $(T_{ib})_m$ and $(T_{ob})_m$. To evaluate these nodal temperatures we write

$$\Delta q_m = C_i \big[(T_{ib})_m - (T_{ib})_{m+1} \big] \tag{9-49}$$

$$\Delta q_m = C_o \big[(T_{ob})_{m+1} - (T_{ob})_m \big] \tag{9-50}$$

The solution of Eqs. (9-46) through (9-50) is quite straightforward. The procedure is broken down as follows: (1) calculate η_m and U_m for $m=1$ at the inlet; (2) calculate Δq_m for $m=1$ from Eq. (9-47); (3) calculate $(T_{ib} - T_{ob})_{m+1}$ for $m=1$ from Eq. (9-46); (4) calculate $(T_{ib})_{m+1}$ and $(T_{ob})_{m+1}$ from Eqs. (9-49) and (9-50); (5) evaluate η_m and U_m for $m=2$; (6) continue steps (1) through (5) for $m=2,3,4,\ldots,M$; and (7) calculate q_c from Eq. (9-48). As in the numerical finite-difference approach to the solution of conduction-heat-transfer problems, to ensure proper accuracy, solutions should be obtained for larger and larger values of M (i.e., smaller and smaller values of Δx) until the calculations for q_c converge.

As indicated in Chaps. 6 and 7, natural convection effects are also sometimes a factor, especially for large temperature differences. This point should be kept in mind when dealing with high heat flux systems.

9-8 CLOSURE

In our introductory study of convective heat exchangers, emphasis has been placed on the development of practical thermal analyses. These are the most common type of heat exchangers used in industry, but it should be noted that various other types of heat exchangers are found in practice. For example, the hot and cold fluids pass alternately over the same surface in systems known as regenerators, and both fluids pass simultaneously through the same passage in mass/heat transfer systems such as cooling towers and distillation columns. For further information on regenerators, mixed fluid systems, and radiative heat exchangers, as well as convective heat exchangers, books by Kays and London [8], Kern [14], and Fraas and Ozisik [15] are recommended.

Finally, it should be recognized that the development of heat exchangers involves mechanical design considerations, fabrication, and testing, as well as thermal analysis. Factors that must be considered in the mechanical design include operating pressure and temperature, stresses produced by thermal expansion, and corrosion. Various aspects of mechanical design, fabrication, and testing of convective heat exchangers are introduced in [2] and [13].

PROBLEMS

9-1. Determine the overall coefficient of heat transfer for a thin-walled double-pipe heat exchanger with inside pipe diameter of 10 cm, outer pipe diameter of 12 cm, and length 1.5 m. Air enters the tube at 250 K with a velocity of 10 m/s and water enters the annulus at 10°C with a velocity of 1 m/s.

9-2. Assuming that the outlet temperatures for the Freon and water of Example 9-1 are 10°C and 50°C, respectively, refine the calculation for h_i and h_o by evaluating the properties at the arithmetic average of the inlet and outlet temperatures.

9-3. Resolve Example 9-1 for the case in which the double-pipe heat exchanger is constructed of 2-mm-thick steel tubing. The inside and outside tube diameters are 2 cm and 3 cm.

9-4. Hot water at 98°C flows at a rate of 25 cm/s through a horizontal steel pipe with 5 cm I.D. and 5 mm wall thickness. The exterior of the pipe is exposed to air at 20°C and 1 atm. Determine the overall heat-transfer coefficient with respect to the inside surface area of the pipe.

9-5. Solve Example 9-3 for the case in which the mass flow rate of the water is 5 kg/s instead of 0.5 kg/s.

9-6. Freon F-12 enters the annulus of a parallel flow double-pipe heat exchanger at -20°C and exits at about 9.75°C. The hot water enters at 98°C and exits at about 49.4°C and the wall temperature at the midpoint is 47.6°C (see Example 9-4). Determine the approximate effect of the temperature-induced property variation of these two fluids on the overall coefficient of heat transfer by evaluating the properties at the arithmetic average of the inlet and outlet temperatures and by utilizing correction factors for the Nusselt numbers.

9-7. Develop expressions for T_{ib} and T_{ob} in a parallel-flow double-pipe heat exchanger for large values of x.

9-8. Develop an expression for q_c in a counterflow heat exchanger for the case in which $C_i = -C_o$ (i.e., for $\eta = 0$). Also calculate and plot T_{ib} and T_{ob} for this special situation.

9-9. Demonstrate that Eqs. (9-35) and (9-36) are equivalent.

9-10. Check the solution to Example 9-4 by computing q_c, T_{i2}, and A_s by the LMTD approach, assuming that $T_{o2} = 9.75°C$.

9-11. Refine the solution of Example 9-4 by evaluating the properties of the fluids at the average temperatures $(T_{i1} + T_{i2})/2$ and $(T_{o1} + T_{o2})/2$, and by utilizing the property correction factor for h_i and h_o given by Eq. (6-42). (See Prob. 9-6.)

9-12. To check the predictions for T_{o1} and T_{i2} in Example 9-5, utilize the LMTD approach to calculate q_c.

9-13. Refine the solution of Example 9-5 by evaluating the properties of the fluids at the average temperatures $(T_{i1} + T_{i2})/2$ and $(T_{o1} + T_{o2})/2$, and by utilizing the property correction factor for h_i and h_o given by Eq. (6-42).

9-14. Demonstrate that Eqs. (9-33) and (9-34) reduce to Eq. (6-74) as C_{min}/C_{max} approaches zero and $\bar{U} = \bar{h}$.

9-15. Solve Example 9-4 for a heat exchanger length of 10 m.

9-16. Solve Example 9-5 for a heat exchanger length of 10 m.

9-17. Water at 20°C flows in the annulus of a double-pipe heat exchanger at a rate of 0.4 kg/s. Hot water at 90°C flows through the tube at a rate of 0.2 kg/s. The heat exchanger is constructed of 1-mm-thick copper tubing with 3 cm inside diameter, 5 cm outside diameter, and 5 m length. Determine the outlet temperatures and the rate of heat transfer for a counterflow arrangement.

9-18. Solve Example 9-6 by the effectiveness approach.

9-19. Water is heated in a double-pipe heat exchanger from 15°C to 40°C. Oil $[c_p = 2.5 \text{ kJ}/(\text{kg} \ °C)]$ with a mass flow of 0.03 kg/s and inlet and outlet temperatures of 80°C and 35°C serves as the heating fluid. Utilize the LMTD approach to determine the required surface area for a counterflow arrangement with $\bar{U} = 300 \text{ W}/(\text{m}^2 \ °C)$. Also determine the mass flow rate of water and the effectiveness of the exchanger.

9-20. Water is heated in a counterflow double-pipe heat exchanger from 35°C to 85°C by an oil with a specific heat of 1.5 kJ/(kg °C) and mass flow rate of 50 kg/min. The oil is cooled from 215°C to 180°C and the overall coefficient of heat transfer is 400 W/(m² °C). Determine the rate of heat transfer, the mass flow rate of the water, and the surface area A_s of the heat exchanger.

9-21. A double-pipe heat exchanger cools water with a mass flow rate of 0.5 kg/s from 50°C to 5°C. A refrigerant enters at $-25°C$ with a capacity rate of 5.0 kW/°C. The mean overall coefficient of heat transfer of the heat exchanger is 1200 W/(m² °C). Determine the surface area required for parallel flow and counterflow operation.

9-22. A double-pipe heat exchanger is utilized to heat oil with a capacity rate of 15 kW/°C from 30°C to 80°C. Hot water enters at 125°C and 2 atm and exits at 95°C. Determine the surface area and rate of heat transfer if $\overline{U}$ is equal to 250 W/(m² °C) for counterflow.

9-23. Water flowing at a rate of 60 kg/min is heated from 35°C to 75°C by oil which undergoes a change from 130°C to 100°C. The overall coefficient of heat transfer is 400 W/(m² °C). Determine the capacity rate of the oil and the surface area of the exchanger for counterflow.

9-24. Calculate and plot the bulk stream temperature distributions for Example 9-6.

9-25. Calculate and plot the bulk stream temperature distributions for Example 9-7.

9-26. Calculate and plot the bulk stream temperature distributions for Example 9-8.

9-27. Solve Example 9-4 for the case in which the heating water is saturated with a quality of 0.5 and a pressure of 1 atm.

9-28. A thin-walled double-pipe heat exchanger is used to heat 55 lb$_m$/min of o'' with a specific heat of 0.525 Btu/(lb$_m$ °F) from 60°F to 100°F. Saturated water at 1 atm enters the exchanger with a quality of 0.9 and a mass flow rate of 30 lb$_m$/min. Estimate the length of the heat exchanger.

9-29. Following the pattern established in the approximate solution developed for the uniform wall temperature boundary condition in Prob. 6-44, the heat transfer in a heat exchanger can be approximated by

$$q_c = \overline{U} A_s \left(\frac{\Delta T_2 + \Delta T_1}{2} \right)$$

As in the analysis of Chap. 6, the substitution of $(\Delta T_2 + \Delta T_1)/2$ for the LMTD relationship for a double-pipe heat exchanger leads to less than 1% error for $0.75 < \Delta T_1 < \Delta T_2 < 1.5$. The coupling of this expression with Eqs. (9-9) and (9-11) provides a simple direct means of approximating q_c and the outlet temperatures. Utilize this approximate approach to estimate the rate of heat transfer and outlet temperature for Example 9-4.

9-30. Utilize the approximate approach of Prob. 9-29 to estimate the rate of heat transfer and outlet temperature for Example 9-5.

9-31. Water flowing at a mass rate of 10 kg/s is to be heated from 15°C to 33°C. The heating fluid enters the shell-and-tube heat exchanger at 75°C with a capacity rate of 25 kW/°C. Assuming that the heat exchanger is rated at an overall coefficient of heat transfer of 1700 W/(m² °C) for these conditions, determine the required surface area A_s. Compare this application to the one of Example 9-8.

9-32. A heat exchanger with one shell pass and two tube passes heats 20,000 lb$_m$/h of water in the tubes from 100°F to 250°F. The heating water enters as saturated liquid at 500°F and exits at 300°F. Determine the rate of heat transfer and the surface area of the heat exchanger if $\overline{U} = 1200$ Btu/(h ft² °F).

9-33. Air at 27°C and 1 atm with a mass velocity of 4 kg/(m² s) is to be heated in a crossflow heat exchanger with 70 3-cm-diameter 2-m-long tubes. The tubes are placed five tubes deep in an in-line array with longitudinal and transverse pitches of 1.25 cm. Hot water at 98°C enters the tubes with a mass flow rate of 1.66 kg/s. Estimate the overall coefficient of heat transfer, the rate of heat transfer q_c, and the exit temperatures of the air and water.

9-34. A crossflow heat exchanger is used to heat 0.6 kg/s of air (mixed) at 15°C. Hot water enters at 45°C with a mass flow rate of 0.5 kg/s. The overall coefficient of heat transfer is 300 W/(m² °C) and the total surface area of the heat exchanger is 5 m². Estimate the exit water temperature and the heat-transfer rate.

9-35. Solve Example 9-9 for the case in which the mass flow rate of the heating water is 5 kg/s instead of 0.5 kg/s.

9-36. Utilize the approximate approach of Prob. 9-29 to estimate the rate of heat-transfer and outlet temperatures for Example 9-9.

9-37. Utilize the approximate approach of Prob. 9-29 to estimate the rate of heat-transfer and outlet temperatures for Example 9-10.

9-38. Water is heated in a shell-and-tube heat exchanger from 35°C to 85°C by an oil with a specific heat of 1.5 kJ/(kg °C) and mass flow rate of 50 kg/min. The oil is to be cooled from 215°C to 180°C and the overall coefficient of heat transfer is 400 W/(m² °C). The water makes one shell pass and the oil makes two tube passes. Estimate the rate of heat transfer, the mass flow rate of the water, and the surface area A_s of the heat exchanger.

9-39. Air flowing at a rate of 2 kg/s is heated in a crossflow heat exchanger from 27°C to 50°C. Water enters the tubes at 90°C and exits at 75°C. Estimate the mass flow rate of the water, the rate of heat transfer, and the surface area of the heat exchanger if $\overline{U}$ is equal to 450 W/(m² °C).

9-40. Utilize the approximate approach of Prob. 9-29 to estimate the rate of heat transfer and outlet temperatures for Example 9-11.

9-41. From a test on a single-shell, two-tube-pass heat exchanger, the following data are available: oil [$c_p = 2$ kJ/(kg °C)] in turbulent flow inside the tubes entered at 70°C at a rate of 0.6 kg/s and exited at 40°C; water flowing on the shell side entered at 15°C and exited at 25°C. A change in service conditions requires the cooling of a similar oil from an initial temperature of 95°C but at two-thirds of the flow rate used in the performance test. Estimate the outlet temperature of the oil for the same water mass flow rate and inlet temperature.

9-42. An unmixed crossflow heat exchanger is to be used to heat air with the hot exhaust gases from a turbine. The air enters with a temperature of 27°C and a mass flow rate of 0.7 kg/s. The temperature, specific heat, and flow rate of the hot gases are 900°C, 1 kJ/(kg °C), and 0.6 kg/s. Determine the surface area of the heat exchanger required to produce an air temperature of 150°C if $\overline{U}$ is equal to 300 W/(m² °C).

9-43. A heat exchanger using 0.5 kg/s of steam at the exhaust of a turbine at a temperature of 50°C and quality of 90% is to be used to heat 3 kg/s of seawater [$c_p = 3.98$ kJ/(kg °C)] from 20°C to 40°C. The heat exchanger is to be sized for one shell pass and four tube passes with 20 parallel tube circuits of 2.50-cm-I.D. and 2.75-cm-O.D. copper tubing. For the clean heat exchanger, the mean heat-transfer coefficients of the steam and water sides are approximately 10,000 W/(m² °C) and 1500 W/(m² °C), respectively. Calculate the tube length required for long-term service.

REFERENCES

[1] PIERSON, O.L., "Experimental Investigation of Influence of Tube Arrangement on Convection Heat Transfer and Flow Resistance in Cross Flow of Gases over Tube Banks," *Trans. ASME*, **59**, 1937, 563–572.

[2] *Steam—Its Generation and Use*, 38th ed. New York: The Babcock and Wilcox Company, 1975.

[3] NAKAYAMA, W., T. DAIKOKU, H. KUWAHARA, and K. KAKIZAKI, "High-Performance Heat Transfer Surface Thermoexcel," *Hitachi Rev.*, **24**, 1975, 329–334.

[4] "Rodbaffle Technology," Phillips Petroleum Company, *Bull. 14896*, 1976.

[5] EPSTEIN, N., "Fouling in Heat Exchangers," *Sixth Int. Heat Transfer Conf., Toronto*, **6**, 1978, 235–253.

[6] *Standards of Tubular Exchanger Manufacturers Association*, 4th ed. New York: Tubular Exchanger Manufacturers Association, Inc., 1959.

[7] MUELLER, A. C., "Thermal Design of Shell-and-Tube Heat Exchangers for Liquid-to-Liquid Heat Transfer," *Eng. Bull., Res. Ser. 121*, Purdue University Engineering Experiment Station, 1954.

[8] KAYS, W. M., and A. L. LONDON, *Compact Heat Exchangers*, 2nd ed. New York: McGraw-Hill Book Company, 1964.

[9] ROHSENOW, W. M., and J. P. HARTNETT, *Handbook of Heat Transfer*. New York: McGraw-Hill Book Company, 1973.

[10] BOWMAN, R. A., A. C. MUELLER, and W. M. NAGLE, "Mean Temperature Difference in Design," *Trans. ASME*, **62**, 1940, 283–294.

[11] HOLMAN, J. P., *Heat Transfer*, 4th ed. New York: McGraw-Hill Book Company, 1976.

[12] KREITH, F., *Principles of Heat Transfer*, 3rd ed. New York: Intext Press, Inc., 1973.

[13] *Standards of Tubular Exchanger Manufacturers Association*, 5th ed. New York: Tubular Exchanger Manufacturers Association, Inc., 1974.

[14] KERN, D. Q., *Process Heat Transfer*. New York: McGraw-Hill Book Company, 1950.

[15] FRAAS, A. P. and M. N. OZISIK, *Heat Exchanger Design*. New York: John Wiley & Sons, Inc., 1965.

CHAPTER 10

CONVECTION HEAT TRANSFER: THEORETICAL ANALYSIS

10-1 INTRODUCTION

In this chapter we introduce the theoretical treatment of both laminar and turbulent convection heat transfer. As indicated in Chap. 5, the complete theoretical analysis of convection-heat-transfer problems requires the use of physical laws in the development of mathematical formulations and solutions for the fluid flow and energy transfer within the fluid. The development of these differential or numerical formulations involves (1) the application of the fundamental principles pertaining to mass, momentum, and energy to a control volume within the flow field; and (2) the use of the particular laws pertaining to fluid shear stress and heat flux. These physical laws, which were presented in Chap. 1, are summarized in Table 10-1. Once the mathematical formulations are developed, solutions are obtained for the velocity and temperature distributions by the various techniques which were introduced in Chap. 3.

We begin our study with laminar flow theory, after which the subject of turbulent flow theory will be taken up. Our introductory study of basic

TABLE 10-1 Fundamental and particular laws: summary

---

1. *Fundamental laws*
   a. Conservation of mass (continuity)
      Rate of creation of mass $= 0$
      $$\Sigma \dot{m}_o - \Sigma \dot{m}_i + \frac{\Delta m_s}{\Delta t} = 0 \qquad\qquad\qquad\text{(i)}$$
   b. Momentum principle (Newton second law of motion)
      Rate of creation of momentum $=$ Sum of forces
      $$\frac{d}{dt}(mu) = \Sigma F_x \qquad\qquad\qquad\qquad\text{(iia)}$$
      $$ma_x = \Sigma F_x \qquad \text{for constant mass} \qquad\text{(iib)}$$
   c. Conservation of energy (first law of thermodynamics)
      Rate of creation of energy $= 0$
      $$\Sigma \dot{E}_o - \Sigma \dot{E}_i + \frac{\Delta E_s}{\Delta t} = 0 \qquad\qquad\qquad\text{(iii)}$$
2. *Particular laws*
   a. Newton law of viscosity, x direction[1]
      $$\tau = \mu \frac{\partial u}{\partial y} \qquad \text{rectangular systems} \qquad\qquad\text{(iv)}$$
      $$\tau = -\mu \frac{\partial u}{\partial r} \qquad \text{cylindrical internal flow system} \qquad\text{(v)}$$
      $$\text{(see Example 10-1)}$$
   b. Fourier law of conduction
      Axial x direction
      $$q_x'' = -k \frac{\partial T}{\partial x} \qquad\qquad\qquad\qquad\text{(vi)}$$
      Direction perpendicular to surface
      $$q_y'' = -k \frac{\partial T}{\partial y} \qquad \text{rectangular system} \qquad\qquad\text{(vii)}$$
      $$q_r'' = k \frac{\partial T}{\partial r} \qquad \text{cylindrical internal flow system} \qquad\text{(viii)}$$
      $$\text{(see Example 10-1)}$$

---

[1]Equations (iv) and (v) are actually boundary layer approximations that neglect the contribution of $\partial v / \partial x$ to τ. Equations (iv) and (v) actually take the more general forms $\tau = \mu(\partial u / \partial y + \partial v / \partial x)$ and $\tau = -\mu(\partial u / \partial r - \partial v / \partial x)$, respectively.

convection-heat-transfer theory will be presented in the context of several classic steady one- and two-dimensional internal and external flows and will feature the use of simple analytical solution techniques.

EXAMPLE 10-1

Write the Newton law of viscosity and the Fourier law of conduction for flow in a circular tube.

Solution

Representative dimensionless velocity and temperature profiles are shown in Fig. E10-1. The usual coordinate for tube flow is r, as shown in this figure. Alternatively, u and T can be expressed in terms of y,

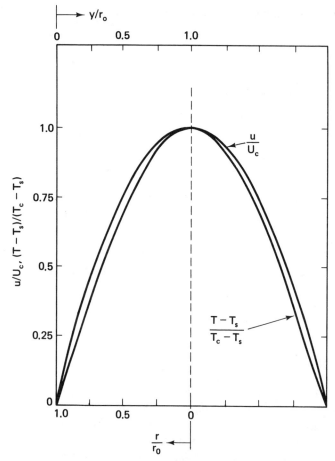

FIGURE E10-1   Velocity and temperature distributions for HFD and TFD laminar flow in a circular tube with uniform wall-flux heating.

where the distance from the wall is defined in terms of r by

$$y = r_0 - r \tag{a}$$

The shear stress can be expressed in terms of the distance from the wall by

$$\tau = \mu \frac{\partial u}{\partial y} \tag{b}$$

It follows from Eq. (a) that

$$= -dr \tag{c}$$

Hence, the Newtonian law of viscosity can also be written as

$$\tau = -\mu \frac{\partial u}{\partial r} \tag{d}$$

Similarly, the heat transfer into the fluid via molecular conduction can be expressed as

$$dq_y = -k \, dA_y \frac{\partial T}{\partial y} \tag{e}$$

or

$$q_y'' = \frac{dq_y}{dA_y} = -k \frac{\partial T}{\partial y} \tag{f}$$

Introducing Eq. (c), the Fourier law of conduction takes the form

$$q_r'' = k \frac{\partial T}{\partial r} \tag{g}$$

Based on this orientation, heat transferred to the fluid is taken as positive.

10-2   LAMINAR FLOW THEORY

10-2-1   Introduction

The general approach to analyzing laminar convection processes involves the development of (1) mathematical formulations for continuity, momentum, and energy transfer within the fluid; (2) solutions for the velocity profiles u and v, and temperature distribution T within the fluid; and (3) predictions for the wall shear stress, wall heat flux (or wall temperature), and/or coefficients of friction and heat transfer. In our study of this fundamental topic, both differential and integral approaches will be used, with consideration given to the major forced convection/natural convection and internal flow/external flow categories. The basic systems to be studied include fully developed flow in tubes and flow over flat plates. The concepts introduced in the study of these classical convection-heat-transfer processes provide a foundation for the theoretical analysis of the more complex problems which are generally encountered in practice.

10-2-2   Fully Developed Flow in Tubes

The problem illustrated in Fig. 10-1 of convection heat transfer for steady laminar flow in a tube is analyzed in this section for fully developed and uniform property conditions, with uniform wall-flux heating. Although emphasis is placed on circular tubes, the concepts introduced in this

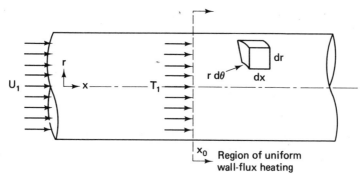

FIGURE 10-1 Laminar flow in a circular tube with uniform wall-flux heating for $x \gg x_0$.

section apply to other internal flow systems with uniform cross-sectional area, such as annuli, channels, and parallel plates.

As mentioned in Chap. 6, hydrodynamic and thermal boundary layers develop in the entrance region for internal flow systems, with the flow becoming fully developed in the region downstream where the hydrodynamic and thermal boundary layers are independent of x. Recall that fully developed conditions are defined by

$$\frac{\partial u}{\partial x} = 0 \qquad \text{HFD} \tag{10-1}$$

$$\frac{\partial}{\partial x}\left(\frac{T - T_s}{T_b - T_s}\right) = 0 \qquad \text{TFD} \tag{10-2}$$

For these conditions, both f and h are independent of x.

In our analysis of this problem, we first develop differential formulations for continuity, momentum, and energy transfer.

Mathematical formulation

To develop the differential formulations for continuity, momentum, and energy, the fundamental laws are applied to a differential volume of fluid $r\, d\theta\, dr\, dx$ shown in Fig. 10-1. The steps required in the development of the differential formulation follow the guidelines established in Chaps. 2 and 3 for conduction heat transfer.

Continuity Applying the principle of conservation of mass to the control volume shown in Figs. 10-1 and 10-2, we obtain

$$\sum \dot{m}_i = \sum \dot{m}_o + \frac{\Delta \cancel{m_s}}{\cancel{\Delta t}}$$

$$d\dot{m}_x + d\dot{m}_r = d\dot{m}_{x+dx} + d\dot{m}_{r+dr} \tag{10-3}$$

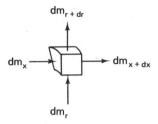

FIGURE 10-2 Conservation of mass relative to differential cylindrical control volume.

where $\Delta m_s / \Delta t = 0$ for steady flow, $d\dot{m}_x = \rho u r \, d\theta \, dr$ and $d\dot{m}_r = \rho v r \, d\theta \, dx$. Utilizing the definition of the partial derivative, this expression takes the form

$$\frac{\partial}{\partial x}(d\dot{m}_x) \, dx + \frac{\partial}{\partial r}(d\dot{m}_r) \, dr = 0 \tag{10-4}$$

or

$$\frac{\partial}{\partial x}(\rho u) + \frac{1}{r}\frac{\partial}{\partial r}(\rho r v) = 0 \tag{10-5}$$

For incompressible flow, the continuity equation becomes

$$\frac{\partial u}{\partial x} + \frac{1}{r}\frac{\partial}{\partial r}(rv) = 0 \tag{10-6}$$

The formulation for continuity is completed by writing the boundary condition for v,

$$v = 0 \quad \text{at} \quad r = r_0 \tag{10-7}$$

(The x boundary condition for u will be taken care of in the formulation for momentum transfer which follows.)

In the HFD region where $\partial u / \partial x = 0$, the continuity equation for incompressible flow reduces to

$$\frac{\partial}{\partial r}(rv) = 0 \tag{10-8}$$

By coupling this simple equation with the boundary condition, we find that $v = 0$ in the HFD region.

Momentum Transfer   To develop the momentum equation, we apply Newton's second law of motion to the element shown in Figs. 10-1 and

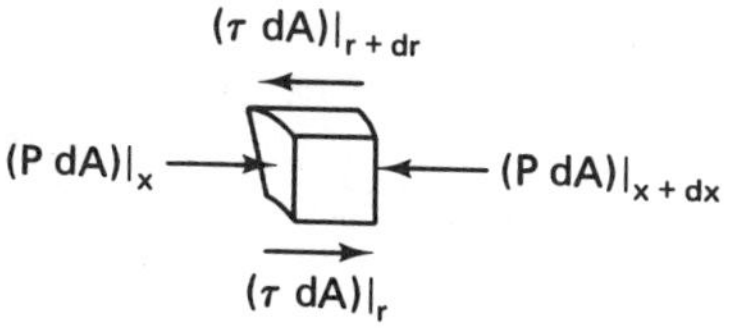

FIGURE 10-3 Summation of forces acting in x direction on differential cylindrical control volume.

10-3; that is,

$$\frac{d}{dt}(mu) = \Sigma F_x \tag{10-9a}$$

where

$$\Sigma F_x = (\tau r \, d\theta \, dx)|_r + (P r \, d\theta \, dr)|_x - (\tau r \, d\theta \, dx)|_{r+dr} - (P r \, d\theta \, dr)|_{x+dx} \tag{10-9b}$$

and

$$\frac{d}{dt}(mu) = dV \frac{d}{dt}(\rho u) \tag{10-9c}$$

where $dV = r \, d\theta \, dr \, dx$. Because ρu can be a function of t, x, and r, we write

$$d(\rho u) = \frac{\partial}{\partial t}(\rho u) dt + \frac{\partial}{\partial x}(\rho u) dx + \frac{\partial}{\partial r}(\rho u) dr \tag{10-10}$$

Hence, $d(mu)/dt$ becomes

$$\frac{d}{dt}(mu) = dV \left[\frac{\partial}{\partial t}(\rho u) + u \frac{\partial}{\partial x}(\rho u) + v \frac{\partial}{\partial r}(\rho u) \right] \tag{10-11}$$

where $\partial(\rho u)/\partial t = 0$ for this steady flow problem.

By combining Eqs. (10-9) and (10-11) and by introducing the definition of the partial derivative, we obtain

$$u \frac{\partial}{\partial x}(\rho u) + v \frac{\partial}{\partial r}(\rho u) = \frac{1}{r} \left[-\frac{\partial}{\partial r}(r\tau) - r \frac{\partial P}{\partial x} \right] \tag{10-12}$$

Utilizing the Newton law of viscosity [Eq. (v) in Table 10-1] and assuming that P is essentially a function of x alone, the momentum equation takes the form

$$u \frac{\partial}{\partial x}(\rho u) + v \frac{\partial}{\partial r}(\rho u) = \frac{1}{r} \frac{\partial}{\partial r}\left(\mu r \frac{\partial u}{\partial r} \right) - \frac{dP}{dx} \tag{10-13}$$

or, for constant properties,

$$u\frac{\partial u}{\partial x} + v\frac{\partial u}{\partial r} = \frac{v}{r}\frac{\partial}{\partial r}\left(r\frac{\partial u}{\partial r}\right) - \frac{1}{\rho}\frac{dP}{dx} \qquad (10\text{-}14)$$

The differential momentum equation is coupled with boundary conditions for u of the form

$$u = U_1 \qquad \text{at} \quad x = 0 \qquad (10\text{-}15)$$

for a uniform distribution at the entrance,

$$u = 0 \qquad \text{at} \quad r = r_0 \qquad (10\text{-}16)$$

for no slip at the wall, and

$$\frac{\partial u}{\partial r} = 0 \qquad \text{at} \quad r = 0 \qquad (10\text{-}17)$$

because of symmetry.

For HFD flow, $\partial u/\partial x = 0$ and $v = 0$ (based on continuity) such that our differential formulation for momentum transfer in a uniform property fluid is given by Eqs. (10-16), (10-17), and

$$\frac{v}{r}\frac{d}{dr}\left(r\frac{du}{dr}\right) - \frac{1}{\rho}\frac{dP}{dx} = 0 \qquad (10\text{-}18)$$

Referring back to Eq. (10-11), we see that the rate of creation of momentum for incompressible flow (ma_x) is equal to zero for HFD conditions, such that the momentum equation actually represents the force balance $\Sigma F_x = 0$.

Energy Transfer   In analyzing the energy transfer for this problem, we assume that the effects of potential energy, kinetic energy, viscous dissipation, axial conduction, and thermal radiation are negligible. (Kinetic energy and/or viscous dissipation are important for viscous fluids such as oil and in high-speed aerodynamic problems, and the effects of axial conduction are sometimes significant for liquid metals.)

Applying the first law of thermodynamics to the control volume dV shown in Fig. 10-4 we write

$$\Sigma \dot{E}_i = \Sigma \dot{E}_o + \frac{\Delta E_s}{\Delta t}$$

$$d\dot{H}_x + d\dot{H}_r + dq_{r+dr} = d\dot{H}_{x+dx} + d\dot{H}_{r+dr} + dq_r \qquad (10\text{-}19)$$

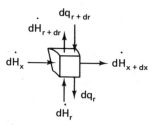

FIGURE 10-4 Conservation of energy relative to differential cylindrical control volume.

where $\Delta E_s/\Delta t = 0$ for steady-state conditions, $d\dot{H}_x = \rho c_p u T r \, d\theta \, dr$ and $d\dot{H}_r = \rho c_p v T r \, d\theta \, dx$ for a single-phase fluid. Notice that the designation of radial conduction heat transfer into the control volume by dq_{r+dr} is consistent with out standard thermodynamic orientation in which heat transferred from the surface into the fluid is taken as positive.

Utilizing the definition of the partial derivative, it follows that

$$\frac{\partial}{\partial r}(dq_r)dr = \frac{\partial}{\partial x}(d\dot{H}_x)dx + \frac{\partial}{\partial r}(d\dot{H}_r)dr \qquad (10\text{-}20)$$

Introducing the Fourier law of conduction [Eq. (viii) in Table 10-1], this equation takes the form

$$\frac{\partial}{\partial x}(\rho c_p u T) + \frac{1}{r}\frac{\partial}{\partial r}(\rho c_p r v T) = \frac{1}{r}\frac{\partial}{\partial r}\left(rk\frac{\partial T}{\partial r}\right) \qquad (10\text{-}21)$$

Expanding this expression, we obtain

$$\rho u \frac{\partial}{\partial x}(c_p T) + \frac{\rho r v}{r}\frac{\partial}{\partial r}(c_p T) + c_p T\left[\frac{\partial}{\partial x}(\rho u) + \frac{1}{r}\frac{\partial}{\partial r}(\rho r v)\right] = \frac{1}{r}\frac{\partial}{\partial r}\left(rk\frac{\partial T}{\partial r}\right) \qquad (10\text{-}22)$$

where the bracketed term is zero because of continuity [Eq. (10-5)]. For constant properties, this equation reduces to

$$u\frac{\partial T}{\partial x} + v\frac{\partial T}{\partial r} = \frac{\alpha}{r}\frac{\partial}{\partial r}\left(r\frac{\partial T}{\partial r}\right) \qquad (10\text{-}23)$$

Assuming that heating is initiated at x_0, the thermal boundary conditions are

$$T = T_1 \quad \text{at} \quad x = x_0 \qquad (10\text{-}24)$$

for a uniform inlet temperature distribution T_1,

$$\frac{\partial T}{\partial r} = 0 \quad \text{at} \quad r = 0 \qquad (10\text{-}25)$$

for symmetrical heating, and

$$k\frac{\partial T}{\partial r} = q_0'' \qquad \text{at} \quad r = r_0 \qquad (10\text{-}26)$$

for a uniform wall-heat flux condition.

Because $v = 0$ in the HFD region, our energy equation for uniform properties becomes

$$u\frac{\partial T}{\partial x} = \frac{\alpha}{r}\frac{\partial}{\partial r}\left(r\frac{\partial T}{\partial r}\right) \qquad (10\text{-}27)$$

For TFD conditions, $\partial T/\partial x$ is prescribed in accordance with the defining equation for TFD flow given by Eq. (10-2). This point will be expanded upon momentarily.

Solution

To solve for the rate of convection heat transfer for laminar flow in a tube, the fluid flow and energy equations must be solved for the velocity and temperature distributions. Once these distributions are known, the friction factor and coefficient of heat transfer can be determined. For problems involving variable property developing flow, the more general continuity, momentum, and energy equations given by Eqs. (10-5), (10-13), and (10-22) must be solved simultaneously. For the somewhat simpler case of uniform property developing flow, Eqs. (10-6) and (10-14) are first solved for the velocity distribution, after which Eq. (10-23) is solved for the temperature profile. Numerical and approximate analytical solutions are available in the literature for these type problems in which the flow is developing [1]–[9].

For our case involving uniform property fully developed laminar flow, an analytical solution is first developed for the fluid-flow aspects of the problem, after which predictions are developed for the temperature distribution and coefficient of heat transfer.

Fluid Flow—HFD Region   For HFD flow, $v = 0$ and the momentum equation and boundary conditions are given by

$$\frac{v}{r}\frac{d}{dr}\left(r\frac{du}{dr}\right) - \frac{1}{\rho}\frac{dP}{dx} = 0 \qquad (10\text{-}18)$$

and

$$u = 0 \quad \text{at} \quad r = r_0 \qquad \frac{du}{dr} = 0 \quad \text{at} \quad r = 0 \qquad (10\text{-}16,17)$$

To solve this problem, we first develop a solution for the dimensionless velocity distribution.

Dimensionless Velocity Distribution. Separating the variables in Eq. (10-18) and integrating, we have

$$r\frac{du}{dr} = \frac{1}{\mu}\frac{dP}{dx}\frac{r^2}{2} + C_1 \tag{10-28}$$

where C_1 is set equal to zero in accordance with Eq. (10-17). Separating the variables and integrating once again, the velocity profile takes the form

$$u = \frac{1}{\mu}\frac{dP}{dx}\frac{r^2}{4} + C_2 \tag{10-29}$$

Utilizing the no slip condition at the wall, C_2 is given by

$$C_2 = -\frac{1}{\mu}\frac{dP}{dx}\frac{r_0^2}{4} \tag{10-30}$$

such that our solution for u is

$$u = -\frac{1}{4\mu}\frac{dP}{dx}r_0^2\left[1 - \left(\frac{r}{r_0}\right)^2\right] \tag{10-31}$$

Alternatively, by setting u equal to the centerline velocity U_c at r equal to zero, we have the somewhat more convenient expression,

$$\frac{u}{U_c} = 1 - \left(\frac{r}{r_0}\right)^2 \tag{10-32}$$

Although we now have a theoretical expression for the velocity profile, our solution is incomplete because U_c and dP/dx are as yet unknown.

Bulk Stream Velocity and Friction Factor. In order to express these unknown parameters in terms of the bulk stream velocity U_b which is equal to U_1, we couple Eq. (10-32) with the defining equation for U_b, Eq. (6-2), as follows:

$$U_b = \frac{2}{r_0^2}\int_0^{r_0} ru\,dr \tag{10-33}$$

$$= \frac{2}{r_0^2}\int_0^{r_0} rU_c\left[1 - \left(\frac{r}{r_0}\right)^2\right]dr$$

$$= \frac{U_c}{2} = -\frac{1}{8\mu}\frac{dP}{dx}r_0^2 \tag{10-34}$$

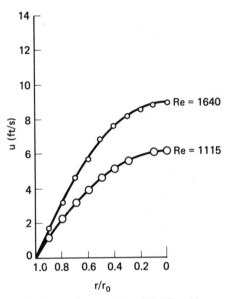

FIGURE 10-5 Comparison of Eq. (10-35) with experimental data for u for HFD laminar flow in circular tube. (Data from Senecal [10] for laminar flow of air in a 0.75-in.-I.D. tube.)

The substitution of this result back into Eq. (10-32) gives

$$u = 2U_b\left[1 - \left(\frac{r}{r_0}\right)^2\right] \tag{10-35}$$

where $U_b = U_1$. This expression is shown in Fig. 10-5 to agree very well with experimental data.

Now that the velocity distribution is fully specified, an expression can be developed for the wall shear stress τ_0 by the use of the Newton law of viscous shear.

$$\tau_0 = \mu\frac{du}{dy}\bigg|_0 = -\mu\frac{du}{dr}\bigg|_{r_0} = \mu 2U_b\left(\frac{2r}{r_0^2}\right)\bigg|_{r_0}$$

$$= \mu 2U_b\left(\frac{2r}{r_0^2}\right)\bigg|_{r_0} = 4\frac{\mu U_b}{r_0} \tag{10-36}$$

It follows that the Fanning friction factor takes the form

$$f = \frac{\tau_0}{\rho U_b^2/2} = \frac{8\mu U_b}{\rho U_b^2 r_0} = 16\frac{\nu}{DU_b} = \frac{16}{\mathrm{Re}} \tag{10-37}$$

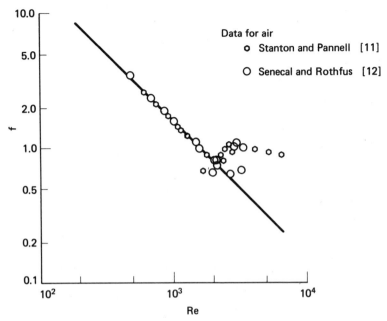

FIGURE 10-6 Comparison of Eq. (10-37) with experimental data for f for HFD laminar flow in circular tubes.

This equation is compared with experimental data in Fig. 10-6. The agreement between theory and experiment is exceptional for Reynolds numbers below the transitional value of approximately 2000.

EXAMPLE 10-2

Show that the shear stress τ varies linearly with r or y for HFD laminar flow in a tube.

Solution

To see the behavior of τ, we write the Newton law of viscosity.

$$\tau = \mu \frac{\partial u}{\partial y} = -\mu \frac{\partial u}{\partial r}$$

With u given by Eq. (10-35),

$$u = 2U_b \left[1 - \left(\frac{r}{r_0} \right)^2 \right] \qquad (10\text{-}35)$$

we have

$$\tau = \mu 2 U_b \left(\frac{2r}{r_0^2} \right)$$

such that

$$\tau_0 = 4 \frac{\mu U_b}{r_0}$$

Thus, τ/τ_0 is given by

$$\frac{\tau}{\tau_0} = \frac{r}{r_0} = 1 - \frac{y}{r_0}$$

τ is indeed linear.

Energy Transfer—TFD Region   With the velocity distribution specified by Eq. (10-35) for HFD conditions, the energy equation becomes

$$2 U_b \left[1 - \left(\frac{r}{r_0} \right)^2 \right] \frac{\partial T}{\partial x} = \frac{\alpha}{r} \frac{\partial}{\partial r} \left(r \frac{\partial T}{\partial r} \right) \qquad (10\text{-}38)$$

For the TFD region, the term $\partial T/\partial x$ can be specified in terms of T, T_s, and T_b by utilizing the defining equation for TFD conditions,

$$\frac{\partial}{\partial x} \left(\frac{T - T_s}{T_b - T_s} \right) = 0 \qquad (10\text{-}2)$$

and by recognizing that the local coefficient of heat transfer h is independent of x. For a uniform wall-heat-flux boundary condition, it follows from the general Newton law of cooling,

$$q_c'' = h(T_s - T_b) \qquad (10\text{-}39)$$

that $T_s - T_b$ is independent of x. Thus,

$$\frac{dT_b}{dx} = \frac{dT_s}{dx} \qquad (10\text{-}40a)$$

such that Eq. (10-2) gives

$$\frac{\partial T}{\partial x} = \frac{dT_s}{dx} = \frac{dT_b}{dx} \qquad (10\text{-}40b)$$

Hence, the energy equation takes the one-dimensional form

$$2 U_b \left[1 - \left(\frac{r}{r_0} \right)^2 \right] \frac{dT_b}{dx} = \frac{\alpha}{r} \frac{d}{dr} \left(r \frac{dT}{dr} \right) \qquad (10\text{-}41)$$

for uniform wall-flux heating.

This equation can be further simplified by returning to the lumped analysis in Chap. 6, in which T_b is given by [from Eq. (6-57)]

$$T_b = T_1 + \frac{q_0'' p x}{\dot{m} c_p} \tag{10-42}$$

This equation actually applies to both the thermal entrance and fully developed regions, and is shown in Fig. 6-10. Hence, for a uniform wall-heat-flux boundary condition, dT_b/dx is given by

$$\frac{dT_b}{dx} = \frac{q_0'' p}{\dot{m} c_p} = \frac{4 q_0''}{\rho c_p U_b D} \tag{10-43}$$

such that the energy equation becomes

$$\frac{4 q_0''}{k} \left(\frac{r}{r_0} \right) \left[1 - \left(\frac{r}{r_0} \right)^2 \right] = \frac{d}{dr} \left(r \frac{dT}{dr} \right) \tag{10-44}$$

This equation is coupled with the r boundary conditions

$$\frac{\partial T}{\partial r} = 0 \qquad \text{at} \quad r = 0 \tag{10-25}$$

$$k \frac{\partial T}{\partial r} = q_0'' \qquad \text{at} \quad r = r_0 \tag{10-26}$$

Following the pattern established in the solution of the fluid-flow problem, we first develop a solution for the dimensionless temperature distribution.

Dimensionless Temperature Distribution. A first integration of Eq. (10-44) from 0 to r and the use of Eq. (10-25) gives

$$r \frac{dT}{dr} = \frac{4 q_0''}{r_0 k} \left(\frac{r^2}{2} - \frac{r^4}{4 r_0^2} \right) \tag{10-45}$$

[This same result is obtained by integrating from r_0 to r with the condition at the wall prescribed by Eq. (10-26).] A second integration gives

$$T = \frac{4 q_0''}{r_0 k} \left(\frac{r^2}{4} - \frac{r^4}{16 r_0^2} \right) + C_1 \tag{10-46}$$

As just indicated, both boundary conditions satisfy Eq. (10-45), but neither can be utilized to obtain C_1 in Eq. (10-46). We circumvent this anomaly by setting T equal to T_s at $r = r_0$. However, it is important to note that T_s is still an unknown function of x that eventually must be evaluated in terms of the specified input q_0''. The use of this intermediate step gives

$$C_1 = T_s - \frac{4q_0''}{r_0 k} \frac{3}{16} r_0^2 \qquad (10\text{-}47)$$

such that the temperature profile becomes

$$T - T_s = -\frac{q_0'' r_0}{k} \left[\frac{3}{4} - \left(\frac{r}{r_0}\right)^2 + \frac{1}{4}\left(\frac{r}{r_0}\right)^4 \right] \qquad (10\text{-}48)$$

To put this expression into a more convenient dimensionless format, T is set equal to T_c at r equal to zero, with the result

$$\frac{T - T_s}{T_c - T_s} = 1 - \frac{4}{3}\left(\frac{r}{r_0}\right)^2 + \frac{1}{3}\left(\frac{r}{r_0}\right)^4 \qquad (10\text{-}49)$$

where $T_c - T_s = -\frac{3}{4} q_0'' r_0 / k$. Whereas the HFD velocity distribution is parabolic (second order), the temperature distribution is seen to be fourth order. (These two profiles are compared in Fig. E10-1.)

Bulk Stream Temperature and Nusselt Number. In order to express T_s in terms of T_b, we utilize the defining expression for T_b given by Eq. (6-11). This equation reduces to the following form for a circular-tube geometry:

$$T_b = \frac{2}{r_0^2 U_b} \int_0^{r_0} uTr \, dr \qquad (10\text{-}50)$$

The substitution of the expressions for u and T given by Eqs. (10-35) and (10-48) into this equation gives (see Prob. 10-9)

$$T_b = T_s - \frac{11}{24} \frac{q_0'' r_0}{k} \qquad (10\text{-}51)$$

Because T_b is also given by Eq. (10-42) which was developed by the simple lumped analysis, Eq. (10-51) specifies T_s.

Eliminating T_s in Eq. (10-48) by the use of Eq. (10-51), we have a final expression for the temperature distribution which takes the form

$$T - T_b = -\frac{q_0'' r_0}{k} \left[\frac{7}{24} - \left(\frac{r}{r_0}\right)^2 + \frac{1}{4}\left(\frac{r}{r_0}\right)^4 \right] \qquad (10\text{-}52)$$

To obtain an expression for the Nusselt number Nu, we merely rearrange Eq. (10-51).

$$h = \frac{q_0''}{T_s - T_b} = \frac{24}{11}\frac{k}{r_0}$$

or

$$\mathrm{Nu} = \frac{hD}{k} = \frac{48}{11} = 4.36 \tag{10-53}$$

This equation is consistent with the limiting predictions developed by Sellars et al. [4] for thermal developing flow.

EXAMPLE 10-3

Consider laminar HFD-plane Couette flow of a very viscous fluid for the case shown in Fig. E10-3, in which the temperature of both walls is maintained at T_0. Develop solutions for the temperature distribution and Nusselt number in the region far downstream.

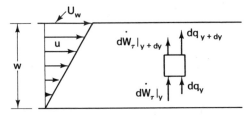

FIGURE E10-3 Differential control volume for laminar HFD-plane Couette flow ($u = U_w y/w$).

Solution

The analysis of the fluid-flow aspects of this problem is suggested as an exercise for the student (see Prob. 10-10). The velocity distribution for laminar HFD-plane Couette flow takes the form

$$u = U_w \frac{y}{w}$$

and

$$U_b = \frac{U_w}{2}$$

The differential formulation for the energy transfer for this situation is similar to the formulation developed earlier in this section for tube flow, except for the fact that the effects of viscous dissipation must now be accounted for because the fluid is very viscous.

Referring to the differential control volume dV shown in Fig. E10-3, the rate of mechanical energy which is converted to thermal energy by the shearing forces within the fluid is expressed by

$$dW_\tau = dF_x \, u = \tau_y u \, dx \, dz$$

The inclusion of this term of our energy balance gives

$$d\dot{H}_x + d\dot{H}_y + dq_y + d\dot{W}_\tau|_{y+dy} = d\dot{H}_{x+dx} + d\dot{H}_{y+dy} + dq_{y+dy} + d\dot{W}_\tau|_y$$

$$\frac{\partial}{\partial x}(d\dot{H}_x) \, dx + \frac{\partial}{\partial y}(d\dot{H}_y) \, dy + \frac{\partial}{\partial y}(dq_y) \, dy - \frac{\partial}{\partial y}(d\dot{W}_\tau) \, dy = 0$$

Because $v = 0$ for HFD conditions, $d\dot{H}_y = 0$. In addition, the asymptotic wall temperature condition requires that $\partial T/\partial x = 0$, such that $d\dot{H}_x = 0$. Hence, our differential energy equation reduces to the form

$$\frac{d}{dy}(dq_y) = \frac{d}{dy}(d\dot{W}_\tau)$$

or

$$\frac{d}{dy}\left(k\frac{dT}{dy}\right) + \frac{d}{dy}\left(\mu u \frac{du}{dy}\right) = 0$$

For uniform properties our formulation for asymptotic uniform wall temperature heating takes the form

$$\frac{d^2T}{dy^2} + \frac{\mu}{k}\frac{d}{dy}\left(u\frac{du}{dy}\right) = 0$$

$$T = T_0 \qquad \text{at} \quad y = 0$$

$$T = T_0 \qquad \text{at} \quad y = w$$

With u set equal to $U_w y/w$, we have

$$\frac{d^2T}{dy^2} + \frac{\mu}{k}\frac{U_w^2}{w^2} = 0$$

Integrating and applying the boundary conditions, the solution for T becomes

$$T - T_0 = \frac{\mu U_w^2}{2k}\left[\frac{y}{w} - \left(\frac{y}{w}\right)^2\right]$$

The rate of heat transfer at $y=0$ is obtained by utilizing the Fourier law of conduction.

$$q_0'' = -k\frac{\partial T}{\partial y}\bigg|_0 = -\frac{\mu U_w^2}{2w}$$

To obtain the coefficient of heat transfer or Nusselt number, we evaluate the bulk stream temperature T_b.

$$T_b - T_0 = \frac{1}{wU_b}\int_0^w \frac{yU_w}{w}(T - T_0)\,dy$$

$$= \frac{1}{wU_b}\frac{2U_b}{w}\int_0^w y\frac{\mu U_w^2}{2k}\left[\frac{y}{w} - \left(\frac{y}{w}\right)^2\right]dy$$

$$= \frac{\mu U_w^2}{12k}$$

Hence, the Nusselt number is given by

$$\text{Nu} = \frac{q_0''}{T_0 - T_b}\frac{2w}{k} = 12$$

10-2-3   Boundary Layer Flow over Flat Plates

We now turn our attention to the problem illustrated in Fig. 10-7 of convection heat transfer associated with steady laminar boundary layer flow over a flat plate with uniform wall-flux heating maintained in the region $x \gg x_0$. We have already seen in Chap. 6 that hydrodynamic and

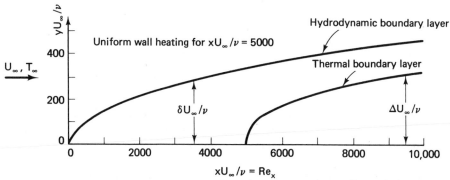

FIGURE 10-7 Development of hydrodynamic and thermal boundary layers for flow over flat plate with constant free stream velocity.

thermal boundary layers develop over the surface, with δ and Δ continually increasing with x.

Because of the simplicity and versatility of the integral approach, this very popular analytical method will be employed in this section. The differential and numerical finite difference approaches are introduced in [13] and [14].

Mathematical formulation

The integral formulation can be developed by either of two methods. The indirect method involves the integration of the differential equations. This approach will be introduced in Example 10-5. In this connection, the differential equations for steady laminar boundary layer flow over a flat plate (assuming uniform properties, and neglecting factors such as potential energy, kinetic energy, viscous dissipation, axial conduction, and thermal radiation) are given by (see Prob. 10-12)

$$\frac{\partial u}{\partial x} + \frac{\partial v}{\partial y} = 0 \tag{10-54}$$

$$u\frac{\partial u}{\partial x} + v\frac{\partial u}{\partial y} = \nu\frac{\partial^2 u}{\partial y^2} - \frac{1}{\rho}\frac{dP}{dx} \tag{10-55}$$

$$u\frac{\partial T}{\partial x} + v\frac{\partial T}{\partial y} = \alpha\frac{\partial^2 T}{\partial y^2} \tag{10-56}$$

where $dP/dx = 0$ for a uniform free stream velocity. The seven accompanying boundary conditions are

$$
\begin{array}{llll}
v = 0 & \text{at } y = 0 & u = 0 & \text{at } y = 0 \\
u = U_\infty & \text{at } x = 0 & u = U_\infty & \text{as } y \to \infty \\
T = T_\infty & \text{at } x = x_0 & T = T_\infty & \text{as } y \to \infty
\end{array}
$$
$$(\text{10-57,58})$$
$$(\text{10-59,60})$$
$$(\text{10-61,62})$$

and

$$-k\frac{\partial T}{\partial y} = q_c'' \quad \text{at } y = 0 \tag{10-63}$$

for the case in which the wall flux is specified. The development of these equations follows the patterns established in Sec. 10-2-2 for developing internal flow and in the following example on natural convection.

EXAMPLE 10-4

Develop the differential formulation for laminar natural convection on a vertical plate.

Solution

Because natural convection flow is caused by temperature induced density gradients within the fluid, we must account for buoyancy in the development of the momentum equation. Since this is the only factor that distinguishes natural convection from forced convection, the continuity and energy equations for these two problems are identical. Therefore, attention is focused on the development of the mathematical formulation for the momentum transfer.

Applying Newton's second law to the differential element shown in Fig. E10-4, we obtain

$$\frac{d}{dt}(mu) = \Sigma F_x$$

where

$$\Sigma F_x = (\tau \, dx \, dz)|_{y+dy} + (P \, dy \, dz)|_x - (P \, dy \, dz)|_{x+dx}$$
$$- (\tau \, dx \, dz)|_y - \rho g \, dx \, dy \, dz$$

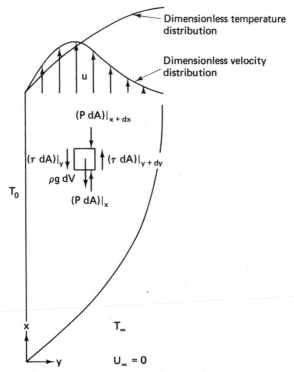

FIGURE E10-4 Laminar natural convection flow over a vertical flat plate.

and

$$\frac{d}{dt}(mu) = dV\left[u\frac{\partial}{\partial x}(\rho u) + v\frac{\partial}{\partial y}(\rho u)\right]$$

for steady-state conditions; the term $\rho g\, dV$ represents the gravitational force. Utilizing the Newton law of viscosity and the defining equation for the derivative, the momentum equation takes the form

$$u\frac{\partial}{\partial x}(\rho u) + v\frac{\partial}{\partial y}(\rho u) = \frac{\partial}{\partial y}\left(\mu\frac{\partial u}{\partial y}\right) - \frac{dP}{dx} - \rho g \qquad \text{(a)}$$

Except for the buoyancy term ρg, this equation is identical to the momentum equation for forced convection. The effect of buoyancy is modeled by accounting for the dependence of the gravitational term ρg on temperature. For isothermal conditions, ρg is constant and no flow occurs.

The boundary conditions that accompany this equation are given by

$$u=0 \quad \text{at} \quad x=0 \qquad u=0 \quad \text{at} \quad y=0 \qquad v=0 \quad \text{at} \quad y=0$$

$$u=0 \quad \text{as} \quad y\to\infty \qquad \text{or at} \quad y=\delta \qquad\qquad\qquad \text{(b)}$$

Based on Eq. (b) and on the physics of the problem, the momentum equation can be written in the following form for the region outside the boundary layer:

$$0 = -\frac{dP}{dx} - \rho_\infty g$$

Substituting this result into Eq. (a), the momentum equation takes the form

$$u\frac{\partial}{\partial x}(\rho u) + v\frac{\partial}{\partial y}(\rho u) = \frac{\partial}{\partial y}\left(\mu\frac{\partial u}{\partial y}\right) + g(\rho_\infty - \rho)$$

For situations in which $T_0 - T_\infty$ is large, the dependence of all of the properties on temperature must be accounted for. However, for moderate temperature differences, the variation of the properties can be neglected, except for the density that appears in the buoyancy term. For this case, we approximate the properties associated with the non-buoyancy terms by their free stream conditions, such that the differential momentum equation reduces to

$$\rho_\infty\left(u\frac{\partial u}{\partial x} + v\frac{\partial u}{\partial y}\right) = \mu_\infty\frac{\partial^2 u}{\partial y^2} + g(\rho_\infty - \rho) \qquad \text{(c)}$$

The dependence of ρ on T is often expressed in terms of the *coefficient of thermal expansion* β which is defined by

$$\beta = \frac{1}{V}\frac{\partial V}{\partial T}\bigg|_P = -\frac{1}{\rho}\frac{\partial \rho}{\partial T}\bigg|_P$$

This property is given in Table A-C-3 in the Appendix for various common fluids. Expanding this defining equation, we obtain (for $P \simeq$ constant)

$$\int_{\rho_\infty}^{\rho} d\rho \simeq -\int_{T_\infty}^{T}\rho\beta\, dT = -\overline{\rho\beta}\,(T - T_\infty)$$

or

$$\rho - \rho_\infty = -\overline{\rho\beta}\,(T - T_\infty)$$

where

$$\overline{\rho\beta} = \frac{1}{T - T_\infty}\int_{T_\infty}^{T}\rho\beta\, dT$$

Substituting this result for ρ into Eq. (c) gives

$$u\frac{\partial u}{\partial x} + v\frac{\partial u}{\partial y} = \nu_\infty\frac{\partial^2 u}{\partial y^2} + \frac{g}{\rho_\infty}\,\overline{\rho\beta}\,(T - T_\infty) \tag{d}$$

For the situation in which the temperature differences are moderate, the terms $\overline{\rho\beta}$ can be approximated by $\rho_\infty\beta_\infty$ such that Eq. (d) reduces to

$$u\frac{\partial u}{\partial x} + v\frac{\partial u}{\partial y} = \nu_\infty\frac{\partial^2 u}{\partial y^2} + g\beta_\infty(T - T_\infty) \tag{e}$$

The differential formulation is completed by writing the continuity, energy, and thermal boundary equations; that is,

$$\frac{\partial u}{\partial x} + \frac{\partial v}{\partial y} = 0$$

$$u\frac{\partial T}{\partial x} + v\frac{\partial T}{\partial y} = \alpha_\infty\frac{\partial^2 T}{\partial y^2}$$

$$T = T_\infty \quad \text{at } x = 0 \qquad T = T_\infty \quad \text{as } y \to \infty \qquad T = T_0 \quad \text{at } y = 0$$

for uniform wall-temperature heating of a single-phase fluid with uniform properties.

In the direct approach to the development of the integral formulation, we apply the fundamental laws pertaining to mass, momentum, and energy transfer to the lumped differential volumes of fluid $\delta\ dx\ dz$ and $\Delta\ dx\ dz$.

Continuity   Applying the principle of conservation of mass to the control volume $\delta\ dx\ dz$ shown in Fig. 10-8, we obtain

$$\Sigma \dot{m}_i = \Sigma \dot{m}_0 + \frac{\Delta \dot{m}_s}{\Delta t}$$

$$\dot{m}_x + d\dot{m}_\delta = \dot{m}_{x+dx} \tag{10-64}$$

where $\Delta m_s/\Delta t = 0$ for steady flow, and the mass flow through the wall is zero $[v(x,0)=0]$; $d\dot{m}_\delta$ is the rate of mass transfer from the free stream into the control volume. $\dot{m}_x$ is related to $d\dot{m}_x\ (=\rho u\ dz\ dy)$ by

$$\dot{m}_x = \int_0^\delta d\dot{m}_x \tag{10-65}$$

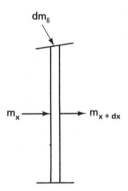

FIGURE 10-8   Conservation of mass relative to lumped differential rectangular control volume; $dV = \delta\ dx\ dz$.

By utilizing the definition of the derivative, Eq. (10-64) reduces to

$$d\dot{m}_\delta = \frac{d\dot{m}_x}{dx} dx$$

$$= \left(\frac{d}{dx} \int_0^\delta d\dot{m}_x \right) dx = \left[\frac{d}{dx} \int_0^\delta (\rho u\ dz)\ dy \right] dx \tag{10-66}$$

Similarly, the differential mass flow rate $d\dot{m}_\Delta$ from the free stream into the thermal boundary layer volume $\Delta\ dx\ dz$ can be represented by

$$d\dot{m}_\Delta = \left(\frac{d}{dx} \int_0^\Delta d\dot{m}_x \right) dx = \left[\frac{d}{dx} \int_0^\Delta (\rho u\ dz)\ dy \right] dx \tag{10-67}$$

Momentum Transfer   Applying the momentum principle to the entire lumped differential volume $\delta\,dx\,dz$, we obtain

$$\frac{d}{dt}(mu) = \Sigma F_x \qquad (10\text{-}68a)$$

where

$$\Sigma F_x = -\tau_0\,dx\,dz + (P\delta\,dz)|_x - (P\delta\,dz)|_{x+dx} \qquad (10\text{-}68b)$$

[see Fig. 10-9(a)], and

$$\frac{d}{dt}(mu) = \dot{M}_0 - \dot{M}_i + \frac{\Delta M_s}{\Delta t} = \dot{M}_{x+dx} - \dot{M}_x - d\dot{M}_\delta \qquad (10\text{-}68c)$$

[see Fig. 10-9(b)]; $\Delta M_s/\Delta t = 0$ for steady flow, $d\dot{M}_\delta = U_\infty\,d\dot{m}_\delta$, and

$$\dot{M}_x = \int_0^\delta u\,d\dot{m}_x \qquad (10\text{-}69)$$

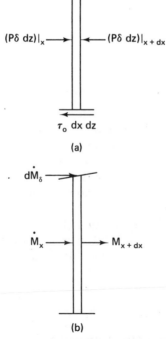

FIGURE 10-9 (a) Summation of forces acting in x direction relative to lumped differential rectangular control volume. (b) Rate of creation of x direction momentum relative to lumped rectangular differential control volume.

Utilizing the definition of the derivative, Eq. (10-68a) reduces to

$$\frac{d\dot{M}_x}{dx} dx - d\dot{M}_\delta = -\tau_0 \, dx \, dz - \frac{dP}{dx} \delta \, dx \, dz \qquad (10\text{-}70)$$

The substitution of Eq. (10-69) and the defining relationship for $d\dot{M}_\delta$ into this expression gives

$$\frac{d}{dx} \int_0^\delta u \, d\dot{m}_x \, dx - U_\infty \, d\dot{m}_\delta = -\tau_0 \, dx \, dz - \frac{dP}{dx} \delta \, dx \, dz \qquad (10\text{-}71)$$

Substituting for $d\dot{m}_x$ and $d\dot{m}_\delta$, this equation becomes

$$\frac{d}{dx} \int_0^\delta u^2 \, dy - U_\infty \frac{d}{dx} \int_0^\delta u \, dy = -\frac{\tau_0}{\rho} - \frac{\delta}{\rho} \frac{dP}{dx} \qquad (10\text{-}72)$$

for incompressible flow. Utilizing the Newton law of viscosity and rearranging, we obtain

$$U_\infty \frac{d}{dx} \int_0^\delta u \, dy - \frac{d}{dx} \int_0^\delta u^2 \, dy = \nu \left. \frac{du}{dy} \right|_0 + \frac{\delta}{\rho} \frac{dP}{dx} \qquad (10\text{-}73)$$

For the case in which $dP/dx = 0$, U_∞ is constant, and this equation takes the convenient form

$$\frac{d}{dx} \int_0^\delta u(U_\infty - u) \, dy = \nu \left. \frac{du}{dy} \right|_0 = \frac{\tau_0}{\rho} \qquad (10\text{-}74)$$

The boundary conditions that accompany the integral momentum equation are given by Eqs. (10-58) through (10-60). Notice that the boundary condition for v given by Eq. (10-57) has already been satisfied in the development of the integral continuity equation, which itself has been incorporated into the integral momentum equation.

Energy Transfer The application of the first law of thermodynamics to the lumped volume $\Delta \, dx \, dz$ shown in Fig. 10-10 gives (for negligible potential and kinetic energy, viscous dissipation, axial conduction, and thermal radiation effects):

$$\Sigma \dot{E}_i = \Sigma \dot{E}_o + \frac{\Delta \dot{E}_s}{\Delta t}$$

$$\dot{H}_x + d\dot{H}_\Delta + dq_c = \dot{H}_{x+dx} \qquad (10\text{-}75)$$

where $\Delta E_s / \Delta t = 0$ for steady-state conditions, $d\dot{H}_\Delta = c_p T_\infty \, d\dot{m}_\Delta$ for no

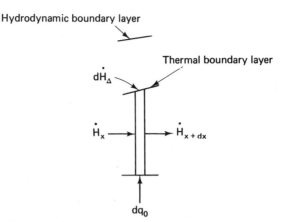

FIGURE 10-10 Conservation of energy relative to lumped differential rectangular control volume; $dV = \Delta \, dx \, dz$.

change in phase, and

$$\dot{H}_x = \int_0^\Delta d\dot{H}_x = \int_0^\Delta c_p T \, d\dot{m}_x \qquad (10\text{-}76)$$

Utilizing the definition of the derivative, we have

$$d\dot{H}_\Delta + dq_c = \frac{d\dot{H}_x}{dx} \, dx \qquad (10\text{-}77)$$

With $\dot{H}_x$ given by Eq. (10-76) and with $d\dot{m}_\Delta$ specified on the basis of continuity by Eq. (10-67), we have

$$(c_p T)_\infty \left(\frac{d}{dx} \int_0^\Delta d\dot{m}_x \right) dx + dq_c = \left(\frac{d}{dx} \int_0^\Delta c_p T \, d\dot{m}_x \right) dx \qquad (10\text{-}78)$$

or

$$\frac{d}{dx} \int_0^\Delta \left[c_p T - (c_p T)_\infty \right] d\dot{m}_x = \frac{dq_c}{dx} \qquad (10\text{-}79)$$

Replacing $d\dot{m}_x$ by $(\rho u \, dz) dy$, we obtain

$$\rho c_p \frac{d}{dx} \int_0^\Delta u (T - T_\infty) dy = \frac{dq_c}{dx \, dz} = q_c'' \qquad (10\text{-}80)$$

for uniform properties.

This integral energy equation is coupled with the thermal boundary conditions given by Eqs. (10-61) through (10-63). For the case in which a specified wall heat flux is maintained, q_c'' is simply replaced by the specified input. On the other hand, for the case in which the wall temperature is specified, the Fourier law of conduction is utilized, with Eq. (10-80) taking the form

$$\rho c_p \frac{d}{dx} \int_0^\Delta u(T - T_\infty)\,dy = -k \left. \frac{\partial T}{\partial y} \right|_0 \qquad (10\text{-}81)$$

Solution

An approximate solution is now developed for steady uniform property laminar boundary layer flow over a flat plate $(dP/dx = 0)$ with specified wall flux heating. Because we are dealing with a uniform property flow, the fluid-flow aspects of the problem are treated first, after which the energy transfer will be handled.

Fluid Flow   The integral momentum equation,

$$\frac{d}{dx} \int_0^\delta u(U_\infty - u)\,dy = \nu \left. \frac{\partial u}{\partial y} \right|_0 \qquad (10\text{-}74)$$

involves the one unknown u. The boundary layer thickness δ is of course dependent upon u. A simple approximate solution to this integral form of the momentum equation can be obtained by treating δ as the unknown, with the velocity profile u being approximated in terms of δ on the basis of the physics of the problem.

Dimensionless Velocity Distribution.   Referring to Fig. 10-11, which depicts the laminar hydrodynamic boundary layer, approximate velocity

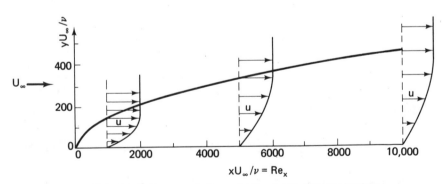

FIGURE 10-11 Representative velocity distribution for laminar boundary layer flow over a flat plate with constant free stream velocity.

profiles are easily sketched in that satisfy the boundary conditions

$$u = 0 \quad \text{at} \quad y = 0 \tag{10-58}$$

$$u = U_\infty \quad \text{at} \quad y \simeq \delta \quad \text{or as} \quad y \to \infty \tag{10-60}$$

and the physical requirement

$$\frac{\partial u}{\partial y} = 0 \quad \text{at} \quad y \simeq \delta \quad \text{or as} \quad y \to \infty \tag{10-82}$$

Many analytical expressions are available that meet these minimal require-ments. One example is the simple second-order polynomial

$$\frac{u}{U_\infty} = 2\frac{y}{\delta} - \left(\frac{y}{\delta}\right)^2 \quad \text{for} \quad y < \delta \tag{10-83}$$

Improvements in our approximation for u can sometimes be made by satisfying higher-order requirements that are called for by the differential momentum equation,

$$u\frac{\partial u}{\partial x} + v\frac{\partial u}{\partial y} = \nu\frac{\partial^2 u}{\partial y^2} \tag{10-55}$$

By evaluating this equation at $y = 0$ and at $y = \delta$, we find

$$\frac{\partial^2 u}{\partial y^2} = 0 \quad \text{at} \quad y = 0 \tag{10-84}$$

$$\frac{\partial^2 u}{\partial y^2} \simeq 0 \quad \text{at} \quad y = \delta \tag{10-85}$$

What is more, we can show that the differential equation is only satisfied when (see Prob. 10-21)

$$\frac{\partial^n u}{\partial y^n} = 0 \quad \text{at} \quad y = 0 \quad \text{and at} \quad y \simeq \delta \quad \text{or as} \quad y \to \infty \tag{10-86}$$

where $n = 3, 4, 5, \ldots$. The exact solution for u satisfies all these require-ments. However, it should be noted that the fulfillment of all of these constants at $y = 0$ and at $y \simeq \delta$ does not guarantee that u satisfies Eq. (10-55) in the region $0 < y < \delta$ (see Prob. 10-24).

As already indicated, the second-order profile given by Eq. (10-83) satisfies the boundary conditions and the physical constraint given by Eq. (10-82). It even satisfies the requirements given by Eq. (10-86), but it fails to satisfy the specifications given by Eqs. (10-84) and (10-85). On the basis of previous studies, more accurate approximations for u have been developed by the use of third-, fourth-, and fifth-order polynomials. The most convenient of these higher-order approximations is obtained on the basis of the third-order polynomial

$$u = a_0 + a_1 y + a_2 y^2 + a_3 y^3 \qquad (10\text{-}87)$$

Since we now have four coefficients, Eqs. (10-58), (10-60), (10-82), and (10-84) can be satisfied. The coupling of Eq. (10-87) with these conditions gives $a_0 = 0$, $2a_2 = 0$, $a_0 + a_1\delta + a_2\delta^2 + a_3\delta^3 = U_\infty$, and $a_1 + 2a_2\delta + 3a_3\delta^2 = 0$. It follows that $a_1 = 3U_\infty/(2\delta)$, and $a_3 = -U_\infty/(2\delta^3)$, such that Eq. (10-87) becomes

$$\frac{u}{U_\infty} = \frac{3}{2}\frac{y}{\delta} - \frac{1}{2}\left(\frac{y}{\delta}\right)^3 \qquad \text{for } y \gtrless \delta \qquad (10\text{-}88)$$

or

$$U_\infty - u = U_\infty\left[1 - \frac{3}{2}\frac{y}{\delta} + \frac{1}{2}\left(\frac{y}{\delta}\right)^3\right] \qquad \text{for } y \gtrless \delta \qquad (10\text{-}89)$$

The velocity profiles shown in Fig. 10-11 are based on this equation.

Boundary Layer Thickness δ and Friction Factor.   Now that reasonable approximations are available for the velocity profile u in terms of δ, we return to the integral momentum equation in order to develop predictions for δ. Utilizing our third-order polynomial approximation, the integral momentum equation becomes

$$U_\infty^2 \frac{d}{dx}\int_0^\delta \left[\frac{3}{2}\left(\frac{y}{\delta}\right) - \frac{1}{2}\left(\frac{y}{\delta}\right)^3\right]\left[1 - \frac{3}{2}\frac{y}{\delta} + \frac{1}{2}\left(\frac{y}{\delta}\right)^3\right]dy = \frac{3}{2}\frac{\nu U_\infty}{\delta}$$

$$(10\text{-}90)$$

The integration of this equation gives

$$\delta\frac{d\delta}{dx} = \frac{140}{13}\frac{\nu}{U_\infty} \qquad (10\text{-}91)$$

This differential equation in δ is coupled with the requirement

$$\delta = 0 \qquad \text{at } x = 0 \qquad (10\text{-}92)$$

Equation (10-91) is readily integrated to give

$$\delta^2 = \frac{280}{13}\frac{\nu x}{U_\infty} \qquad (10\text{-}93)$$

This equation was actually used in the construction of Figs. 10-7 and 10-11.

Substituting this result for δ into Eq. (10-88), we obtain an expression for the velocity distribution of the form

$$\frac{u}{U_\infty} = \frac{3}{2}\left[\sqrt{\frac{13}{280}}\,\frac{y}{\sqrt{xv/U_\infty}}\right] - \frac{1}{2}\left[\sqrt{\frac{13}{280}}\,\frac{y}{\sqrt{xv/U_\infty}}\right]^3 \qquad \text{for } y \leqslant \delta$$

$$= 0.323\eta - 0.005\eta^3 \qquad \text{for } \eta \leqslant 4.64 \tag{10-94}$$

where $\eta = y/\sqrt{xv/U_\infty} = \sqrt{280/13}\,y/\delta$. This equation is compared with experimental data and the exact solution in Fig. 10-12. With u expressed in terms of η (or y/δ), we find that the velocity profiles are geometrically similar at all values of x. Consequently, η is known as a *similarity coordinate.*

To obtain the wall shear stress, we employ the Newton law of viscosity; that is,

$$\tau_0 = \mu\,\frac{\partial u}{\partial y}\bigg|_0 = \frac{3}{2}\mu\frac{U_\infty}{\delta} = \rho U_\infty^2\frac{f_x}{2} \tag{10-95}$$

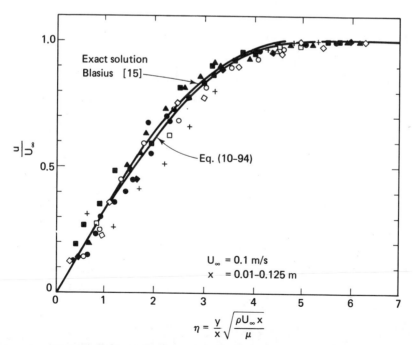

FIGURE 10-12 Velocity distribution for laminar boundary layer flow over a flat plate. (Data from Hansen [8].)

Hence, f_x becomes

$$f_x = \frac{3}{\text{Re}_\delta} = 3\sqrt{\frac{13}{280}} \; \frac{1}{\sqrt{\text{Re}_x}} = \frac{0.646}{\sqrt{\text{Re}_x}} \tag{10-96}$$

This expression is only 3% below the exact solution, Eq. (6-82), which was compared with experimental data in Fig. 6-15.

   Energy Transfer   Approximate integral solutions can be fairly easily developed for either specified wall flux or step uniform wall temperature boundary conditions. We will focus our attention on the more general but simpler of these two problems, specified wall flux heating. The step uniform wall-temperature problem, which is treated in most other introductory heat-transfer texts, will be considered in the context of laminar natural convection in Example 10-5, as well as in Prob. 10-23.

   To develop an approximate solution for the thermal boundary layer thickness Δ which appears in the integral energy equation,

$$\frac{d}{dx} \int_0^\Delta u(T - T_\infty)\, dy = \frac{q_c''}{\rho c_p} \tag{10-80}$$

we must develop an approximation for the temperature distribution.

   Dimensionless Temperature Distribution.   Following the pattern of our fluid-flow analysis, T is approximated by the third-order polynomial

$$T = b_0 + b_1 y + b_2 y^2 + b_3 y^3 \tag{10-97}$$

where the four coefficients are selected in accordance with the following four conditions for specified wall-flux heating:

$$T = T_\infty \quad \text{at } y = \Delta \quad \text{or as } y \to \infty \qquad -k\frac{\partial T}{\partial y} = q_c'' \quad \text{at } y = 0$$

$$\tag{10-62, 63}$$

$$\frac{\partial T}{\partial y} = 0 \quad \text{at } y = \Delta \quad \text{or as } y \to \infty \qquad \frac{\partial^2 T}{\partial y^2} = 0 \quad \text{at } y = 0 \qquad \text{(10-98a, b)}$$

[Equation (10-98b) is obtained by evaluating the differential energy equation, Eq. (10-56), at the wall, where u and v are both zero.] Following through with the evaluation of the coefficients b_n, we obtain

$$T - T_\infty = \frac{2}{3}\frac{q_c'' \Delta}{k}\left[1 - \frac{3}{2}\frac{y}{\Delta} + \frac{1}{2}\left(\frac{y}{\Delta}\right)^3\right] \qquad \text{for } y \leqslant \Delta \tag{10-99}$$

By setting $y = 0$, the interrelation between the two unknowns T_s and Δ is found to be

$$T_s - T_\infty = \frac{2q_c'' \Delta}{3k} \qquad (10\text{-}100)$$

such that T can also be put into the dimensionless form

$$\frac{T - T_s}{T_\infty - T_s} = \frac{3}{2}\frac{y}{\Delta} - \frac{1}{2}\left(\frac{y}{\Delta}\right)^3 \qquad (10\text{-}101)$$

Note the similarity between this dimensionless expression for T and the third-order profile for u given by Eq. (10-88).

Thermal Boundary Layer Thickness Δ and Nusselt Number. Substituting our third-order polynomial approximations for u and T [Eqs. (10-88) and (10-99)] into the integral energy equation for the case in which Δ is less than δ, we have

$$\frac{d}{dx}\int_0^\Delta \left\{ U_\infty\left[\frac{3}{2}\frac{y}{\delta} - \frac{1}{2}\left(\frac{y}{\delta}\right)^3\right]\right\}\left\{\left(\frac{2}{3}\frac{q_c''\Delta}{k}\right)\left[1 - \frac{3}{2}\frac{y}{\Delta} + \frac{1}{2}\left(\frac{y}{\Delta}\right)^3\right]\right\}dy = \frac{q_c''}{\rho c_p}$$

$$(10\text{-}102)$$

where δ is specified by Eq. (10-93). Carrying through with the integration with respect to y, we obtain

$$\frac{d}{dx}\left\{\left(\frac{U_\infty q_c'' \delta^2}{k}\right)\left[\frac{1}{10}\left(\frac{\Delta}{\delta}\right)^3 - \frac{1}{140}\left(\frac{\Delta}{\delta}\right)^5\right]\right\} = \frac{q_c''}{\rho c_p} \qquad (10\text{-}103)$$

Noting that $\Delta = 0$ at $x = 0$, this equation is easily integrated with respect to x to obtain

$$\frac{U_\infty q_c'' \delta^2}{k}\left[\frac{1}{10}\left(\frac{\Delta}{\delta}\right)^3 - \frac{1}{140}\left(\frac{\Delta}{\delta}\right)^5\right] = \frac{1}{\rho c_p}\int_0^x q_c''\, dx$$

or

$$\frac{1}{10}\left(\frac{\Delta}{\delta}\right)^3 - \frac{1}{140}\left(\frac{\Delta}{\delta}\right)^5 = \frac{\alpha}{U_\infty \delta^2}\frac{1}{q_c''}\int_0^x q_c''\, dx \qquad (10\text{-}104)$$

Utilizing Eq. (10-93) to eliminate δ^2 on the right-hand side, we have

$$\left(\frac{\Delta}{\delta}\right)^3 - \frac{1}{14}\left(\frac{\Delta}{\delta}\right)^5 = \frac{13}{28\,\mathrm{Pr}\,x}\frac{1}{q_c''}\int_0^x q_c''\, dx \qquad (10\text{-}105)$$

For step uniform wall-flux heating,

$$q_c'' = 0 \quad \text{for} \quad x < x_0$$
$$q_c'' = q_0'' \quad \text{for} \quad x \gg x_0 \tag{10-106}$$

Eq. (10-105) reduces to

$$\left(\frac{\Delta}{\delta}\right)^3 - \frac{1}{14}\left(\frac{\Delta}{\delta}\right)^5 = \frac{13}{28\,\text{Pr}}\left(1 - \frac{x_0}{x}\right) \tag{10-107}$$

For this and other specified wall-flux conditions, calculations can be obtained for Δ/δ vs. x by taking Δ/δ as the independent variable. As an alternative, a convenient but approximate explicit expression can be obtained for Δ/δ by neglecting the fifth-order term in Eq. (10-107); that is,

$$\frac{\Delta}{\delta} \simeq \left[\frac{13}{28}\frac{1}{\text{Pr}}\left(1 - \frac{x_0}{x}\right)\right]^{1/3} \tag{10-108}$$

This equation was utilized in the development of Fig. 10-7 with $\text{Pr} = 0.70$ and $\text{Re}_{x_0} = 5000$. For $x_0 = 0$, Eq. (10-108) reduces to

$$\frac{\Delta}{\delta} = \left(\frac{13}{28\,\text{Pr}}\right)^{1/3} \tag{10-109}$$

Because u was specified by Eq. (10-88) in the evaluation of the integral in Eq. (10-80), this solution is limited to the case for which Δ is less than δ. (Note that $u = U_\infty$ in the region where Δ is greater than δ.) To see the conditions for which our simple approximate solution for step uniform wall-flux heating satisfies this restriction, Δ/δ in Eq. (10-108) is set equal to or less than unity; i.e.,

$$\frac{13}{28}\frac{1}{\text{Pr}}\left(1 - \frac{x_0}{x}\right) \leqslant 1 \tag{10-110}$$

or

$$\text{Pr} \geqslant \frac{13}{28}\left(1 - \frac{x_0}{x}\right) \tag{10-111}$$

This equation, which represents the minimum values of Pr for which our analysis is valid, is shown in Fig. 10-13. Based on this result, we conclude that this analysis does not apply to low-Prandtl-number fluids, except for small values of $x - x_0$.

Based on Eqs. (10-93), (10-99), and (10-108), predictions can be obtained for the temperature profile and wall temperature or coefficient of

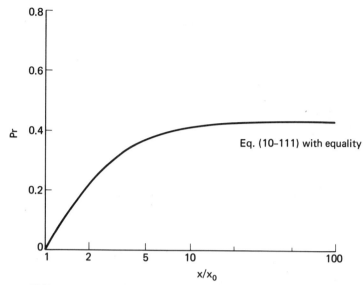

FIGURE 10-13 Criterion for applicability of approximate integral solution for heat transfer.

heat transfer for step uniform wall-flux heating. For example, the wall temperature T_s is obtained by combining these three equations with y set equal to zero; that is,

$$T_s - T_\infty = \frac{2q_0'' \delta}{3k} \left[\frac{13}{28} \frac{1}{Pr} \left(1 - \frac{x_0}{x} \right) \right]^{1/3}$$

$$= 2.4 \frac{q_0''}{k} \sqrt{\frac{x\nu}{U_\infty}} \left[\frac{1}{Pr} \left(1 - \frac{x_0}{x} \right) \right]^{1/3} \qquad (10\text{-}112)$$

Rearranging this expression, we have

$$Nu_x = \frac{q_0''}{T_s - T_\infty} \frac{x}{k} = 0.417 \frac{\sqrt{Re_x} \; Pr^{1/3}}{(1 - x_0/x)^{1/3}} \qquad (10\text{-}113)$$

This equation reduces to

$$Nu_x = 0.417 \sqrt{Re_x} \; Pr^{1/3} \qquad (10\text{-}114)$$

for $x_0 = 0$ or for large values of x/x_0. Equations (10-113) and (10-114) are shown in Fig. 10-14 in terms of $Nu_x/(\sqrt{Re_x} \; Pr^{1/3})$ vs. Re_x.

For step uniform wall-temperature heating, the approximate integral approach with third-order polynomial profiles gives rise to an expression

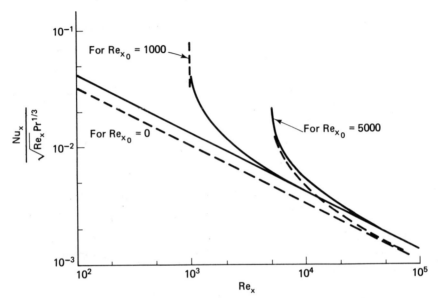

FIGURE 10-14 Theoretical predictions for Nusselt number for
laminar boundary layer flow over a flat plate.

for the Nusselt number of the form (see Prob. 10-23)

$$Nu_x = \frac{0.323 \sqrt{Re_x} \; Pr^{1/3}}{\left[1 - \left(\dfrac{x_0}{x}\right)^{3/4}\right]^{1/3}} \tag{10-115}$$

For large values of x/x_0, this equation takes the form

$$Nu_x = 0.323 \sqrt{Re_x} \; Pr^{1/3} \tag{10-116}$$

These expressions are also shown in Fig. 10-14. It should be noted that Eq.
(10-116) for uniform wall-temperature heating can be expressed in terms of
f_x by utilizing Eq. (10-96).

$$Nu_x = \frac{f_x}{2} Re_x \; Pr^{1/3} \tag{10-117}$$

EXAMPLE 10-5

Develop an approximate integral solution for laminar natural convection flow over a vertical flat plate with uniform wall-temperature heating.

Solution

The differential formulation for laminar free convection flow over a vertical flat plate with moderate temperature differences is given in Example 10-4 as

$$\frac{\partial u}{\partial x} + \frac{\partial v}{\partial y} = 0 \tag{a}$$

$$u \frac{\partial u}{\partial x} + v \frac{\partial u}{\partial y} = \nu_\infty \frac{\partial^2 u}{\partial y^2} + g\beta_\infty (T - T_\infty) \tag{b}$$

$$u \frac{\partial T}{\partial x} + v \frac{\partial T}{\partial y} = \alpha_\infty \frac{\partial^2 T}{\partial y^2} \tag{c}$$

$$
\begin{array}{llll}
u = 0 & \text{and} & T = T_\infty & \text{at} \quad x = 0 \\
u = 0, v = 0 & \text{and} & T = T_0 & \text{at} \quad y = 0 \\
u = 0 & \text{and} & T = T_\infty & \text{as} \quad y \to \infty
\end{array} \tag{d}
$$

To develop the integral momentum equation for laminar natural convection boundary layer flow, we integrate the differential momentum equation across the boundary layer. First, the differential equation is rewritten as follows:

$$\frac{\partial u^2}{\partial x} - u \frac{\partial u}{\partial x} + \frac{\partial}{\partial y}(uv) - u \frac{\partial v}{\partial y} = \nu_\infty \frac{\partial^2 u}{\partial y^2} + g\beta_\infty (T - T_\infty)$$

Utilizing the continuity equation, this expression reduces to the form

$$\frac{\partial u^2}{\partial x} + \frac{\partial}{\partial y}(uv) = \nu_\infty \frac{\partial^2 u}{\partial y^2} + g\beta_\infty (T - T_\infty)$$

Integrating each term of this equation with respect to y, we obtain

$$\int_0^y \frac{\partial u^2}{\partial x} dy + \int_0^y \frac{\partial}{\partial y}(uv) dy = \int_0^y \nu_\infty \frac{\partial^2 u}{\partial y^2} dy + \int_0^y g\beta_\infty (T - T_\infty) dy$$

or

$$\frac{\partial}{\partial x} \int_0^y u^2 \, dy + (uv) \Big|_0^y = \nu_\infty \frac{\partial u}{\partial y} \Big|_0^y + \int_0^y g\beta_\infty (T - T_\infty) dy \tag{e}$$

Setting y equal to δ and noting that $u=0$ at $y=0$, Eq. (e) reduces to

$$\frac{d}{dx}\int_0^{\delta} u^2\,dy = -\nu_{\infty}\frac{\partial u}{\partial y}\Big|_0 + \int_0^{\delta} g\beta_{\infty}(T-T_{\infty})\,dy \qquad \text{(f)}$$

The integral formulation is completed by writing the integral energy equation

$$\frac{d}{dx}\int_0^{\Delta} u(T-T_{\infty})\,dy = -\alpha_{\infty}\frac{\partial T}{\partial y}\Big|_0 \qquad \text{(g)}$$

The thermal boundary conditions are given by Eqs. (d) for uniform wall-temperature heating.

As in the analysis of forced convection boundary layer flow, u and T are approximated by third-order polynomials; that is,

$$u = a_0 + a_1 y + a_2 y^2 + a_3 y^3$$
$$T = b_0 + b_1 y + b_2 y^2 + b_3 y^3$$

In regard to the conditions that should be satisfied, the evaluation of Eq. (b) at $y=0$ indicates that

$$\frac{\partial^2 u}{\partial y^2} = -\frac{g\beta_{\infty}}{\nu_{\infty}}(T_0-T_{\infty}) \qquad \text{at} \quad y=0$$

In addition, u must be equal to zero as y approaches large values. Aside from these two constraints, the other conditions for u and T utilized in the analysis for forced convection apply for uniform wall-temperature heating. The evaluation of the coefficients a_n and b_n in accordance with these conditions gives

$$u = \frac{g\beta_{\infty}(T_0-T_{\infty})}{\nu_{\infty}}\frac{\delta^2}{4}\left[\frac{y}{\delta} - 2\left(\frac{y}{\delta}\right)^2 + \left(\frac{y}{\delta}\right)^3\right] \qquad \text{(h)}$$

and

$$\frac{T-T_0}{T_{\infty}-T_0} = \frac{3}{2}\frac{y}{\Delta} - \frac{1}{2}\left(\frac{y}{\Delta}\right)^3 \qquad \text{(i)}$$

Following the path taken in the analysis of forced convection boundary layer flow, these equations can be substituted into the integral momentum and energy equations, Eqs. (f) and (g), to produce two equations in the unknowns δ and Δ. However, because both Eqs. (f) and

(g) involve u, T, δ, and Δ, the resulting expressions become quite unwieldy and are difficult to solve. Fortunately, the physics of the problem allows us to circumvent this difficulty. Because the flow is established as a result of the temperature potential $T - T_\infty$, u becomes small as T approaches T_∞. Therefore, δ and Δ can be assumed to be of the same order; that is,

$$\delta \simeq \Delta$$

Based on this simplifying assumption, the single unknown Δ can be obtained by the approximate solution of either the integral momentum or energy equation. Because Eq. (g) is simpler than Eq. (f), the integral energy equation will be utilized as our basis for evaluating δ and Δ.

The substitution of Eqs. (h) and (i) into the integral energy equation with δ set equal to Δ gives rise to an expression of the form

$$\frac{g\beta_\infty(T_0 - T_\infty)}{4\nu_\infty} \int_0^\Delta \Delta^2 \left[\frac{y}{\Delta} - 2\left(\frac{y}{\Delta}\right)^2 + \left(\frac{y}{\Delta}\right)^3 \right] \left[1 - \frac{3}{2}\frac{y}{\Delta} + \frac{1}{2}\left(\frac{y}{\Delta}\right)^3 \right] dy$$

$$= \frac{3}{2}\frac{\alpha_\infty}{\Delta}$$

or

$$\frac{0.038}{4} \frac{g\beta_\infty(T_0 - T_\infty)}{\nu_\infty} \frac{d\Delta^3}{dx} = \frac{3}{2}\frac{\alpha_\infty}{\Delta}$$

This equation can be rewritten as

$$\Delta^3 \frac{d\Delta}{dx} = \frac{2}{0.038} \frac{\nu_\infty^2}{\mathrm{Pr}_\infty\, g\beta_\infty(T_0 - T_\infty)} \tag{j}$$

With $\Delta = 0$ at $x = 0$, the solution for Δ is

$$\frac{\Delta}{x} = 3.81 \mathrm{Ra}_x^{-0.25} \tag{k}$$

where the *Raleigh number* Ra_x is equal to $\mathrm{Pr}\,\mathrm{Gr}_x$ and the *Grashof number* Gr_x is defined by

$$\mathrm{Gr}_x = \frac{gx^3\beta}{\nu^2}(T_0 - T_\infty) \tag{l}$$

The properties are evaluated at the free stream temperature T_∞. (As mentioned in Chap. 7, the Grashof number is a key parameter in natural convection flows. Based on experimental observations, the transition from laminar to turbulent flow occurs at a value of Gr_x of the order of 10^9.)

A solution of the integral momentum equation with the assumptions $\delta = \Delta$ and $\beta = \beta_\infty$ gives rise to a solution for Δ that is exactly the same for $Pr = 1.18$ and that is within about 10% of Eq. (k) for $0.75 \gtrsim Pr \gtrsim 1.75$ (see Prob. 10-25). Therefore, the analysis should be reasonably accurate within this range of Prandtl numbers.

With δ and Δ given by Eq. (k), the velocity and temperature profiles can be obtained from Eqs. (h) and (i). These profiles are plotted in Fig. E10-5. Note that Eq. (h) can be written as

$$\frac{ux}{\nu} = \sqrt{\frac{Gr_x}{Pr}} \; \frac{(3.81)^2}{4} \left[\frac{y}{\delta} - 2\left(\frac{y}{\delta}\right)^2 + \left(\frac{y}{\delta}\right)^3 \right]$$

An expression is obtained for the wall heat flux as follows:

$$q_0'' = -k \left. \frac{\partial T}{\partial y} \right|_0 = \frac{3}{2} \frac{k}{\Delta} (T_0 - T_\infty) = 0.394 \frac{k}{x} (T_0 - T_\infty) Ra_x^{0.25} \qquad \text{(m)}$$

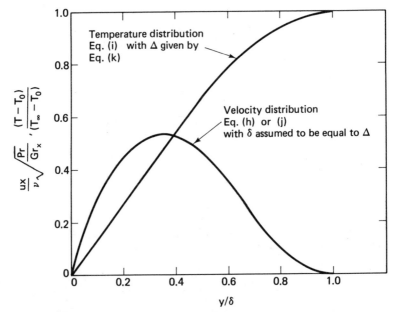

FIGURE E10-5   Predictions for velocity and temperature distributions for laminar natural convection flow over a vertical flat plate.

Because the wall heat flux for natural convection is often expressed in terms of the Nusselt number, Eq. (m) is rewritten in the form

$$Nu_x = \frac{h_x x}{k} = 0.394 \, Ra_x^{0.25} \tag{n}$$

Applying the defining equation for $\bar{h}$, we obtain

$$\overline{Nu} = \frac{\bar{h} x}{k} = 0.525 \, Ra_x^{0.25} \tag{o}$$

Referring back to Fig. 7-3, this equation lies only 5 to 6% below the empirical correlation given by Eq. (7-8). This basic agreement between Eq. (o) and experiment underwrites the acceptability of the assumptions such as $\delta \simeq \Delta$ and $\beta \simeq \beta_\infty$ which were utilized in our analysis for moderate Prandtl number fluids. Further, for air with $Pr = 0.72$, Eq. (o) is within 0.5% of a numerical solution by Schmidt and Beckmann [16] and within about 10% of an exact solution by Schuh [17] for the Prandtl number range $0.25 \gtrsim Pr \gtrsim 5.0$.

10-3 TURBULENT FLOW THEORY

10-3-1 Introduction

As indicated in Chap. 5, turbulent flow is characterized by nonstreamline unsteady random motion of fluid elements within the flow stream. This chaotic churning action is brought about by minute disturbances within the system which become unstable for values of the Reynolds number above a certain critical value. For Reynolds numbers below this value, disturbances still occur but are stabilized or dampened by the viscous properties of the fluid, such that laminar conditions prevail.

In the following sections, attention is given to general characteristics of turbulent convection processes, the classical approach to modeling turbulence, and the analysis of two basic turbulent convection processes.

10-3-2 Characteristics of Turbulent Convection Processes

The unsteady nature of turbulent flow

Extensive experimental flow visualization and anemometer studies have been conducted over the past few years which provide us with a qualitative description of the mechanism associated with turbulent convection processes. First, these studies demonstrate the unsteady character of

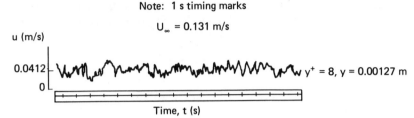

FIGURE 10-15 Measurements of instantaneous axial velocity u for fully turbulent boundary layer flow over a flat plate. (From Runstadler et al. [18].)

the entire flow field, including the region very close to the wall. This point is illustrated by Fig. 10-15, in which fluctuations in the instantaneous axial velocity u at a point very close to the wall are shown for turbulent flow over a flat plate.

A fairly realistic description of the turbulent transport mechanism is shown in Fig. 10-16, which pictures a burst process that involves the intermittent exchange of fluid between the turbulent core and wall region. As mentioned by Kays [5], relatively large elements of low-velocity fluid adjacent to the wall surface periodically lift off the surface and eventually break up out in the turbulent core. Of course, continuity considerations require that the fluid ejected from the wall region by the burst or lift-off process must be replaced by inrushing fluid with properties, such as axial velocity and temperature, which are associated with the turbulent core.

Because turbulence is characterized by fluctuating transport properties, the time-variant velocity profile u is often expressed in terms of mean $\bar{u}$ and fluctuating u' values; that is,

$$u = \bar{u} + u' \tag{10-118}$$

where

$$\bar{u} = \frac{1}{t} \int_0^t u \, dt \tag{10-119}$$

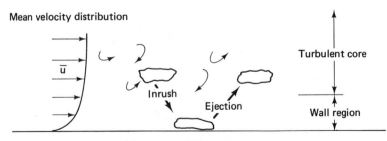

FIGURE 10-16 Turbulent burst process.

and

$$\overline{u'} = \frac{1}{t} \int_0^t u' \, dt = 0 \qquad (10\text{-}120)$$

The flow is said to be steady on a time-average basis if $\partial\bar{u}/\partial t = 0$. u, $\bar{u}$, and u' are shown in Fig. 10-17 for a situation in which the flow undergoes a change from laminar to turbulent conditions. We see that $u' = 0$ for laminar flow and $u' \neq 0$ for turbulent flow. For developing flows, we have $\partial\bar{u}/\partial x \neq 0$, whereas HFD internal flow is defined for turbulent conditions by

$$\frac{\partial\bar{u}}{\partial x} = 0 \qquad (10\text{-}121)$$

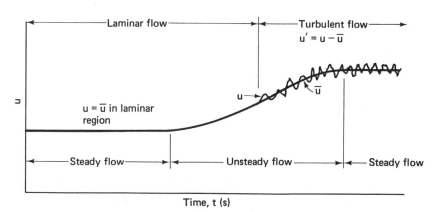

FIGURE 10-17 Representative instantaneous axial velocity u for process that undergoes a change from laminar to turbulent flow.

v, T, and P are also expressed in terms of mean and fluctuating values. In regard to the temperature profile, TFD conditions are defined for turbulent flow in terms of the time-average temperature distribution $\bar{T}$ by

$$\frac{\partial}{\partial x}\left(\frac{\bar{T} - T_s}{T_b - T_s}\right) = 0 \qquad (10\text{-}122)$$

Unlike the problem of laminar flow, predictions for the instantaneous unsteady transport properties for turbulent flow cannot be obtained. However, reliable predictions can be developed for the mean transport properties by the use of existing theoretical approaches to the analysis of turbulent flow and by the use of certain empirical inputs.

The fundamental and particular laws utilized in the analysis of laminar convection processes also apply to turbulent flow. Hence, the principles of continuity, momentum, and energy, together with the Newton law of viscosity and the Fourier law of conduction, will provide the backbone for our study of turbulent flow. But, it is the unsteady and time-average forms of these principles that must be used in the analysis of turbulence.

The consequences of turbulent flow

It should also be recalled that the lumped approach developed in Chap. 6 applies to turbulent flow as well as to laminar flow. However, the appropriate coefficients of friction and heat transfer for turbulent flow must be utilized in these rather simple analyses. Based on the physical picture of turbulence portrayed by Fig. 10-16, we expect that the macroscopic exchange of fluid between the wall and turbulent core regions will bring about a larger transport of momentum and heat than for laminar flow. The coefficients of friction and heat transfer given in Chap. 6 for laminar and turbulent flow substantiate this qualitative observation. This point is reinforced by Figs. 10-6 and 6-5 in which experimental measurements for the friction factors associated with laminar and turbulent HFD flow in circular tubes and flow over flat plates are shown. Notice the increase in friction factor as the flow transcends from laminar to turbulent conditions.

The consequence of this unsteady exchange of fluid between the wall and core regions is also vividly portrayed by the velocity profiles shown in Fig. 10-18 for channel flow. Although the Reynolds number is equal to 4000 for both these profiles, one is laminar and the other is (transitional) turbulent. Notice that much larger wall-region velocities are found in turbulent flow than in laminar flow. Hence, the transport of momentum from the bulk steam to the wall region is indeed most effective for turbulent flow.

Similarly, convection heat and mass transfer are much more effective for turbulent flow than for laminar flow. Therefore, industrial convection heat and mass transfer equipment is often operated in the turbulent flow regime in order to achieve high transfer rates.

10-3-3   Classical Approach to Modeling Turbulence

The key modeling concepts involved in the classical analysis of turbulent convection processes are introduced in this section in the context of the differential formulation approach. In addition, the very useful integral formulation will be presented.

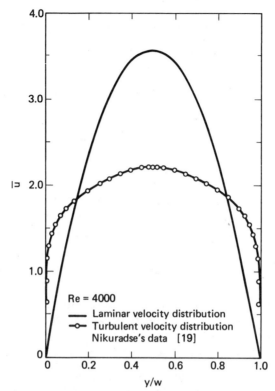

FIGURE 10-18 Velocity distributions for both laminar and (transitional) turbulent HFD channel flow. (From *Fluid Dynamics and Heat Transfer* by J. G. Knudsen and D. L. Katz. Copyright 1958 by McGraw-Hill Book Company. Used with permission of McGraw-Hill Book Company.)

Differential formulation

Fluid Flow In the classical approach to turbulence, the continuity and momentum equations are first written in terms of the instantaneous unsteady velocity distribution, after which equations are developed for the time-average velocity distribution.

Instantaneous Equations. Applying the principles of continuity and momentum to an unsteady fluctuating flow field, we obtain (incompressible flow, Cartesian coordinates)

Continuity

$$\frac{\partial u}{\partial x} + \frac{\partial v}{\partial y} = 0 \qquad (10\text{-}123)$$

Momentum

$$\frac{\partial u}{\partial t} + u\frac{\partial u}{\partial x} + v\frac{\partial u}{\partial y} = \frac{1}{\rho}\frac{\partial}{\partial y}\left(\mu\frac{\partial u}{\partial y}\right) - \frac{1}{\rho}\frac{\partial P}{\partial x} \qquad (10\text{-}124)$$

The development of these equations follows step by step the formulation of the differential continuity and momentum equations for unsteady laminar flow.

Because of the complex random unsteady nature of turbulence, Eqs. (10-123) and (10-124) cannot be solved for the instantaneous fluctuating profiles u and v. However, these equations provide a theoretical basis for the development of time-average equations which, when coupled with certain empirical inputs, can be solved for the important mean velocity distributions $\bar{u}$ and $\bar{v}$. Once we know $\bar{u}$ and $\bar{v}$, the mean wall shear stress can be determined from

$$\bar{\tau}_0 = \mu\frac{\partial \bar{u}}{\partial y}\bigg|_0 \qquad (10\text{-}125)$$

and predictions can eventually be obtained for the mean temperature distribution $\bar{T}$ and Nusselt number by solving the time-average energy equation. (For variable property conditions, the time average fluid flow and energy equations must be solved simultaneously.)

Time-Average Equations. To develop the time-average continuity equation, u and v are replaced by their mean and fluctuation components; that is,

$$\frac{\partial \bar{u}}{\partial x} + \frac{\partial u'}{\partial x} + \frac{\partial \bar{v}}{\partial y} + \frac{\partial v'}{\partial y} = 0 \qquad (10\text{-}126)$$

Then, taking the time average of each term, we obtain

$$\overline{\frac{\partial \bar{u}}{\partial x}} + \overline{\frac{\partial u'}{\partial x}} + \overline{\frac{\partial \bar{v}}{\partial y}} + \overline{\frac{\partial v'}{\partial y}} = 0 \qquad (10\text{-}127)$$

or simply,

$$\frac{\partial \bar{u}}{\partial x} + \frac{\partial \bar{v}}{\partial y} = 0 \qquad (10\text{-}128)$$

To develop the time-average momentum equation, we first rewrite Eq. (10-124) as

$$\rho\left[\frac{\partial u}{\partial t} + \frac{\partial u^2}{\partial x} + \frac{\partial}{\partial y}(uv) - u\left(\frac{\partial u}{\partial x} + \frac{\partial v}{\partial y}\right)\right] = \frac{\partial}{\partial y}\left(\mu\frac{\partial u}{\partial y}\right) - \frac{\partial P}{\partial x}$$

$$(10\text{-}129)$$

By introducing the mean and fluctuating properties into this equation and taking the time average of each term, we obtain (for $\partial \bar{u}/\partial t = 0$)

$$\rho\left[\frac{\partial \overline{u^2}}{\partial x} + \frac{\partial}{\partial y}(\overline{uv}) + \frac{\partial \overline{u'^2}}{\partial x} + \frac{\partial}{\partial y}(\overline{u'v'})\right] = \frac{\partial}{\partial y}\left(\mu\frac{\partial \bar{u}}{\partial y}\right) - \frac{\partial \bar{P}}{\partial x} \qquad (10\text{-}130)$$

Although the fluctuation terms of the left-hand side of this equation actually represent the transport of momentum by turbulent eddy motion, this equation is traditionally coupled with the continuity equation and put into the form

$$\rho\left(\bar{u}\frac{\partial \bar{u}}{\partial x} + \bar{v}\frac{\partial \bar{u}}{\partial y}\right) = \frac{\partial}{\partial y}\left(\mu\frac{\partial \bar{u}}{\partial y} - \rho\overline{u'v'}\right) - \frac{\partial}{\partial x}(\bar{P} + \overline{u'^2}) \qquad (10\text{-}131)$$

where $\partial\overline{u'^2}/\partial x$ is generally assumed to be negligible on the basis of experience. Because the term $\mu\,\partial\bar{u}/\partial y$ represents the mean shear stress associated with molecular transport, the term $-\rho\overline{u'v'}$ has come to be called the *apparent turbulent shear stress* $\overline{\tau_t}$; that is,

$$\overline{\tau_t} = -\rho\,\overline{u'v'} \qquad (10\text{-}132)$$

$\overline{\tau_t}$ is also referred to as the *Reynolds shear stress* in honor of the man who first introduced this concept [21]. The actual mean shear stress, $\mu\,\partial\bar{u}/\partial y$, is designated by $\overline{\tau_y}$. Based on this convention, the two terms $\overline{\tau_t}$ and $\overline{\tau_y}$ taken together can be thought of as an *apparent total shear stress* which we shall denote by $\bar{\tau}$.

$$\bar{\tau} = \overline{\tau_y} + \overline{\tau_t} = \mu\frac{\partial \bar{u}}{\partial y} - \rho\overline{u'v'} \qquad (10\text{-}133)$$

Based on these definitions for $\bar{\tau}$ and $\overline{\tau_t}$ and assuming negligible pressure gradients in the y direction, the time-average momentum equation, Eq. (10-131), takes the compact and useful form

$$\rho\left(\bar{u}\frac{\partial \bar{u}}{\partial x} + \bar{v}\frac{\partial \bar{u}}{\partial y}\right) = \frac{\partial \bar{\tau}}{\partial y} - \frac{d\bar{P}}{dx} \qquad (10\text{-}134)$$

With the Reynolds shear stress term equal to zero for laminar conditions, this equation is seen to reduce to the momentum equation for laminar flow, Eq. (10-55).

The differential formulation for turbulent fluid flow is completed by writing the boundary conditions and by specifying the turbulence parameter $\overline{\tau_t}$. The boundary conditions are similar in form to the boundary conditions for laminar flow. For example, we have the no-slip condition at the wall

$$\bar{u} = 0 \qquad \text{at} \quad y = 0 \tag{10-135}$$

Specification of Reynolds Stress. The distinctive feature of these turbulent fluid-flow equations is the fact that the Reynolds stress $\bar{\tau}_t$ must be specified by the use of theoretical and/or empirical inputs. Although $\bar{\tau}_t$ can be evaluated on the basis of experimental measurements for u' and v', the usual analytical approach, known as the mean-field method, is to relate this turbulence parameter to the local mean axial velocity $\bar{u}$ by

$$\bar{\tau}_t = \mu_t \frac{\partial \bar{u}}{\partial y} \tag{10-136}$$

where the *eddy viscosity* μ_t must be specified; $\nu_t (\equiv \mu_t / \rho)$ is the *eddy diffusivity* or *eddy kinematic viscosity*. (The symbol ε_m is often used instead of ν_t in the turbulence literature.) Notice the similarity between this equation and the Newton law of viscosity,

$$\tau_y = \mu \frac{\partial u}{\partial y} \qquad \text{or} \qquad \bar{\tau}_y = \mu \frac{\partial \bar{u}}{\partial y} \tag{10-137, 138}$$

With the Reynolds stress $\bar{\tau}_t$ expressed in terms of the eddy viscosity μ_t, the apparent total shear stress becomes

$$\bar{\tau} = \bar{\tau}_y + \bar{\tau}_t = (\mu + \mu_t) \frac{\partial \bar{u}}{\partial y} \tag{10-139}$$

or

$$\frac{\bar{\tau}}{\rho} = (\nu + \nu_t) \frac{\partial \bar{u}}{\partial y} \tag{10-140}$$

The differential time-average momentum equation can be written in terms of μ_t as

$$\rho \left(\bar{u} \frac{\partial \bar{u}}{\partial x} + \bar{v} \frac{\partial \bar{u}}{\partial y} \right) = \frac{\partial}{\partial y} \left[(\mu + \mu_t) \frac{\partial \bar{u}}{\partial y} \right] - \frac{d\bar{P}}{dx} \tag{10-141}$$

Representative measurements for ν_t are shown in Fig. 10-19 for fully turbulent flow in terms of the dimensionless distances y^+ and y/δ; the dimensionless distance y^+ is given by

$$y^+ = \frac{U^* y}{\nu} \tag{10-142}$$

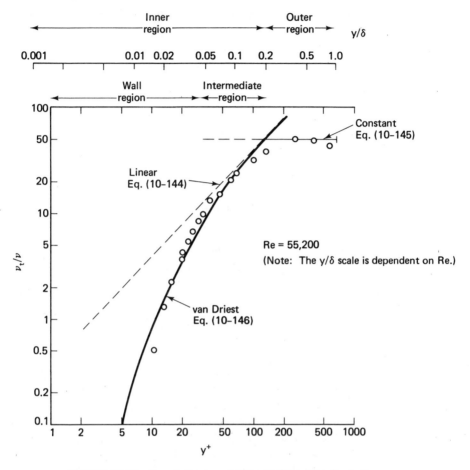

FIGURE 10-19 Correlations and experimental data for turbulent eddy diffusivity ν_t. (Data by Hussain and Reynolds [23].)

The *friction velocity* U^* is defined by

$$U^* = \sqrt{\frac{\bar{\tau}_0}{\rho}} \qquad (10\text{-}143)$$

and δ represents the channel half-width, tube radius, or boundary layer thickness. Although these data are for channel flow, data for fully turbulent flow in tubes and over flat plates follow the same pattern, such that the same correlations for ν_t are generally applied to any fully turbulent flow field.

For practical purposes, a fully turbulent flow field can be broken into the following zones:

$$
\begin{array}{ll}
\text{Inner region} & y/\delta \gtrsim y_1/\delta \\
\text{Wall region} & y^+ \lesssim 50 \\
\text{Intermediate region} & y^+ \gtrsim 50 \\
\text{Outer region} & y/\delta \gtrsim y_1/\delta
\end{array}
$$

where y_1/δ generally lies between 0.15 and 0.2. The domain outside the wall region is also known as the turbulent core. Referring to Fig. 10-19, we approximate ν_t/ν in the intermediate and outer regions (i.e., in the turbulent core) by expressions of the form

$$
\frac{\nu_t}{\nu} = \kappa y^+ \qquad
\begin{array}{c}
\text{Intermediate region} \\
y^+ \gtrsim 50, \quad y/\delta \gtrsim y_1/\delta
\end{array}
\qquad (10\text{-}144)
$$

and

$$
\frac{\nu_t}{\nu} = \alpha_1 \delta^+ \qquad
\begin{array}{c}
\text{Outer region} \\
y/\delta \gtrsim y_1/\delta
\end{array}
\qquad (10\text{-}145)
$$

where the empirical constant κ is approximately 0.4 for fully turbulent HFD internal flow and $\alpha_1 (= \kappa y_1/\delta)$ is usually between 0.06 and 0.08. (Other approximations are also found in the literature for ν_t.) In the wall region, ν_t/ν is clearly nonlinear.

One of the objectives in the analysis of turbulence is to minimize the number of empirical inputs by employing reasonable theoretical models of the actual turbulent transport process. These efforts have been most productive in treating the region adjacent to the wall. Among the various mathematical models for the behavior of ν_t within the wall region for fully turbulent flow, the formulation by van Driest [24] is perhaps the most popular. This classical analysis gives rise to an expression for ν_t/ν within the entire inner region of the form

$$
\frac{\nu_t}{\nu} = (\kappa y^+)^2 \left[1 - \exp\left(-\frac{y^+}{a}\right)\right]^2 \frac{du^+}{dy^+} \qquad (10\text{-}146)
$$

where a is an empirical constant which is approximately equal to 26. This equation reduces to Eq. (10-144) as y^+ approaches a value of the order of 50. Equation (10-146) is shown to correlate the experimental data in Fig. 10-19 quite well within both the wall and intermediate regions. This basic approach has been used extensively by Cebeci and Smith [25] and many others.

An alternative nonclassical approach to modeling wall turbulence has recently been developed [26]–[28] which is somewhat akin to the van Driest approach. This *surface renewal* approach to the analysis of turbulent convection transport processes will be touched upon in the closing section of this chapter.

Other empirical correlations for ν_t within the inner region have been developed by Rotta [29], Reichardt [30], Deissler [31], and Spalding [32].

EXAMPLE 10-6

A popular alternative mean-field representation for $\bar{\tau}_t$ was proposed by Prandtl [22] in 1910 of the form

$$\bar{\tau}_t = \rho \ell^2 \left(\frac{\partial \bar{u}}{\partial y} \right)^2 \tag{a}$$

where the mixing length ℓ is imagined to be the small transverse distance that fluid elements move during a turbulent fluctuation. In the intermediate region where $\nu_t/\nu = \kappa y^+$, ℓ has been found to be equal to κy. Develop a relationship between ν_t and ℓ.

Solution

To relate ℓ to ν_t, we compare Eq. (a) with Eq. (10-136),

$$\tau_t = \mu_t \frac{\partial \bar{u}}{\partial y} = \rho \ell^2 \left(\frac{\partial \bar{u}}{\partial y} \right)^2$$

$$\nu_t = \ell^2 \frac{\partial \bar{u}}{\partial y} \tag{b}$$

or

$$\frac{\nu_t}{\nu} = \ell^{+2} \frac{\partial u^+}{\partial y^+} \tag{c}$$

where $\ell^+ = \ell U^*/\nu$.

Energy Transfer Following the pattern established in the development of the fluid-flow equations, the time-average energy equation can be written as (for $\partial \bar{T}/\partial t = 0$)

$$\rho c_p \left(\bar{u} \frac{\partial \bar{T}}{\partial x} + \bar{v} \frac{\partial \bar{T}}{\partial y} \right) = -\frac{\partial \overline{q''}}{\partial y} \tag{10-147}$$

where $\overline{q''}$ is the *apparent total heat flux*. $\overline{q''}$ is defined in terms of the actual mean heat flux $\overline{q_y''}$ and the *apparent turbulent heat flux* $\overline{q_t''}$ as follows:

$$\overline{q''} = \overline{q_y''} + \overline{q_t''} \tag{10-148}$$

where

$$\overline{q_y''} = -k\frac{\partial \overline{T}}{\partial y} \qquad \text{mean heat flux} \qquad (10\text{-}149)$$

$$\overline{q_t''} = \rho c_p \overline{v'T'} \qquad \text{apparent turbulent heat flux} \qquad (10\text{-}150)$$

Notice that Eq. (10-147) reduces to the energy equation for laminar flow, Eq. (10-56), as the turbulent fluctuation term approaches zero.

The time-average energy equation must, of course, be coupled with appropriate boundary conditions. As an example, for turbulent flow over a surface with uniform wall flux heating, we have

$$-k\frac{\partial \overline{T}}{\partial y} = \overline{q_0''} \qquad \text{at} \quad y=0 \qquad (10\text{-}151)$$

Finally, our formulation is closed by specifying the turbulence parameter $\overline{q_t''}$. The apparent turbulent heat flux is generally expressed in terms of $\overline{T}$ by

$$\overline{q_t''} = -k_t\frac{\partial \overline{T}}{\partial y} \qquad (10\text{-}152)$$

or

$$\frac{\overline{q_t''}}{\rho c_p} = -\alpha_t\frac{\partial \overline{T}}{\partial y} \qquad (10\text{-}153)$$

where k_t is the *eddy thermal conductivity* and α_t is the *eddy thermal diffusivity*. (The symbol ε_H is often used in place of α_t in the turbulence literature.)

With $\overline{q_t''}$ expressed in this mean-field format, the differential time-average energy equation takes the form

$$\rho c_p\left(\overline{u}\frac{\partial \overline{T}}{\partial x} + \overline{v}\frac{\partial \overline{T}}{\partial y}\right) = \frac{\partial}{\partial y}\left[(k+k_t)\frac{\partial \overline{T}}{\partial y}\right] \qquad (10\text{-}154)$$

Although ν_t and α_t are not fluid properties, they are often expressed in the form of a dimensionless ratio called the *turbulent Prandtl number* Pr_t,

$$\text{Pr}_t = \frac{c_p \mu_t}{k_t} = \frac{\nu_t}{\alpha_t} \qquad (10\text{-}155)$$

As early as 1874, Reynolds postulated that Pr_t should be approximately unity because $\overline{q_t''}$ and $\overline{\tau_t}$ both result from the same mechanism; that is, $\alpha_t \simeq \nu_t$ or $Pr_t \simeq 1$. This simple assumption has been utilized with quite good results for basic fully turbulent flow of moderate to high Prandtl number fluids. However, recent experimental evidence for moderate Pr fluids indicates that Pr_t is greater than unity in at least a portion of the wall region, falls to a value of the order of 0.9 in the intermediate region out to about $y/\delta = 0.4$, and eventually falls to very small values in the outermost part of the boundary layer [33, 34]. The net effect of this Pr_t variation with y^+ is evidently balanced out across the boundary layer, such that the $Pr_t = 1$ approximation provides reasonable engineering predictions. However, the turbulent Prandtl number cannot be assumed to be unity for low-Prandtl-number liquid metals. Formulations have been developed for Pr_t for liquid metals by Azer and Chao [35], Jenkins [36], and others.

Integral formulation

Following the pattern established in Example 10-5, we integrate the differential time-average continuity, momentum, and energy equations to obtain the integral formulation for turbulent developing flow with $d\overline{P}/dx = 0$. The resulting time-average integral momentum and energy equations take the forms

$$\frac{\partial}{\partial x} \int_0^\delta \bar{u}(U_\infty - \bar{u}) \, dy = \frac{\overline{\tau_0}}{\rho} \tag{10-156}$$

$$\frac{\partial}{\partial x} \int_0^\Delta \bar{u}(\overline{T} - T_\infty) \, dy = \frac{\overline{q_c''}}{\rho c_p} \tag{10-157}$$

for systems with uniform properties and negligible potential and kinetic energy, viscous dissipation, axial conduction, and thermal radiation effects.

The boundary conditions for the integral formulation are identical to those which are utilized in the differential formulation. Notice that these integral equations do *not* involve the turbulent fluctuation terms $\overline{\tau_t}$ and $\overline{q_t''}$! Although the integral formulation for turbulent flow is of the same form as for laminar flow [see Eqs. (10-74) and (10-80) and the boundary conditions], it should be observed that these modeling equations for turbulence involve $\bar{u}$, $\overline{T}$, $\overline{\tau_0}$, and $\overline{q_c''}$ rather than u, T, τ_0, and q_c''. Thus, the turbulent fluctuating momentum and energy transport mechanism that the parameters $\overline{\tau_t}$ and $\overline{q_t''}$ represent still manages to play its role in the formulation. In this regard, the rather well-behaved nature of the profiles for laminar flow can be attributed to the fact that the momentum and energy transfer mechanisms operate evenhandedly across the entire flow field. On the other hand, the fact that the mean fluctuation momentum and energy transfer (or apparent turbulent shear stress and heat flux, $\overline{\tau_t}$ and $\overline{q_t''}$) associated with turbulent flow vary with y in a nonlinear fashion produces turbulent profiles that possess quite distinctive and different characteristics in the wall region, the intermediate region, and the outer region. The approximation of these profiles by the use of polynomials and physical insight alone is simply not possible. To circumvent this difficulty, we turn to simplified differential analyses and empirical correlations for the mean profiles in order to develop integral solutions for the local mean wall shear stress and heat transfer. This basic approach to the analysis of turbulence will be developed in the context of fully turbulent boundary layer flow over a flat plate.

10-3-4   Solutions

The classical time-average differential continuity, momentum, and energy equations for turbulent convection heat transfer can be solved by differential/numerical and integral techniques. These methods of solution follow the patterns established in the analysis of laminar flow, except for complications that arise because of the turbulent eddy transport mechanism. To demonstrate the mechanics involved in developing solutions for turbulent convection-heat-transfer processes, consideration is given to fully turbulent TFD internal flow and fully turbulent boundary layer flow over a flat plate, with uniform wall-flux heating maintained in both cases.

Fully turbulent flow in tubes

To minimize the mathematical details, the analysis of convection heat transfer for fully developed, fully turbulent internal flow is presented in the context of the parallel-plate system shown in Fig. 10-20. As indicated in Chap. 6, convection heat and momentum transfer are fairly insensitive to the geometry for fully turbulent flow in internal flow systems with uniform cross-sectional area. Hence, predications for coefficients of friction and heat transfer which are obtained for fully turbulent flow between parallel plates can also be used for the standard circular tube geometry, as well as other geometries with uniform cross-sectional area.

Fluid Flow—HFD Region   Our primary objectives in this section are to develop predictions for the mean velocity profile $\bar{u}$ and the mean wall shear stress $\bar{\tau}_0$ or friction factor f. This information will then be utilized in the following section to develop solutions for the mean temperature distribution and the Nusselt number.

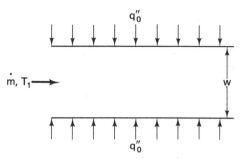

FIGURE 10-20 Fully developed turbulent flow between parallel plates with uniform wall-flux heating.

For HFD turbulent flow between parallel plates, the classical time-average momentum equation reduces to the form

$$\frac{d}{dy}\left[(\mu+\mu_t)\frac{d\bar{u}}{dy}\right] - \frac{d\bar{P}}{dx} = 0 \qquad (10\text{-}158)$$

The boundary conditions are

$$\bar{u}=0 \quad \text{at} \quad y=0 \qquad \frac{d\bar{u}}{dy}=0 \quad \text{at} \quad y=\frac{w}{2} \qquad (10\text{-}159,160)$$

Our formulation is closed by the specification of μ_t. As indicated in the previous section, ν_t can be approximated by relationships such as those by van Driest or Spalding in the wall region, and by

$$\frac{\nu_t}{\nu} = \kappa y^+ \qquad \begin{array}{l}\text{intermediate region}\\ y^+\gtrsim 50, \quad y/\delta\gtrsim 0.2\end{array} \qquad (10\text{-}144)$$

and

$$\frac{\nu_t}{\nu} = \alpha_1\delta^+ \qquad \begin{array}{l}\text{outer region}\\ y/\delta\gtrsim 0.2\end{array} \qquad (10\text{-}145)$$

where $\kappa=0.4$, $\alpha_1\simeq0.08$, and $\delta=w/2$ for HFD channel flow.

Dimensionless Mean Velocity Distribution. Equation (10-158) is separated and integrated to obtain

$$(\mu+\mu_t)\frac{d\bar{u}}{dy} = \frac{d\bar{P}}{dx}y + C_1 \qquad (10\text{-}161)$$

Utilizing the boundary condition given by Eq. (10-160), we have

$$(\mu + \mu_t)\frac{d\bar{u}}{dy} = \frac{d\bar{P}}{dx}\left(y - \frac{w}{2}\right) \tag{10-162}$$

A somewhat more convenient form of this equation can be obtained by eliminating the pressure gradient term in favor of the mean wall shear stress $\bar{\tau}_0$. (This step can also be taken in our analysis of laminar flow instead of expressing dP/dx in terms of U_c. See Prob. 10-35).

$$\frac{d\bar{P}}{dx} = -\frac{2}{w}(\mu + \mu_t)\frac{d\bar{u}}{dy}\bigg|_0 = -\frac{2}{w}\bar{\tau}_0 \tag{10-163}$$

The coupling of this result with Eq. (10-162) gives

$$(\mu + \mu_t)\frac{d\bar{u}}{dy} = \bar{\tau}_0\left(1 - \frac{y}{w/2}\right) \tag{10-164}$$

or

$$\bar{\tau} = \bar{\tau}_0\left(1 - \frac{y}{w/2}\right) \tag{10-165}$$

Hence, the apparent total shear stress $\bar{\tau}$ is seen to vary linearly with y, as was found to be the case for laminar flow (Example 10-2). Equation (10-164) takes the dimensionless form

$$\left(1 + \frac{\nu_t}{\nu}\right)\frac{du^+}{dy^+} = 1 - \frac{y}{w/2} \tag{10-166}$$

where y/w is also equal to y^+/w^+.

To produce predictions for the dimensionless velocity distribution u^+, we again separate the variables and integrate to obtain

$$u^+ = \int_0^{y^+} \frac{1 - y/w/2}{1 + \nu_t/\nu}\, dy^+ \tag{10-167}$$

Although this equation can be integrated numerically as it stands, the use of several practical approximations simplifies our computational task.

Focusing attention on the inner region ($y/\delta = y/w/2 \gtrsim 0.2$), where the $y/w/2$ term is small, Eq. (10-166) can be approximated by

$$\left(1 + \frac{\nu_t}{\nu}\right)\frac{du^+}{dy^+} = 1 \qquad \begin{array}{c}\text{inner region}\\[4pt] \dfrac{y}{w/2} \gtrsim 0.2\end{array} \tag{10-168}$$

Referring back to Eq. (10-164) or (10-165), we see that this approximation is equivalent to setting $\bar{\tau}$ equal to $\bar{\tau}_0$. In fact, the inner region can be defined as the zone in which $\bar{\tau} \simeq \bar{\tau}_0$. Separating the variables and integrating this equation, we have

$$u^+ = \int_0^{y^+} \frac{1}{1 + \nu_t/\nu} \, dy^+ \qquad \frac{y}{w/2} \gtrsim 0.2 \qquad \text{inner region} \qquad (10\text{-}169)$$

The numerical integration of this equation in the wall region where ν_t/ν can be approximated by the van Driest equation, Eq. (10-146), is discussed in Example 10-7. The van Driest calculations are shown to compare very favorably with experimental data for u^+ in Fig. 10-21.

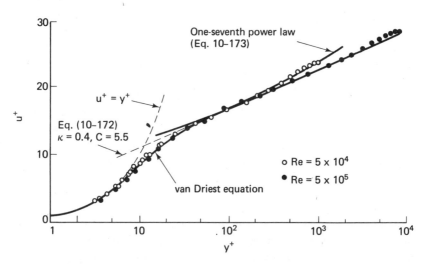

FIGURE 10-21 Dimensionless velocity distributions for HFD fully turbulent flow in circular tubes. (Data from Lindgren [37].)

In the innermost part of the wall region, known as the viscous sublayer, where ν_t is much less than ν, Eq. (10-169) reduces to

$$u^+ = y^+ \qquad (10\text{-}170)$$

The experimental data in Fig. 10-21 are in good agreement with this limiting equation.

In the intermediate region in which ν_t/ν is approximately equal to κy^+ and where ν_t is much greater than ν, Eq. (10-169) reduces to

$$u^+ = \int \frac{dy^+}{\kappa y^+} + C \qquad (10\text{-}171)$$

such that u^+ is given by

$$u^+ = \frac{1}{\kappa} \ln y^+ + C \qquad (10\text{-}172)$$

The empirical constants are generally given by $\kappa = 0.4$ and $C = 5.5$ for fully turbulent HFD internal flows. As shown in Example 10-7, the constant C and the constant a in the van Driest equation are related. The specification of one predetermines the other. This famous logarithmic equation is compared with experimental data in Fig. 10-21. The agreement between Eq. (10-172) and the data is exceptional within the intermediate region. This equation is even in reasonable agreement with the data in the outer region, which is surprising when one considers the fact that $\bar{\tau} \neq \bar{\tau}_0$ and $\nu_t / \nu \neq \kappa y^+$ in this region. But as expected, Eq. (10-172) fails to correlate the data in the wall region.

As shown in Fig. 10-21, this logarithmic equation can be approximated by

$$u^+ = 8.7 y^{+\,1/7} \qquad (10\text{-}173)$$

between $y^+ = 30$ and 500. This simple one-seventh power law has been found to correlate experimental data fairly well in the intermediate and outer regions for Reynolds numbers in the important 10^4 to 10^5 region.

Because the theoretical basis for Eq. (10-172) and others like it, such as the van Driest equation, is restricted to the inner region, we will refer to such equations as *laws of the inner region*. (These types of equations are frequently referred to as *laws of the wall* in turbulence literature.) It is our good fortune that these equations also correlate the experimental data reasonably well in the outer region. However, the development of more accurate theoretically based expressions for u^+ in the outer region requires that Eq. (10-167) be solved. The development of a simple analytical solution for u^+ within this region is considered in Prob. 10-39.

We now turn our attention to the development of predictions for the friction factor. Once f is known, the mean velocity distribution $\bar{u}$ can be obtained by utilizing the defining equation for u^+; that is,

$$u^+ = \frac{\bar{u}}{U^*}$$

$$\bar{u} = u^+ U^* = u^+ \sqrt{\frac{\bar{\tau}_0}{\rho}} = u^+ U_b \sqrt{\frac{f}{2}} \qquad (10\text{-}174)$$

EXAMPLE 10-7

Develop an expression for u^+ within the entire inner region for fully turbulent HFD flow by utilizing the van Driest equation for ν_t.

Solution

The van Driest expression for ν_t within the inner region is given by

$$\frac{\nu_t}{\nu} = (\kappa y^+)^2 \left[1 - \exp\left(-\frac{y^+}{a}\right)\right]^2 \frac{du^+}{dy^+} \qquad (10\text{-}146)$$

The relationship between ν_t and du^+/dy^+ can be obtained from the momentum equation as follows:

$$\left(1 + \frac{\nu_t}{\nu}\right) \frac{du^+}{dy^+} = 1 \qquad (10\text{-}168)$$

$$\frac{du^+}{dy^+} = \frac{1}{1 + \nu_t/\nu} \qquad (a)$$

Hence, Eq. (10-146) can be put in the form

$$\frac{\nu_t}{\nu} = (\kappa y^+)^2 \left[1 - \exp\left(-\frac{y^+}{a}\right)\right]^2 \frac{1}{1 + \nu_t/\nu} \qquad (b)$$

or

$$\left(\frac{\nu_t}{\nu}\right)^2 + \frac{\nu_t}{\nu} - (\kappa y^+)^2 \left[1 - \exp\left(-\frac{y^+}{a}\right)\right]^2 = 0 \qquad (c)$$

The solution to this equation is

$$\frac{\nu_t}{\nu} = \frac{-1 + \sqrt{1 + 4(\kappa y^+)^2 [1 - \exp(-y^+/a)]^2}}{2} \qquad (d)$$

We are now ready to obtain an expression for u^+. Substituting this expression for ν_t into the momentum equation, Eq. (10-168), and integrating, we have

$$u^+ = \int_0^{y^+} \frac{1}{1 + \dfrac{-1 + \{1 + 4(\kappa y^+)^2 [1 - \exp(-y^+/a)]^2\}^{1/2}}{2}} \, dy^+ \qquad (e)$$

van Driest integrated this equation numerically with a set equal to 26. His results are shown in Fig. 10-21 to be in excellent agreement with experimental data. A simple FORTRAN program for performing this numerical integration is presented in Example 10-10.

EXAMPLE 10-8

Develop solutions for the dimensionless velocity profile u^+ in the inner region for HFD fully turbulent flow in a circular tube.

Solution

The time-average differential momentum equation for turbulent flow in a circular tube is given by

$$\rho\left(\bar{u}\frac{\partial\bar{u}}{\partial x}+\bar{v}\frac{\partial\bar{u}}{\partial r}\right)=-\frac{1}{r}\frac{\partial}{\partial r}(r\bar{\tau})-\frac{\partial\bar{P}}{\partial x} \tag{a}$$

For HFD, this equation reduces to

$$\frac{d}{dr}(r\bar{\tau})=-r\frac{d\bar{P}}{dx} \tag{b}$$

Integrating once, we obtain

$$r\bar{\tau}=-\frac{r^2}{2}\frac{d\bar{P}}{dx}+C_1 \tag{c}$$

Because $\bar{\tau}=0$ at $r=0$, $C_1=0$. Hence, we have

$$\bar{\tau}=-\frac{r}{2}\frac{d\bar{P}}{dx} \tag{d}$$

Setting $\bar{\tau}=\bar{\tau}_0$ at $r=r_0$, we see that

$$\frac{d\bar{P}}{dx}=-\frac{2\bar{\tau}_0}{r_0} \tag{e}$$

such that Eq. (d) also takes the form

$$\bar{\tau}=\bar{\tau}_0\frac{r}{r_0} \tag{f}$$

With r replaced by r_0-y, this equation takes the form

$$\bar{\tau}=\bar{\tau}_0\left(1-\frac{y}{r_0}\right) \tag{g}$$

This equation for turbulent flow in a circular tube is seen to be of the same form as Eq. (10-165) for turbulent flow in a channel, except for the fact that $w/2$ is replaced by r_0. Therefore, the expressions for u^+ which were developed for HFD flow between parallel plates also apply to

HFD tube flow, with $w/2$ being replaced by r_0. It follows that u^+ is approximated by Eq. (e) of Example 10-7 for the wall region, and by Eq. (10-172) for the region $y^+ \gtrsim 50$ and $y/\delta \gtrsim 0.2$. Equation (10-172) can also be utilized as a first approximation in the outer region. (See Prob. 10-40 for the development of a more accurate expression for u^+ in the outer region.)

Bulk Stream Velocity and Friction Factor. As in our analysis of laminar internal flow, the final step in obtaining predictions for the velocity distribution $\bar{u}$ and the friction factor involves the evaluation of the bulk stream velocity U_b. By definition, the bulk stream velocity for turbulent flow between parallel plates is given by

$$U_b = \frac{1}{w} \int_0^w \bar{u} \, dy \tag{10-175}$$

With $\bar{u}$ expressed in terms of u^+, we obtain

$$U_b = \frac{U^*}{w^+} \int_0^{w^+} u^+ \, dy^+ \tag{10-176}$$

where $w^+ = U^* w / \nu$. Because $U^* = U_b \sqrt{f/2}$, we have

$$\sqrt{\frac{2}{f}} = \frac{2}{w^+} \int_0^{w^+/2} u^+ \, dy^+ \tag{10-177}$$

Therefore, with u^+ known, calculations can be obtained for f.

As a first approximation, we will utilize the simple one-seventh power law given by Eq. (10-173) throughout the entire flow field. Although this equation fails to correlate the data for u^+ in the wall region, the volume flow rate in this region is negligibly small when compared to the volume flow rate in the intermediate and outer regions. Hence, the error introduced by this approximation is small. Substituting the one-seventh power law into Eq. (10-177), we have

$$\sqrt{\frac{2}{f}} = \frac{2}{w^+} \int_0^{w^+/2} 8.7 y^{+1/7} dy^+$$

$$\sqrt{\frac{2}{f}} = \frac{7}{8}(8.7)\left(\frac{w^+}{2}\right)^{1/7} \tag{10-178}$$

where

$$w^+ = \frac{wU^*}{\nu} = \frac{wU_b}{\nu}\sqrt{\frac{f}{2}} = \frac{\text{Re}}{2}\sqrt{\frac{f}{2}} \tag{10-179}$$

Solving for f, we obtain

$$f = 0.081 \, \text{Re}^{-0.25} \qquad (10\text{-}180)$$

This equation lies only $2\frac{1}{2}\%$ above the well-known Blasius equation,

$$f = 0.079 \, \text{Re}^{-0.25} \qquad (10\text{-}181)$$

which was obtained on the basis of an empirical curve fit to data for HFD turbulent flow in circular tubes and channels in the low Reynolds number range from 5000 to 10^5.

In order to obtain predictions for f for larger values of the Reynolds number, the logarithmic correlation for u^+ can be utilized. As shown in Example 10-9, the use of this more accurate correlation for u^+ gives rise to an implicit expression for f of the form

$$\sqrt{\frac{2}{f}} = C - \frac{1}{\kappa} + \frac{1}{\kappa} \ln\left(\frac{\text{Re}}{4} \sqrt{\frac{f}{2}} \right) \qquad (10\text{-}182)$$

for flow between parallel plates. Setting $\kappa = 0.4$ and $C = 5.5$, we have

$$\sqrt{\frac{2}{f}} = 2.5 \ln\left(\text{Re} \sqrt{\frac{f}{2}} \right) - 0.466 \qquad (10\text{-}183)$$

Rearranging this equation, an explicit expression is obtained for Re in terms of f.

$$\text{Re} = 1.2 \sqrt{\frac{2}{f}} \, \exp\left(0.4 \sqrt{\frac{2}{f}} \right) \qquad (10\text{-}184)$$

An expression is also developed in Example 10-9 for HFD turbulent flow in a circular tube which takes the form

$$\sqrt{\frac{2}{f}} = 2.5 \ln\left(\text{Re} \sqrt{\frac{f}{2}} \right) + 0.0171 \qquad (10\text{-}185)$$

This expression was first derived by Prandtl in 1935. However, Prandtl adjusted the constants on the basis of experimental measurements by Nikuradse [19] in order to obtain better agreement with the data in the low-Reynolds-number range. The final Prandtl–Nikuradse equation is given by

$$\sqrt{\frac{2}{f}} = 2.46 \ln\left(\text{Re} \sqrt{\frac{f}{2}} \right) + 0.292 \qquad (10\text{-}186)$$

This equation was shown to be in excellent agreement with the experimental data in Fig. 6-5 over the entire turbulent flow range. The equation for flow between parallel plates, Eq. (10-183), is approximately 20% above Eq. (10-186). However, this difference is of the order of the scatter in the experimental data.

For values of Re $\gtrsim 10^4$, Eq. (10-186) can be approximated by a simple explicit correlation for f of the form

$$f = 0.046 \, \text{Re}^{-0.2} \qquad \text{Re} \gtrsim 10^4 \tag{10-187}$$

This simple equation was shown to compare very well with the data in Fig. 6-5.

EXAMPLE 10-9

Utilize the logarithmic inner law to develop expressions for the Fanning friction factor f for fully turbulent HFD flow (a) between parallel plates, and (b) in circular tubes.

Solution

For flow between parallel plates, f is given by Eq. (10-177),

$$\sqrt{\frac{2}{f}} = \frac{2}{w^+} \int_0^{w^+/2} u^+ \, dy^+ \tag{10-177}$$

Utilizing the logarithmic inner law for u^+, we have

$$\sqrt{\frac{2}{f}} = \frac{2}{w^+} \int_0^{w^+/2} \left(C + \frac{1}{\kappa} \ln y^+ \right) dy^+$$

$$= \frac{2}{w^+} \left[Cy^+ + \frac{1}{\kappa} (y^+ \ln y^+ - y^+) \right]\Bigg|_0^{w^+/2}$$

$$= C - \frac{1}{\kappa} + \frac{1}{\kappa} \ln\left(\frac{\text{Re}}{4} \sqrt{\frac{f}{2}} \right) \tag{10-182}$$

With $C = 5.5$ and $\kappa = 0.4$, we have

$$\sqrt{\frac{2}{f}} = 3 + 2.5 \ln\left(\frac{\text{Re}}{4} \sqrt{\frac{f}{2}} \right)$$

$$= 2.5 \ln\left(\text{Re} \sqrt{\frac{f}{2}} \right) - 0.466 \tag{10-183}$$

For a circular-tube geometry, U_b is defined by

$$U_b = \frac{2}{r_0^2}\int_0^{r_0} r\bar{u}\,dr = \frac{2}{r_0^2}\int_0^{r_0}(r_0-y)\bar{u}\,dy$$

Introducing the friction velocity, this equation takes the form

$$\sqrt{\frac{2}{f}} = \frac{2}{r_0^+}\int_0^{r_0^+}\left(1-\frac{y^+}{r_0^+}\right)u^+\,dy^+$$

With u^+ specified by the logarithmic law, we have

$$\sqrt{\frac{2}{f}} = \frac{2}{r_0^+}\int_0^{r_0^+}\left(C+\frac{1}{\kappa}\ln y^+ - C\frac{y^+}{r_0^+} - \frac{y^+}{\kappa r_0^+}\ln y^+\right)dy^+$$

$$= \frac{2}{r_0^+}\left[Cr_0^+ + \frac{1}{\kappa}(r_0^+\ln r_0^+ - r_0^+) - \frac{C}{2}r_0^+ - \frac{1}{\kappa r_0^+}\left(\frac{r_0^{+2}}{2}\ln r_0^+ - \frac{r_0^{+2}}{4}\right)\right]$$

$$= \frac{1}{\kappa}\ln r_0^+ + C - \frac{3}{2\kappa}$$

$$= 2.5\ln\left(\frac{\mathrm{Re}}{2}\sqrt{\frac{f}{2}}\right) + 1.75 = 2.5\ln\left(\mathrm{Re}\sqrt{\frac{f}{2}}\right) + 0.0171$$

Energy Transfer—TFD Region   We now seek to develop predictions for the mean temperature profile $\bar{T}$ and the Nusselt number Nu for fully turbulent TFD flow between parallel plates with uniform wall flux heating. Our analysis will be seen to involve steps that are similar to those that were utilized in the solution of the laminar flow problem and that are analogous to those that were employed in the analysis of the turbulent momentum transfer.

The classical differential formulation for the mean energy transfer for TFD turbulent channel flow is given by

$$\rho c_p \bar{u}\frac{\partial \bar{T}}{\partial x} = \frac{\partial}{\partial y}\left[(k+k_t)\frac{\partial \bar{T}}{\partial y}\right] \tag{10-188}$$

$$\frac{\partial \bar{T}}{\partial y} = 0 \qquad \text{at } y = \frac{w}{2} \tag{10-189}$$

and

$$-k\frac{\partial \bar{T}}{\partial y} = \overline{q_0''} \qquad \text{at } y = 0 \tag{10-190}$$

for a uniform wall heat flux. For this uniform wall-heat-flux boundary condition, $\partial \overline{T}/\partial x$ is evaluated by utilizing the defining equation for TFD flow, Eq. (10-122), and a lumped energy balance, with the result (see development, Chap. 6)

$$\frac{\partial \overline{T}}{\partial x} = \frac{dT_b}{dx} = \frac{dT_s}{dx} = \frac{\overline{q_0''}\, P}{\dot{m} c_p} = \frac{4\,\overline{q_0''}}{\rho c_p U_b D_H} \tag{10-191}$$

where $D_H = 2w$ for flow between parallel plates. Hence, the energy equation takes the simpler form

$$\rho c_p \bar{u} \frac{dT_b}{dx} = \frac{d}{dy}\left[(k+k_t)\frac{d\overline{T}}{dy}\right] \tag{10-192}$$

or

$$\frac{4\,\overline{q_0''}}{D_H}\frac{\bar{u}}{U_b} = \frac{d}{dy}\left[(k+k_t)\frac{d\overline{T}}{dy}\right] \tag{10-193}$$

As indicated in Sec. 10-2, the turbulent thermal conductivity k_t can be expressed in terms of ν_t by

$$\alpha_t = \frac{k_t}{\rho c_p} = \frac{\nu_t}{\mathrm{Pr}_t} \tag{10-194}$$

where the turbulent Prandtl number Pr_t is approximately equal to unity for moderate Prandtl number fluids.

Dimensionless Mean Temperature Distribution. Recognizing that

$$-\left[(k+k_t)\frac{d\overline{T}}{dy}\right]\bigg|_0 = -k\frac{d\overline{T}}{dy}\bigg|_0 = \overline{q_0''} \tag{10-195}$$

Eq. (10-193) is separated and integrated to obtain

$$(k+k_t)\frac{d\overline{T}}{dy} + \overline{q_0''} = \frac{4\,\overline{q_0''}}{U_b D_H}\int_0^y \bar{u}\,dy \tag{10-196}$$

or

$$-(k+k_t)\frac{d\overline{T}}{dy} = \overline{q_0''}\left(1 - \frac{4}{U_b D_H}\int_0^y \bar{u}\,dy\right) \tag{10-197}$$

This equation is put into the dimensionless form

$$\left(\frac{\alpha}{\nu} + \frac{\alpha_t}{\nu}\right)\frac{dT^+}{dy^+} = 1 - \frac{4}{\text{Re}}\int_0^{y^+} u^+ \, dy^+ \qquad (10\text{-}198)$$

where T^+ is defined in terms of the unknown wall temperature T_s by

$$T^+ = \frac{(T_s - \bar{T})\rho c_p U^*}{\overline{q_0''}} \qquad (10\text{-}199)$$

and the integral $\int_0^{y^+} u^+ dy^+$ can be evaluated by specifying u^+ in accordance with our fluid-flow analysis.

Separating the variables and integrating once again, we write

$$T^+ = \int_0^{y^+} \left[\frac{1 - (4/\text{Re})\int_0^{y^+} u^+ \, dy^+}{\dfrac{1}{\text{Pr}} + \dfrac{\nu_t/\nu}{\text{Pr}_t}}\right] dy^+ \qquad (10\text{-}200)$$

With ν_t, Pr_t, and u^+ specified, this equation can be numerically integrated. However, to simplify our problem, we break this integral into two parts. In the inner region $(y/w/2 \gtrsim 0.2)$, where $\int_0^{y^+} u^+ \gtrsim dy^+$ is small, we write

$$T^+ = \int_0^{y^+} \frac{1}{\dfrac{1}{\text{Pr}} + \dfrac{\nu_t/\nu}{\text{Pr}_t}} dy^+ \qquad \begin{array}{c}\text{inner region}\\[4pt] \dfrac{y}{w/2} \gtrsim 0.2\end{array} \qquad (10\text{-}201)$$

In the outer region where the velocity profile is quite flat, we approximate $\bar{u}$ by U_b, with the result that

$$\frac{4}{\text{Re}}\int_0^{y^+} u^+ \, dy^+ = \frac{4}{2wU_b/\nu}\frac{U_b}{U^*}\frac{yU^*}{\nu} = \frac{2}{w}y \qquad (10\text{-}202)$$

and

$$T^+ = \int \frac{1 - y/w/2}{\dfrac{1}{\text{Pr}} + \dfrac{\nu_t/\nu}{\text{Pr}_t}} dy^+ + B \qquad \begin{array}{c}\text{outer region}\\[4pt] \dfrac{y}{w/2} \gg 0.2\end{array} \qquad (10\text{-}203)$$

where B is a constant of integration. With Pr and Pr_t set equal to unity, Eqs. (10-201) and (10-203) are seen to be equivalent to the relationships for u^+ given by Eqs. (10-169) and (10-167), respectively.

Concerning the wall region, Eq. (10-201) has been solved numerically by investigators for various inputs for ν_t/ν and Pr_t. To illustrate, the numerical integration of this equation with Pr_t set equal to unity and with ν_t/ν given by the van Driest equation, Eq. (10-146), is discussed in Example 10-10. Predictions obtained in Example 10-10 for T^+ are shown in Fig. 10-22 to be in excellent agreement with experimental data for fluids with moderate values of the Prandtl number. In the viscous sublayer where ν_t is much less than ν, Eq. (10-201) gives rise to the limiting result

$$T^+ = Pr\, y^+ \qquad \text{for} \quad y^+ \lesssim 5 \tag{10-204}$$

This equation is shown to be consistent with the experimental data and with the van Driest calculations in Fig. 10-22.

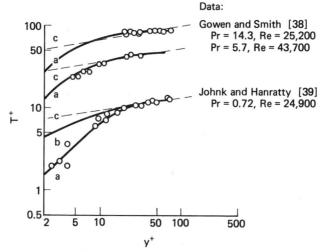

FIGURE 10-22 Dimensionless temperature distributions for TFD fully turbulent flow in circular tubes. Curves: (a) van Driest equation (from Example 10-7); (b) logarithmic law; (c) one-seventh power law.

In the intermediate part of the inner region where ν_t/ν is approximately equal to κy^+ and where α_t is much greater than α for fluids other than liquid metals, Eq. (10-201) reduces to the form

$$T^+ = \int \frac{dy^+}{\kappa y^+} + B = \frac{1}{\kappa}\ln y^+ + B \tag{10-205}$$

where the constant B is a function of the Prandtl number. As shown in Example 10-10, the numerical integration of Eq. (10-201) with ν_t/ν properly specified within the wall region provides us with the relationship

TABLE 10-2   Calculations for B and B_1

Pr	B Example 10-10	Eq. (10-206)	B_1
0.5	0.217	0.689	5.71
0.72	2.74	2.93	7.15
1.0	5.50	5.50	8.73
2.0	13.4	13.2	13.2
3.0	20.0	19.7	17.0
5.0	31.2	30.9	23.4
5.7	34.7	34.5	25.4
7.5	43.3	43.1	30.3
10.0	54.1	53.6	36.5
14.3	71.2	70.8	46.3

between B and Pr. B is given in Table 10-2 for representative values of Pr. White [13] has proposed a useful correlation for B of the form

$$B = 12.8 \ Pr^{0.68} - 7.3 \qquad (10\text{-}206)$$

This correlation is shown to be quite reasonable for values of the Prandtl number of the order of unity and greater.

Equation (10-205) is compared with experimental data in Fig. 10-22 for turbulent TFD flow in circular tubes. The agreement between this logarithmic inner law and the data are seen to be quite good throughout the intermediate region. As in the case for the u^+ correlations, we find that our inner law for T^+ can also be used to approximate the data in the outer region.

Equation (10-205) can also be approximated by a simple correlation of the form

$$T^+ = B_1 y^{+1/7} \qquad (10\text{-}207)$$

B_1 is obtained by equating Eqs. (10-205) and (10-207) at some value of y^+. For example, if we equate these expressions at $y^+ = 50$, then $B_1 = (9.78 + B)/1.75$. Calculations for B_1 based on this relationship are given in Table 10-2 for several values of Pr. This one-seventh power law is shown in Fig. 10-22 to correlate the data for T^+ quite well in the intermediate region. Similar to the restriction imposed on the one-seventh power law for u^+, Eq. (10-207) is best suited for values of the Reynolds number between 10^4 and 10^5.

Concerning the theoretical treatment of the outer region, a simple analytical solution to Eq. (10-203) is given in Prob. 10-47. However, as indicated above, the dimensionless temperature distribution within the outer region can be approximated by our simple inner laws.

Next, we develop predictions for the Nusselt number. With Nu and f known, calculations can be obtained for the mean temperature distribution $\overline{T}$ by utilizing the defining equation for T^+; that is,

$$T^+ = \frac{T_s - \overline{T}}{q_0''} \rho c_p U^*$$

or

$$\frac{\overline{T} - T_s}{T_b - T_s} = \frac{\overline{q_0''}}{T_s - T_b} \frac{T^+}{\rho c_p U^*} = \frac{\text{Nu } T^+}{\text{Re Pr} \sqrt{f/2}} \tag{10-208}$$

EXAMPLE 10-10

Develop predictions for T^+ and u^+ within the inner region for TFD fully turbulent flow. Utilize the van Driest equation for ν_t and the Reynolds assumption $\alpha_t = \nu_t$ (i.e., $\text{Pr}_t = 1$).

Solution

The solution for T^+ within the inner region is given by Eq. (10-201),

$$T^+ = \int_0^{y^+} \frac{1}{\dfrac{1}{\text{Pr}} + \dfrac{\nu_t/\nu}{\text{Pr}_t}} \, dy^+ \tag{10-201}$$

We have seen in Example 10-7 that van Driest's equation for ν_t can be written as

$$\frac{\nu_t}{\nu} = \frac{-1 + \sqrt{1 + 4(\kappa y^+)^2 [1 - \exp(-y^+/a)]^2}}{2} \tag{a}$$

Setting Pr_t equal to unity and utilizing the van Driest equation for ν_t, we have

$$T^+ = \int_0^{y^+} \frac{1}{\dfrac{1}{\text{Pr}} + \dfrac{-1 + \{1 + 4(\kappa y^+)^2 [1 - \exp(-y^+/a)]^2\}^{1/2}}{2}} \, dy^+ \tag{b}$$

A FORTRAN program designed to integrate Eq. (b) numerically is presented in Fig. E-A-5.

Predictions for T^+ obtained by running this program on a digital computer are presented in Fig. 10-22 for various values of Prandtl number with $\kappa = 0.4$ and $a = 27.4$. The predictions for each value of Pr approach the limiting curve

$$T^+ = B + \frac{1}{\kappa} \ln y^+ \tag{c}$$

To determine the value of B, we set T^+ in Eq. (b) equal to Eq. (c) for values of $y^+ \gtrsim 100$. This simple step is also included in the FORTRAN program. Calculations for B obtained by running this program are given in Table 10-2.

By comparing Eq. (e) in Example 10-7 with Eq. (b) above, we see that the velocity profile u^+ is identical to the temperature distribution T^+ for $Pr = 1$ and $Pr_t = 1$.

Bulk Stream Temperature and Nusselt Number. In the classical approach, predictions are obtained for the Nusselt number by utilizing the defining equation for the bulk stream temperature T_b. (It should be recalled that this same approach was utilized in our analysis of the laminar flow problem.) For turbulent channel flow, T_b is defined by

$$T_b = \frac{1}{wU_b} \int_0^w \bar{u}\bar{T} \, dy = \frac{2}{wU_b} \int_0^{w/2} \bar{u}\bar{T} \, dy \tag{10-209}$$

or, in terms of T^+,

$$T_b^+ = \frac{T_s - T_b}{q_0'' / (\rho c_p U^*)} = \frac{2U^*}{w^+ U_b} \int_0^{w^+/2} u^+ T^+ \, dy^+ \tag{10-210}$$

Solving for $\overline{q_0''}/(T_s - T_b)$, we obtain

$$\frac{\overline{q_0''}}{T_s - T_b} = \frac{\rho c_p U_b}{\dfrac{2}{w^+} \displaystyle\int_0^{w^+/2} u^+ T^+ \, dy^+} \tag{10-211}$$

The Nusselt number is therefore given by

$$Nu = \left(\frac{\overline{q_0''}}{T_s - T_b} \right) \frac{2w}{k} = \frac{Re \, Pr}{\dfrac{2}{w^+} \displaystyle\int_0^{w^+/2} u^+ T^+ \, dy^+} \tag{10-212}$$

Based on the solutions for u^+ and T^+ which have just been developed, this equation can be integrated by analytical or numerical methods. Perhaps the simplest approach to this problem is to approximate u^+ and T^+ by the one-seventh power laws. A solution based on this approach is given in Prob. 10-49. Another simple approach is to approximate $\bar{u}$ by U_b, and to approximate T^+ by the logarithmic law, Eq. (10-205). Utilizing these approximations, we obtain an expression for Nu as follows:

$$
\begin{aligned}
\text{Nu} &= \frac{\text{Re Pr}}{\dfrac{U_b}{U^*}\dfrac{2}{w^+}\displaystyle\int_0^{w^+/2}\left(\frac{1}{\kappa}\ln y^+ + B\right)dy^+} \\[2ex]
&= \frac{\sqrt{f/2}\ \text{Re Pr}}{\dfrac{2}{w^+}\left[\dfrac{1}{\kappa}(y^+\ln y^+ - y^+) + By^+\right]_0^{w^+/2}} \\[2ex]
&= \frac{\sqrt{f/2}\ \ \text{Re Pr}}{\dfrac{1}{\kappa}\left[\ln\left(\dfrac{\text{Re}}{4}\sqrt{\dfrac{f}{2}}\right) - 1\right] + B}
\end{aligned}
\tag{10-213}
$$

where B is given in Table 10-2. In order to put this equation into a more manageable form, we utilize the relationship between f and Re given by Eq. (10-182); that is,

$$
\sqrt{\frac{2}{f}} = C - \frac{1}{\kappa} + \frac{1}{\kappa}\ln\left(\frac{\text{Re}}{4}\sqrt{\frac{f}{2}}\right)
\tag{10-182}
$$

By coupling Eqs. (10-213) and (10-182) to eliminate the $\ln(\text{Re }\sqrt{f/2}/4)$ term, we obtain

$$
\text{Nu} = \frac{(f/2)\text{Re Pr}}{1 + \sqrt{f/2}\,(B - C)}
\tag{10-214}
$$

This equation is compared with experimental data for air in Fig. 10-23 by substituting for B and f. For moderate values of the Prandtl number, this equation can be reasonably well approximated by a simple correlation of the form

$$
\text{Nu} = \frac{f}{2}\text{Re Pr}^{0.45}
\tag{10-215}
$$

By substituting White's correlation for B, Eq. (10-206), into Eq. (10-214), we obtain

$$
\text{Nu} = \frac{(f/2)\text{Re Pr}}{1 + 12.8\sqrt{f/2}\,(\text{Pr}^{0.68} - 1)}
\tag{10-216}
$$

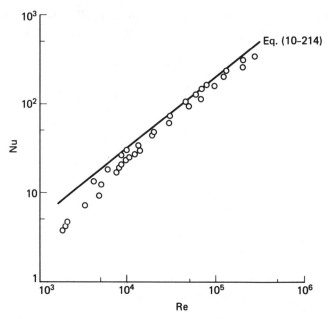

FIGURE 10-23 Comparison of predictions for Nusselt number with representative experimental data. (Data from Sams and Desmon [40], Deissler and Eian [41], and Barnes and Jackson [42].)

This same equation was obtained by White [13] for tube flow. Furthermore, Eq. (10-216) is almost identical to the popular Petukhov-Kirillov [43] correlation given by Eq. (6-33).

Fully turbulent boundary layer flow over flat plates

We now consider turbulent boundary layer flow over a flat plate with uniform wall-flux heating maintained over the surface as shown in Fig. 10-24. To simplify matters, the integral approach will be featured in our solution of this problem.

Fluid Flow The classical differential formulation for the fluid flow associated with turbulent boundary layer flow is given by Eqs. (10-128) and (10-141),

$$\frac{\partial \bar{u}}{\partial x} + \frac{\partial \bar{v}}{\partial y} = 0 \tag{10-217}$$

$$\rho\left(\bar{u}\frac{\partial \bar{u}}{\partial x} + \bar{v}\frac{\partial \bar{u}}{\partial y}\right) = \frac{\partial}{\partial y}\left[(\mu + \mu_t)\frac{\partial \bar{u}}{\partial y}\right] - \frac{d\bar{P}}{dx} \tag{10-218}$$

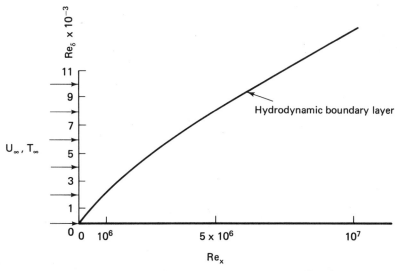

FIGURE 10-24 Turbulent boundary layer flow over a flat plate with constant free stream velocity and uniform wall-flux heating.

where $d\bar{P}/dx = 0$ for flow over a flat plate with constant free stream velocity, and

$$\bar{u} = U_\infty \quad \text{at } x = 0 \qquad \bar{u} = 0 \quad \text{at } y = 0 \qquad (10\text{-}219, 220)$$

$$\bar{u} = U_\infty \quad \text{at } y = \delta \qquad \bar{v} = 0 \quad \text{at } y = 0 \qquad (10\text{-}221, 222)$$

Our formulation is completed by the specification of μ_t.[1] As indicated in Sec. 10-3-3, the eddy diffusivity can be approximated by

$$\frac{\nu_t}{\nu} = \kappa y^+ \qquad \begin{array}{l} \text{intermediate region} \\ y^+ \gtrsim 50, \quad y/\delta \gtrsim 0.2 \end{array} \qquad (10\text{-}144)$$

and

$$\frac{\nu_t}{\nu} = \alpha_1 \delta^+ \qquad \begin{array}{l} \text{outer region} \\ y/\delta \gtrsim 0.2 \end{array} \qquad (10\text{-}145)$$

These approximations are utilized in many of the modern differential/numerical approaches to analyzing turbulent boundary layer flow processes. In the wall region where ν_t/ν is nonlinear, the van Driest type of equation is frequently employed.

---

[1]Alternatively, the Reynolds stress for turbulent boundary layer flow is often expressed in terms of the Prandtl mixing length ℓ. This popular modeling concept is introduced in Example 10-6.

These equations provide the basis for the development of analytical and numerical solutions for the mean velocity distribution and mean wall shear stress for turbulent boundary layer flow. Our emphasis will be placed on the development of simplified analytical solutions for u^+ and f_x for fully turbulent boundary layer flow over a flat plate.

Dimensionless Mean Velocity Distribution. Following the path taken in the classical analysis of HFD fully turbulent channel flow, we integrate Eq. (10-218) to obtain

$$(\mu + \mu_t)\frac{d\bar{u}}{dy} = \bar{\tau}_0 + \int_0^y \rho\left(\bar{u}\frac{\partial \bar{u}}{\partial x} + \bar{v}\frac{\partial \bar{u}}{\partial y}\right)dy \qquad (10\text{-}223)$$

or

$$\bar{\tau} = \bar{\tau}_0 + \int_0^y \rho\left(\bar{u}\frac{\partial \bar{u}}{\partial x} + \bar{v}\frac{\partial \bar{u}}{\partial y}\right)dy \qquad (10\text{-}224)$$

Whereas $\bar{\tau}$ was found to be a simple linear function of y for turbulent fully developed internal flow, Eq. (10-224) indicates that the relationship between $\bar{\tau}$ and y is nonlinear and quite complex for boundary layer flow. However, in the inner region ($y/\delta \gtrsim 0.2$) where $\bar{\tau} \approx \bar{\tau}_0$, Eq. (10-223) takes the simple form

$$(\mu + \mu_t)\frac{d\bar{u}}{dy} = \bar{\tau}_0 \qquad (10\text{-}225)$$

or

$$\left(1 + \frac{\nu_t}{\nu}\right)\frac{du^+}{dy^+} = 1 \qquad (10\text{-}226)$$

This equation is identical to Eq. (10-168) for the inner region of a turbulent HFD internal flow. Therefore, the solutions for u^+ that were developed in Sec. 10-3-4 for the inner region of internal flows also apply to this external flow problem. For example, in the intermediate region where $\nu_t/\nu = \kappa y^+$, u^+ is given by the logarithmic law

$$u^+ = \frac{1}{\kappa}\ln y^+ + C \qquad \begin{matrix}\text{intermediate region} \\ y^+ \gtrsim 50, \quad y/\delta \gtrsim 0.2\end{matrix} \qquad (10\text{-}227)$$

where the constants can be approximated by the same values of κ and C that are used in fully turbulent HFD internal flows (i.e., $\kappa = 0.4$ and $C = 5.5$). (These empirical constants are frequently given by $\kappa = 0.41$ and $C = 5.0$ for turbulent boundary layer flow.) As an alternative to this equation, the one-seventh power law,

$$u^+ = 8.7y^{+1/7} \qquad (10\text{-}228)$$

can be used for Reynolds numbers Re_x ranging between 10^5 and 10^7. In the wall region where these two equations do not apply, relationships such as those developed on the basis of the van Driest or Spalding inputs for ν_t can be used.

Equations (10-227) and (10-228) and the van Driest equation have been found to be in quite adequate agreement with experimental data, even in the outer region.

The development of a solution for u^+ in the outer region requires that the convective terms $\bar{u}\,\partial\bar{u}/\partial x + \bar{v}\,\partial\bar{u}/\partial y$ be retained in Eq. (10-223). This, in turn, requires that the continuity equation, Eq. (10-217), be satisfied. Even though μ_t is essentially constant in the outer region, these are both complicating issues that are difficult to overcome via analytical techniques. However, these equations can be solved numerically. For example, numerical solutions have been developed for u^+ within the outer region for fully turbulent boundary layer flow over a flat plate with uniform free stream velocity that are in good agreement with the celebrated empirical "law of the wake" by Coles [44], which is given by

$$u^+ = C + \frac{1}{\kappa}\ln y^+ + \frac{1}{\kappa}\sin^2\frac{\pi y}{2\delta} \qquad (10\text{-}229)$$

Although this expression correlates the data very well, the simpler inner laws are often used as a first approximation for fully turbulent flow over a flat plate. More general laws are available for u^+ for flows with nonzero pressure gradients.

As in the case of HFD internal flow, these expressions for u^+ must be coupled with predictions for the friction factor in order to obtain the mean velocity profile $\bar{u}$.

Integral Solution for Friction Factor. Predictions can be developed for the friction factor f_x for turbulent boundary layer flow by coupling correlations for u^+ with the integral momentum equation. Although we are presently considering the case for which $d\bar{P}/dx = 0$, this powerful solution approach can also be extended to more general problems involving nonzero pressure gradients.

It should be mentioned that integral analyses have also been developed for predicting f_x which involve the use of empirical inputs for $\bar{u}$ rather than u^+. This approach is presented in many introductory heat-transfer texts. Although this approach can be credited with early successes in the analysis of fully turbulent boundary layer flow over a flat plate, it does not lend itself to the solution of the more general cases for which $d\bar{P}/dx \neq 0$, nor is it as accurate as the more modern approach that utilizes inputs for u^+.

To develop predictions for f_x based on the modern integral approach, we first write the integral momentum equation in terms of the dimensionless velocity u^+ as follows:

$$\frac{d}{dx}\int_0^\delta \bar{u}(U_\infty - \bar{u})dy = \frac{\tau_0}{\rho} \tag{10-156}$$

$$\nu\frac{d}{dx}\int_0^{\delta^+} u^+(U_\infty - u^+ U^*)dy^+ = U_\infty^2 \frac{f_x}{2}$$

$$\frac{d}{d\mathrm{Re}_x}\left[\int_0^{\delta^+} u^+\left(1 - u^+\sqrt{\frac{f_x}{2}}\right)dy^+\right] = \frac{f_x}{2} \tag{10-230}$$

By specifying u^+, Eq. (10-230) can be solved for f_x or δ^+.

To demonstrate, u^+ is approximated by the one-seventh power law, Eq. (10-228). Substituting this correlation for u^+ into the integral momentum equation, we obtain

$$\frac{d}{d\mathrm{Re}_x}\left[\int_0^{\delta^+} 8.7y^{+1/7}\left(1 - 8.7y^{+1/7}\sqrt{\frac{f_x}{2}}\right)dy^+\right] = \frac{f_x}{2}$$

$$\frac{d}{d\mathrm{Re}_x}\left[\frac{7}{8}(8.7)\delta^{+8/7} - \frac{7}{9}(8.7)^2\delta^{+9/7}\sqrt{\frac{f_x}{2}}\right] = \frac{f_x}{2} \tag{10-231}$$

This gives us one equation in the two unknowns f_x and δ. A second independent relationship between these two unknowns is obtained by setting $\bar{u} = U_\infty$ at $y = \delta$ in our power law correlation; that is,

$$\sqrt{\frac{2}{f_x}} = 8.7\delta^{+1/7} \tag{10-232}$$

Utilizing this expression, Eq. (10-231) reduces to

$$\frac{7}{72(8.7)^7}\frac{d}{d\mathrm{Re}_x}\left(\frac{2}{f_x}\right)^4 = \frac{f_x}{2} \tag{10-233}$$

Separating the variables and rearranging, we have

$$\frac{2}{f_x}d\left(\frac{2}{f_x}\right)^4 = \frac{72}{7}(8.7)^7\,d\mathrm{Re}_x \tag{10-234}$$

or

$$4\left(\frac{2}{f_x}\right)^4 d\left(\frac{2}{f_x}\right) = \frac{72}{7}(8.7)^7\,d\mathrm{Re}_x \tag{10-235}$$

As a first approximation, this equation is assumed to apply over the whole plate length with $2/f_x = 0$ at $\mathrm{Re}_x = 0$. Utilizing this boundary condition, the solution for f_x is

$$\frac{4}{5}\left(\frac{2}{f_x}\right)^5 = \frac{72}{7}(8.7)^7 \, \mathrm{Re}_x$$

or

$$f_x = 0.0581 \, \mathrm{Re}_x^{-0.2} \tag{10-236}$$

This expression lies only $1\frac{1}{2}\%$ below a commonly used empirical correlation which takes the form

$$f_x = 0.059 \, \mathrm{Re}_x^{-0.2} \tag{10-237}$$

Equation (10-237) was shown to be in good agreement with experimental data in Fig. 6-15 for values of the Reynolds number from transition up to about 10^7.

Integral solutions for f_x have been developed which correlate the data over a broader Reynolds number range by utilizing more accurate inputs for u^+ such as the Spalding inner law, the logarithmic inner law, and the Coles law of the wake. For example, the use of the logarithmic inner law gives (Prob. 10-57)

$$\mathrm{Re}_x = \frac{e^{-\kappa C}}{\kappa^3}\left[e^{\kappa \sqrt{2/f_x}}\left(\kappa^2 \frac{2}{f_x} - 4\kappa\sqrt{\frac{2}{f_x}} + 6\right) - 6 - 2\kappa\sqrt{\frac{2}{f_x}}\,\right] \tag{10-238}$$

This equation correlates the data quite well over the entire turbulent Reynolds number range. The use of Spalding's more involved inner law leads to predictions that are within 1% of this equation. White [13] has developed an explicit correlation for f_x that is in very good agreement with Eq. (10-238), which takes the form

$$f_x = \frac{0.455}{\ln^2(0.06 \, \mathrm{Re}_x)} \tag{10-239}$$

This equation was shown to be in good agreement with experimental data in Fig. 6-15.

Energy Transfer   The classical differential formulation for energy transfer associated with fully turbulent boundary layer flow over a flat plate with uniform heating maintained over the plate is given by Eqs. (10-154) and the boundary conditions; that is,

$$\rho c_p \left(\bar{u} \frac{\partial \bar{T}}{\partial x} + \bar{v} \frac{\partial \bar{T}}{\partial y} \right) = \frac{\partial}{\partial y} \left[(k + k_t) \frac{\partial \bar{T}}{\partial y} \right] \qquad (10\text{-}240)$$

$$\bar{T} = T_\infty \quad \text{at} \quad x = 0 \qquad \bar{T} = T_\infty \quad \text{as} \quad y \to \infty \qquad (10\text{-}241, 242)$$

and

$$-k \frac{\partial \bar{T}}{\partial y} = \overline{q_0''} \qquad \text{at} \quad y = 0 \qquad (10\text{-}243)$$

The formulation is closed by the specification of k_t or α_t. For fluids with moderate values of Prandtl number, we set Pr_t equal to unity, as a first approximation; that is,

$$\alpha_t = \frac{k_t}{\rho c_p} = \nu_t \qquad (10\text{-}244)$$

Dimensionless Mean Temperature Distribution. Integrating Eq. (10-240), we write

$$(\alpha + \alpha_t) \frac{\partial \bar{T}}{\partial y} = \int_0^y \left(\bar{u} \frac{\partial \bar{T}}{\partial x} + \bar{v} \frac{\partial \bar{T}}{\partial y} \right) dy - \frac{\overline{q_0''}}{\rho c_p} \qquad (10\text{-}245)$$

In the inner region in which the integral on the right-hand side of this equation is small, this equation reduces to

$$-(\alpha + \alpha_t) \frac{d\bar{T}}{dy} = \frac{\overline{q_0''}}{\rho c_p} \qquad \begin{array}{c} \text{inner region} \\ y/\delta \gtrsim 0.2 \end{array} \qquad (10\text{-}246)$$

By expressing $\bar{T}$ in terms of the dimensionless temperature profile T^+, we obtain

$$\left(\frac{\alpha}{\nu} + \frac{\alpha_t}{\nu} \right) \frac{dT^+}{dy^+} \simeq 1 \qquad \begin{array}{c} \text{inner region} \\ y/\delta \gtrsim 0.2 \end{array} \qquad (10\text{-}247)$$

or

$$T^+ = \int_0^{y^+} \frac{1}{\dfrac{1}{\text{Pr}} + \dfrac{\nu_t/\nu}{\text{Pr}_t}} \, dy^+ \qquad (10\text{-}248)$$

which is identical to our result within the inner region for TFD fully turbulent flow between parallel plates, Eq. (10-201). As we have already seen, the solution to this equation in the intermediate region takes the form

$$T^+ = \frac{1}{\kappa}\ln y^+ + B \qquad \begin{array}{c} \text{intermediate region} \\ y^+ \gtrsim 50, \quad y/\delta \gtrsim 0.2 \end{array} \qquad (10\text{-}249)$$

where B is dependent on Pr. This logarithmic equation can be approximated by the one-seventh power law

$$T^+ = B_1 y^{+1/7} \qquad (10\text{-}250)$$

where $B_1 = (9.78 + B)/1.75$. In the wall region, T^+ can be obtained by the numerical integration of Eq. (10-248) with ν_t specified by equations, such as the van Driest expression, that apply in the zone $y^+ \gtrsim 50$. It is the solution of Eq. (10-248) within this region that provides us with predictions for B. B and B_1 are given in Table 10-2 for representative values of the Prandtl number with $\kappa = 0.4$ and $C = 5.5$.

Equations (10-249) and (10-250) and the van Driest equation are compared with an empirical correlation of experimental data for turbulent boundary layer flow of air in Fig. 10-25. These expressions are seen to be in good agreement with the data within both the inner and outer regions.

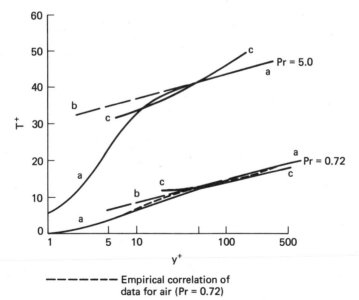

— — — — — Empirical correlation of
data for air (Pr = 0.72)

FIGURE 10-25 Dimensionless temperature distribution for fully turbulent boundary layer flow over flat plates. Curves: (a) van Driest equation; (b) logarithmic law; (c) one-seventh power law.

The inner laws for T^+ are generally utilized to approximate the dimensionless temperature distribution in the outer region. To obtain more accurate predictions within this region, the full energy equation, Eq. (10-240), is generally solved numerically.

Integral Solution for Nusselt Number.   Until recently, no true integral type solution to this problem had been developed. However, fresh thinking on the subject by White and his co-workers [47], [48] has been very fruitful. The general approach by White and his associates has recently been simplified and extended to nonuniform wall heating conditions [49], [50].

To develop this integral solution for the local mean Nusselt number for fully turbulent boundary layer flow, we start with the integral energy equation for turbulent flow, Eq. (10-157),

$$\frac{d}{dx}\int_0^\Delta \bar{u}\left(\bar{T}-T_\infty\right)dy = \frac{\overline{q_c''}}{\rho c_p} \tag{10-157}$$

where $\overline{q_c''}=\overline{q_0''}$. Following the approach which was utilized in our integral solution for f_x, this equation is written in terms of u^+ and T^+ as follows:

$$\nu\frac{d}{dx}\int_0^{\Delta^+} u^+\left[\frac{(T-T_0)-(T_\infty-T_0)}{\overline{q_0''}/(\rho c_p U^*)}\right]\frac{\overline{q_0''}}{\rho c_p U^*}\,dy^+ = \frac{\overline{q_0''}}{\rho c_p}$$

or

$$\frac{d}{d\mathrm{Re}_x}\left[\frac{\overline{q_0''}}{\sqrt{f_x/2}}\int_0^{\Delta^+} u^+(T_\infty^+ - T^+)dy^+\right] = \overline{q_0''} \tag{10-251}$$

Because we are dealing with a uniform wall-heat flux boundary condition, this equation reduces to the simpler form

$$\frac{d}{d\mathrm{Re}_x}\left[\sqrt{\frac{2}{f_x}}\int_0^{\Delta^+} u^+(T_\infty^+ - T^+)dy^+\right] = 1 \tag{10-252}$$

With uniform wall flux heating initiated at $x=0$, our integral formulation for the time-average energy transfer is closed by writing

$$\Delta^+ = 0 \quad \text{at} \quad \mathrm{Re}_x = 0 \tag{10-253}$$

The solution to Eq. (10-252) is simply

$$\sqrt{\frac{2}{f_x}}\int_0^{\Delta^+} u^+(T_\infty^+ - T^+)dy^+ = \mathrm{Re}_x \tag{10-254}$$

With u^+ and T^+ specified, this equation provides us with predictions for Δ^+ in terms of Re_x. Then setting $\overline{T} = T_\infty$ at $y = \Delta$, our defining equation for T^+ gives

$$T_\infty^+ = (T_0 - T_\infty) \frac{\rho c_p U^*}{q_0''} \qquad (10\text{-}255)$$

or

$$\text{Nu}_x = \frac{q_0''}{T_0 - T_\infty} \frac{x}{k} = \frac{\sqrt{f_x/2}\ \text{Re}_x \text{Pr}}{T_\infty^+} \qquad (10\text{-}256)$$

For example, with u^+ and T^+ approximated by the power laws, Eqs. (10-228) and (10-250), Eq. (10-254) gives rise to

$$\int_0^{\Delta^+} 8.7 y^{+1/7} (B_1 \Delta^{+1/7} - B_1 y^{+1/7}) dy^+ = \sqrt{\frac{f_x}{2}}\ \text{Re}_x$$

$$8.7 B_1 \left(\frac{7}{8} \Delta^{+1/7} \Delta^{+8/7} - \frac{7}{9} \Delta^{+9/7} \right) = \sqrt{\frac{f_x}{2}}\ \text{Re}_x$$

$$\Delta^+ = \left(\frac{\sqrt{f_x/2}\ \text{Re}_x}{0.846\, B_1} \right)^{7/9} \qquad (10\text{-}257)$$

Substituting this result together with the power law for T^+ into Eq. (10-256), we obtain an expression for Nu_x of the form

$$\text{Nu}_x = \frac{\sqrt{f_x/2}\ \text{Re}_x\ \text{Pr}}{B_1 \left(\dfrac{\sqrt{f_x/2}\ \text{Re}_x}{0.846\, B_1} \right)^{1/9}} = 0.982 \left(\frac{\sqrt{f_x/2}\ \text{Re}_x}{B_1} \right)^{8/9} \text{Pr} \qquad (10\text{-}258)$$

By utilizing Eq. (10-237), Nu_x can be written as

$$\text{Nu}_x = \frac{0.209}{B_1^{8/9}} \text{Re}_x^{0.8}\ \text{Pr} \qquad (10\text{-}259)$$

or

$$\text{Nu}_x = \frac{7.08}{B_1^{8/9}} \frac{f_x}{2} \text{Re}_x\ \text{Pr} \qquad (10\text{-}260)$$

With B_1 given in Table 10-2, Eq. (10-259) is compared in Fig. 10-26 with experimental data for turbulent boundary layer flow of air over a flat

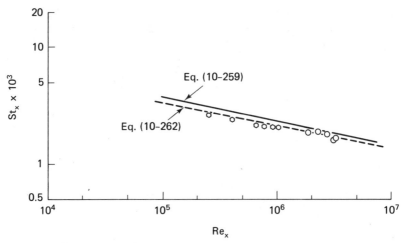

FIGURE 10-26 Comparison of predictions for Stanton number with experimental data. (From Reynolds et al. [51].)

plate with uniform wall temperature heating. The agreement between our theoretical relationship for uniform wall flux heating and the uniform wall temperature data is seen to be fairly good. This result reinforces our earlier conclusion that the Nusselt number for fully turbulent conditions is essentially independent of the form of the thermal boundary condition.

By substituting the values of B_1 given in Table 10-2 for moderate Prandtl numbers into Eq. (10-260), we obtain predictions for Nu_x that are very well correlated by

$$Nu_x = \frac{f_x}{2} Re_x\, Pr^{0.45} \tag{10-261}$$

The correlation by White [13] given by Eq. (6-98),

$$Nu_x = \frac{(f_x/2)Re_x\, Pr}{1 + 12.8\sqrt{f_x/2}\ (Pr^{0.68} - 1)} \tag{10-262}$$

is also shown in Fig. 10-26. This equation is in slightly better agreement with the data than Eq. (10-259) [or Eq. (10-260)].

10-4 CLOSURE

For purposes of analysis, the two primary categories in convection-heat-transfer systems include *laminar flow* and *turbulent flow* conditions. Whereas the mathematical formulations for laminar flow systems are generally quite straightforward, the solution of the resulting equations can be quite complex. For the simplest problems, as typified by the fully developed tube flow problem introduced in Sec. 10-2-2, analytical solutions can be readily obtained. Analytical solutions to a number of simple laminar convection problems are presented in textbooks by White [13] and others. Somewhat more complex laminar flow problems involving hydrodynamic and/or thermal development, such as the flat-plate boundary layer flow problem considered in Sec. 10-2-3, can be fairly easily solved by means of the integral approach. As illustrated in Prob. 10-29, analytical solutions can even be developed for some of these types of problems by introducing similarity coordinates. These problems of intermediate complexity are also sometimes solved numerically. The most complex problems involving compressibility effects, variable properties, buoyancy, axial conduction, strong adverse and favorable pressure gradients, mass transfer through a wall, unsteady conditions, and other complications are almost always handled numerically.

The analysis of turbulent convection heat transfer involves the conceptual problem of how to mathematically model the complex turbulent transport process itself. The classical mean-field approach introduced in this chapter is the best-known method of attack. Once the eddy diffusivity ν_t and eddy thermal diffusivity α_t (or turbulent Prandtl number Pr_t) are known, the solution of turbulent convection-heat-transfer problems follows the same basic pattern as for laminar flow. Again, analytical techniques can be utilized for the simplest problems. But numerical and integral techniques are generally called upon to handle the more complex problems, with numerical approaches becoming more and more dominant. One advanced program, known as STAN-5, developed by a Stanford group [53], is capable of handling both internal and external turbulent flow convection-heat-transfer processes. Comprehensive general-purpose computer programs have also been developed by a group at the Imperial College of London [54], [55] and others which are widely used in industries and universities.

Although the basic classical mean-field approach introduced in this chapter can be utilized to solve a fairly broad range of turbulent convection processes, it must be pointed out that this general approach has produced rather mixed results when applied to some of the more complex processes [56]. This point is particularly well documented by the 1968 Stanford Conference [57], in which many numerical and integral solution techniques were tested against a fairly broad range of turbulent boundary layer flows. Whereas mean-field models are very useful in engineering analysis, they fail to handle some important effects, such as strong accelerations and separation. Furthermore, recent theoretical and experimental studies involving combined forced and natural convection of gases [58], supercritical fluids [59], and liquid metals [60] in vertical tubes reveal classical predictions for heat transfer which are in direct conflict with the experimental findings. It is for complex problems such as these that our lack of reliable information pertaining to the distribution in ν_t and α_t (or Pr_t) is most critical. In fact, it is because of this lack of information on ν_t and α_t that turbulent convection processes involving variable property and buoyancy effects, hydrodynamic unsteady conditions, and other complications have not to date been successfully handled by the classical approach.

In an attempt to overcome some of the inadequacies of the classical mean-field approach, rather extensive study has been directed toward the development of higher-order analyses in which the Reynolds stress $\bar{\tau}_t$ is expressed in terms of turbulence quantities such as turbulent kinetic energy and turbulent dissipation. In this classical turbulent-field approach, one or more partial differential equations are written for turbulence quantities. Higher-order analyses of this type have achieved considerable success, with one-equation and two-equation models being used in some engineering industries.

Because of the limitations in the classical mean-field approach to turbulence, the recent development of an alternative method of modeling turbulent convection transport processes has taken on increased significance. As pointed out by White [13], a growing number of analyses have been developed in recent years on the basis of the principle of surface renewal which treat wall turbulence as a true convective phenomenon, without the need for eddy diffusivity ν_t or turbulent Prandtl number Pr_t assumptions. (Although this approach does not require the use of ν_t or Pr_t, predictions have been developed for these classical turbulence parameters on the basis of the surface renewal approach [26], [27].) The general surface renewal concept was first put forth by Danckwerts [62] in 1951. This pioneering work was followed up by Einstein and Li [63] and Hanratty [64] in 1956 and numerous others in more recent years [26]–[28], [59], [65]–[72]. In this approach, the actual unsteady transport processes associated with the turbulent burst process pictured back in Fig. 10-17 is mathematically modeled. Rather than to go into the details of the surface renewal modeling approach, perhaps it will suffice to say that this simple method has been utilized with some measure of success in the analysis of complex turbulent transport problems that have heretofore defied solution. For example, the surface renewal modeling concept has been utilized to analyze turbulent convective heat transfer in supercritical fluids both with and without buoyancy influence [59], [77]. But it should be understood that this model of wall turbulence is considered to be a supplement to, not a replacement for, the classical approach. In this regard, coupled surface renewal/classical models have evolved which embody the strengths of both approaches [73], [74].

In closing, we reiterate the point that the classical mean-field approach to the analysis of turbulent convective heat transfer which has been featured in this chapter has proven to adequately handle many turbulent convection problems. Computer programs based on this approach are generally utilized in industry. However, new approaches are required for the analysis of certain of the more complex turbulent convection-heat-transfer processes. The classical turbulent-field method and the surface renewal method are two such new approaches which will possibly become standard tools in the future.

PROBLEMS

10-1. Utilize Eq. (10-18) to develop the following equation for HFD laminar flow in a tube:

$$\frac{\tau}{\tau_0} = \frac{r}{r_0}$$

10-2. Develop a solution for the velocity distribution and Fanning friction factor for HFD laminar flow between horizontal parallel plates. Compare your solution with the results for f presented in Chap. 6.

10-3. Develop a solution for the velocity distribution and Fanning friction factor for HFD laminar flow between the inclined parallel plates shown in Fig. P10-3. Compare your solution with the results of Prob. 10-2.

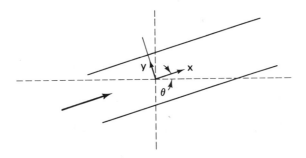

FIGURE P10-3

10-4. Develop a solution for the temperature distribution and Nusselt number for laminar fully developed flow between parallel plates with uniform wall flux heating maintained on both plates. Compare your solution with the results for Nu presented in Chap. 6.

10-5. Repeat Prob. 10-4 for the case in which one plate is insulated and a uniform wall flux q_0'' is maintained at the other surface.

10-6. Repeat Prob. 10-5 for the case in which $q''_w = 2q''_0$.

10-7. Experimental velocity profile data for HFD laminar flow in a concentric annulus are shown in Fig. P10-7. Develop a solution for the velocity distribution and compare your predictions with these data. Also develop an expression for the Fanning friction factor. Compare your solution for f with the results presented in Fig. 6-2.

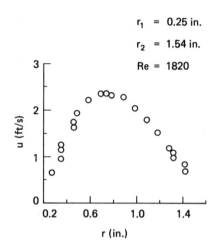

$$r_1 = 0.25 \text{ in.}$$
$$r_2 = 1.54 \text{ in.}$$
$$Re = 1820$$

FIGURE P10-7 Data for axial velocity distribution for HFD laminar flow in concentric annulus. (From Rothfus [52].)

10-8. Develop a solution for the temperature distribution and Nusselt number for laminar fully developed flow in a concentric annulus with uniform heating maintained along the inner surface and with the outer surface insulated. Compare your solution to the results for Nu presented in Chap. 6.

10-9. Integrate Eq. (10-50) to obtain Eq. (10-51).

10-10. Develop a solution for the velocity distribution in laminar fully developed plane Couette flow, for which case $u=0$ at $y=0$, $u=U_w$ at $y=w$, and $dP/dx=0$. Also obtain an expression for the Fanning friction factor.

10-11. Develop a solution for the velocity distribution in laminar fully developed Couette flow with $dP/dx \neq 0$. Plot your results for both positive and negative values of dP/dx [i.e., $-w^2 \, dP/dx/(4\mu U_w) = -2, -0.5, 0, 0.5, 2$].

10-12. Integrate Eq. (10-90) to obtain Eq. (10-91). Then obtain Eq. (10-93).

10-13. Demonstrate that the temperature distribution for laminar TFD flow in a circular tube with uniform wall flux heating can be expressed as

$$\frac{T - T_s}{T_b - T_s} = \frac{24}{11}\left[\frac{3}{4} - \left(\frac{r}{r_0}\right)^2 + \frac{1}{4}\left(\frac{r}{r_0}\right)^4 \right]$$

or

$$T - T_1 = \frac{4q_0'' r_0}{k} \left[\frac{2x/D}{\text{Re Pr}} - \frac{7}{96} + \frac{1}{4} \left(\frac{r}{r_0} \right)^2 - \frac{1}{16} \left(\frac{r}{r_0} \right)^4 \right]$$

10-14. Develop the differential formulation for convection heat transfer associated with steady laminar boundary layer flow over a flat plate. Neglect property variation, buoyancy, potential energy, kinetic energy, viscous dissipation, axial conduction, and thermal radiation.

10-15. Repeat Prob. 10-14, but include the effects of property variation.

10-16. Repeat Prob. 10-14, but include viscous dissipation.

10-17. Repeat Prob. 10-14, but include axial conduction.

10-18. Repeat Prob. 10-14, but include potential and kinetic energy.

10-19. Repeat Prob. 10-14, but include buoyancy effects.

10-20. Develop the integral formulation for momentum transfer associated with steady laminar flow in the entrance region of a channel. Also outline the steps required to solve these equations for f.

10-21. Demonstrate the validity of Eq. (10-86).

10-22. Develop an approximate integral solution for the hydrodynamic boundary layer thickness δ, velocity distribution u, wall shear stress τ_0, and friction factor f_x for laminar boundary layer flow over a flat plate. (a) Utilize a second-order polynomial approximation for u. (b) Utilize a fourth-order polynomial approximation for u.

10-23. Repeat Prob. 1-22 by utilizing a fourth-order polynomial approximation for u. Explain why the accuracy of this higher order polynomial approximation for u is lower than for the third-order polynomial.

10-24. Develop an approximate integral solution for the thermal boundary layer thickness Δ, temperature distribution, and Nusselt number for laminar boundary layer flow over a flat plate with uniform wall temperature heating maintained for $x \gg x_0$. Show that for $x_0 = 0$, the Nusselt number is given by Eq. (10-115). Utilize third-order polynomial approximations for u and T.

10-25. Referring to Example 10-5, develop the solution of the integral momentum equation for $\delta = \Delta$ and $\beta = \beta_\infty$. Compare this solution for Δ to the solution given by Eq. (k). Also develop an expression for Nu_x and $\overline{\text{Nu}}$ based on this result.

10-26. Develop an approximate integral solution for the Nusselt number Nu_x for laminar boundary layer flow with the wall heat flux given by $q_c'' = q_0''[1 + \sin (\pi x/L)]$ for $x \gg 0$.

10-27. Develop an approximate integral solution for the Nusselt number for thermal developing HFD laminar flow between parallel plates with uniform wall-flux heating at $y = 0$ and with the surface at $y = w$ insulated. Utilize a fourth-order polynomial approximation for T.

10-28. Develop an implicit numerical finite-difference formulation for laminar convection heat transfer for boundary layer flow over a flat plate.

10-29. The similarity coordinate η ($\equiv y\sqrt{U_\infty/(x\nu)}$) was established by means of the integral approach for laminar boundary layer flow over a flat plate. Show that the differential continuity and momentum equations, Eqs. (10-54) and (10-55), can be put into the form

$$\frac{d}{d\eta}\left(\frac{u''}{u'}\right) + \frac{1}{2}\frac{u}{U_\infty} = 0$$

where $u' = du/d\eta$ and $u'' = d^2u/d\eta^2$. Show that the three boundary conditions in x and y reduce to the following two conditions in η: $u(0) = 0$ and $u(\infty) = U_\infty$.

10-30. The ordinary differential equation given in Prob. 10-29 has been solved by Blasius [15], Kays [5], and others. Based on the classical solution by Blasius, $u'(0)$ is equal to $0.332 U_\infty$. Based on this result, show that the Fanning friction factor takes the form

$$f_x = 0.664\, \mathrm{Re}_x^{-1/2}$$

Also develop an expression for $\bar{f}$.

10-31. Utilize the van Driest equation for ν_t to obtain an expression for the mixing length ℓ in the wall region for fully turbulent flow.

10-32. Following the pattern established in the development of the classical momentum equation for turbulent flow, Eq. (10-134), develop the classical energy equation given by Eq. (10-147).

10-33. Develop the integral momentum equation for turbulent flow, Eq. (10-156), by integrating Eq. (10-134) across the boundary layer δ for $dP/dx = 0$.

10-34. Develop the integral energy equation for turbulent flow, Eq. (10-157), by integrating Eq. (10-147) across the thermal boundary layer Δ.

10-35. Returning to the analysis of HFD laminar tube flow, show that Eq. (10-31) or (10-32) can be written in terms of τ_0 as

$$u = \frac{\tau_0}{\rho}\frac{r_0}{2\nu}\left[1 - \left(\frac{r}{r_0}\right)^2\right]$$

or

$$u^+ = \frac{r_0^+}{2}\left[1 - \left(\frac{r^+}{r_0^+}\right)^2\right]$$

Then obtain an expression for the Fanning friction factor by integrating to obtain U_b (see Example 10-9); that is,

$$U_b = \frac{2}{r_0^2}\int_0^{r_0} ru\, dr$$

or

$$\sqrt{\frac{2}{f}} = \frac{2}{r_0^+} \int_0^{r_0^+} r^+ u^+ \, dr^+$$

10-36. Show that Eq. (10-169) can be put into the form

$$y^+ = \int_0^{u^+} \left(1 + \frac{\nu_t}{\nu}\right) du^+$$

10-37. Spalding's correlation for ν_t takes the form

$$\nu_t = \nu\kappa e^{-\kappa c}\left[e^{\kappa u^+} - 1 - \kappa u^+ - \frac{(\kappa u^+)^2}{2}\right]$$

Combine this equation with the expression given in Prob. 10-36 to obtain an expression for u^+. Compare this result with the van Driest equation and the experimental data of Fig. 10-21.

10-38. Show graphically that the van Driest result for u^+ in the inner region can be reasonably well approximated by the simple equation

$$u^+ = 16\left[1 - \exp\left(\frac{-y^+}{16}\right)\right]$$

for $y^+ < 46.8$, and

$$u^+ = 2.5 \ln y^+ + 5.5$$

for $y^+ > 46.8$. (These equations are based on the simple surface renewal approach to turbulence. See [28].)

10-39. Develop an expression for u^+ in the outer region for fully developed fully turbulent flow between parallel plates by setting ν_t/ν equal to 0.08 $(w^+/2)$ in this region (i.e., $y/w/2 > 0.2$) and by including the $y/w/2$ term in Eq. (10-167).

10-40. Repeat Prob. 10-39 for a circular-tube geometry.

10-41. The famous logarithmic equation, Eq. (10-172), was obtained by setting ν_t/ν equal to κy^+ and by assuming that $\nu_t >> \nu$ in Eq. (10-169). Develop an equation for u^+ in the intermediate region by integrating Eq. (10-169) without this simplification; that is,

$$u^+ = \int \frac{dy^+}{1 + \kappa y^+} + C$$

Compare your result to Eq. (10-172).

10-42. Combine Eqs. (10-178) and (10-179) to obtain Eq. (10-180).

10-43. Show that Eq. (10-201) can be written in the form

$$T^+ = \int_0^{u^+} \frac{1 + \nu_t/\nu}{\dfrac{1}{Pr} + \dfrac{\nu_t/\nu}{Pr_t}} \, du^+$$

Notice that this equation is more convenient when ν_t/ν is specified by the Spalding equation given in Prob. 10-37.

10-44. Verify Eq. (10-191) by developing the appropriate lumped analysis.

10-45. Verify Eqs. (10-202) and (10-203).

10-46. Equation (10-205) is obtained from Eq. (10-201) by setting $\nu_t/\nu = \kappa y^+$, $\mathrm{Pr}_t = 1$, and by assuming that $\nu_t/\nu > > 1/\mathrm{Pr}$. Show that this equation can be more accurately solved for these values of ν_t/ν and Pr_t to obtain

$$T^+ = \frac{1}{\kappa}\ln\left(\frac{1}{\mathrm{Pr}} + \kappa y^+\right) + B$$

10-47. Develop an expression for T^+ in the outer region for fully developed fully turbulent flow between parallel plates by setting ν_t/ν equal to 0.08 $(w^+/2)$ and Pr_t equal to unity in this region and by including the term $y/w/2$ in Eq. (10-203).

10-48. Repeat Prob. 10-47 for a circular-tube geometry.

10-49. Develop an expression for the Nusselt number for fully developed and fully turbulent flow between parallel plates by integrating Eq. (10-212) with u^+ and T^+ specified by the one-seventh power laws.

10-50. Verify Eq. (10-213).

10-51. Demonstrate that Eq. (10-214) can be approximated by Eq. (10-215) in the moderate Prandtl number range 0.5 to 5.0.

10-52. Determine the maximum percent difference between Eq. (10-216) and the Petukhov/Kirillov equation, Eq. (6-33), for $\mathrm{Re} = 10^5$ and $0.5 < \mathrm{Pr} < 50$.

10-53. Verify Eqs. (10-233) and (10-236).

10-54. Verify Eqs. (10-251) and (10-257).

10-55. The analogy approach is perhaps the simplest way of developing predictions for the Nusselt number for turbulent boundary layer flow over a flat plate. In this approach, the turbulent Prandtl number Pr_t is set equal to unity and the thermal boundary layer thickness Δ is approximated by the hydrodynamic boundary layer thickness δ. By utilizing the analogy concept with u^+ and T^+ approximated by the logarithmic laws, show that an expression can be developed for Nu_x for uniform wall-temperature heating and fully turbulent flow along the entire length of the plate of the form

$$\mathrm{Nu}_x = \frac{(f_x/2\,\mathrm{Re}_x)\mathrm{Pr}}{1 + \sqrt{f_x/2}\,(B-C)}$$

10-56. Repeat Example 10-10 by utilizing the Spalding expression for ν_t given in Prob. 10-37.

10-57. Develop Eq. (10-238).

10-58. Develop a law for u^+ in the wake region for turbulent boundary layer flow over a flat plate by approximating the mean shear stress distribution by a simple third-order polynomial. (Recall that t can be taken as constant in the outer region.)

10-59. Develop a law for T^+ in the wake region for turbulent boundary layer flow over a flat plate by approximating the mean heat flux distribution by a simple third-order polynomial.

10-60. The displacement thickness δ^* is often utilized in laminar and turbulent boundary layer analyses. δ^* is defined by

$$\delta^* = \int_0^\infty \left(1 - \frac{\bar{u}}{U}\right) dy \simeq \int_0^\infty \left(1 - \frac{\bar{u}}{U}\right) dy$$

(a) Develop the relationship between δ^* and δ for laminar boundary layer flow over a flat plate by utilizing a third-order polynomial approximation for u. (b) Repeat part (a) for turbulent flow by utilizing the one-seventh power law for u^+.

REFERENCES

[1] LANGHAAR, H. L., "Steady Flow in the Transition Length of a Straight Tube," *J. Appl. Mech.*, **9**, 1942, A55–A58.

[2] SEIGEL, R., E. M. SPARROW, and T. M. HALLMAN, "Steady Laminar Heat Transfer in Circular Tube with Prescribed Wall Heat Flux" *Appl. Sci. Res.*, **A7**, 1958, 386–392.

[3] HEATON, H. S., W. C. REYNOLDS, and W. M. KAYS, *Int. J. Heat Mass Transfer*, **7**, 1964, 763.

[4] SELLARS, J. R., M. TRIBUS, and J. S. KLEIN, "Heat Transfer to Laminar Flows in a Round Tube or Flat Conduit, The Graetz Problem Extended," *Trans. ASME*, **78**, 1956, 441–448.

[5] KAYS, W. M., *Convective Heat and Mass Transfer*. New York: McGraw-Hill Book Company, 1966.

[6] KAYS, W. M., "Numerical Solution for Laminar Flow Heat Transfer in Circular Tubes," *Trans. ASME*, **77**, 1955, 1265–1274.

[7] GOLDBERG, P., "A Digital Computer Solution for Laminar Flow Heat Transfer in Circular Tubes, M. S. Thesis, Mechanical Engineering Department, Massachusetts Institute of Technology, 1958.

[8] HANSEN, M., "Velocity Distribution in the Boundary Layer of a Submerged Plate," *NACA TM 585*, 1930.

[9] GRAETZ, L., Über die Wärmeleitfähigkeit von Flüssigkeiten," *Ann. Phys. Chem.*, **25**, 1885, 337.

[10] SENECAL, V. E., "Characteristics of Transition Flow in Smooth Tubes," Ph.D. Thesis, Carnegie Institute of Technology, 1952.

[11]   STANTON, T. E., and J. R. PANNELL, "Similarity of Motion in Relation to the Surface Friction of Fluids," *Trans. Roy. Soc. (London),* **A214**, 1914, 199.

[12]   SENECAL, V. E., and R. R. ROTHFUS, *Chem. Eng. Progr.,* **49**, 1953, 533.

[13]   WHITE, F. M., *Viscous Fluid Flow.* New York: McGraw-Hill Book Company, 1974.

[14]   PANTANKAR, S. V., *Numerical Heat Transfer and Fluid Flow.* Washington, D.C.: Hemisphere Publishing Co., 1980.

[15]   BLASIUS, H., "Grenzschichten in Flüssigkeiten mit kleiner Reibung," *Z. Math. Phys.,* **56**, 1908, 1–37; English translation in *NACA TM 1256.*

[16]   SCHMIDT, E., and W. BECKMANN, "Das Temperatur- und Geschwindigkeitsfeld von einer licher Wandtemperatur," *Forsch. Gebiete Ingenieurw.,* **1**, 1930, 391.

[17]   SCHUH, H., "Einige Probleme bei freier Strömung zäher Flüssigkeiten," *Göttinger Monogr. Bd. B., Grenzschichten,* 1946.

[18]   RUNSTADLER, P. W., S. J. KLINE, and W. C. REYNOLDS, "An Experimental Investigation of the Flow Structure of the Turbulent Boundary Layer," *Rept. MD-8,* Stanford University, 1963.

[19]   NIKURADSE, J., "Widerstandsgesetz und Geschwindigkeit von turbulenten Wasserströmungen in glatten und rauhen Rohren," *Proc. Third Int. Cong. Appl. Mech.,* **1**, 1930, 239.

[20]   KNUDSEN, J. G., and D. L. KATZ, *Fluid Dynamics and Heat Transfer.* New York: McGraw-Hill Book Company, 1958.

[21]   REYNOLDS, O., *Scientific Papers of Osborne Reynolds,* Vol. II. London: Cambridge University Press, 1901.

[22]   PRANDTL, L., "Eine Beziehung zwischen Wärmeaustausch und Strömungswiderstand der Flüssigkeiten," *Phys. Zeit.,* **11**, 1910, 1072.

[23]   HUSSAIN, A. K. M. F., and W. C. REYNOLDS, "Measurements in Fully Developed Turbulent Channel Flow," *J. Fluids Eng.,* **97**, 1975, 569.

[24]   VAN DRIEST, E. R., "Turbulent Boundary Layer in Compressible Fluids," *J. Aero. Sci.,* **18**, 1951, 145–161.

[25]   CEBECI, T., and A. M. O. SMITH, *Analysis of Turbulent Boundary Layers.* New York: Academic Press, Inc., 1974.

[26]   THOMAS, L. C., "A Formulation for ϵ_m and ϵ_H Based on the Surface Renewal Principle," *AIChE J.,* **24**, 1978, 101.

[27]   RAJAGOPAL, R., and L. C. THOMAS, "The Formulation of Relationship for ϵ_m and ϵ_H Based on a Physically Realistic Model of Turbulence, *Fifth Int. Heat Transfer Conf., Tokyo,* 1974.

[28]   THOMAS, L. C., "The Turbulent Burst Phenomenon: Inner Laws for u^+ and T^+," NATO Advanced Study Institute on Turbulence, Istanbul, Turkey, 1978.

[29]   ROTTA, J., "Das in Wandnähe gültige Geschwindigkeitsgegetz turbulenter Strömungen," *Ing. Arch.,* **18**, 1950, 277–279.

[30]   REICHARDT, H., "Vollständige Darstellung der turbulenten Geschwindigkeitssverteilung in glatten Leitungen," *ZAMM,* **31**, 1951, 208–219.

[31] DEISSLER, R. G., "Analysis of Turbulent Heat Transfer, Mass Transfer and Friction in Smooth Tubes at High Prandtl and Schmidt Numbers," *NACA TN 3145*, 1954.

[32] SPALDING, D. B., "A Single Formula for the Law of the Wall," *J. Appl. Mech.*, **28**, 1961, 455–457.

[33] BRADSHAW, P., "Compressible Turbulent Shear Layers," in *Annual Review of Fluid Mechanics*, Vol. 9. Palo Alto, Calif.: Annual Reviews, Inc., 1977.

[34] SIMPSON, R. L., D. G. WHITTEN, and R. J. MOFFAT, "An Experimental Study of the Turbulent Prandtl Number of Air with Injection and Suction," *Int. J. Heat Mass Transfer*, **13**, 1970, 125–143.

[35] AZER, N. Z., and B. T. CHAO, "Turbulent Heat Transfer in Liquid Metals— Fully Developed Pipe Flow with Constant Wall Temperature," *Int. J. Heat Mass Transfer*, **3**, 1961, 77–83.

[36] JENKINS, R., "Variation of Eddy Conductivity with Prandtl Modulus and Its Use in Prediction of Turbulent Heat Transfer Coefficients," Heat Transfer and Fluid Mechanics Institute, Stanford University, 1951.

[37] LINDGREN, E. R., Oklahoma State University, Civil Engineering Department, *Rept. IAD621071*, 1965.

[38] GOWEN, R. A., and J. W. SMITH, "The Effects of the Prandtl Number on Temperature Profiles for Heat Transfer in Turbulent Pipe Flow," *Chem. Eng. Sci.*, **22**, 1967, 1701–1711.

[39] JOHNK, R. E., and T. J. HANRATTY, "Temperature Profiles for Turbulent Flow of Air in a Pipe—II," *Chem. Eng. Sci.*, **17**, 1962, 802.

[40] SAMS, E. W., and L. G. DESMON, "Heat Transfer from High Temperature Surfaces to Fluids," *NACA Memo E9D12*, 1949.

[41] DEISSLER, R. G., and C. S. EIAN, "Analytical and Experimental Investigation of Heat Transfer with Variable Fluid Properties," *NACA TN 2629*, 1952.

[42] BARNES, J. F., and J. D. JACKSON, "Heat Transfer to Air, Carbon Dioxide and Helium Flowing through Smooth Circular Tubes under Conditions of Large Surface/Gas Temperature Ratio," *J. Mech. Eng. Sci.*, **3**, 1961, 303.

[43] PETUKHOV, B. S., and V. V. KIRILLOV, "Heat Exchange for Turbulent Flow of Liquid in Tubes, *Teploenergetika*, **5**, 1958, 63–68.

[44] COLES, D., "The Law of the Wake in the Turbulent Boundary Layer," *J. Fluid Mech.*, **1**, 1956, 191–225.

[45] WHITE, F. M., "A New Integral Method for Analyzing the Turbulent Boundary Layer with Arbitrary Pressure Gradient," *J. Basic Eng.*, **91**, 1969, 371–378.

[46] BRAND, R. S., and L. N. PERSEN, "Implications of the Law of the Wall for Turbulent Boundary Layers," *Acta Polytechnica Scandinavica*, Physics Including Nucleonics Series No. 30, Trondheim, 1964, 61.

[47] WHITE, F. M., R. C. LESSMANN, and G. H. CHRISTOPH, "A Simplified Approach to the Analysis of Turbulent Boundary Layers in Two and Three Dimensions," *Tech. Rept. AFFDL-TR-136*, Air Force Flight Dynamics Laboratory, Wright-Patterson Air Force Base, November 1972.

[48] CHRISTOPH, G. H., R. C. LESSMANN, and F. M. WHITE, "Calculations of Turbulent Heat Transfer and Skin Friction," *AIAA J.*, **11**, 1973, 1046.

[49] THOMAS, L. C., "A Simple Integral Approach to Turbulent Thermal Boundary Layer Flow," *J. Heat Transfer*, **100**, 1978, 744.

[50] AL-SHARIF, M., "A Modern Integral Approach to Turbulent Thermal Boundary Layer Flow," M.S. Thesis, University of Petroleum and Minerals, Dhahran, Saudi Arabia, 1979.

[51] REYNOLDS, W. C., W. M. KAYS, and S. J. KLINE, "Heat Transfer in the Turbulent Incompressible Boundary Layer—III: Arbitrary Wall Temperature and Heat Flux, *NASA Memo 12-3-58W*, 1958.

[52] ROTHFUS, R. R., "Velocity Distribution and Fluid Friction in Concentric Annuli," Ph.D. Thesis, Carnegie Institute of Technology, 1948.

[53] CRAWFORD, M. E., and W. M. KAYS, "STAN-5—A Program for Numerical Computation of Two-Dimensional Internal/External Boundary Layer Flows," *Stanford Univ. Mech. Eng. Rept. HMT-23*, 1975.

[54] SPALDING, D. B., *GENMIX—A General Computer Program for Two-Dimensional Parabolic Phenomena.* Oxford: Pergamon Press, 1977.

[55] PATANKAR, S. V., and D. B. SPALDING, "Numerical predictions of three dimensional flows," Imperial College of London, Mech. Eng. Dept., Rept. No. HTS/72/4, 1972.

[56] WHITE, F. M., and G. H. CHRISTOPH, "A Simple Theory for the Two-Dimensional Compressible Turbulent Boundary Layer," *J. Basic Eng.*, **1972**, 636–642.

[57] KLINE, S. J., D. J. COCKRELL, M. V. MORKOVIN, and G. SOVRAN, *Proc. Comput. Turbulent Boundary Layer*, Vol. I, Department of Mechanical Engineering, Stanford University, 1968.

[58] BATES, J. A., R. A. SCHMALL, G. A. HASEN, and D. M. McELIGOT, "Effects of Buoyant Body Forces on Forced Convection in Heated Laminarizing Flows," *Fifth Int. Heat Transfer Conf., Tokyo, II, FC4.4*, 1974, 141.

[59] KAKARALA, C. R., "Development of Turbulent Wall Layer Models for Momentum and Heat Transfer in Tube Flow," Ph.D. Dissertation, University of Akron, Akron, Ohio, 1976.

[60] BUHR, H. D., A. D. CARR, and R. E. BALZHISER, "Temperature Profiles in Liquid Metals and the Effects of Superimposed Free Convection in Turbulent Flow," *Int. J. Heat Mass Transfer*, **11**, 1968, 641.

[61] REYNOLDS, W. C., "Computation of Turbulent Flows," *in Annual Review of Fluid Mechanics, Vol. 8*, Palo Alto, Calif.: Annual Reviews, Inc., 1978.

[62] DANCKWERTS, P. V., "Significance of Liquid-Film Coefficients in Gas Absorption," *I and E C*, **43**, 1951, 1460.

[63] EINSTEIN, H. A., and H. L. LI, "The Viscous Sublayer along a Smooth Boundary," *ASCE, J. Mech. Div.*, **82**, 1956, 293.

[64] HANRATTY, T. J., "Turbulent Exchange of Mass and Momentum with a Boundary," *AIChE J.*, **2**, 1956, 359.

[65] HARRIOTT, P., "A Random Eddy Modification of the Penetration Theory," *Chem. Eng. Sci.*, **17**, 1962, 149.

[66] Thomas, L. C., "Temperature Profiles for Liquid Metals and Moderate Prandtl Number Fluids," *J. Heat Transfer*, **92**, 1970, 565.

[67] Meek, R. L., and A. D. Baer, "The Periodic Viscous Sublayer in Turbulent Flow," *AIChE J.*, **16**, 1970, 841.

[68] Chung, B. T. F., L. T. Fan, and C. L. Hwang, "General Mathematical Models of Transport Processes with and without Chemical Reactions," *Can. J. Chem. Eng.*, **49**, 1971, 340.

[69] Rajagopal, R., "Adaptation of Surface Renewal and Penetration Type Models to Turbulent Momentum and Heat Transfer for Newtonian and Non-Newtonian Fluids," Ph.D. Dissertation, University of Akron, Akron, Ohio, 1973.

[70] Katsibas, P., and R. J. Gordon, "Momentum and Energy Transfer in Turbulent Pipe Flow: The Penetration Model Revisited," *AIChE J.*, **20**, 1974, 191.

[71] Kakarala, C. R., and L. C. Thomas, "A Theoretical Analysis of Turbulent Convective Heat Transfer for Supercritical Fluids," *Fifth Int. Heat Transfer Conf., Tokyo*, 1974.

[72] Sideman, S., and W. V. Pinczewski, "Turbulent Heat and Mass Transfer at Interfaces: Transport Models and Mechanisms," *Topics in Transport Phenomena*. Washington, D.C.: Hemisphere Publishing Corporation, 1975.

[73] Ibrahim, M. B., and L. C. Thomas, "A Surface Renewal/Classical Analysis of Turbulent Momentum Transfer," *Sixteenth Midwestern Conf. on Mechanics*, Manhattan, Kansas, 1979.

[74] Thomas, L. C., and M. B. Ibrahim, "Theoretical Predictions for the Turbulent Prandtl number based on a model of the turbulent burst phenomenon," *ASME Winter Annual Metting*, New York, 1979.

A MATHEMATICAL CONCEPTS

A-1 The Calculus

The importance of the calculus in the study of heat transfer cannot be overstated. Therefore, we want to briefly review fundamental concepts pertaining to the calculus.

The derivative

For cases in which the dependent variable ψ is a function of only one independent variable such as x, we are reminded of the following definition for the derivative:

$$\frac{d\psi}{dx} = \lim_{\Delta x \to 0} \frac{\psi(x + \Delta x) - \psi(x)}{\Delta x} \qquad \text{(A-1)}$$

Because dx itself is infinitesimal, this equation can be written in the form

$$\frac{d\psi}{dx} = \frac{\psi(x+dx)-\psi(x)}{dx} \tag{A-2}$$

Hence, $\psi(x+dx)$ can be written in terms of $\psi(x)$ and $d\psi/dx$ as

$$\psi(x+dx) = \psi(x) + \frac{d\psi}{dx}dx \tag{A-3}$$

This equation will be found to be very important in the analysis of one-dimensional conduction heat transfer in Chap. 2.

If ψ is a function of more than one independent variable such as x, y, z or t, the partial derivative of ψ with respect to x at the location x, y, z and at the instant t is defined as

$$\frac{\partial\psi}{\partial x} = \lim_{\Delta x \to 0} \frac{\psi(x+\Delta x)-\psi(x)}{\Delta x} \qquad \text{at} \quad x, y, z \text{ and } t \tag{A-4}$$

or

$$\frac{\partial\psi}{\partial x} = \frac{\psi(x+dx)-\psi(x)}{dx} \qquad \text{at} \quad x, y, z \text{ and } t \tag{A-5}$$

For this more general situation, $\psi(x+dx)$ is expressed in terms of $\psi(x)$ and the partial derivative; that is,

$$\psi(x+dx) = \psi(x) + \frac{\partial\psi}{\partial x}dx \tag{A-6}$$

Similar expressions can be written which involve partial derivatives with respect to the other independent variables. For example, the partial derivative of ψ with respect to y can be written as

$$\frac{\partial\psi}{\partial y} = \frac{\psi(y+dy)-\psi(y)}{dy} \qquad \text{at} \quad x, y, z \text{ and } t \tag{A-7}$$

Differential equations

Both ordinary and partial differential equations are encountered in the study of heat transfer. For situations in which the equations are *linear*, analytical solution techniques often can be used. Second-order linear ordinary differential equations take the form

$$\psi + a(x)\frac{d\psi}{dx} + b(x)\frac{d^2\psi}{dx^2} = c(x) \tag{A-8}$$

where the coefficients a, b, and c are functions of the independent variable x only. This equation in *nonlinear* if any one of the coefficients a, b, or c is a function of ψ. Similarly, partial differential equations which involve coefficients that are functions of only the independent variables are linear.

The number of boundary conditions in an independent variable required in the solution of any ordinary or partial differential equation is equal to the highest-order differential in that variable. Similar to the definition for a linear differential equation, linear boundary conditions take the general form

$$\psi + d(x)\frac{d\psi}{dx} + e(x)\frac{d^2\psi}{dx^2} = f(x) \quad \text{at} \quad x = x_1 \quad \text{(A-9)}$$

A differential equation or boundary condition that is satisfied by a function ψ is said to be *homogeneous* if it is also satisfied by $C\psi$, where C is an arbitrary constant. Thus, the linear ordinary differential equation given by Eq. (A-8) is homogeneous if $c(x)$ is zero and is *nonhomogeneous* if $c(x) \neq 0$. Similarly, the boundary condition given by Eq. (A-9) is homogeneous if $f(x)$ is zero.

As an example of a problem involving linear nonhomogeneous equations, consider Eq. A-8 with boundary conditions of the form

$$\psi + d_1(x)\frac{d\psi}{dx} = f_1(x) \quad \text{at} \quad x = x_1 \quad \text{(A-10)}$$

$$\psi + d_2(x)\frac{d\psi}{dx} = f_2(x) \quad \text{at} \quad x = x_2 \quad \text{(A-11)}$$

These types of boundary conditions are often encountered in heat-transfer problems. Notice that this problem involves three nonhomogeneous terms, $c(x)$, $f_1(x)$, and $f_2(x)$. The solution to this problem takes the form

$$\psi = \psi_0 + C_1\psi_1 + C_2\psi_2 \quad \text{(A-12)}$$

where ψ_0 is the particular solution of the nonhomogeneous differential equation and $C_1\psi_1 + C_2\psi_2$ is the solution of the homogeneous differential equation. The constants of integration C_1 and C_2 are evaluated on the basis of the boundary conditions.

A-2 Coordinate Systems

The foregoing review of basic mathematical concepts has been presented in the context of the *Cartesian coordinate system* shown in Fig. A-A-1. Our study will also require the use of the *cylindrical coordinate system* shown in Fig. A-A-2, and, to a lesser extent, the *spherical coordinate system* shown in Fig. A-A-3.

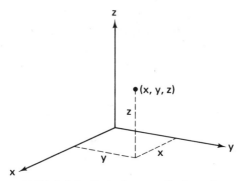

FIGURE A-A-1   Cartesian coordinate system.

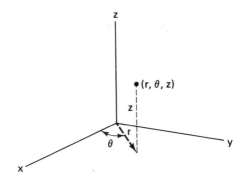

FIGURE A-A-2   Cylindrical coordinate system.

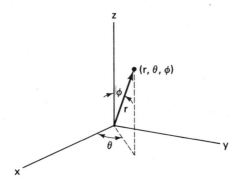

FIGURE A-A-3   Spherical coordinate system.

In general, the temperature distribution can be a function of all three spatial coordinates (e.g., x, y, and z) and a single time variable t. We will refer to such multidimensional heat-transfer systems as four-dimensional.[1] Of course, three-dimensional (e.g., t,x,y or x,y,z), two-dimensional (e.g.,

---

[1]Many investigators prefer to utilize the spatial dimensions alone in classifying the dimensionality of problems.

t, x or x, y), and one-dimensional (e.g., t or x) systems are also often encountered. Systems in which the temperature is a function of time will be referred to as *unsteady*. *Steady-state* conditions prevail when the temperature distribution within a system is independent of time.

B DIMENSIONS, UNITS, AND SIGNIFICANT FIGURES

Until the transition from English to metric (the international system, SI) units is complete, it will be necessary for us to be conversant with both systems. Consequently, both SI and English units will be used in our study, with emphasis given to the SI system.

The units for the four key fundamental dimensions—*time, length, mass*, and *temperature*—in these systems are summarized as follows:

System	Time, t	Length, L	Mass, m	Temperature, T
SI	second, s	meter, m	kilogram, kg	Kelvin, K
				Celsius, °C
English	second, s	foot, ft	pound mass, $1b_m$	Rankine, °R
				Fahrenheit, °F

Focusing attention on the units for temperature, in the SI system the *Kelvin* K is used for the absolute temperature scale[2] and the *degree Celsius* °C is used for the Celsius temperature scale. These two SI units for temperature are related by

$$T(K) = T(°C) + 273.2 \, K \qquad (B-1)$$

Note that an increment °C is equal to an increment of 1 K; i.e.,

$$°C = 1 \, K \qquad (B-2)$$

The Rankine °R and Fahrenheit °F units for temperature scale in the English system are related by

$$T(°R) = T(°F) + 459.7°R \qquad (B-3)$$

and

$$°F = 1°R \qquad (B-4)$$

---

[2]Note that the SI unit for the absolute temperature scale is Kelvin K and not degree Kelvin K.

The units for the SI and English systems are related by

$$T(°F) = \frac{9°F}{5°C} T(°C) + 32°F \tag{B-5}$$

$$°R = \frac{5}{9} K \tag{B-6}$$

or

$$°F = \frac{5}{9} °C \tag{B-7}$$

We are also reminded of the following units for the derived dimensions *force*, *energy* and *power*:

System	Force, F	Energy, E	Power, P
SI	newton, N	joule, J	watt, W
English	pound force, $1b_f$	British thermal unit, Btu	Btu/h

By definition, we have

$$1\ N = 1\ kg\ m/s \tag{B-8}$$

$$1\ J = 1\ N\ m \tag{B-9}$$

$$1\ W = 1\ J/s \tag{B-10}$$

Standard symbols and basic units for parameters encountered in heat transfer are presented in Table A-B-1 together with conversion factors.

In regard to heat transfer, the following units generally will be utilized in this text:

q—heat-transfer rate, W

Q—total heat transfer, J

The total heat transfer over an increment of time Q is related to q by the equation

$$q = \frac{\partial Q}{\partial t} \tag{B-11}$$

Hence, Q can be obtained from q by writing

$$Q = \int_0^t q\ dt \tag{B-12}$$

at any point in space.

Three significant figures will be maintained for most of our calculations.

TABLE A-B-1 Standard symbols and units for heat transfer parameters

Parameter	Symbol	SI units	English units	Conversion (English to SI)	Conversion (SI to English)
Time	t	s	s	—	—
Length	L	m	ft	1 ft = 0.3049 m	1 m = 3.281 ft
Mass	m	kg	lb_m	1 lb_m = 0.4359 kg	1 kg = 2.205 lb_m
Temperature	T	°C	°F	°F = (5/9)°C	°C = (9/5)°F
Temperature, absolute	T	K	°R	1°R = (5/9) K	K = (9/5) °R
Area	A	m²	ft²	1 ft² = 0.09290 m²	1 m² = 10.76 ft²
Volume	V	m³	ft³	1 ft³ = 0.002831 m³	1 m³ = 35.31 ft³
Velocity	v	m/s	ft/s	1 ft/s = 0.3048 m/s	1 m/s = 3.281 ft/s
Density	ρ	kg/m³	lb_m/ft³	1 lb_m/ft³ = 16.02 kg/m³	1 kg/m³ = 0.06243 lb_m/ft³
Force	F	N	lb_f	1 lb_f = 4.448 N	1 N = 0.2248 lb_f
Pressure	P	N/m²	lb_f/in.²	1 lb_f/in.² = 6895 N/m²	1 N/m² = 1.450×10⁻⁴ lb_f/in.²
Dynamic viscosity	μ	kg/(m s)	lb_m/(ft s)	1 lb_m/(ft s) = 1.488 kg/(m s)	1 kg/(m s) = 0.6720 lb_m/(ft s)
Kinematic viscosity	ν	m²/s	ft²/s	1 ft²/s = 0.0290 m²/s	1 m²/s = 10.76 ft²/s
Thermal diffusivity	α	m²/s	ft²/s	1 ft²/s = 0.09290 m²/s	1 m²/s = 10.76 ft²/s
Specific heat	c_p, c_o	J/(kg °C)	Btu/(lb_m °F)	1 Btu/(lb_m °F) = 4.187 kJ/(kg °C)	1 kJ/(kg °C) = 0.2388 Btu/(lb_m °F)
Thermal conductivity	k	W/(m °C)	Btu/(h ft °F)	1 Btu/(h ft °F) = 1.731 W/(m °C)	1 W/(m °C) = 0.5778 Btu/(h ft °F)
Coefficient of convection heat transfer	h	W/(m² °C)	Btu/(h ft² °F)	1 Btu/(h ft² °F) = 5.179 W/(m² °C)	1 W/(m² °C) = 0.1761 Btu/(h ft² °F)
Heat transfer (energy)	Q	J	Btu	1 Btu = 1.055 kJ	1 kJ = 0.9478 Btu
Heat transfer rate (power)	q	W	Btu/h	1 Btu/h = 0.2931 W	1 W = 3.412 Btu/h
Heat transfer per unit area (flux)	q''	W/m²	Btu/(h ft²)	1 Btu/(h ft²) = 3.154 W/m²	1 W/m² = 0.3171 Btu/(h ft²)
Energy generation per unit volume	$\dot{q}$	W/m³	Btu/(h ft³)	1 Btu/(h ft³) = 10.35 W/m³	1 W/m³ = 0.0966 Btu/(h ft³)

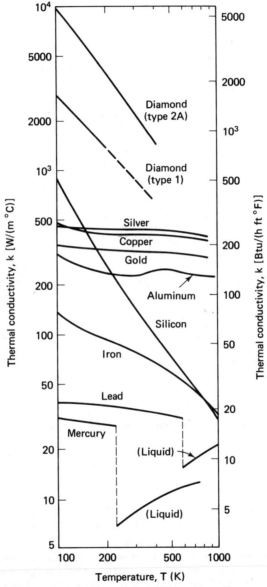

FIGURE A-C-1   Dependence of thermal conductivity on temperature: metals and other good heat conductors—moderate temperature zone. (From Touloukian et al., [2] and [3] of Chap. 1.)

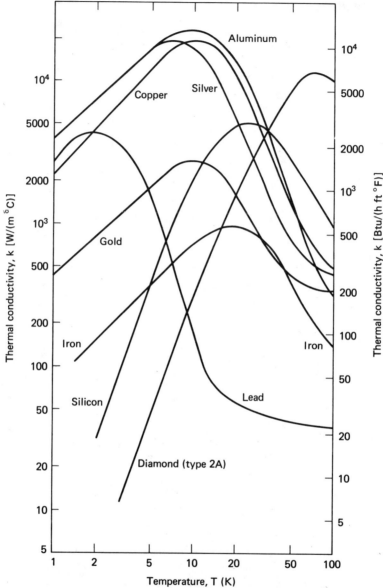

FIGURE A-C-2 Dependence of thermal conductivity on temperature: metals and other good heat conductors—cryogenic temperature zone. (From Touloukian et al., [2] and [3] of Chap. 1.)

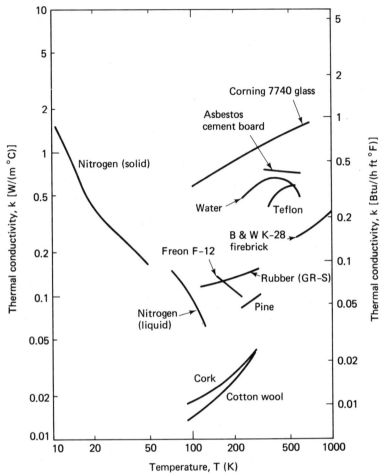

FIGURE A-C-3 Dependence of thermal conductivity on temperature: nonmetallic solids and saturated liquids. (From Touloukian et al., [3] and [4] of Chap. 1.)

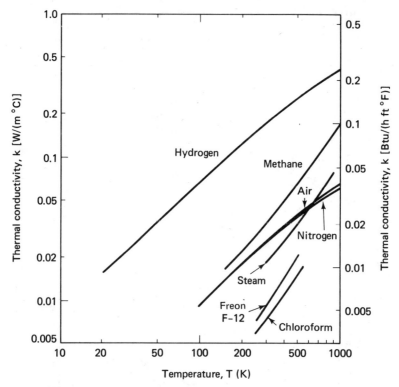

FIGURE A-C-4 Dependence of thermal conductivity on temperature: gases. (From Touloukian et al., [4] of Chap. 1.)

TABLE A-C-1 Property values of solid metals[a]

Metal	Properties at 20°C				Thermal conductivity, k [$W/(m\,K)$]									
	ρ (kg/m³)	c_p [kJ/(kg·°C)]	k [W/(m·°C)]	$\alpha \times 10^5$ (m²/s)	−100°C −148°F	0°C 32°F	100°C 212°F	200°C 392°F	300°C 572°F	400°C 752°F	600°C 1112°F	800°C 1472°F	1000°C 1832°F	1200°C 2192°F
Aluminum:														
Pure	2,707	0.896	204	8.418	215	202	206	215	228	249				
Al-Cu (Duralumin) 94–96% Al, 3–5% Cu, trace Mg	2,787	0.883	164	6.676	126	159	182	194						
Al-Mg (Hydronalium) 91–95% Al, 5–9% Mg	2,611	0.904	112	4.764	93	109	125	142						
Al-Si (Silumin) 87% Al, 13% Si	2,659	0.871	164	7.099	149	163	175	185						
Al-Si (Silumin, copper-bearing) 86.5% Al, 1% Cu	2,659	0.867	137	5.933	119	137	144	152	161					
Al-Si (Alusil) 78–80% Al, 20–22% Si	2,627	0.854	161	7.172	144	157	168	175	178					
Al-Mg-Si 97% Al, 1% Mg, 1% Si, 1% Mn	2,707	0.892	177	7.311		175	189	204						
Lead	11,373	0.130	35	2.343	36.9	35.1	33.4	31.5	29.8					
Iron:														
Pure	7,897	0.452	73	2.034	87	73	67	62	55	48	40	36	35	36
Wrought iron (C H 0.5%)	7,849	0.46	59	1.626		59	57	52	48	45	36	33	33	33
Cast iron (C≈4%)	7,272	0.42	52	1.703										
Steel (C max≈1.5%)														
Carbon steel C≈0.5%	7,833	0.465	54	1.474		55	52	48	45	42	35	31	29	31
1.0%	7,801	0.473	43	1.172		43	43	42	40	36	33	29	28	29
1.5%	7,753	0.486	36	0.970		36	36	36	35	33	31	28	28	29

TABLE A-C-1 (continued)

Metal	Properties at 20°C ρ (kg/m³)	c_p [kJ/(kg·°C)]	k [W/(m·°C)]	α×10⁵ (m²/s)	Thermal conductivity, k [W/(m K)] −100°C −148°F	0°C 32°F	100°C 212°F	200°C 392°F	300°C 572°F	400°C 752°F	600°C 1112°F	800°C 1472°F	1000°C 1832°F	1200°C 2192°F
Nickel steel														
Ni≈0%	7,897	0.452	73	2.026										
10%	7,945	0.46	26	0.720										
20%	7,993	0.46	19	0.526										
30%	8,073	0.46	12	0.325										
40%	8,169	0.46	10	0.279										
50%	8,266	0.46	14	0.361										
60%	8,378	0.46	19	0.493										
70%	8,506	0.46	26	0.666										
80%	8,618	0.46	35	0.872										
90%	8,762	0.46	47	1.156										
100%	8,906	0.448	90	2.276										
Invar Ni=36%	8,137	0.46	10.7	0.286										
Chrome Steel														
Cr=0%	7,897	0.452	73	2.026	87	73	67	62	55	48	40	36	35	36
1%	7,865	0.46	61	1.665		62	55	52	47	42	36	33	33	
2%	7,865	0.46	52	1.443		54	48	45	42	38	33	31	31	
5%	7,833	0.46	40	1.110		40	38	36	36	33	29	29	29	
10%	7,785	0.46	31	0.867		31	31	31	29	29	28	28	29	
20%	7,689	0.46	22	0.635		22	22	22	22	24	24	26	29	
30%	7,625	0.46	19	0.542										
Cr-Ni (chrome-nickel):														
15% Cr, 10% Ni	7,865	0.46	19	0.526										
18% Cr, 8% Ni (V2A)	7,817	0.46	16.3	0.444		16.3	17	17	19	19	22	26	31	
20% Cr, 15% Ni	7,833	0.46	15.1	0.415										
25% Cr, 20% Ni	7,865	0.46	12.8	0.361										
Ni-Cr (nickel-chrome):														
80% Ni, 15% Cr	8,522	0.46	17	0.444										
60% Ni, 15% Cr	8,266	0.46	12.8	0.333										
40% Ni, 15% Cr	8,073	0.46	11.6	0.305										
20% Ni, 15% Cr	7,865	0.46	14.0	0.390		14.0	15.1	15.1	16.3	17	19	22		

Thermal conductivity, k [$W/(m\,K)$]

Metal	Properties at 20°C ρ (kg/m³)	c_p [kJ/(kg °C)]	k [W/(m °C)]	$\alpha \times 10^5$ (m²/s)	−100°C −148°F	0°C 32°F	100°C 212°F	200°C 392°F	300°C 572°F	400°C 752°F	600°C 1112°F	800°C 1472°F	1000°C 1832°F	1200°C 2192°F
Cr-Ni-Al:														
6% Cr, 1.5% Al, 0.55% Si (Sicromal 8)	7,721	0.490	22	0.594										
24% Cr, 2.5% Al, 0.55% Si (Sicromal 12)	7,673	0.494	19	0.501										
Manganese steel														
Mn = 0%	7,897	0.494	73	1.863										
1%	7,865	0.46	50	1.388										
2%	7,865	0.46	38	1.050		38	36	36	36	35	33			
5%	7,849	0.46	22	0.637										
10%	7,801	0.46	17	0.483										
Tungsten steel														
W = 0%	7,897	0.452	73	2.026										
1%	7,913	0.448	66	1.858										
2%	7,961	0.444	62	1.763		62	59	54	48	45	36			
5%	8,073	0.435	54	1.525										
10%	8,314	0.419	48	1.391										
20%	8,826	0.389	43	1.249										
Silicon steel														
Si = 0%	7,897	0.452	73	2.026										
1%	7,769	0.46	42	1.164										
2%	7,673	0.46	31	0.888										
5%	7,417	0.46	19	0.555										
Copper:														
Pure	8,954	0.3831	386	11.234	407	386	379	374	369	363	353			
Aluminum bronze 95% Cu, 5% Al	8,666	0.410	83	2.330										
Bronze 75% Cu, 25% Sn	8,666	0.343	26	0.859										
Red Brass 85% Cu, 9% Sn, 6% Zn	8,711	0.385	61	1.804		59	71							
Brass 70% Cu, 30% Zn	8,522	0.385	111	3.412	88		128	144	147	147				
German silver 62% Cu, 15% Ni, 22% Zn	8,618	0.394	24.9	0.733	19.2		31	40	45	48				
Constantan 60% Cu, 40% Ni	8,922	0.410	22.7	0.612	21		22.2	26						

TABLE A-C-1 (continued)

Metal	Properties at 20°C				Thermal conductivity, k [$W/(m\,K)$]									
	ρ (kg/m³)	c_p [kJ/(kg°C)]	k [W/(m°C)]	$\alpha \times 10^5$ (m²/s)	$-100°C$ $-148°F$	$0°C$ $32°F$	$100°C$ $212°F$	$200°C$ $392°F$	$300°C$ $572°F$	$400°C$ $752°F$	$600°C$ $1112°F$	$800°C$ $1472°F$	$1000°C$ $1832°F$	$1200°C$ $2192°F$
Magnesium:														
Pure	1,746	1.013	171	9.708	178	171	168	163	157					
Mg-Al (electrolytic)														
6–8% Al, 1–2% Zn	1,810	1.00	66	3.605		52	62	74	83					
Mg-Mn 2% Mn	1,778	1.00	114	6.382	93	111	125	130						
Molybdenum	10,220	0.251	123	4.790	138	125	118	114	111	109	106	102	99	92
Nickel:														
Pure (99.9%)	8,906	0.4459	90	2.266	104	93	83	73	64	59				
Impure (99.2%)	8,906	0.444	69	1.747		69	64	59	55	52	55	62	67	69
Ni-Cr 90% Ni, 10% Cr	8,666	0.444	17	0.444		17.1	18.9	20.9	22.8	24.6	22.5			
80% Ni, 20% Cr	8,314	0.444	12.6	0.313		12.3	13.8	15.6	17.1	18.9				
Silver:														
Purest	10,524	0.2340	419	17.004	419	417	415	412						
Pure (99.9%)	10,524	0.2340	407	16.563	419	410	415	374	362					
Tungsten	19,350	0.1344	163	6.271		166	151	142	133	126	112			
Zinc, pure	7,144	0.3843	112.2	4.106	114	112	109	106	100	93		76		
Tin, pure	7,304	0.2265	64	3.884	74	65.9	59	57						

Source: From *Heat and Mass Transfer* by E.R.G. Eckert and R.M. Drake. Copyright 1972 by McGraw-Hill Book Company. Used with permission of McGraw-Hill Book Company.

[a] 1 kg/m³ = 0.06243 lb$_m$/ft³ for ρ, 1 kJ/(kg °C) = 0.2388 Btu/(lb$_m$ °F) for c_p, 1 W/(m°C) = 0.5778 Btu/(h ft °F) for k, 1 m²/s = 10.76 ft²/s for α, °C = 1 K and °F = 1 °R.

TABLE A-C-2 Property values of solid nonmetals[a]

Substance	T (°C)	k [$W/(m\ °C)$]	ρ (kg/m^3)	c_v [$kJ/(kg\ °C)$]	$\alpha \times 10^7$ (m^2/s)
Structural and heat-resistant materials					
Asphalt	20–55	0.74–0.76			
Brick:					
Building brick	20	0.69	1600	0.84	5.2
Carborundum brick	600	18.5			
	1400	11.1			
Chrome brick	200	2.32	3000	0.84	9.2
	550	2.47			9.8
	900	1.99			7.9
Diatomaceous earth, molded	200	0.24			
and fired	870	0.31			
Fireclay brick, burnt 2426°F	500	1.04	2000	0.96	5.4
	800	1.07			
	1100	1.09			
Burnt 2642°F	500	1.28	2300	0.96	5.8
	800	1.37			
	1100	1.40			
Missouri	200	1.00	2600	0.96	4.0
	600	1.47			
	1400	1.77			
Magnesite	200	3.81		1.13	
	650	2.77			
	1200	1.90			
Cement, mortar	23	1.16			
Concrete:					
Cinder	23	0.76			
Stone 1-2-4 mix	20	1.37	1900–2300	0.88	8.2–6.8
Glass:					
Window	20	0.78 (avg)	2700	0.84	3.4
Corosilicate	30–75	1.09	2200		
Plaster, gypsum	20	0.48	1440	0.84	4.0
Stone:					
Granite		1.73–3.98	2640	0.82	8–18
Limestone	100–300	1.26–1.33	2500	0.90	5.6–5.9
Marble		2.07–2.94	2500–2700	0.80	10–13.6
Sandstone	40	1.83	2160–2300	0.71	11.2–11.9
Wood (across the grain):					
Balsa	30	0.055	140		
Cypress	30	0.097	460		
Fir	23	0.11	420	2.72	0.96
Maple or oak	30	0.166	540	2.4	1.28
Yellow pine	23	0.147	640	2.8	0.82
White pine	30	0.112	430		

TABLE A-C-2 (continued)

Substance	T ($°C$)	k $[W/(m\,°C)]$	ρ (kg/m^3)	c_v $[kJ/(kg\,°C)]$	$\alpha \times 10^7$ (m^2/s)
		Insulating material			
Asbestos:					
Loosely packed	−45	0.149			
	0	0.154	470–570	0.816	3.3–4
	100	0.161			
Asbestos-cement boards	20	0.74			
Sheets	51	0.166			
Felt, 40 laminations/in.	38	0.057			
	150	0.069			
	260	0.083			
20 laminations/in.	38	0.078			
	150	0.095			
	260	0.112			
Corrugated, 4 plies/in.	38	0.087			
	93	0.100			
	150	0.119			
Asbestos cement	⋯	2.08			
Balsam wool	32	0.04	35		
Cardboard, corrugated	⋯	0.064			
Corkboard	30	0.043	160		
Cork:					
Regranulated	32	0.045	45–120	1.88	2–5.3
Ground	32	0.043	150		
Diatomaceous earth (Sil-o-cel)	0	0.061	320		
Felt:					
Hair	30	0.036	130–200		
Wool	30	0.052	330		
Fiber, insulating board	20	0.048	240		
Glass wool	23	0.038	24	0.7	22.6
		0.144			
Kapok	30	0.035			
Magnesia, 85%	38	0.067	270		
	93	0.071			
	150	0.074			
	204	0.080			
Rock wool	32	0.040	160		
Loosely packed	150	0.067	64		
	260	0.087			
Sawdust	23	0.059			
Silica aerogel	32	0.024	140		
Wood shavings	23	0.059			

Source: Reproduced from *International Critical Tables* with the permission of the National Academy of Sciences, Washington, D.C.

[a]1 $W/(m\,°C) = 0.5778$ Btu/(h ft °F) for k, 1 $kg/m^3 = 0.06243$ lb_m/ft^3 for ρ,
1 $kJ/(k_b\,°C) = 0.2388$ Btu/(lb_m °F) for c_v, 1 $m^2/s = 10.76$ ft²/s for α, °C=1 K and °F= 1 °R.

TABLE A-C-3   Property values of saturated liquids[a]

T ($°C$)	ρ (kg/m^3)	c_p [$kJ/(kg\ °C)$]	$\nu \times 10^6$ (m^2/s)	k [$W/(m\ °C)$]	$\alpha \times 10^7$ (m^2/s)	Pr	$\beta \times 10^3$ ($1/°C$)
			Water, H_2O				
0	1,002.28	4.2178	1.788	0.552	1.308	13.6	−0.67
20	1,000.52	4.1818	1.006	0.597	1.430	7.02	0.20
40	994.59	4.1784	0.658	0.628	1.512	4.34	
60	985.46	4.1843	0.478	0.651	1.554	3.02	
80	974.08	4.1964	0.364	0.668	1.636	2.22	
100	960.63	4.2161	0.294	0.680	1.680	1.74	7.2
120	945.25	4.250	0.247	0.685	1.708	1.446	
140	928.27	4.283	0.214	0.684	1.724	1.241	
160	909.69	4.342	0.190	0.680	1.729	1.099	
180	889.03	4.417	0.173	0.675	1.724	1.004	
200	866.76	4.505	0.160	0.665	1.706	0.937	14.
220	842.41	4.610	0.150	0.652	1.680	0.891	
240	815.66	4.756	0.143	0.635	1.639	0.871	
260	785.87	4.949	0.137	0.611	1.577	0.874	
281	752.55	5.208	0.135	0.580	1.481	0.910	
300	714.26	5.728	0.135	0.540	1.324	1.019	21.
			Ammonia, NH_3				
−50	703.69	4.463	0.435	0.547	1.742	2.60	
−40	691.68	4.467	0.406	0.547	1.775	2.28	
−30	679.34	4.476	0.387	0.549	1.801	2.15	
−20	666.69	4.509	0.381	0.547	1.819	2.09	
−10	653.55	4.564	0.378	0.543	1.825	2.07	
0	640.10	4.635	0.373	0.540	1.819	2.05	2.2
10	626.16	4.714	0.368	0.531	1.801	2.04	2.3
20	611.75	4.798	0.359	0.521	1.775	2.02	2.5
30	596.37	4.890	0.349	0.507	1.742	2.01	
40	580.99	4.999	0.340	0.493	1.701	2.00	
50	564.33	5.116	0.330	0.476	1.654	1.99	
			Carbon dioxide, CO_2				
−50	1,156.34	1.84	0.119	0.0855	0.4021	2.96	
−40	1,117.77	1.88	0.118	0.1011	0.4810	2.46	
−30	1,076.76	1.97	0.117	0.1116	0.5272	2.22	
−20	1,032.39	2.05	0.115	0.1151	0.5445	2.12	
−10	983.38	2.18	0.113	0.1099	0.5133	2.20	
0	926.99	2.47	0.108	0.1045	0.4578	2.38	
10	860.03	3.14	0.101	0.0971	0.3608	2.80	
20	772.57	5.0	0.091	0.0872	0.2219	4.10	14.
30	597.81	36.4	0.080	0.0703	0.0279	28.7	

T ($°C$)	ρ (kg/m^3)	c_p [$kJ/(kg\ °C)$]	$\nu \times 10^6$ (m^2/s)	k [$W/(m\ °C)$]	$\alpha \times 10^7$ (m^2/s)	Pr	$\beta \times 10^3$ ($1/°C$)
			Sulfur dioxide, SO_2				
−50	1,560.84	1.3595	0.484	0.242	1.141	4.24	
−40	1,536.81	1.3607	0.424	0.235	1.130	3.74	
−30	1,520.64	1.3616	0.371	0.230	1.117	3.31	
−20	1,488.60	1.3624	0.324	0.225	1.107	2.93	
−10	1,463.61	1.3628	0.288	0.218	1.097	2.62	
0	1,438.46	1.3636	0.257	0.211	1.081	2.38	
10	1,412.51	1.3645	0.232	0.204	1.066	2.18	
20	1,386.40	1.3653	0.210	0.199	1.050	2.00	1.94
30	1,359.33	1.3662	0.190	0.192	1.035	1.83	
40	1,329.22	1.3674	0.173	0.185	1.019	1.70	
50	1,299.10	1.3683	0.162	0.177	0.999	1.61	
			Methyl chloride, CH_3Cl				
−50	1,052.58	1.4759	0.320	0.215	1.388	2.31	
−40	1,033.35	1.4826	0.318	0.209	1.368	2.32	
−30	1,016.53	1.4922	0.314	0.202	1.337	2.35	
−20	999.39	1.5043	0.309	0.196	1.301	2.38	
−10	981.45	1.5194	0.306	0.187	1.257	2.43	
0	962.39	1.5378	0.302	0.178	1.213	2.49	
10	942.36	1.5600	0.297	0.171	1.166	2.55	
20	923.31	1.5860	0.293	0.163	1.112	2.63	
30	903.12	1.6161	0.288	0.154	1.058	2.72	
40	883.10	1.6504	0.281	0.144	0.996	2.83	
50	861.15	1.6890	0.274	0.133	0.921	2.97	
			Dichlorodifluoromethane (Freon F-12), CCl_2F_2				
−50	1,546.75	0.8750	0.310	0.067	0.501	6.2	
−40	1,518.71	0.8847	0.279	0.069	0.514	5.4	
−30	1,489.56	0.8956	0.253	0.069	0.526	4.8	1.9
−20	1,460.57	0.9073	0.235	0.071	0.539	4.4	
−10	1,429.49	0.9203	0.221	0.073	0.550	4.0	
0	1,397.45	0.9345	0.214	0.073	0.557	3.8	3.1
10	1,364.30	0.9496	0.203	0.073	0.560	3.6	
20	1,330.18	0.9659	0.198	0.073	0.560	3.5	
30	1,295.10	0.9835	0.194	0.071	0.560	3.5	
40	1,257.13	1.0019	0.191	0.069	0.555	3.5	4.4
50	1,215.96	1.0216	0.190	0.067	0.545	3.5	
			Eutectic calcium chloride solution, 29.9% $CaCl_2$				
−50	1,319.76	2.608	36.35	0.402	1.166	312.	
−40	1,314.96	2.6356	24.97	0.415	1.200	208.	
−30	1,310.15	2.6611	17.18	0.429	1.234	139.	
−20	1,305.51	2.688	11.04	0.445	1.267	87.1	
−10	1,300.70	2.713	6.96	0.459	1.300	53.6	

TABLE A-C-3   (continued)

T ($°C$)	ρ (kg/m^3)	c_p [$kJ/(kg\ °C)$]	$\nu \times 10^6$ (m^2/s)	k [$W/(m\ °C)$]	$\alpha \times 10^7$ (m^2/s)	Pr	$\beta \times 10^3$ ($1/°C$)
0	1,296.06	2.738	4.39	0.472	1.332	33.0	
10	1,291.41	2.763	3.35	0.485	1.363	24.6	
20	1,286.61	2.788	2.72	0.498	1.394	19.6	
30	1,281.96	2.814	2.27	0.511	1.419	16.0	
40	1,277.16	2.839	1.92	0.523	1.445	13.3	
50	1,272.51	2.868	1.65	0.535	1.468	11.3	

Glycerin, $C_3H_5(OH)_3$

0	1,276.03	2.261	0.00831	0.282	0.983	84.7	
10	1,270.11	2.319	0.00300	0.284	0.965	31.0	
20	1,264.02	2.386	0.00118	0.286	0.947	12.5	0.50
30	1,258.09	2.445	0.00050	0.286	0.929	5.38	
40	1,252.01	2.512	0.00022	0.286	0.914	2.45	
50	1,244.96	2.583	0.00015	0.287	0.893	1.63	

Ethylene glycol, $C_2H_4(OH)_2$

0	1,130.75	2.294	57.53	0.242	0.934	615	
20	1,116.65	2.382	19.18	0.249	0.939	204	0.65
40	1,101.43	2.474	8.69	0.256	0.939	93	
60	1,087.66	2.562	4.75	0.260	0.932	51	
80	1,077.56	2.650	2.98	0.261	0.921	32.4	
100	1,058.50	2.742	2.03	0.263	0.908	22.4	

Engine oil (unused)

0	899.12	1.796	4280	147	0.911	47,100	
20	888.23	1.880	900	145	0.872	10,400	0.70
40	876.05	1.964	240	144	0.834	2,870	
60	864.04	2.047	83.9	140	0.800	1,050	
80	852.02	2.131	37.5	138	0.769	490	
100	840.01	2.219	20.3	137	0.738	276	
120	828.96	2.307	12.4	135	0.710	175	
140	816.94	2.395	8.0	133	0.686	116	
160	805.89	2.483	5.6	132	0.663	84	

Mercury, Hg

0	13,628.22	0.1403	0.124	8.20	42.99	0.0288	
20	13,579.04	0.1394	0.114	8.69	46.06	0.0249	0.182
50	13,505.84	0.1386	0.104	9.40	50.22	0.0207	
100	13,384.58	0.1373	0.0928	10.51	57.16	0.0162	
150	13,264.28	0.1365	0.0853	11.49	63.54	0.0134	
200	13,144.94	0.1570	0.0802	12.34	69.08	0.0116	
250	13,025.60	0.1357	0.0765	13.07	74.06	0.0103	
315.5	12,847.00	0.134	0.0673	14.02	81.5	0.0083	

Source: From *Heat and Mass Transfer* by E.R.G. Eckert and R.M. Drake. Copyright 1972 by McGraw-Hill Book Company. Used with permission of McGraw-Hill Book Company.
[a]1 $kg/m^3 = 0.06243\ lb_m/ft^3$ for ρ, 1 $kJ/(kg\ °C) = 0.2388\ Btu/(lb_m\ °F)$ for c_p, 1 $m^2/s = 10.76\ ft^2/s$ for ν and α, 1 $W/(m\ °C) = 0.5778\ Btu/(h\ ft\ °F)$ for k, $°C = 1\ K$ and $°F = 1\ °R$.

TABLE A-C-4 Property values of common liquid metals[a]

Metal	Melting point (°C)	Normal boiling point (°C)	T (°C)	$\rho \times 10^{-3}$ (kg/m^3)	$\mu \times 10^3$ $[kg/(m\ s)]$	c_p $[kJ/(kg\ °C)]$	k $[W/(m\ °C)]$	Pr
Bismuth	271	1480	316	10.01	1.62	0.144	16.4	0.014
			760	9.47	0.79	0.165	15.6	0.0084
Lead	327	1740	371	10.5	2.40	0.159	16.1	0.024
			704	10.1	1.37	0.155	14.9	0.016
Lithium	179	1320	204	0.51	0.60	4.19	38.1	0.065
			982	0.44	0.42	4.19		
Mercury	−38.9	357	10	13.6	1.59	0.138	8.1	0.027
			316	12.8	0.86	0.134	14.0	0.0084
Potassium	63.9	760	149	0.81	0.37	0.796	45.0	0.0066
			704	0.67	0.14	0.754	33.1	0.0031
Sodium	97.8	883	204	0.90	0.43	1.34	80.3	0.0072
			704	0.78	0.18	1.26	59.7	0.0038
Sodium–potassium:								
22% Na	19	826	93.3	0.848	0.49	0.946	24.4	0.019
			760	0.69	0.146	0.883		
56% Na	−11.1	784	93.3	0.89	0.58	1.13	25.6	0.026
			760	0.74	0.16	1.04	28.9	0.058
Lead–bismuth:								
44.5% Pb	125	1670	288	10.3	1.76	0.147	10.7	0.024
			649	9.84	1.15			

Source: Adapted to SI units from J. G. Knudsen and D. L. Katz, *Fluid Dynamics and Heat Transfer*, McGraw-Hill Book Company, New York, 1958.
[a] $1\ kg/m^3 = 0.06243\ 1b_m/ft^3$ for ρ, $1\ kg/(m\ s) = 0.672\ 1b_m/(ft\ s)$ for μ, $1\ kJ/(kg\ °C) = 0.2388$ $Btu/(1b_m\ °F)$ for c_p, $1\ W/(m\ °C) = 0.5778\ Btu/(h\ ft\ °F)$ for k, $°C = 1\ K$, and $°F = 1\ °R$.

TABLE A-C-5 Property values of gases at atmospheric pressure[a]

T (k)	ρ (kg/m^3)	c_p $[kJ/(kg\ °C)]$	$\mu \times 10^5$ $[kg/(m\ s)]$	$\nu \times 10^6$ (m^2/s)	k $[W/(m\ °C)]$	$\alpha \times 10^4$ (m^2/s)	Pr
			Air				
100	3.6010	1.0266	0.6924	1.923	0.009246	0.02501	0.770
150	2.3675	1.0099	1.0283	4.343	0.013735	0.05745	0.753
200	1.7684	1.0061	1.3289	7.490	0.01809	0.10165	0.739
250	1.4128	1.0053	1.488	9.49	0.02227	0.13161	0.722
300	1.1774	1.0057	1.846	15.68	0.02624	0.22160	0.708
350	0.9980	1.0090	2.075	20.76	0.03003	0.2983	0.697
400	0.8826	1.0140	2.286	25.90	0.03365	0.3760	0.689
450	0.7833	1.0207	2.484	28.86	0.03707	0.4222	0.683
500	0.7048	1.0295	2.671	37.90	0.04038	0.5564	0.680
550	0.6423	1.0392	2.848	44.34	0.04360	0.6532	0.680
600	0.5879	1.0551	3.018	51.34	0.04659	0.7512	0.680
650	0.5430	1.0635	3.177	58.51	0.04953	0.8578	0.682
700	0.5030	1.0752	3.332	66.25	0.05230	0.9672	0.684
750	0.4709	1.0856	3.481	73.91	0.05509	1.0774	0.686
800	0.4405	1.0978	3.625	82.29	0.05779	1.1951	0.689
850	0.4149	1.1095	3.765	90.75	0.06028	1.3097	0.692
900	0.3925	1.1212	3.899	99.3	0.06279	1.4271	0.696
950	0.3716	1.1321	4.023	108.2	0.06525	1.5510	0.699
1000	0.3524	1.1417	4.152	117.8	0.06752	1.6779	0.702
1100	0.3204	1.160	4.44	138.6	0.0732	1.969	0.704
1200	0.2947	1.179	4.69	159.1	0.0782	2.251	0.707
1300	0.2707	1.197	4.93	182.1	0.0837	2.583	0.705
1400	0.2515	1.214	5.17	205.5	0.0891	2.920	0.705
1500	0.2355	1.230	5.40	229.1	0.0946	3.262	0.705
1600	0.2211	1.248	5.63	254.5	0.100	3.609	0.705
1700	0.2082	1.267	5.85	280.5	0.105	3.977	0.705
1800	0.1970	1.287	6.07	308.1	0.111	4.379	0.704
1900	0.1858	1.309	6.29	338.5	0.117	4.811	0.704
2000	0.1762	1.338	6.50	369.0	0.124	5.260	0.702
2100	0.1682	1.372	6.72	399.6	0.131	5.715	0.700
2200	0.1602	1.419	6.93	432.6	0.139	6.120	0.707
2300	0.1538	1.482	7.14	464.0	0.149	6.540	0.710
2400	0.1458	1.574	7.35	504.0	0.161	7.020	0.718
2500	0.1394	1.688	7.57	543.5	0.175	7.441	0.730
			Helium				
200	0.2435	5.200	1.566	64.38	0.1177	0.9288	0.694
255	0.1906	5.200	1.817	95.50	0.1357	1.3675	0.70
366	0.13280	5.200	2.305	173.6	0.1691	2.449	0.71
477	0.10204	5.200	2.750	269.3	0.197	3.716	0.72
589	0.08282	5.200	3.113	375.8	0.225	5.215	0.72
700	0.07032	5.200	3.475	494.2	0.251	6.661	0.72
800	0.06023	5.200	3.817	634.1	0.275	8.774	0.72
900	0.05286	5.200	4.136	781.3	0.298	10.834	0.72

T (k)	ρ (kg/m^3)	c_p $[kJ/(kg\,°C)]$	$\mu \times 10^5$ $[kg/(m\,s)]$	$\nu \times 10^6$ (m^2/s)	k $[W/(m\,°C)]$	$\alpha \times 10^4$ (m^2/s)	Pr
			Hydrogen				
30	0.84722	10.840	0.1606	1.895	0.0228	0.02493	0.759
50	0.50955	10.501	0.2516	4.880	0.0362	0.0676	0.721
100	0.24572	11.229	0.4212	17.14	0.0665	0.2408	0.712
150	0.16371	12.602	0.5595	34.18	0.0981	0.475	0.718
200	0.12270	13.540	0.6813	55.53	0.1282	0.772	0.719
250	0.09819	14.059	0.7919	80.64	0.1561	1.130	0.713
300	0.08185	14.314	0.8963	109.5	0.182	1.554	0.706
350	0.07016	14.436	0.9954	141.9	0.206	2.031	0.697
400	0.06135	14.491	1.086	177.1	0.228	2.568	0.690
450	0.05462	14.499	1.178	215.6	0.251	3.164	0.682
500	0.04918	14.507	1.264	257.0	0.272	3.817	0.675
550	0.04469	14.532	1.348	301.6	0.292	4.516	0.668
600	0.04085	14.537	1.429	349.7	0.315	5.306	0.664
700	0.03492	14.574	1.589	455.1	0.351	6.903	0.659
800	0.03060	14.675	1.740	569	0.384	8.563	0.664
900	0.02723	14.821	1.878	690	0.412	10.217	0.676
1000	0.02451	14.968	2.016	822	0.440	11.997	0.686
1100	0.02227	15.165	2.146	965	0.464	13.726	0.703
1200	0.02050	15.366	2.275	1107	0.488	15.484	0.715
1300	0.01890	15.575	2.408	1273	0.512	17.394	0.733
1333	0.01842	15.638	2.444	1328	0.519	18.013	0.736
			Oxygen				
100	3.9918	0.9479	0.7768	1.946	0.00903	0.023876	0.815
150	2.6190	0.9178	1.149	4.387	0.01367	0.05688	0.773
200	1.9559	0.9131	1.485	7.593	0.01824	0.10214	0.745
250	1.5618	0.9157	1.787	11.45	0.02259	0.15794	0.725
300	1.3007	0.9203	2.063	15.86	0.02676	0.22353	0.709
350	1.1133	0.9291	2.316	20.80	0.03070	0.2968	0.702
400	0.9755	0.9420	2.554	26.18	0.03461	0.3768	0.695
450	0.8682	0.9567	2.777	31.99	0.03828	0.4609	0.694
500	0.7801	0.9722	2.991	38.34	0.04173	0.5502	0.697
550	0.7096	0.9881	3.197	45.05	0.04517	0.6441	0.700
600	0.6504	1.0044	3.392	52.15	0.04832	0.7399	0.704
			Nitrogen				
100	3.4808	1.0722	0.6862	1.971	0.009450	0.025319	0.786
200	1.7108	1.0429	1.295	7.568	0.01824	0.10224	0.747
300	1.1421	1.0408	1.784	15.63	0.02620	0.22044	0.713
400	0.8538	1.0459	2.198	25.74	0.03335	0.3734	0.691
500	0.6824	1.0555	2.570	37.66	0.03984	0.5530	0.684
600	0.5687	1.0756	2.911	51.19	0.04580	0.7486	0.686
700	0.4934	1.0969	3.213	65.13	0.05123	0.9466	0.691
800	0.4277	1.1225	3.484	81.46	0.05609	1.1685	0.700
900	0.3796	1.1464	3.749	91.06	0.06070	1.3946	0.711
1000	0.3412	1.1677	4.400	117.2	0.06475	1.6250	0.724
1100	0.3108	1.1857	4.228	136.0	0.06850	1.8591	0.736
1200	0.2851	1.2037	4.450	156.1	0.07184	2.0932	0.748

TABLE A-C-5 (continued)

T (k)	ρ (kg/m^3)	c_p $[kJ/(kg\ °C)]$	$\mu \times 10^5$ $[kg/(m\ s)]$	$\nu \times 10^6$ (m^2/s)	k $[W/(m\ °C)]$	$\alpha \times 10^4$ (m^2/s)	Pr
			Carbon dioxide				
220	2.4733	0.783	1.111	4.490	0.010805	0.05920	0.818
250	2.1657	0.804	1.259	5.813	0.012884	0.07401	0.793
300	1.7973	0.871	1.496	8.321	0.016572	0.10588	0.770
350	1.5362	0.900	1.721	11.19	0.02047	0.14808	0.755
400	1.3424	0.942	1.932	14.39	0.02461	0.19463	0.738
450	1.1918	0.980	2.134	17.90	0.02897	0.24813	0.721
500	1.0732	1.013	2.326	21.67	0.03352	0.3084	0.702
550	0.9739	1.047	2.508	25.74	0.03821	0.3750	0.685
600	0.8938	1.076	2.683	30.02	0.04311	0.4383	0.668
			Carbon monoxide				
220	1.55363	1.0429	1.383	8.903	0.01906	0.11760	0.758
250	0.8410	1.0425	1.540	11.28	0.02144	0.15063	0.750
300	1.13876	1.0421	1.784	15.67	0.02525	0.21280	0.737
350	0.97425	1.0434	2.009	20.62	0.02883	0.2836	0.728
400	0.85363	1.0484	2.219	25.99	0.03226	0.3605	0.722
450	0.75848	1.0551	2.418	31.88	0.0436	0.4439	0.718
500	0.68223	1.0635	2.606	38.19	0.03863	0.5324	0.718
550	0.62024	1.0756	2.789	44.97	0.04162	0.6240	0.721
600	0.56850	1.0877	2.960	52.06	0.04446	0.7190	0.724
			Ammonia, NH_3				
220	0.3828	2.198	0.7255	19.0	0.0171	0.2054	0.93
273	0.7929	2.177	0.9353	11.8	0.0220	0.1308	0.90
323	0.6487	2.177	1.104	17.0	0.0270	0.1920	0.88
373	0.5590	2.236	1.289	23.0	0.0327	0.2619	0.87
423	0.4934	2.315	1.467	29.7	0.0391	0.3432	0.87
473	0.4405	2.395	1.649	37.4	0.0467	0.4421	0.84
			Steam (H_2O vapor)				
380	0.5863	2.060	1.271	21.6	0.0246	0.2036	1.060
400	0.5542	2.014	1.344	24.2	0.0261	0.2338	1.040
450	0.4902	1.980	1.525	31.1	0.0299	0.307	1.010
500	0.4405	1.985	1.704	38.6	0.0339	0.387	0.996
550	0.4005	1.997	1.884	47.0	0.0379	0.475	0.991
600	0.3652	2.026	2.067	56.6	0.0422	0.573	0.986
650	0.3380	2.056	2.247	66.4	0.0464	0.666	0.995
700	0.3140	2.085	2.426	77.2	0.0505	0.772	1.000
750	0.2931	2.119	2.604	88.8	0.0549	0.883	1.005
800	0.2739	2.152	2.786	102.	0.0592	1.001	1.010
850	0.2579	2.186	2.969	115.	0.0637	1.130	1.019

Source: From *Heat and Mass Transfer* by E.R.G. Eckert and R.M. Drake. Copyright 1972 by McGraw-Hill Book Company. Used with permission of McGraw-HIll Book Company.
[a] 1 $kg/m^3 = 0.06243$ lb_m/ft^3 for ρ, 1 $kJ/(kg\ °C) = 0.2388$ Btu/(lb_m °F) for c_p, 1 kg/(m s) $= 0.6720$ $lb_m/(ft\ s)$ for μ, 1 $m^2/s = 10.76$ ft^2/s for ν and α, 1 W/(m °C) $= 0.5778$ Btu/(h ft °F) for k, °C $= 1$ K and °F $= 1$ °R. The values of μ, k, c_p, and Pr are fairly insensitive to pressure for He, H_2, O_2, N_2, and air.

TABLE A-C-6 Liquid–vapor surface tension of various fluids

	Saturation temperature		Surface tension	
Liquid	°F	°C	$\alpha \times 10^4$ (lb_f/ft)	$\alpha \times 10^3$ (N/m)
Water	32	0	51.8	75.6
Water	60	15.6	50.2	73.2
Water	100	37.8	47.8	69.7
Water	200	93.3	41.2	60.1
Water	212	100	40.3	58.8
Water	320	160	31.6	46.1
Water	440	227	21.9	31.9
Water	560	293	11.1	16.2
Water	680	360	1.0	1.46
Water	705.4	374	0	0
Sodium	1618	881	77	11.2
Potassium	1400	760	43	62.7
Rubidium	1270	688	30	43.8
Cesium	1260	682	20	29.2
Mercury	675	357	27	39.4
Benzene (C_6H_6)	176	80	19	27.7
Ethyl alcohol (C_2H_5OH)	173	78.3	15	21.9
Freon 11	112	44.4	5.8	8.5

TABLE A-C-7 Emissivities of various surfaces

Surface	Emissivity ε at various temperatures			Solar temperature
Metals:				
Aluminum, polished	0.049 at 400 K	0.054 at 523 K	0.060 at 600 K	0.30
Chromium, polished	0.080 at 310 K	0.17 at 530 K	0.26 at 810 K	0.49
Copper				
Polished	0.041 at 340 K	0.034 at 560 K	0.050 at 1200 K	0.50
Oxidized at 1030 K	0.50 at 590 K	0.53 at 700 K	0.86 at 1000 K	
Gold, polished	0.019 at 130 K	0.022 at 251 K	0.068 at 1100 K	
Iron (Armco ingot),				
polished	0.32 at 470 K	0.41 at 670 K	0.57 at 1100 K	
Iron oxide	0.96 at 310 K	0.85 at 810 K		0.74
Nickel, polished	0.090 at 270 K	0.12 at 670 K	0.17 at 1100 K	0.50
Silver, polished	0.020 at 300 K	0.030 at 670 K	0.040 at 1100 K	0.11
Stainless steel:				
Type 18-8	0.25 at 310 K	0.42 at 580 K	0.99 at 1100 K	
AISI Type 310	0.26 at 810 K	0.29 at 1000 K	0.29 at 1400 K	
AISI Type 310 oxidized at 1255 K	0.46 at 360 K	0.36 at 500 K	0.64 at 760 K	
Tungsten filament	0.034 at 400 K	0.11 at 1000 K	0.32 at 3000 K	0.35
Zinc, polished	0.020 at 310 K	0.030 at 530 K	0.060 at 1600 K	0.46
Building and insulating materials:				
Asbestos paper	0.93 at 310 K	0.93 at 530 K		0.93
Asphalt	0.93 at 310 K	0.90 at 810 K		0.93
Brick–fire clay	0.90 at 310 K	0.70 at 810 K	0.75 at 1600 K	
Paints:				
Parsons black	0.98 at 240 K	0.98 at 460 K		0.98 (N)
Acrylic white	0.90 at 300 K			0.26 (N)
White (ZnO)	0.82 at 1100 K	0.81 at 1200 K	0.82 at 1300 K	0.013 (N)
Lampblack paint	0.96 at 310 K	0.97 at 530 K	0.97 at 1600 K	0.97
Other:				
Glass:				
Corning	0.68 at 139 K	0.90 at 210 K	0.82 at 260 K	
Pyrex	0.85 at 82 K	0.88 at 420 K	0.75 at 1100 K	0.19(N)
Carbon (graphite)	0.65 at 2000 K	0.69 at 2300 K	0.74 at 2600 K	0.86(N)

(N) designates normal emissivity ε_N

D ANALYTICAL SOLUTION FOR STEADY TWO-DIMENSIONAL CONDUCTION HEAT TRANSFER

The product solution approach is utilized in this section to solve for the temperature distribution in the steady two-dimensional rectangular plate shown in Fig. 3-5. The differential formulation is given by

$$\frac{\partial^2 \psi}{\partial x^2} + \frac{\partial^2 \psi}{\partial y^2} = 0 \qquad (3\text{-}48)$$

$$\psi = 0 \quad \text{at} \quad x = 0 \qquad \psi = 0 \qquad \text{at} \quad x = L \qquad (3\text{-}49, 50)$$

$$\psi = 0 \quad \text{at} \quad y = 0 \qquad \psi = F(x) \quad \text{at} \quad y = w \qquad (3\text{-}51, 52)$$

where $\psi = T - T_1$ and $F(x) = f(x) - T_1$. Because only one nonhomogeneous condition appears in this system of equations, we are ready to employ the product solution approach.

We now assume a product solution of the form

$$\psi(x,y) = X(x)Y(y) \qquad (D\text{-}1)$$

where $X(x)$ and $Y(y)$ are functions of x and y, respectively. Substituting this assumed product solution into Eqs. (3-48) through (3-52), we obtain

$$Y\frac{d^2X}{dx^2} + X\frac{d^2Y}{dy^2} = 0 \qquad (D\text{-}2)$$

$$X(0)Y(y) = 0 \qquad X(L)Y(y) = 0 \qquad (D\text{-}3, 4)$$

$$X(x)Y(0) = 0 \qquad X(x)Y(w) = F(x) \qquad (D\text{-}5, 6)$$

Rearranging Eq. (D-2) with the purpose of separating the variables, we obtain

$$\frac{1}{X}\frac{d^2X}{dx^2} = -\frac{1}{Y}\frac{d^2Y}{dy^2} \qquad (D\text{-}7)$$

The left-hand side of this equation is a function of x alone and the right-hand side is only a function of y. It follows that in order to satisfy Eq. (D-7), both sides of the equation must be equal to a constant; that is,

$$\frac{1}{X}\frac{d^2X}{dx^2} = -\frac{1}{Y}\frac{d^2Y}{dy^2} = \gamma_n \qquad (D\text{-}8)$$

where γ_n represents all constants for which Eqs. (D-2) through (D-6) are satisfied. The value of γ_n must be selected on the basis of a consideration of the boundary conditions.

670

Equation (D-8) gives us two equations of the forms

$$\frac{d^2X}{dx^2} - \gamma_n X = 0 \tag{D-9}$$

and

$$\frac{d^2Y}{dy^2} + \gamma_n Y = 0 \tag{D-10}$$

Both of these equations are satisfied by exponential functions. The assumptions

$$X = Ae^{ax} \quad \text{and} \quad Y = Be^{by} \tag{D-11, 12}$$

lead to values of a and b of the forms

$$a^2 = +\gamma_n \qquad a = \pm\sqrt{\gamma_n} \tag{D-13a, b}$$

$$b^2 = -\gamma_n \qquad b = \pm\sqrt{-\gamma_n} \tag{D-14a, b}$$

Hence, the solutions become

$$X = A_1 e^{\sqrt{\gamma_n}\,x} + A_2 e^{-\sqrt{\gamma_n}\,x} \tag{D-15}$$

$$Y = B_1 e^{\sqrt{-\gamma_n}\,y} + B_2 e^{-\sqrt{-\gamma_n}\,y} \tag{D-16}$$

The constants A_1, A_2, B_1, B_2, and γ_n are determined by use of the boundary conditions. We observe that the boundary conditions for X are both homogeneous. Thus, x is the homogeneous coordinate in this problem and y is the nonhomogeneous coordinate. The boundary conditions in the homogeneous coordinate are considered first. Equations (D-3) and (D-4) give

$$A_1 + A_2 = 0 \tag{D-17}$$

and

$$A_1 e^{\sqrt{\gamma_n}\,L} + A_2 e^{-\sqrt{\gamma_n}\,L} = 0 \tag{D-18}$$

or

$$A_1\left(e^{\sqrt{\gamma_n}\,L} - e^{-\sqrt{\gamma_n}\,L}\right) = 0 \tag{D-19}$$

If A_1 is taken as zero, we have the trivial solution $\psi = 0$. Therefore, the term in parentheses must be zero; that is,

$$e^{\sqrt{\gamma_n}\, L} - e^{-\sqrt{\gamma_n}\, L} = 0 \tag{D-20}$$

Our problem is to find all values of γ_n for which this equation is satisfied. To simplify matters, we introduce the substitution

$$\gamma_n = \pm\lambda_n^2 \tag{D-21}$$

For positive values of γ_n, Eq. (D-20) becomes

$$e^{\lambda_n L} - e^{-\lambda_n L} = 0 \tag{D-22}$$

or

$$\sinh(\lambda_n L) = 0 \tag{D-23}$$

This equation is only satisfied for $\lambda_n L = 0$ or $\lambda_n = 0$. The substitution of $\lambda_n = 0$ (or $\gamma_n = 0$) into Eqs. (D-15) and (D-16) is seen to produce the trivial result that X is independent of x and Y is independent of y. Therefore, we conclude that γ_n is not positive.

For negative values of γ_n, Eq. (D-20) takes the form

$$e^{i\lambda_n L} - e^{-i\lambda_n L} = 0 \tag{D-24}$$

or

$$\sin(\lambda_n L) = 0 \tag{D-25}$$

Because

$$\sin(n\pi) = 0 \qquad \text{for} \qquad n = 0, 1, 2, \ldots \tag{D-26}$$

we conclude that Eq. (D-20) is satisfied for

$$\lambda_n L = n\pi \qquad \text{for} \qquad n = 0, 1, 2, \ldots \tag{D-27}$$

Thus, γ_n is given by

$$\gamma_n = -\lambda_n^2 = -\left(\frac{n\pi}{L}\right)^2 \qquad \text{for} \qquad n = 0, 1, 2, \ldots \tag{D-28}$$

and our solution for X and Y for any integer n takes the form

$$X_n = A_1 e^{i(n\pi/L)x} + A_2 e^{-i(n\pi/L)x} \tag{D-29}$$

or (since $A_2 = -A_1$)

$$X_n = 2A_1 i \sin \frac{n\pi x}{L} \tag{D-30}$$

and

$$Y_n = B_1 e^{(n\pi/L)y} + B_2 e^{-(n\pi/L)y} \tag{D-31}$$

We now utilize the y-boundary conditions. Setting $Y=0$ at $y=0$ in accordance with Eq. (D-5), we have

$$0 = B_1 + B_2 \tag{D-32}$$

Thus, Eq. (D-31) takes the form

$$Y_n = B_1 \left[e^{(n\pi/L)y} - e^{-(n\pi/L)y} \right] \tag{D-33}$$

or

$$Y_n = 2B_1 \sinh \frac{n\pi y}{L} \tag{D-34}$$

Substituting this result together with Eq. (D-30) into Eq. (D-1) gives

$$\psi_n = X_n Y_n = \left(2A_1 i \sin \frac{n\pi x}{L} \right) \left(2B_1 \sinh \frac{n\pi y}{L} \right)$$
$$= C_n \sinh \frac{n\pi y}{L} \sin \frac{n\pi x}{L} \tag{D-35}$$

for any integer n, where $C_n = 2iA_1 B_1$. Because ψ_n represents a solution for each integer, the full solution is given by

$$\psi = \sum_{n=1}^{\infty} \psi_n = \sum_{n=1}^{\infty} C_n \sinh \frac{n\pi y}{L} \sin \frac{n\pi x}{L} \tag{D-36}$$

The coefficient C_n is evaluated by use of the final nonhomogeneous y-boundary condition by substituting Eq. (D-36) into Eq. (D-6); that is,

$$F(x) = \sum_{n=1}^{\infty} C_n \sinh \frac{n\pi w}{L} \sin \frac{n\pi x}{L}$$

or

$$F(x) = \sum_{n=1}^{\infty} b_n \sin \frac{n\pi x}{L} \tag{D-37}$$

where $b_n = C_n \sinh(n\pi w/L)$. This equation is recognized as *the Fourier sine series* expansion of the function $F(x)$, where b_n can be expressed as

$$b_n = \frac{2}{L} \int_0^L F(x) \sin\frac{n\pi x}{L}\, dx \qquad \text{(D-38)}$$

Thus, our final solution for the temperature distribution takes the form

$$T - T_1 = \psi = \sum_{n=1}^{\infty} b_n \frac{\sinh(n\pi y/L)}{\sinh(n\pi w/L)} \sin\frac{n\pi x}{L} \qquad \text{(D-39)}$$

E NUMERICAL COMPUTATIONS

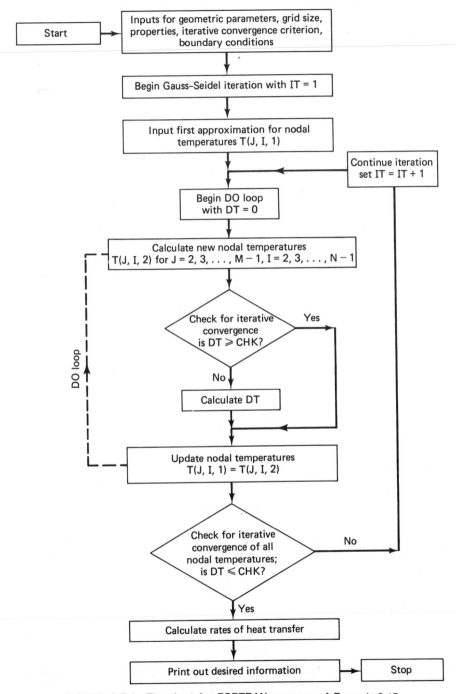

FIGURE A-E-1 Flowchart for FORTRAN program of Example 3-17.

```
C          EXAMPLE 3-17
CCC1       REAL K
CCC2       REAL L
CCC3       DIMENSION T(100,100,2)
CCC4       DIMENSION FX(100)
C
C          INPUTS FOR GEOMETRIC PARAMETERS.
CCC5       L=1.0
CCC6       W=L
CCC7       D=0.01
C
C          INPUTS FOR GRID SIZE.
CCC8       M=10
CCC9       DX=L/(M-1)
CC10       N=M
CC11       M1=M-1
CC12       N1=N-1
C
C          INPUTS FOR PROPERTIES.
CC13       K=100.
C
C          INPUTS FOR ITERATIVE CONVERGENCE CRITERION.
CC14       CHK=1.E-4
C
C          INPUTS FOR BOUNDARY CONDITIONS.
CC15       T1=0.0
C
C          DO LOOP 1 INPUTS B.C. AT SURFACES E AND D.
CC16       DO 1 I=1,N
CC17       T(1,I,1)=T1
CC18       T(1,I,2)=T(1,I,1)
CC19       T(M,I,1)=T1
CC20     1 T(M,I,2)=T(M,I,1)
C
C          DO LOOP 2 INPUTS B.C. AT SURFACES A AND C.
CC21       DO 2 J=2,M1
CC22       T(J,1,1)=T1
CC23       T(J,1,2)=T(J,1,1)
CC24       FX(J)=100.*SIN(3.14160*(J-1)*DX/L)
CC25       T2X=FX(J)
CC26       T(J,N,1)=T2X
CC27     2 T(J,N,2)=T(J,N,1)
C
C          BEGIN GAUSS-SEIDEL ITERATION WITH IT=1.
CC28       IT=1
C
C          INPUT FIRST APPROXIMATION FOR NODAL TEMPERATURES.
CC29       DO 3 J=2,M1
CC30       DO 3 I=2,N1
CC31     3 T(J,I,1)=1.
C
C          BEGIN DO LOOP WITH DT=0.
CC32     4 DT=0.0
C
C          CALCULATE NEW NODAL TEMPERATURES T(J,I,2).
CC33       DO 5 J=2,M1
CC34       DO 5 I=2,N1
CC35       T(J,I,2)=(T(J+1,I,1)+T(J-1,I,1)+T(J,I+1,1)+T(J,I-1,1))/4.
C
C          CHEK FOR ITERATIVE CONVERGENCE.
CC36       IF (DT .GE. CHK) GO TO 5
C
C          CALCULATE DT.
CC37       DT=ABS((T(J,I,2)-T(J,I,1))/T(J,I,1))
C
C          UPDATE NODAL TEMPERATURES.
CC38     5 T(J,I,1)=T(J,I,2)
C
C          CHECK FOR ITERATIVE CONVERGENCE OF ALL NODAL TEMPERATURES.
CC39       IF (DT .LE. CHK) GO TO 6
C
```

FIGURE A-E-2 FORTRAN program for Example 3-17.

```
                   CONTINUE ITERATION SET IT=IT+1.
C040               IT=IT+1
C041               GO TO 4
               C
               C   CALCULATE RATES OF HEAT TRANSFER.
C042           6   SGA=0.0
C043               SQC=0.0
C044               DO 7 J=2,M1
C045               DQA=K*DX*(T(J,N1,2)+T(J-1,N,2)/2.+T(J+1,N,2)/2.-2.*T(J,N,2))
C046               SQA=SQA+DQA
C047               DQC=-K*DX*(T(J,2,2)+T(J-1,1,2)/2.+T(J+1,1,2)/2.-2.*T(J,1,2))
C048           7   SQC=SQC+DQC
               C
               C   PRINT OUT DESIRED INFORMATION.
C049           8   PRINT 9
C050           9   FORMAT (1H1)
C051               DO 10 II=1,N
C052               I=N-II+1
C053               PRINT 11,(T(J,I,2),J=1,M)
C054          10   CONTINUE
C055               PRINT 12,M1
C056               PRINT 13,IT
C057               PRINT 14,SQA,SQC
C058          11   FORMAT (1H0,5X,10(F8.2,3X))
C059          12   FORMAT (1H0,' M = ',I5)
C060          13   FORMAT (1H0,' NUMBER OF ITERATIONS REQUIRED FOR SPECIFIED ACCURACY
                  X ',I9)
C061          14   FORMAT (1H0,' HEAT TRANSFER RATE THRU SURFACE A = ',F10.5/' HEAT
                  XTRANSFER RATE THRU SURFACE C = ',F10.5)
C062               STOP
C063               END
```

FIGURE A-E-2 (continued)

```
             C        EXAMPLE 3-20
0001                  REAL L
0002                  REAL K
0003                  DIMENSION T(100,2,2)
0004                  L=C.004
0005                  K=50.0
0006                  ALPHA=2.E-5
0007                  M=10
0008                  M1=M-1
0009                  DX=L/M1
0010                  S=2.0
0011                  DTIME=S*DX**2./ALPHA
0012                  CHK=1.E-4
0013                  CSS=1.F-4
0014                  TI=1.E-5
0015                  DO 1 J=2,M1
0016                  T(J,1,1)=TI
0017                1 T(J,1,2)=TI
0018                  T1=1.E-5
0019                  T(1,1,1)=T1
0020                  T(1,2,1)=T1
0021                  T(1,1,2)=T1
0022                  T(1,2,2)=T1
0023                  TM=100.
0024                  T(M,1,1)=TM
0025                  T(M,2,1)=TM
0026                  T(M,1,2)=TM
0027                  T(M,2,2)=TM
0028                  QFL1=1.0
0029                  TAU=1
0030                2 CONTINUE
0031                  IT=1
0032                  DO 3 J=2,M1
0033                3 T(J,2,1)=T(J,1,2)
0034                4 DT=0
0035                  DO 5 J=2,M1
0036                  T(J,2,2)=(T(J+1,2,1)+T(J-1,2,1)+T(J,1,2)/S)/(2.+1./S)
0037                  IF (DT .GE. CHK) GO TO 5
0038                  DT=ABS((T(J,2,2)-T(J,2,1))/T(J,2,2))
0039                5 T(J,2,1)=T(J,2,2)
0040                  IF (DT .LE. CHK) GO TO 6
0041                  IT=IT+1
0042                  GO TO 4
0043                6 CFL=-K/DX*(T(M,2,2)-T(M1,2,2))
0044                  TIME=TAU*DTIME
0045               80 FORMAT(75X,2(E12.5,5X))
0046                  PRINT 80,TAU,TIME
0047               90 FORMAT(3X,I5,3X,F7.5,3X,I5,3X,E15.7,3X,E12.5)
0048                  PRINT 90,M,CHK,IT,QFL,T(M1,2,2)
0049                  DQFL=ABS((CFL-QFL1)/QFL1)
0050                  IF(DQFL .LE. CSS) GO TO 8
0051                  CFL1=QFL
0052                  TAU=TAU+1
0053                  DO 7 J=2,M1
0054                7 T(J,1,2)=T(J,2,2)
0055                  GC TO 2
0056                8 STCP
0057                  END
```

FIGURE A-E-3 FORTRAN program for Example 3-20.

```
                C       EXAMPLE 6-11
0001            REAL L
0C02            CIMENSICN T(1000)
0003            D=C.1
0004            L=10.
CC05            MDCT=10
0006            M=1C1
0007            M1=M-1
0008            DX=L/M1
OC09            H=4820.
0010            F=3.1416*D
0011            CP=4180.
0012            T(1)=20.
CC13            CC 1 J=2,M
0014            X=(J-1)*DX
0015            TS=20.+80.*(1.-CUS(4.*3.1416*X/L))
0016            T(J)=(H*P/(MDCT*CP))*DX*TS+T(J-1)*(1.-(H*P/(MDCT*CP)*DX))
0017            PRINT 9,X,T(J)
0018          1 CONTINUE
CC19            CC=MDCT*CP*(T(J)-T(1))
0020            PRINT 9,CO
0021          9 FORMAT (3X,2E15.7)
0022            STCP
0023            ENC
```

FIGURE A-E-4 FORTRAN program for Example 6-11.

```
                C       EXAMPLE 1C-10
0C01            DIMENSICN F(1800)
00C2            PR=C.72
0003            AK=0.4
0004            AA=27.4
0005            DO 55 IYP=5,11C,5
0006            YP=IYP
0007            EXPO=EXP(-YP/AA)
0008            UT1=((1.C+4.0*(AK*YP)**2.0*(1.C-EXPC)**2.0)**0.5-1.0)/2.0
0009            DYP=0.1
0010            M=YP/DYP+1
0011            YPD=0.0
0012            DO 5 I=1,M
0013            F(I)=2./((-1.+(1.+4.*(AK*YPD)**2.*(1.-EXF(-YPD/AA))**2.)**.5)+
                X2./PR)
0014          5 YPD=YPD+DYP
0015            M1=M-1
0016            SUM=0.0
0017            DC 95 N=2,M1
0018         95 SUM=SUM+F(N)
0019            AINT=DYP/2.*(F(1)+F(N)+2.*SUM)
0020            TP=AINT
0021            IF(YP .LE. 1CC.) GC TO 96
0022            A=TP-(ALOG(YP))/AK
0023         96 WRITE(6,15)TP,YP,LT1,A
0024         55 CONTINUE
0025         15 FORMAT(4(5 X,E13.7))
0026            STOP
0027            ENC
```

FIGURE A-E-5 FORTRAN program for Example 10-10.

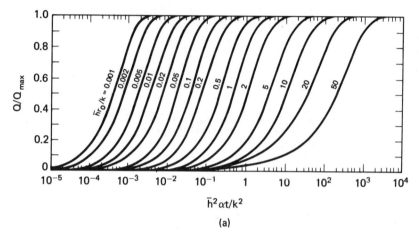

(a)

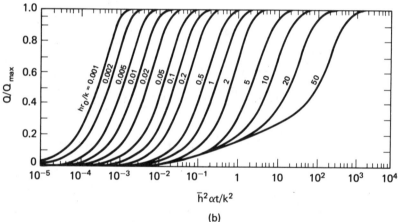

(b)

FIGURE A-F-1   Charts for accumulative heat transfer in solid with surface suddenly exposed to fluid. (a) Long cylinder. (b) Sphere $[q_{max} = c_v V(T_i - T_F)]$. (From *Fundamentals of Heat Transfer* by Grober et al. Copyright 1961 by McGraw-Hill Book Company. Used with permission of McGraw-Hill Book Company.)

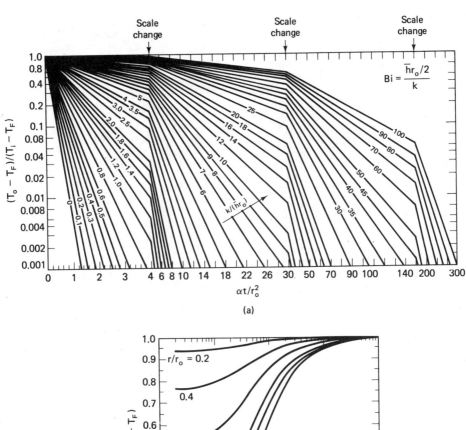

(a)

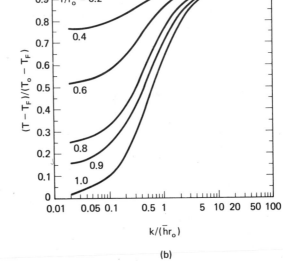

(b)

FIGURE A-F-2 Heisler temperature charts for a solid circular cylinder with surface suddenly exposed to a fluid. (a) Instantaneous centerline temperature T_o. (b) Instantaneous temperature distribution T as a function of centerline temperature T_o. (From M. P. Heisler, *Trans. ASME.* **69**, 1947, 227–236.)

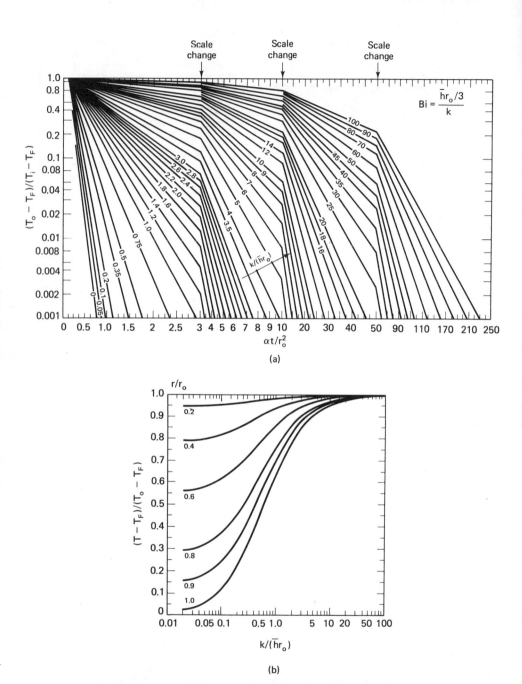

FIGURE A-F-3 Heisler temperature charts for a solid sphere with surface suddenly exposed to a fluid. (a) Instantaneous centerline temperature T_o. (b) Instantaneous temperature distribution T as a function of centerline temperature T_o. (From M. P. Heisler, *Trans. ASME*, **69**, 1947, 227–236.)

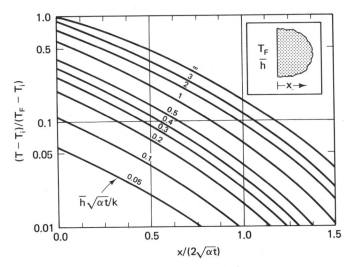

FIGURE A-F-4 Chart for instantaneous temperature distribution in a semi-infinite solid with surface suddenly exposed to a fluid. (From *Conduction Heat Transfer* by P. J. Schneider; Fig. 10-12, p. 266. Copyright 1955 by Addison-Wesley Publishing Company, Inc., Reading, Mass. Used with permission.)

G TABLE FOR RADIATION FUNCTIONS

TABLE A-G-1 Radiation functions

λT_s		$E_{b\lambda}/T_s^5$		
$\mu m \, °R$	$\mu m \, K$	$\dfrac{Btu}{h \, ft^2 \, °R^5 \, \mu m} \times 10^{15}$	$\dfrac{W}{m^2 \, K^5 \, \mu m} \times 10^{11}$	$\dfrac{E_{b0 \to \lambda}}{\sigma T_s^4}$
1,000	555.6	0.000671	0.400×10^{-5}	0.170×10^{-7}
1,200	666.7	0.0202	0.120×10^{-3}	0.756×10^{-6}
1,400	777.8	0.204	0.00122	0.106×10^{-4}
1,600	888.9	1.057	0.00630	0.738×10^{-4}
1,800	1,000.0	3.544	0.02111	0.321×10^{-3}
2,000	1,111.1	8.822	0.05254	0.00101
2,200	1,222.2	17.776	0.10587	0.00252
2,400	1,333.3	30.686	0.18275	0.00531
2,600	1,444.4	47.167	0.28091	0.00983
2,800	1,555.6	66.334	0.39505	0.01643
3,000	1,666.7	87.047	0.51841	0.02537
3,200	1,777.8	108.14	0.64404	0.03677
3,400	1,888.9	128.58	0.76578	0.05059
3,600	2,000.0	147.56	0.87878	0.06672

TABLE A-G-1 (continued)

λT_s		$E_{b\lambda/T_s^5}$		
$\mu m \,°R$	$\mu m \, K$	$\dfrac{Btu}{h\,ft^2\,°R^5\,\mu m}$ $\times 10^{15}$	$\dfrac{W}{m^2\,K^5\,\mu m}$ $\times 10^{11}$	$\dfrac{E_{b0\to\lambda}}{\sigma T_s^4}$
3,800	2,111.1	164.49	0.97963	0.08496
4,000	2,222.2	179.04	1.0663	0.10503
4,200	2,333.3	191.05	1.1378	0.12665
4,400	2,444.4	200.51	1.1942	0.14953
4,600	2,555.6	207.55	1.2361	0.17337
4,800	2,666.7	212.32	1.2645	0.19789
5,000	2,777.8	215.06	1.2808	0.22285
5,200	2,888.9	216.00	1.2864	0.24803
5,400	3,000.0	215.39	1.2827	0.27322
5,600	3,111.1	213.46	1.2713	0.29825
5,800	3,222.2	210.43	1.2532	0.32300
6,000	3,333.3	206.51	1.2299	0.34734
6,200	3,444.4	201.88	1.2023	0.37118
6,400	3,555.6	196.69	1.1714	0.39445
6,600	3,666.7	191.09	1.1380	0.41708
6,800	3,777.8	185.18	1.1029	0.43905
7,000	3,888.9	179.08	1.0665	0.46031
7,200	4,000.0	172.86	1.0295	0.48085
7,400	4,111.1	166.60	0.99221	0.50066
7,600	4,222.2	160.35	0.95499	0.51974
7,800	4,333.3	154.16	0.91813	0.53809
8,000	4,444.4	148.07	0.88184	0.55573
8,200	4,555.6	142.10	0.84629	0.57267
8,400	4,666.7	136.28	0.81163	0.58891
8,600	4,777.8	130.63	0.77796	0.60449
8,800	4,888.9	125.15	0.74534	0.61941
9,000	5,000.0	119.86	0.71383	0.63371
9,200	5,111.1	114.76	0.68346	0.64740
9,400	5,222.2	109.85	0.65423	0.66051
9,600	5,333.3	105.14	0.62617	0.67305
9,800	5,444.4	100.62	0.59925	0.68506
10,000	5,555.6	96.289	0.57346	0.69655
10,200	5,666.7	92.145	0.54877	0.70754
10,400	5,777.8	88.181	0.52517	0.71806
10,600	5,888.9	84.394	0.50261	0.72813
10,800	6,000.0	80.777	0.48107	0.73777
11,000	6,111.1	77.325	0.46051	0.74700
11,200	6,222.2	74.031	0.44089	0.75583
11,400	6,333.3	70.889	0.42218	0.76429
11,600	6,444.4	67.892	0.40434	0.77238
11,800	6,555.6	65.036	0.38732	0.78014
12,000	6,666.7	62.313	0.37111	0.78757

λT_s		$E_{b\lambda/T_s^5}$		
$\mu m \, °R$	$\mu m \, K$	$\dfrac{Btu}{h \, ft^2 \, °R^5 \, \mu m}$ $\times 10^{15}$	$\dfrac{W}{m^2 \, K^5 \, \mu m}$ $\times 10^{11}$	$\dfrac{E_{b0 \to \lambda}}{\sigma T_s^4}$
12,200	6,777.8	59.717	0.35565	0.79469
12,400	6,888.9	57.242	0.34091	0.80152
12,600	7,000.0	54.884	0.32687	0.80806
12,800	7,111.1	52.636	0.31348	0.81433
13,000	7,222.2	50.493	0.30071	0.82035
13,200	7,333.3	48.450	0.28855	0.82612
13,400	7,444.4	46.502	0.27695	0.83166
13,600	7,555.6	44.645	0.26589	0.83698
13,800	7,666.7	42.874	0.25534	0.84209
14,000	7,777.8	41.184	0.24527	0.84699
14,200	7,888.9	39.572	0.23567	0.85171
14,400	8,000.0	38.033	0.22651	0.85624
14,600	8,111.1	36.565	0.21777	0.86059
14,800	8,222.2	35.163	0.20942	0.86477
15,000	8,333.3	33.825	0.20145	0.86880
16,000	8,888.9	27.977	0.16662	0.88677
17,000	9,444.4	23.301	0.13877	0.90168
18,000	10,000.0	19.536	0.11635	0.91414
19,000	10,555.6	16.484	0.09817	0.92462
20,000	11,111.1	13.994	0.08334	0.93349
21,000	11,666.7	11.949	0.07116	0.94104
22,000	12,222.2	10.258	0.06109	0.94751
23,000	12,777.8	8.852	0.05272	0.95307
24,000	13,333.3	7.676	0.04572	0.95788
25,000	13,888.9	6.687	0.03982	0.96207
26,000	14,444.4	5.850	0.03484	0.96572
27,000	15,000.0	5.139	0.03061	0.96892
28,000	15,555.6	4.532	0.02699	0.97174
29,000	16,111.1	4.012	0.02389	0.97423
30,000	16,666.7	3.563	0.02122	0.97644
40,000	22,222.2	1.273	0.00758	0.98915
50,000	27,777.8	0.560	0.00333	0.99414
60,000	33,333.3	0.283	0.00168	0.99649
70,000	38,888.9	0.158	0.940×10^{-3}	0.99773
80,000	44,444.4	0.0948	0.564×10^{-3}	0.99845
90,000	50,000.0	0.0603	0.359×10^{-3}	0.99889
100,000	55,555.6	0.0402	0.239×10^{-3}	0.99918

From R. V. Dunkle, "Thermal Radiation Tables and Applications," *Trans. ASME* **76**, 1954, 549.

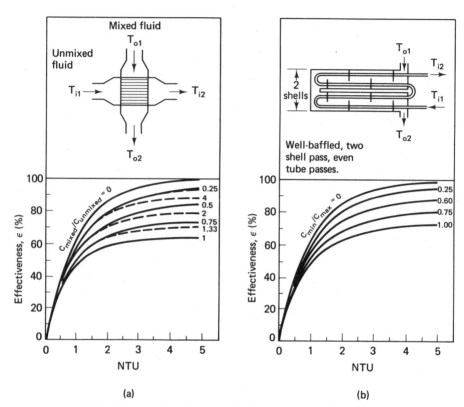

FIGURE A-H-1 Effectiveness of heat exchangers. (a) Crossflow system. (b) Shell-and-tube system. (From *Compact Heat Exchangers* (2nd ed.). by W. M. Kays and A. L. London. Copyright 1964 by McGraw-Hill Book Company. Used with permission of McGraw-Hill Book Company.)

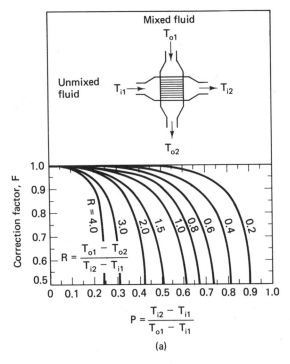

(a)

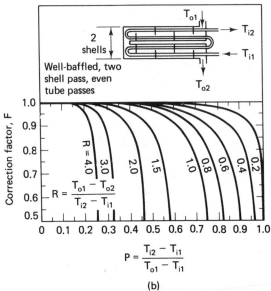

(b)

FIGURE A-H-2 LMTD correction factors for heat exchangers. (a) Crossflow system. (From R. A. Bowman et al., *Trans. ASME*, **62**, 1940, 283–294.) (b) Shell-and-tube system. (Courtesy of Tubular Exchanger Manufacturers Association.)

I SYMBOLS[1]

a	Acceleration, m/s^2
A	Area, m^2
A_N	Area of body normal to direction of flow, m^2
c	Speed of light, m/s
c_o	Speed of light in vacuum, m/s
c_p	Specific heat at constant pressure, $J/(kg\ °C)$
c_v	Specific heat at constant volume, $J/(kg\ °C)$
C	Thermal capacitance, $J/°C$; heat-exchanger capacity rate ($\equiv \dot{m}c_p$), $W/°C$
C_e	Electrical capacitance, F
C_D	Drag coefficient
D	Diameter, m
D_H	Hydraulic diameter ($\equiv 4A/p$), m
D_N	Height of body for crossflow, m
e_λ	Monochromatic emissive power ($\equiv dE_{0\to\lambda}/d\lambda$), $W/m^2/\mu m$
E	Energy, J; total emissive power, W/m^2
E_e	Electrical voltage, V
f	Fanning friction factor
F	Force, N; correction factor for LMTD in heat exchangers
F_a	Arrangement factor for tube bank
F_d	Depth factor for tube bank
F_n	Heat-transfer depth factor for tube bank
F_f	Fouling factor, $m^2\ °C/W$
F_{s-R}	Thermal radiation shape factor
$\mathcal{F}_{s-R}$	Radiation factor
g	Acceleration of gravity, m/s^2
G	Mass velocity ($\equiv \dot{m}/A$), $kg/(m^2\ s)$; thermal irradiation, W/m^2
h	Coefficient of heat transfer for convection, $W/(m^2\ °C)$
h_{fv}	Specific enthalpy of vaporization, J/kg
h_{tc}	Thermal contact coefficient, $W/(m^2\ °C)$
h_R	Coefficient of heat transfer for thermal radiation, $W/(m^2\ °C)$
H	Enthalpy, J
I	Intensity of thermal radiation, W/m^2
I_e	Electrical current, A
J	Radiosity, W/m^2
k	Thermal conductivity, $W/(m\ °C)$
k_e	Effective thermal conductivity for natural convection in an enclosed space, $W/(m\ °C)$
k_t	Eddy thermal conductivity for turbulent flow, $W/(m\ °C)$

[1]Symbols for dimensions are given in Table A-B-1.

KE	Kinetic energy, J
L	Length, m
ℓ	Characteristic length ($\equiv V/A_s$), m
LMTD	Logarithmic mean temperature difference, °C
m	Mass, kg; convective fin parameter $\left[\equiv \sqrt{\overline{h}p/(kA)}\ \right]$, 1/m
m_R	Radiative fin parameter $\left[\equiv \sqrt{\sigma p F_{s-R}/(kA)}\ \right]$, 1/m
$\dot{m}$	Mass flow rate, kg/s
M	Molecular weight; number of finite-difference nodes in x direction; number of heat-flow paths; momentum, kg m/s
n	Index of refraction ($\equiv c/c_o$)
N	Number of finite-difference nodes in y direction; total number of tubes; number of curvilinear squares in a heat-flow path; number of surfaces in a thermal radiation enclosure
N_n	Number of tube columns wide
N_p	Number of tube rows deep
NTU	Number of transfer units [$\equiv \overline{h}A_s/(\dot{m}c_p)$ or $\overline{U}A_s/(\dot{m}c_p)_{\min}$]
P	Perimeter, m
P	Pressure, N/m^2
PE	Potential energy, J
q	Heat-transfer rate, W
q''	Heat-transfer rate per unit area (i.e., heat flux), W/m^2
q_{crit}	Critical (burnout) heat flux associated with boiling, W/m^2
$\overline{q''}$	Apparent mean total heat flux for turbulent flow ($\equiv \overline{q_y''} + \overline{q_t''}$), W/m^2
$\overline{q_t''}$	Apparent mean turbulent heat flux for turbulent flow ($\equiv \rho c_p \overline{v'T'}$), W/m^2
$\dot{q}$	Rate of internal energy generation per unit volume, W/m^3
Q	Heat transfer, J
r	Radial direction, m
r_c	Critical radius, m
R	Thermal resistance, °C/W
R_e	Electrical resistance, Ω
S	Conduction shape factor, m; tube bank pitch, m
t	Time, s
T	Temperature, °C or K
u	Axial velocity distribution, m/s
u^+	Dimensionless mean axial velocity distribution ($\equiv \bar{u}/U^*$)
U	Internal energy, J; velocity, m/s; overall coefficient of heat transfer, W/(m^2 °C)
U^*	Friction velocity ($\equiv \sqrt{\tau_0/\rho}$), m/s
v	Transverse velocity distribution, m/s
V	Volume, m^3

w	Length, m
W	Work, J
$\dot{W}$	Power, W
x,y,z	Space coordinates in Cartesian system, m
x_c	Location of transition from laminar to turbulent flow, m
X	Quality of two-phase fluid
y^+	Dimensionless distance from wall $(\equiv yU^*/\nu)$
Z	Number of finite-difference subvolumes with unknown nodal temperature
Z_s	Total number of finite-difference subvolumes
α	Thermal diffusivity $[\equiv k/(\rho c_p)]$, m²/s; thermal radiation absorptivity
α_e	Temperature coefficient of resistance, 1/°C
α_t	Eddy thermal diffusivity for turbulent flow, m²/s
β	Volume coefficient of expansion, 1/°C
β_T	Temperature coefficient of thermal conductivity, 1/°C
δ	Hydrodynamic boundary layer thickness, m; characteristic length, m
Δ	Thermal boundary layer thickness, m; finite-difference increment
ε	Emissivity; heat-exchanger effectiveness
ε_H	Eddy thermal diffusivity for turbulent flow (often used instead of α_t), m²/s
η	Similarity coordinate, m; heat-exchanger parameter $(\equiv 1/C_i + 1/C_o$ for parallel flow, $\equiv 1/C_i - 1/C_o$ for counterflow), kg/J
η_F	Fin efficiency
θ	Angle, rad
λ	Thermal radiation wavelength, μm; Darcy friction factor $(\equiv 4f)$
μ	Dynamic viscosity, kg/(m s)
μ_t	Eddy viscosity for turbulent flow, kg/(m s)
ν	Kinematic viscosity $(\equiv \mu/\rho)$, m²/s; frequency in electromagnetic wave theory, 1/s
ν_t	Eddy diffusivity for turbulent flow, m²/s
ρ	Density, kg/m³; thermal radiation reflectivity
ρ_e	Electrical resistivity, m Ω
σ	Stefan–Boltzmann constant for thermal radiation $[\equiv 5.67 \times 10^{-8}$ W/(m² K⁴)]; surface tension of liquid–vapor interface, N/m
τ	Shear stress, N/m²; thermal radiation transmissivity
$\bar{\tau}$	Apparent total mean shear stress for turbulent flow $(\equiv \bar{\tau}_y + \bar{\tau}_t)$, N/m²
$\bar{\tau}_t$	Apparent mean turbulent shear stress $(\equiv -\rho\overline{u'v'})$, N/m²
ϕ	Angle, rad

ψ	Relative temperature $(T - T_F,\ T - T_i$, etc.), °C; temperature gradient $(\equiv dT/dx)$, °C/m
$d\omega$	Solid angle $(\equiv dA_r/r^2)$
ξ	General coordinate direction (can represent x, y, z, or r), m

Subscripts

b	Bulk fluid conditions; thermal radiation blackbody surface
B	Boiling
f	Film conditions
F	Fluid
c	Convection; centerline
cp	Constant property
C	Condensation
e	Electrical
g	Graybody in thermal radiation
HS	Heat sink
i	Initial; tube side; inlet
j	Node index; surface index number
k	Conduction
ℓ	Saturated liquid
L	Length, m
m	Mean; finite-difference index for increments in x direction
n	Finite-difference index for increments in y direction
o	Annular or shell side; outlet
o	Uniform wall condition
r	Radial direction
R	Thermal radiation
s	Surface; storage
sat	Saturation condition
ss	Steady-state
T	Transistor
v	Saturated vapor
x, y, z	Directions in Cartesian coordinate system
∞	Free stream conditions
ξ	General coordinate direction

Superscripts

$-$	Mean
$\cdot$	Rate
$'$	Turbulent fluctuating component

$''$	Per unit area
i	Iteration index number
m,n	Coefficients in empirical correlations
τ	Finite-difference index for time

Dimensionless Groups

Bi	Biot number ($\equiv R_k/R_s$, $= \bar{h}\ell/k$ for convection)
Gr	Grashof number [$\equiv g\beta\delta^3(T_s - T_F)/\nu^2$]
Nu	Nusselt number ($\equiv hD_H/k$ or $h\delta/k$)
$\overline{\text{Nu}}$	Mean Nusselt number ($\equiv \bar{h}D_H/k$ or $\bar{h}\delta/k$)
Pe	Peclet number ($\equiv$ Re Pr)
Pr	Prandtl number ($\equiv \mu c_p/k = \nu/\alpha$)
Pr_t	Turbulent Prandtl number ($\equiv \nu_t/\alpha_t$)
Ra	Raleigh number ($\equiv$ Gr Pr)
Re	Reynolds number ($\equiv U_b D_H/\nu$, $U_\infty x/\nu$, or $U_\infty \delta/\nu$)
$\text{Re}_{\max}$	Reynolds number for tube bank flow ($\equiv U_{b,\max}D/\nu$)
Re_f	Reynolds number for filmwise condensation
St	Stanton number [$\equiv$ Nu/(Re Pr)]

INDEX

693